STATA BASE REFERENCE MANUAL
VOLUME 2
K-Q
RELEASE 9

A Stata Press Publication
StataCorp LP
College Station, Texas

Stata Press, 4905 Lakeway Drive, College Station, Texas 77845

Copyright © 1985–2005 by StataCorp LP
All rights reserved
Version 9
Typeset in TEX
Printed in the United States of America

10 9 8 7 6 5 4 3 2 1

ISBN 1-881228-94-0 (volumes 1–3)
ISBN 1-881228-91-6 (volume 1)
ISBN 1-881228-92-4 (volume 2)
ISBN 1-881228-93-2 (volume 3)

For copyright information about the software, type `help copyright` within Stata.

The suggested citation for this software is

StataCorp. 2005. *Stata Statistical Software: Release 9*. College Station, TX: StataCorp LP.

Title

> **kappa** — Interrater agreement

Syntax

Interrater agreement, two unique raters

> kap *varname*$_1$ *varname*$_2$ $\big[$*if*$\big]$ $\big[$*in*$\big]$ $\big[$*weight*$\big]$ $\big[$, *options* $\big]$

Weights for weighting disagreements

> kapwgt *wgtid* $\big[$ 1 \ # 1 $\big[$ \ # # 1 ... $\big]$ $\big]$

Interrater agreement, nonunique raters

> kap *varname*$_1$ *varname*$_2$ *varname*$_3$ $\big[$...$\big]$ $\big[$*if*$\big]$ $\big[$*in*$\big]$ $\big[$*weight*$\big]$

Interrater agreement, nonunique raters, variables record frequency of ratings

> kappa *varlist* $\big[$*if*$\big]$ $\big[$*in*$\big]$

options	description
Main	
<u>t</u>ab	display table of assessments
<u>w</u>gt(*wgtid*)	specify how to weight disagreements; see *Options* for alternatives
<u>a</u>bsolute	treat rating categories as absolute

fweights are allowed; see [U] **11.1.6 weight**.

Description

kap (first syntax) calculates the kappa-statistic measure of interrater agreement when there are two unique raters and two or more ratings.

kapwgt defines weights for use by kap in measuring the importance of disagreements.

kap (second syntax) and kappa calculate the kappa-statistic measure when there are two or more (nonunique) raters and two outcomes, more than two outcomes when the number of raters is fixed, and more than two outcomes when the number of raters varies. kap (second syntax) and kappa produce the same results; they merely differ in how they expect the data to be organized.

kap assumes that each observation is a subject. *varname*$_1$ contains the ratings by the first rater, *varname*$_2$ by the second rater, and so on.

kappa also assumes that each observation is a subject. The variables, however, record the frequencies with which ratings were assigned. The first variable records the number of times the first rating was assigned, the second variable records the number of times the second rating was assigned, and so on.

1

Options

___Main___

`tab` displays a tabulation of the assessments by the two raters.

`wgt`(*wgtid*) specifies that *wgtid* be used to weight disagreements. You can define your own weights using `kapwgt`; in that case, `wgt()` specifies the name of the user-defined matrix. For instance, you might define

    . `kapwgt mine 1 \ .8 1 \ 0 .8 1 \ 0 0 .8 1`

and then

    . `kap rata ratb, wgt(mine)`

In addition, two prerecorded weights are available.

`wgt`(w) specifies weights $1 - |i - j|/(k - 1)$, where i and j index the rows and columns of the ratings by the two raters and k is the maximum number of possible ratings.

`wgt`(w2) specifies weights $1 - \{(i - j)/(k - 1)\}^2$.

`absolute` is relevant only if `wgt()` is also specified; see `wgt()` above. Option `absolute` modifies how i, j, and k are defined and how corresponding entries are found in a user-defined weighting matrix. When `absolute` is not specified, i and j refer to the row and column index, not to the ratings themselves. Say that the ratings are recorded as $\{0, 1, 1.5, 2\}$. There are 4 ratings; $k = 4$, and i and j are still 1, 2, 3, and 4 in the formulas above. Index 3, for instance, corresponds to rating = 1.5. This is convenient but can, with some data, lead to difficulties.

When `absolute` is specified, all ratings must be integers, and they must be coded from the set $\{1, 2, 3, \ldots\}$. Not all values need be used; integer values that do not occur are simply assumed to be unobserved.

Remarks

Remarks are presented under the headings

        The case of two raters
        The case of more than two raters

The kappa-statistic measure of agreement is scaled to be 0 when the amount of agreement is what would be expected to be observed by chance and 1 when there is perfect agreement. For intermediate values, Landis and Koch (1977a, 165) suggest the following interpretations:

below 0.0	Poor
0.00–0.20	Slight
0.21–0.40	Fair
0.41–0.60	Moderate
0.61–0.80	Substantial
0.81–1.00	Almost Perfect

The case of two raters

▷ Example 1

Consider the classification by two radiologists of 85 xeromammograms as normal, benign disease, suspicion of cancer, or cancer (a subset of the data from Boyd et al. 1982 and discussed in the context of kappa in Altman 1991, 403–405).

```
. use http://www.stata-press.com/data/r9/rate2
(Altman p. 403)
. tabulate rada radb
```

Radiologist A's assessment	Radiologist B's assessment				Total
	normal	benign	suspect	cancer	
normal	21	12	0	0	33
benign	4	17	1	0	22
suspect	3	9	15	2	29
cancer	0	0	0	1	1
Total	28	38	16	3	85

Our dataset contains two variables: `rada`, radiologist A's assessment and `radb`, radiologist B's assessment. Each observation is a patient.

We can obtain the kappa measure of interrater agreement by typing

```
. kap rada radb
```

Agreement	Expected Agreement	Kappa	Std. Err.	Z	Prob>Z
63.53%	30.82%	0.4728	0.0694	6.81	0.0000

If each radiologist had made his determination randomly (but with probabilities equal to the overall proportions), we would expect the two radiologists to agree on 30.8% of the patients. In fact, they agreed on 63.5% of the patients, or 47.3% of the way between random agreement and perfect agreement. The amount of agreement indicates that we can reject the hypothesis that they are making their determinations randomly.

◁

▷ Example 2: Weighted kappa, prerecorded weight w

There is a difference between two radiologists disagreeing as to whether a xeromammogram indicates cancer or the suspicion of cancer and disagreeing as to whether it indicates cancer or is normal. The weighted kappa attempts to deal with this. `kap` provides two "prerecorded" weights, `w` and `w2`:

```
. kap rada radb, wgt(w)
Ratings weighted by:
    1.0000    0.6667    0.3333    0.0000
    0.6667    1.0000    0.6667    0.3333
    0.3333    0.6667    1.0000    0.6667
    0.0000    0.3333    0.6667    1.0000
```

Agreement	Expected Agreement	Kappa	Std. Err.	Z	Prob>Z
86.67%	69.11%	0.5684	0.0788	7.22	0.0000

The w weights are given by $1 - |i - j|/(k - 1)$, where i and j index the rows of columns of the ratings by the two raters and k is the maximum number of possible ratings. The weighting matrix is printed above the table. In our case, the rows and columns of the 4×4 matrix correspond to the ratings normal, benign, suspicious, and cancerous.

A weight of 1 indicates that an observation should count as perfect agreement. The matrix has 1s down the diagonals—when both radiologists make the same assessment, they are in agreement. A weight of, say, 0.6667 means that they are in two-thirds agreement. In our matrix, they get that score if they are "one apart"—one radiologist assesses cancer and the other is merely suspicious, or one is suspicious and the other says benign, and so on. An entry of 0.3333 means that they are in one-third agreement, or, if you prefer, two-thirds disagreement. That is the score attached when they are "two apart". Finally, they are in complete disagreement when the weight is zero, which happens only when they are three apart—one says cancer and the other says normal.

◁

▷ Example 3: Weighted kappa, prerecorded weight w2

The other prerecorded weight is w2, where the weights are given by $1 - \{(i - j)/(k - 1)\}^2$:

```
. kap rada radb, wgt(w2)
Ratings weighted by:
    1.0000    0.8889    0.5556    0.0000
    0.8889    1.0000    0.8889    0.5556
    0.5556    0.8889    1.0000    0.8889
    0.0000    0.5556    0.8889    1.0000
```

| | Expected | | | | |
Agreement	Agreement	Kappa	Std. Err.	Z	Prob>Z
94.77%	84.09%	0.6714	0.1079	6.22	0.0000

The w2 weight makes the categories even more alike and is probably inappropriate here.

◁

▷ Example 4: Weighted kappa, user-defined weights

In addition to using prerecorded weights, we can define our own weights with the kapwgt command. For instance, we might feel that suspicious and cancerous are reasonably similar, that benign and normal are reasonably similar, but that the suspicious/cancerous group is nothing like the benign/normal group:

```
. kapwgt xm 1 \ .8 1 \ 0 0 1 \ 0 0 .8 1
. kapwgt xm
1.0000
0.8000 1.0000
0.0000 0.0000 1.0000
0.0000 0.0000 0.8000 1.0000
```

We name the weights xm, and after the weight name, we enter the lower triangle of the weighting matrix, using \ to separate rows. In our example, we have four outcomes, so we continued entering numbers until we had defined the fourth row of the weighting matrix. If we type kapwgt followed by a name and nothing else, it shows us the weights recorded under that name. Satisfied that we have entered them correctly, we now use the weights to recalculate kappa:

```
. kap rada radb, wgt(xm)
Ratings weighted by:
    1.0000   0.8000   0.0000   0.0000
    0.8000   1.0000   0.0000   0.0000
    0.0000   0.0000   1.0000   0.8000
    0.0000   0.0000   0.8000   1.0000
              Expected
  Agreement   Agreement    Kappa   Std. Err.        Z    Prob>Z
    80.47%      52.67%     0.5874     0.0865      6.79    0.0000
```

◁

❏ Technical Note

In addition to using weights for weighting the differences in categories, you can specify Stata's traditional weights for weighting the data. In the examples above, we have 85 observations in our dataset—one for each patient. If we only knew the table of outcomes—that there were 21 patients rated normal by both radiologists, etc.—it would be easier to enter the table into Stata and work from it. The easiest way to enter the data is with `tabi`; see [R] **tabulate twoway**.

```
. tabi 21 12 0 0 \ 4 17 1 0 \ 3 9 15 2 \ 0 0 0 1, replace
           |                      col
       row |       1         2         3         4  |   Total
-----------+--------------------------------------- +----------
         1 |      21        12         0         0  |      33
         2 |       4        17         1         0  |      22
         3 |       3         9        15         2  |      29
         4 |       0         0         0         1  |       1
-----------+--------------------------------------- +----------
     Total |      28        38        16         3  |      85
          Pearson chi2(9) =  77.8111   Pr = 0.000
```

`tabi` reported the Pearson χ^2 for this table, but we do not care about it. The important thing is that, with the `replace` option, `tabi` left the table in memory:

```
. list in 1/5

     | row   col   pop |
     |-----------------|
  1. |   1     1    21 |
  2. |   1     2    12 |
  3. |   1     3     0 |
  4. |   1     4     0 |
  5. |   2     1     4 |
```

The variable `row` is radiologist A's assessment, `col` is radiologist B's assessment, and `pop` is the number so assessed by both. Thus

```
. kap row col [freq=pop]
              Expected
  Agreement   Agreement    Kappa   Std. Err.        Z    Prob>Z
    63.53%      30.82%     0.4728     0.0694      6.81    0.0000
```

If we are going to keep these data, the names `row` and `col` are not indicative of what the data reflect. We could (see [U] **12.6 Dataset, variable, and value labels**)

```
. rename row rada
. rename col radb
. label var rada "Radiologist A's assessment"
. label var radb "Radiologist B's assessment"
. label define assess 1 normal 2 benign 3 suspect 4 cancer
. label values rada assess
. label values radb assess
. label data "Altman p. 403"
```

kap's `tab` option, which can be used with or without weighted data, shows the table of assessments:

```
. kap rada radb [freq=pop], tab
```

Radiologist A's assessment	Radiologist B's assessment normal	benign	suspect	cancer	Total
normal	21	12	0	0	33
benign	4	17	1	0	22
suspect	3	9	15	2	29
cancer	0	0	0	1	1
Total	28	38	16	3	85

Agreement	Expected Agreement	Kappa	Std. Err.	Z	Prob>Z
63.53%	30.82%	0.4728	0.0694	6.81	0.0000

❑

❑ Technical Note

You have data on individual patients. There are two raters, and the possible ratings are 1, 2, 3, and 4, but neither rater ever used rating 3:

```
. use http://www.stata-press.com/data/r9/rate2no3
. tabulate ratera raterb
```

ratera	raterb 1	2	4	Total
1	6	4	3	13
2	5	3	3	11
4	1	1	26	28
Total	12	8	32	52

In this case, kap would determine that the ratings are from the set $\{1, 2, 4\}$ because those were the only values observed. kap would expect a user-defined weighting matrix to be 3×3, and if it were not, kap would issue an error message. In the formula-based weights, the calculation would be based on $i, j = 1, 2, 3$ corresponding to the three observed ratings $\{1, 2, 4\}$.

Specifying the `absolute` option would make it clear that the ratings are 1, 2, 3, and 4; it just so happens that rating = 3 was never assigned. If a user-defined weighting matrix were also specified, kap would expect it to be 4×4 or larger (larger because we can think of the ratings being 1, 2, 3, 4, 5, ... and it just so happens that ratings 5, 6, ... were never observed, just as rating = 3 was not observed). In the formula-based weights, the calculation would be based on $i, j = 1, 2, 4$.

```
. kap ratera raterb, wgt(w)
```
Ratings weighted by:
```
     1.0000    0.5000    0.0000
     0.5000    1.0000    0.5000
     0.0000    0.5000    1.0000
```

Agreement	Expected Agreement	Kappa	Std. Err.	Z	Prob>Z
79.81%	57.17%	0.5285	0.1169	4.52	0.0000

```
. kap ratera raterb, wgt(w) absolute
```
Ratings weighted by:
```
     1.0000    0.6667    0.0000
     0.6667    1.0000    0.3333
     0.0000    0.3333    1.0000
```

Agreement	Expected Agreement	Kappa	Std. Err.	Z	Prob>Z
81.41%	55.08%	0.5862	0.1209	4.85	0.0000

If all conceivable ratings are observed in the data, specifying `absolute` makes no difference. For instance, if rater A assigns ratings $\{1, 2, 4\}$ and rater B assigns $\{1, 2, 3, 4\}$, the complete set of assigned ratings is $\{1, 2, 3, 4\}$, the same that `absolute` would specify. Without `absolute`, it makes no difference whether the ratings are coded $\{1, 2, 3, 4\}$, $\{0, 1, 2, 3\}$, $\{1, 7, 9, 100\}$, $\{0, 1, 1.5, 2.0\}$, or otherwise.   ❏

The case of more than two raters

In the case of more than two raters, the mathematics are such that the two raters are not considered unique. For instance, if there are three raters, there is no assumption that the three raters who rate the first subject are the same as the three raters that rate the second. Although we call this the "more than two raters" case, it can be used with two raters when the raters' identities vary.

The nonunique rater case can be usefully broken down into three subcases: (a) there are two possible ratings, which we will call positive and negative; (b) there are more than two possible ratings, but the number of raters per subject is the same for all subjects; and (c) there are more than two possible ratings, and the number of raters per subject varies. kappa handles all these cases. To emphasize that there is no assumption of constant identity of raters across subjects, the variables specified contain counts of the number of raters rating the subject into a particular category.

Jacob Cohen (1923–1998) was born in New York City. After studying psychology at City College of New York and New York University, he worked as a medical psychologist and then from 1959 at New York University. He made many contributions to research methods, including the kappa measure. He persistently emphasized the value of multiple regression and the importance of power and of measuring effects rather than testing significance.

(Continued on next page)

▷ Example 5: Two ratings

Fleiss, Levin and Paik (2003, 612) offers the following hypothetical ratings by different sets of raters on 25 subjects:

Subject	No. of raters	No. of pos. ratings	Subject	No. of raters	No. of pos. ratings
1	2	2	14	4	3
2	2	0	15	2	0
3	3	2	16	2	2
4	4	3	17	3	1
5	3	3	18	2	1
6	4	1	19	4	1
7	3	0	20	5	4
8	5	0	21	3	2
9	2	0	22	4	0
10	4	4	23	3	0
11	5	5	24	3	3
12	3	3	25	2	2
13	4	4			

We have entered these data into Stata, and the variables are called `subject`, `raters`, and `pos`. `kappa`, however, requires that we specify variables containing the number of positive ratings and negative ratings, that is, `pos` and `raters-pos`:

```
. use http://www.stata-press.com/data/r9/p612
. gen neg = raters-pos
. kappa pos neg
Two-outcomes, multiple raters:
```

Kappa	Z	Prob>Z
0.5415	5.28	0.0000

We would have obtained the same results if we had typed `kappa neg pos`.

◁

▷ Example 6: More than two ratings, constant number of raters, kappa

Each of ten subjects is rated into one of three categories by five raters (Fleiss, Levin, and Paik 2003, 615):

```
. use http://www.stata-press.com/data/r9/p615
. list
```

	subject	cat1	cat2	cat3
1.	1	1	4	0
2.	2	2	0	3
3.	3	0	0	5
4.	4	4	0	1
5.	5	3	0	2
6.	6	1	4	0
7.	7	5	0	0
8.	8	0	4	1
9.	9	1	0	4
10.	10	3	0	2

We obtain the kappa statistic:

```
. kappa cat1-cat3
      Outcome │     Kappa          Z      Prob>Z
─────────────┼──────────────────────────────────
         cat1 │    0.2917       2.92      0.0018
         cat2 │    0.6711       6.71      0.0000
         cat3 │    0.3490       3.49      0.0002
─────────────┼──────────────────────────────────
     combined │    0.4179       5.83      0.0000
```

The first part of the output shows the results of calculating kappa for each of the categories separately against an amalgam of the remaining categories. For instance, the `cat1` line is the two-rating kappa, where positive is `cat1` and negative is `cat2` or `cat3`. The test statistic, however, is calculated differently (see *Methods and Formulas*). The combined kappa is the appropriately weighted average of the individual kappas. Note that there is considerably less agreement about the rating of subjects into the first category than there is for the second.

◁

▷ Example 7: More than two ratings, constant number of raters, kap

Now suppose that we have the same data as in the previous example, but that the data are organized differently:

```
. use http://www.stata-press.com/data/r9/p615b
. list
```

	subject	rater1	rater2	rater3	rater4	rater5
1.	1	1	2	2	2	2
2.	2	1	1	3	3	3
3.	3	3	3	3	3	3
4.	4	1	1	1	1	3
5.	5	1	1	1	3	3
6.	6	1	2	2	2	2
7.	7	1	1	1	1	1
8.	8	2	2	2	2	3
9.	9	1	3	3	3	3
10.	10	1	1	1	3	3

In this case, we would use `kap` rather than `kappa` since the variables record ratings for each rater.

```
. kap rater1 rater2 rater3 rater4 rater5
There are 5 raters per subject:
      Outcome │     Kappa          Z      Prob>Z
─────────────┼──────────────────────────────────
            1 │    0.2917       2.92      0.0018
            2 │    0.6711       6.71      0.0000
            3 │    0.3490       3.49      0.0002
─────────────┼──────────────────────────────────
     combined │    0.4179       5.83      0.0000
```

Note that it does not matter which rater is which when there are more than two raters.

◁

▷ Example 8: More than two ratings, varying number of raters, kappa

In this unfortunate case, kappa can be calculated, but there is no test statistic for testing against $\kappa > 0$. We do nothing differently—kappa calculates the total number of raters for each subject, and, if it is not a constant, kappa suppresses the calculation of test statistics.

```
. use http://www.stata-press.com/data/r9/rvary
. list
```

	subject	cat1	cat2	cat3
1.	1	1	3	0
2.	2	2	0	3
3.	3	0	0	5
4.	4	4	0	1
5.	5	3	0	2
6.	6	1	4	0
7.	7	5	0	0
8.	8	0	4	1
9.	9	1	0	2
10.	10	3	0	2

```
. kappa cat1-cat3
```

Outcome	Kappa	Z	Prob>Z
cat1	0.2685	.	.
cat2	0.6457	.	.
cat3	0.2938	.	.
combined	0.3816	.	.

```
note:  Number of ratings per subject vary; cannot calculate test
       statistics.
```

◁

▷ Example 9: More than two ratings, varying number of raters, kap

This case is similar to the previous example, but the data are organized differently:

```
. use http://www.stata-press.com/data/r9/rvary2
. list
```

	subject	rater1	rater2	rater3	rater4	rater5
1.	1	1	2	2	.	2
2.	2	1	1	3	3	3
3.	3	3	3	3	3	3
4.	4	1	1	1	1	3
5.	5	1	1	1	3	3
6.	6	1	2	2	2	2
7.	7	1	1	1	1	1
8.	8	2	2	2	2	3
9.	9	1	3	.	.	3
10.	10	1	1	1	3	3

In this case, we specify kap instead of kappa since the variables record ratings for each rater.

```
. kap rater1-rater5
There are between 3 and 5 (median = 5.00) raters per subject:
```

Outcome	Kappa	Z	Prob>Z
1	0.2685	.	.
2	0.6457	.	.
3	0.2938	.	.
combined	0.3816	.	.

```
note:  Number of ratings per subject vary; cannot calculate test
       statistics.
```

◁

Saved Results

kap and kappa save in r():

Scalars

r(N)	number of subjects (kap only)	r(kappa)	kappa
r(prop_o)	observed proportion of agreement (kap only)	r(z)	z statistic
r(prop_e)	expected proportion of agreement (kap only)	r(se)	standard error for kappa statistic

Methods and Formulas

kap, kapwgt, and kappa are implemented as ado-files.

The kappa statistic was first proposed by Cohen (1960). The generalization for weights reflecting the relative seriousness of each possible disagreement is due to Cohen (1968). The analysis-of-variance approach for $k = 2$ and $m \geq 2$ is due to Landis and Koch (1977b). See Altman (1991, 403–409) or Dunn (2000, chapter 2) for an introductory treatment and Fleiss, Levin, and Paik (2003, chapter 18) for a more detailed treatment. All formulas below are as presented in Fleiss, Levin, and Paik (2003). Let m be the number of raters, and let k be the number of rating outcomes.

kap: m = 2

Define w_{ij} $(i = 1, \ldots, k, j = 1, \ldots, k)$ as the weights for agreement and disagreement (wgt()), or, if the data are not weighted, define $w_{ii} = 1$ and $w_{ij} = 0$ for $i \neq j$. If wgt(w) is specified, $w_{ij} = 1 - |i - j|/(k - 1)$. If wgt(w2) is specified, $w_{ij} = 1 - \left\{ (i - j)/(k - 1) \right\}^2$.

The observed proportion of agreement is

$$p_o = \sum_{i=1}^{k} \sum_{j=1}^{k} w_{ij} p_{ij}$$

where p_{ij} is the fraction of ratings i by the first rater and j by the second. The expected proportion of agreement is

$$p_e = \sum_{i=1}^{k} \sum_{j=1}^{k} w_{ij} p_{i\cdot} p_{\cdot j}$$

where $p_{i\cdot} = \sum_j p_{ij}$ and $p_{\cdot j} = \sum_i p_{ij}$.

Kappa is given by $\widehat{\kappa} = (p_o - p_e)/(1 - p_e)$.

The standard error of $\widehat{\kappa}$ for testing against 0 is

$$\widehat{s}_0 = \frac{1}{(1 - p_e)\sqrt{n}} \left(\left[\sum_i \sum_j p_{i\cdot} p_{\cdot j} \{ w_{ij} - (\overline{w}_{i\cdot} + \overline{w}_{\cdot j}) \}^2 \right] - p_e^2 \right)^{1/2}$$

where n is the number of subjects being rated, $\overline{w}_{i\cdot} = \sum_j p_{\cdot j} w_{ij}$, and $\overline{w}_{\cdot j} = \sum_i p_{i\cdot} w_{ij}$. The test statistic $Z = \kappa/s_0$ is assumed to be distributed $N(0, 1)$.

kappa: m > 2, k = 2

Each subject i, $i = 1, \ldots, n$ is found by x_i of m_i raters to be positive (the choice as to what is labeled positive is arbitrary).

The overall proportion of positive ratings is $\overline{p} = \sum_i x_i/(n\overline{m})$, where $\overline{m} = \sum_i m_i/n$. The between-subjects mean square is (approximately)

$$B = \frac{1}{n} \sum_i \frac{(x_i - m_i\overline{p})^2}{m_i}$$

and the within-subject mean square is

$$W = \frac{1}{n(\overline{m} - 1)} \sum_i \frac{x_i(m_i - x_i)}{m_i}$$

Kappa is then defined as

$$\widehat{\kappa} = \frac{B - W}{B + (\overline{m} - 1)W}$$

The standard error for testing against 0 (Fleiss and Cuzick 1979) is approximately equal to and is calculated as

$$\widehat{s}_0 = \frac{1}{(\overline{m} - 1)\sqrt{n\overline{m}_H}} \left\{ 2(\overline{m}_H - 1) + \frac{(\overline{m} - \overline{m}_H)(1 - 4\overline{pq})}{\overline{m}\,\overline{pq}} \right\}^{1/2}$$

where $\overline{m}_H$ is the harmonic mean of m_i and $\overline{q} = 1 - \overline{p}$.

The test statistic $Z = \widehat{\kappa}/\widehat{s}_0$ is assumed to be distributed N(0, 1).

kappa: m > 2, k > 2

Let x_{ij} be the number or ratings on subject i, $i = 1, \ldots, n$ into category j, $j = 1, \ldots, k$. Define $\bar{p}_j$ as the overall proportion of ratings in category j, $\bar{q}_j = 1 - \bar{p}_j$, and let $\widehat{\kappa}_j$ be the kappa statistic given above for $k = 2$ when category j is compared with the amalgam of all other categories. Kappa is (Landis and Koch 1977b)

$$\bar{\kappa} = \frac{\sum_j \bar{p}_j \bar{q}_j \widehat{\kappa}_j}{\sum_j \bar{p}_j \bar{q}_j}$$

In the case where the number of raters per subject $\sum_j x_{ij}$ is a constant m for all i, Fleiss, Nee, and Landis (1979) derived the following formulas for the approximate standard errors. The standard error for testing $\widehat{\kappa}_j$ against 0 is

$$\widehat{s}_j = \left\{ \frac{2}{nm(m-1)} \right\}^{1/2}$$

and the standard error for testing $\bar{\kappa}$ is

$$\bar{s} = \frac{\sqrt{2}}{\sum_j \bar{p}_j \bar{q}_j \sqrt{nm(m-1)}} \left\{ \left(\sum_j \bar{p}_j \bar{q}_j \right)^2 - \sum_j \bar{p}_j \bar{q}_j (\bar{q}_j - \bar{p}_j) \right\}^{1/2}$$

References

Abramson, J. H. and Z. H. Abramson. 2001. *Making Sense of Data: A Self-Instruction Manual on the Interpretation of Epidemiological Data*. 3rd ed. New York: Oxford University Press.

Altman, D. G. 1991. *Practical Statistics for Medical Research*. London: Chapman & Hall.

Boyd, N. F., C. Wolfson, M. Moskowitz, T. Carlile, M. Petitclerc, H. A. Ferri, E. Fishell, A. Gregoire, M. Kiernan, J. D. Longley, I. S. Simor, and A. B. Miller. 1982. Observer variation in the interpretation of xeromammograms. *Journal of the National Cancer Institute* 68: 357–63.

Campbell, M. J. and D. Machin. 1999. *Medical Statistics: A Commonsense Approach*. 3rd ed. New York: Wiley.

Cohen, J. 1960. A coefficient of agreement for nominal scales. *Educational and Psychological Measurement* 20: 37–46.

———. 1968. Weighted kappa: Nominal scale agreement with provision for scaled disagreement or partial credit. *Psychological Bulletin* 70: 213–220.

Dunn, G. 2000. *Statistics in Psychiatry*. London: Arnold.

Fleiss, J. L. and J. Cuzick. 1979. The reliability of dichotomous judgments: Unequal numbers of judges per subject. *Applied Psychological Measurement* 3: 537–542.

Fleiss, J. L., B. Levin, and M. C. Paik. 2003. *Statistical Methods for Rates and Proportions*. 3rd ed. New York: Wiley.

Fleiss, J. L., J. C. M. Nee, and J. R. Landis. 1979. Large sample variance of kappa in the case of different sets of raters. *Psychological Bulletin* 86: 974–977.

Gould, W. 1997. stata49: Interrater agreement. *Stata Technical Bulletin* 40: 2–8. Reprinted in *Stata Technical Bulletin Reprints*, vol. 7, pp. 20–28.

Landis, J. R. and G. G. Koch. 1977a. The measurement of observer agreement for categorical data. *Biometrics* 33: 159–174.

———. 1977b. A one-way components of variance model for categorical data. *Biometrics* 33: 671–679.

Reichenheim, M. E. 2000. sxd3: Sample size for the kappa statistic of interrater agreement. *Stata Technical Bulletin* 58: 41–45. Reprinted in *Stata Technical Bulletin Reprints*, vol. 10, pp. 382–387.

———. 2004. Confidence intervals for the kappa statistic. *Stata Journal* 4: 421–428.

Shrout, P. E. 2001. Jacob Cohen (1923–1998). *American Psychologist* 56: 166.

Steichen, T. J. and N. J. Cox. 1998a. sg84: Concordance correlation coefficient. *Stata Technical Bulletin* 43: 35–39. Reprinted in *Stata Technical Bulletin Reprints*, vol. 8, pp. 137–143.

——. 1998b. sg84.1: Concordance correlation coefficient, revisited. *Stata Technical Bulletin* 45: 21–23. Reprinted in *Stata Technical Bulletin Reprints*, vol. 8, pp. 143–145.

——. 2000a. sg84.2: Concordance correlation coefficient: Update for Stata 6. *Stata Technical Bulletin* 54: 25–26. Reprinted in *Stata Technical Bulletin Reprints*, vol. 9, pp. 169–170.

——. 2000b. sg84.3: Concordance correlation coefficient: Minor corrections. *Stata Technical Bulletin* 58: 9. Reprinted in *Stata Technical Bulletin Reprints*, vol. 10, pp. 137.

——. 2002. A note on the concordance correlation coefficient. *Stata Journal* 2: 183–189.

Also See

Related: [R] **tabulate twoway**

Title

kdensity — Univariate kernel density estimation

Syntax

kdensity *varname* [*if*] [*in*] [*weight*] [, *options*]

options	description
Main	
<u>epan</u>echnikov	Epanechnikov kernel function; the default
epan2	alternative Epanechnikov kernel function
<u>bi</u>weight	biweight kernel function
<u>cos</u>ine	cosine trace
<u>gau</u>ssian	Gaussian kernel function
<u>par</u>zen	Parzen kernel function
<u>rec</u>tangle	rectangle kernel function
<u>tri</u>angle	triangle kernel function
<u>g</u>enerate(*newvar$_x$ newvar$_d$*)	store the estimation points in *newvar$_x$* and the density estimate in *newvar$_d$*
<u>w</u>idth(#)	halfwidth of kernel
n(#)	estimate density using # points; default is min(N, 50)
at(*var$_x$*)	estimate density using the values specified by *var$_x$*
<u>nog</u>raph	suppress graph
Kernel plot	
cline_options	affect rendition of the plotted kernel density estimate
Density plots	
<u>normal</u>	add normal density to the graph
<u>normo</u>pts(*cline_options*)	affect rendition of normal density
<u>stu</u>dent(#)	add Student's t density with # degrees of freedom to the graph
<u>sto</u>pts(*cline_options*)	affect rendition of the Student's t density
Add plot	
<u>addp</u>lot(*plot*)	add other plots to the generated graph
Y-Axis, X-Axis, Title, Caption, Legend, Overall	
twoway_options	any options other than by() documented in [G] *twoway_options*

fweights and aweights are allowed; see [U] **11.1.6 weight**.

Description

kdensity produces kernel density estimates and graphs the result.

Options

> **Main**

epanechnikov, epan2, biweight, cosine, gaussian, parzen, rectangle, and triangle specify the kernel. By default, epanechnikov, specifying the Epanechnikov kernel, is used.

generate(*newvar_x newvar_d*) stores the results of the estimation. *newvar_x* will contain the points at which the density is estimated. *newvar_d* will contain the density estimate.

width(#) specifies the halfwidth of the kernel, the width of the density window around each point. If w() is not specified, the "optimal" width is calculated and used. The optimal width is the width that would minimize the mean integrated squared error if the data were Gaussian and a Gaussian kernel were used, so it is not optimal in any global sense. In fact, for multimodal and highly skewed densities, this width is usually too wide and oversmooths the density (Silverman 1986).

n(#) specifies the number of points at which the density estimate is to be evaluated. The default is $\min(N, 50)$, where N is the number of observations in memory.

at(*var_x*) specifies a variable that contains the values at which the density should be estimated. This option allows you to more easily obtain density estimates for different variables or different subsamples of a variable and then overlay the estimated densities for comparison.

nograph suppresses the graph. This option is often used in combination with the generate() option.

> **Kernel plot**

cline_options affect the rendition of the plotted kernel density estimate. See [G] ***connect_options***.

> **Density plots**

normal requests that a normal density be overlaid on the density estimate for comparison.

normopts(*cline_options*) specifies details about the rendition of the normal curve, such as the color and style of line used. See [G] ***connect_options***.

student(#) specifies that a Student's t density with # degrees of freedom be overlaid on the density estimate for comparison.

stopts(*cline_options*) affect the rendition of the Student's t density. See [G] ***connect_options***.

> **Add plot**

addplot(*plot*) provides a way to add other plots to the generated graph. See [G] ***addplot_option***.

> **Y-Axis, X-Axis, Title, Caption, Legend, Overall**

twoway_options are any of the options documented in [G] ***twoway_options***, excluding by(). These include options for titling the graph (see [G] ***title_options***) and options for saving the graph to disk (see [G] ***saving_option***).

Remarks

Kernel density estimators approximate the density $f(x)$ from observations on x. Histograms do this, too, and the histogram itself is a kind of kernel density estimate. The data are divided into nonoverlapping intervals, and counts are made of the number of data points within each interval. Histograms are bar graphs that depict these frequency counts—the bar is centered at the midpoint of each interval—and its height reflects the average number of data points in the interval.

In more general kernel density estimates, the range is still divided into intervals, and estimates of the density at the center of intervals are produced. One difference is that the intervals are allowed to overlap. We can think of sliding the interval—called a window—along the range of the data and collecting the center-point density estimates. The second difference is that, rather than merely counting the number of observations in a window, a kernel density estimator assigns a weight between 0 and 1—based on the distance from the center of the window—and sums the weighted values. The function that determines these weights is called the kernel.

Kernel density estimates have the advantages of being smooth and of being independent of the choice of origin (corresponding to the location of the bins in a histogram).

See Salgado-Ugarte, Shimizu, and Taniuchi (1993) and Fox (1990) for discussions of kernel density estimators that stress their use as exploratory data-analysis tools.

▷ Example 1: Histogram and kernel density estimate

Goeden (1978) reports data consisting of 316 length observations of coral trout. We wish to investigate the underlying density of the lengths. To begin on familiar ground, we might draw a histogram. In [R] **histogram**, we suggest setting the bins to $\min(\sqrt{n}, 10 \cdot \log_{10} n)$, which for $n = 316$ is roughly 18:

```
. use http://www.stata-press.com/data/r9/trocolen

. histogram length, bin(18)
(bin=18, start=226, width=19.777778)
```

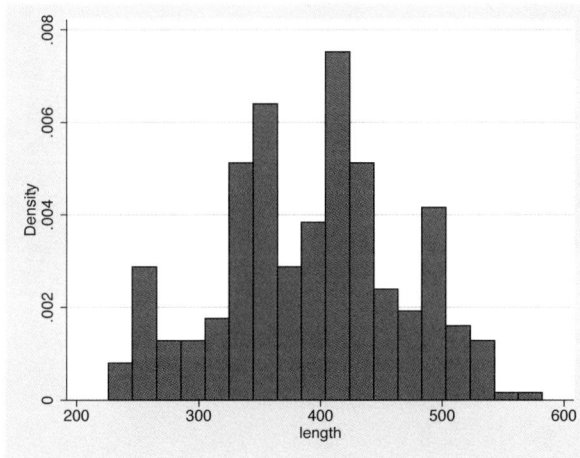

The kernel density estimate, on the other hand, is smooth.

(*Continued on next page*)

```
. kdensity length
```

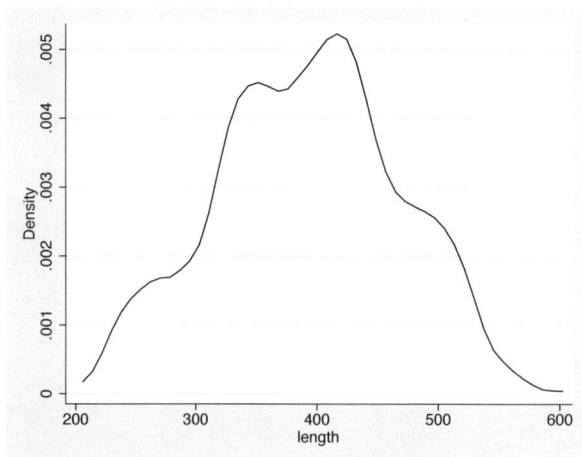

Kernel density estimators are, however, sensitive to an assumption, just as are histograms. In histograms, we specify a number of bins. For kernel density estimators, we specify a width. In the graph above, we used the default width. `kdensity` is smarter than `twoway histogram` in that its default width is not a fixed constant. Even so, the default width is not necessarily best.

`kdensity` saves the width in the returned scalar `width`, so typing `display r(width)` reveals it. Doing this, we discover that the width is approximately 20.

Widths are similar to the inverse of the number of bins in a histogram in that smaller widths provide more detail. The units of the width are the units of x, the variable being analyzed. The width is specified as a halfwidth, meaning that the kernel density estimator with halfwidth 20 corresponds to sliding a window of size 40 across the data.

We can specify halfwidths for ourselves using the `width()` option. Smaller widths do not smooth the density as much:

```
. kdensity length, w(10)
```

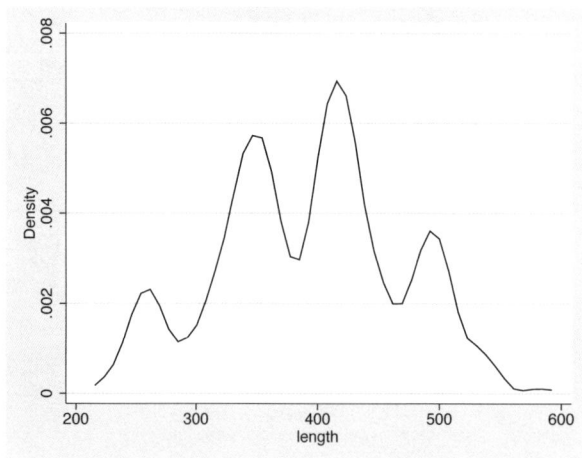

```
. kdensity length, w(15)
```

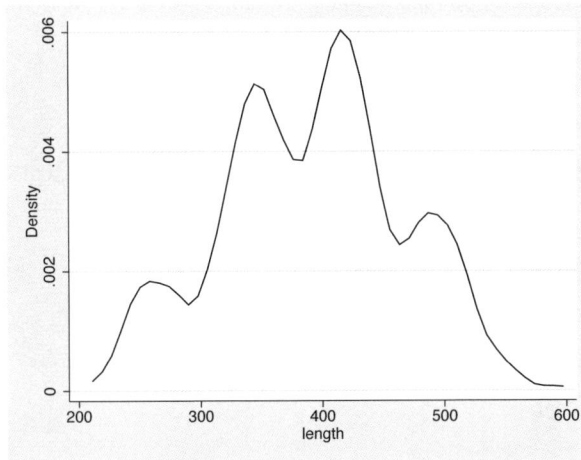

◁

▷ Example 2: Different kernels can produce different results

When widths are held constant, different kernels can produce surprisingly different results. This is really an attribute of the kernel and width combination; for a given width, some kernels are more sensitive than others at identifying peaks in the density estimate.

We can see this when using a dataset with lots of peaks. In the automobile dataset, we characterize the density of `weight`, the weight of the vehicles. Below we compare the Epanechnikov and Parzen kernels.

```
. use http://www.stata-press.com/data/r9/auto, clear
(1978 Automobile Data)
. kdensity weight, epan nogr g(x epan)
. kdensity weight, parzen nogr g(x2 parzen)
. label var epan "Epanechnikov density estimate"
. label var parzen "Parzen density estimate"
. line epan parzen x, sort ytitle(Density) legend(cols(1))
```

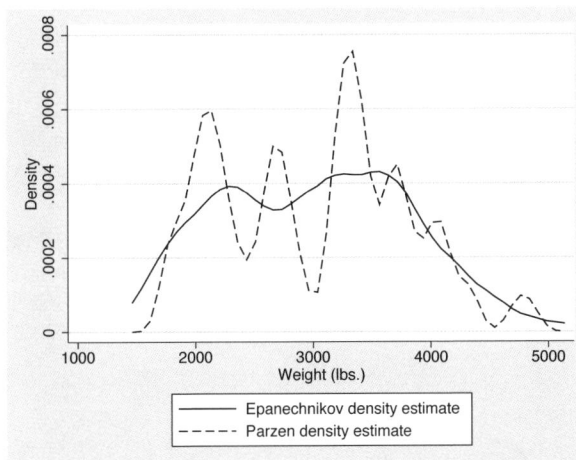

We did not specify a width, so we obtained the default width. That width is not a function of the selected kernel, but of the data. See *Methods and Formulas* for the calculation of the optimal width.

◁

▷ Example 3: Density with overlaid normal density

In examining the density estimates, we may wish to overlay a normal density or a Student's t density for comparison. Using automobile weights, we can get an idea of the distance from normality by using the `normal` option.

```
. kdensity weight, epan normal
```

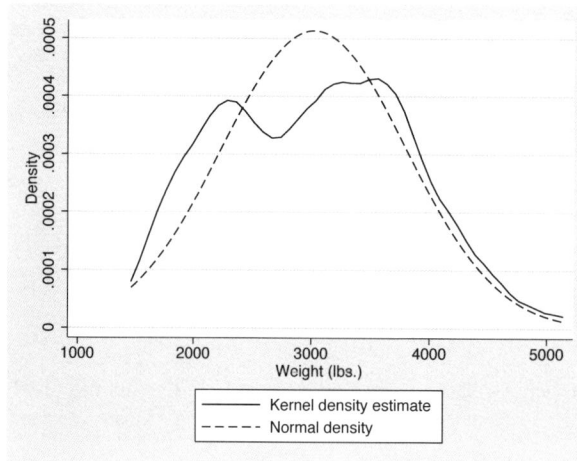

◁

▷ Example 4: Compare two densities

We also may want to compare two or more densities. In this example, we will compare the density estimates of the weights for the foreign and domestic cars.

```
. use http://www.stata-press.com/data/r9/auto, clear
(1978 Automobile Data)
. kdensity weight, nogr gen(x fx)
. kdensity weight if foreign==0, nogr gen(fx0) at(x)
. kdensity weight if foreign==1, nogr gen(fx1) at(x)
. label var fx0 "Domestic cars"
. label var fx1 "Foreign cars"
```

```
. line fx0 fx1 x, sort ytitle(Density)
```

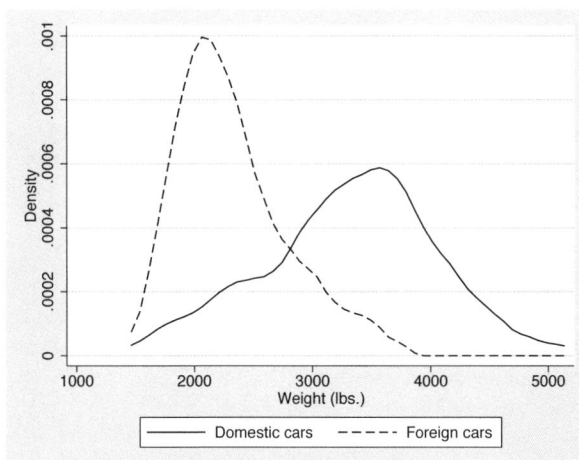

☐ Technical Note

Although all the examples we included had densities of less than 1, the density may exceed 1.

The probability density $f(x)$ of a continuous variable, x, has the units and dimensions of the reciprocal of x. If x is measured in meters, $f(x)$ has units 1/meter. Thus the density is not measured on a probability scale, so it is possible for $f(x)$ to exceed 1.

To see this, think of a uniform density on the interval 0 to 1. The area under the density curve is 1: this is the product of the density, which is constant at 1, and the range, which is 1. If the variable is then transformed by doubling, the area under the curve remains 1 and is the product of the density, constant at 0.5, and the range, which is 2. Conversely, if the variable is transformed by halving, the area under the curve also remains at 1 and is the product of the density, constant at 2, and the range, which is 0.5. (Strictly, the range is measured in certain units, and the density is measured in the reciprocal of those units, so the units cancel on multiplication.)

☐

Saved Results

kdensity saves in r():

Scalars
r(width)	kernel bandwidth
r(n)	number of points at which the estimate was evaluated
r(scale)	density bin width

Macros
r(kernel)	name of kernel

Methods and Formulas

kdensity is implemented as an ado-file.

A kernel density estimate is formed by summing the weighted values calculated with the kernel function K, as in

$$\widehat{f}_K = \frac{1}{nh} \sum_{i=1}^{n} K\left(\frac{x - X_i}{h}\right)$$

where we may define various kernel functions. kdensity includes seven different kernel functions. The Epanechnikov is the default function if no other kernel is specified and is the most efficient in minimizing the mean integrated squared error.

Kernel	Formula	
Biweight	$K[z] = \begin{cases} \frac{15}{16}(1 - z^2)^2 \\ 0 \end{cases}$	if $|z| < 1$ otherwise
Cosine	$K[z] = \begin{cases} 1 + \cos(2\pi z) \\ 0 \end{cases}$	if $|z| < 1/2$ otherwise
Epanechnikov	$K[z] = \begin{cases} \frac{3}{4}(1 - \frac{1}{5}z^2)/\sqrt{5} \\ 0 \end{cases}$	if $|z| < \sqrt{5}$ otherwise
Epan2	$K[z] = \begin{cases} \frac{3}{4}(1 - z^2) \\ 0 \end{cases}$	if $|z| < 1$ otherwise
Gaussian	$K[z] = \frac{1}{\sqrt{2\pi}} e^{-z^2/2}$	
Parzen	$K[z] = \begin{cases} \frac{4}{3} - 8z^2 + 8|z|^3 \\ 8(1 - |z|)^3/3 \\ 0 \end{cases}$	if $|z| \leq 1/2$ if $1/2 < |z| \leq 1$ otherwise
Rectangular	$K[z] = \begin{cases} 1/2 \\ 0 \end{cases}$	if $|z| < 1$ otherwise
Triangular	$K[z] = \begin{cases} 1 - |z| \\ 0 \end{cases}$	if $|z| < 1$ otherwise

From the definitions given in the table, we can see that the choice of h will drive how many values are included in estimating the density at each point. This value is called the *window width* or *bandwidth*. If the window width is not specified, it is determined as

$$m = \min\left(\sqrt{\text{variance}_x}, \ \frac{\text{interquartile range}_x}{1.349}\right)$$

$$h = \frac{0.9m}{n^{1/5}}$$

where x is the variable for which we wish to estimate the kernel and n is the number of observations.

Most researchers agree that the choice of kernel is not as important as the choice of bandwidth. There is a great deal of literature on choosing bandwidths under various conditions; see, for example, Parzen (1962) or Tapia and Thompson (1978). Also see Newton (1988) for a comparison with sample spectral density estimation in time-series applications.

Acknowledgments

We gratefully acknowledge the previous work by Isaías H. Salgado-Ugarte of Universidad Nacional Autónoma de México, and Makoto Shimizu and Toru Taniuchi of the University of Tokyo; see Salgado-Ugarte, Shimizu, and Taniuchi (1993). Their article provides a good overview of the subject of univariate kernel density estimation and presents arguments for its use in exploratory data analysis.

References

Fiorio, C. V. 2004. Confidence intervals for kernel density estimation. *Stata Journal* 4: 168–179.

Fox, J. 1990. Describing univariate distributions. In *Modern Methods of Data Analysis*, ed. J. Fox and J. S. Long, 58–125. Newbury Park, CA: Sage.

Goeden, G. B. 1978. A monograph of the coral trout, *Plectropomus leopardus* (Lacépède). *Res. Bull. Fish. Serv. Queensl.* 1: 42 p.

Newton, H. J. 1988. *TIMESLAB: A Time Series Analysis Laboratory*. Belmont, CA: Wadsworth & Brooks/Cole.

Parzen, E. 1962. On estimation of a probability density function and mode. *Annals of Mathematical Statistics* 32: 1065–1076.

Salgado-Ugarte, I. H. and M. A. Pérez-Hernández. 2003. Exploring the use of variable bandwidth kernel density estimators. *Stata Journal* 3: 133–147.

Salgado-Ugarte, I. H., M. Shimizu, and T. Taniuchi. 1993. snp6: Exploring the shape of univariate data using kernel density estimators. *Stata Technical Bulletin* 16: 8–19. Reprinted in *Stata Technical Bulletin Reprints*, vol. 3, pp. 155–173.

——. 1995a. snp6.1: ASH, WARPing, and kernel density estimation for univariate data. *Stata Technical Bulletin* 26: 23–31. Reprinted in *Stata Technical Bulletin Reprints*, vol. 5, pp. 161–172.

——. 1995b. snp6.2: Practical rules for bandwidth selection in univariate density estimation. *Stata Technical Bulletin* 27: 5–19. Reprinted in *Stata Technical Bulletin Reprints*, vol. 5, pp. 172–190.

——. 1997. snp13: Nonparametric assessment of multimodality for univariate data. *Stata Technical Bulletin* 38: 27–35. Reprinted in *Stata Technical Bulletin Reprints*, vol. 7, pp. 232–243.

Scott, D. W. 1992. *Multivariate Density Estimation: Theory, Practice, and Visualization*. New York: Wiley.

Silverman, B. W. 1986. *Density Estimation for Statistics and Data Analysis*. London: Chapman & Hall.

Simonoff, J. S. 1996. *Smoothing Methods in Statistics*. New York: Springer.

Steichen, T. J. 1998. gr33: Violin plots. *Stata Technical Bulletin* 46: 13–18. Reprinted in *Stata Technical Bulletin Reprints*, vol. 8, pp. 57–65.

Tapia, R. A. and J. R. Thompson. 1978. *Nonparametric Probability Density Estimation*. Baltimore: Johns Hopkins University Press.

Van Kerm, P. 2003. Adaptive kernel density estimation. *Stata Journal* 3: 148–156.

Wand, M. P. and M. C. Jones. 1995. *Kernel Smoothing*. London: Chapman & Hall.

Also See

Related:	[R] **histogram**, [R] **lowess**
Background:	*Stata Graphics Reference Manual*

Title

> **ksmirnov** — Kolmogorov–Smirnov equality-of-distributions test

Syntax

One-sample Kolmogorov–Smirnov test

>   ksmirnov *varname* = *exp* [*if*] [*in*]

Two-sample Kolmogorov–Smirnov test

>   ksmirnov *varname* [*if*] [*in*] , by(*groupvar*) [<u>e</u>xact]

Description

ksmirnov performs one- and two-sample Kolmogorov–Smirnov tests of the equality of distributions. In the first syntax, *varname* is the variable whose distribution is being tested, and *exp* must evaluate to the corresponding (theoretical) cumulative. In the second syntax, *groupvar* must take on two distinct values. The distribution of *varname* for the first value of *groupvar* is compared with that of the second value.

When testing for normality, please see [R] **sktest** and [R] **swilk**.

Options for two-sample test

> Main

by(*groupvar*) is required. It specifies a binary variable that identifies the two groups.

exact specifies that the exact p-value be computed. This may take a long time if $n > 50$.

Remarks

▷ Example 1: Two-sample test

Say that we have data on x that resulted from two different experiments, labeled as group==1 and group==2. Our data contain

```
. list, sep(4)
```

	group	x
1.	2	2
2.	1	0
3.	2	3
4.	1	4
5.	1	5
6.	2	8
7.	2	10

We wish to use the two-sample Kolmogorov–Smirnov test to determine if there are any differences in the distribution of x for these two groups:

```
. ksmirnov x, by(group)
Two-sample Kolmogorov-Smirnov test for equality of distribution functions:
Smaller group        D       P-value  Corrected
1:                0.5000     0.424
2:               -0.1667     0.909
Combined K-S:     0.5000     0.785       0.735
```

The first line tests the hypothesis that x for group 1 contains *smaller* values than group 2. The largest difference between the distribution functions is 0.5. The approximate p-value for this is 0.424, which is not significant.

The second line tests the hypothesis that x for group 1 contains *larger* values than group 2. The largest difference between the distribution functions in this direction is 0.1667. The approximate p-value for this small difference is 0.909.

Finally, the approximate p-value for the combined test is 0.785, corrected to 0.735. The p-values ksmirnov calculates are based on the asymptotic distributions derived by Smirnov (1939). These approximations are not very good for small samples ($n < 50$). They are too conservative—real p-values tend to be substantially smaller. We have also included a less conservative approximation for the nondirectional hypothesis based on an empirical continuity correction—the 0.735 reported in the third column.

That number, too, is only an approximation. An exact value can be calculated using the exact option:

```
. ksmirnov x, by(group) exact
Two-sample Kolmogorov-Smirnov test for equality of distribution functions:
Smaller group        D       P-value    Exact
1:                0.5000     0.424
2:               -0.1667     0.909
Combined K-S:     0.5000     0.785     0.657
```

◁

▷ Example 2: One-sample test

Let's now test whether x in the example above is distributed normally. Kolmogorov–Smirnov is not a particularly powerful test in testing for normality, and we do not endorse such use of it; see [R] **sktest** and [R] **swilk** for better tests.

In any case, we will test against a normal distribution with the same mean and standard deviation:

```
. summarize x
    Variable |      Obs       Mean   Std. Dev.       Min        Max
           x |        7   4.571429   3.457222          0         10
. ksmirnov x = normal((x-4.571429)/3.457222)
One-sample Kolmogorov-Smirnov test against theoretical distribution
         normal((x-4.571429)/3.457222)
Smaller group        D       P-value  Corrected
x:                0.1650     0.683
Cumulative:      -0.1250     0.803
Combined K-S:     0.1650     0.991       0.978
```

Since Stata has no way of knowing that we based this calculation on the calculated mean and standard deviation of x, the test statistics will be slightly conservative in addition to being approximations. Nevertheless, they clearly indicate that the data cannot be distinguished from normally distributed data.

◁

Saved Results

ksmirnov saves in r():

Scalars

r(D_1)	D from line 1	r(D)	combined D
r(p_1)	p-value from line 1	r(p)	combined p-value
r(D_2)	D from line 2	r(p_cor)	corrected combined p-value
r(p_2)	p-value from line 2	r(p_exact)	exact combined p-value

Macros

r(group1)	name of group from line 1	r(group2)	name of group from line 2

Methods and Formulas

ksmirnov is implemented as an ado-file.

In general, the Kolmogorov–Smirnov test (Kolmogorov 1933; Smirnov 1939; also see Conover 1999, 428–465) is not very powerful against differences in the tails of distributions. In return for this, it is fairly powerful for alternative hypotheses that involve lumpiness or clustering in the data.

The directional hypotheses are evaluated with the statistics

$$D^+ = \max_x \left\{ F(x) - G(x) \right\}$$
$$D^- = \min_x \left\{ F(x) - G(x) \right\}$$

where $F(x)$ and $G(x)$ are the empirical distribution functions for the sample being compared. The combined statistic is

$$D = \max \left(|D^+|, |D^-| \right)$$

The p-value for this statistic may be obtained by evaluating the asymptotic limiting distribution. Let m be the sample size for the first sample, and let n be the sample size for the second sample. Smirnov (1939) shows that

$$\lim_{m,n \to \infty} \Pr\left\{ \sqrt{mn/(m+n)} D_{m,n} \le z \right\} = 1 - 2 \sum_{i=1}^{\infty} (-1)^{i-1} \exp\left(-2i^2 z^2 \right)$$

The first five terms form the approximation P_a used by Stata. The exact p-value is calculated by a counting algorithm; see Gibbons (1971, 127–131). A corrected p-value was obtained by modifying the asymptotic p-value using a numerical approximation technique

$$Z = \Phi^{-1}(P_a) + 1.04/\min(m,n) + 2.09/\max(m,n) - 1.35/\sqrt{mn/(m+n)}$$
$$p\text{-value} = \Phi(Z)$$

where $\Phi()$ is the cumulative normal distribution.

Andrei Nikolayevich Kolmogorov (1903–1987), of Russia, was one of the great mathematicians of the twentieth century, making outstanding contributions in many different branches, including set theory, measure theory, probability and statistics, approximation theory, functional analysis, classical dynamics, and theory of turbulence. He was a faculty member at Moscow State University for over 60 years.

Nikolai Vasilyevich Smirnov (1900–1966) was a Russian statistician whose work included contributions in nonparametric statistics, order statistics, and goodness of fit. After army service and the study of philosophy and philology, he turned to mathematics and eventually rose to be head of Mathematical Statistics at the Steklov Mathematical Institute in Moscow.

References

Aivazian, S. A. 1997. Smirnov, Nikolai Vasilyevich. In *Leading Personalities in Statistical Sciences from the Seventeenth Century to the Present*, ed. N. L. Johnson and S. Kotz, 208–210. New York: Wiley.

Conover, W. J. 1999. *Practical Nonparametric Statistics*. 3rd ed. New York: Wiley.

Gibbons, J. D. 1971. *Nonparametric Statistical Inference*. New York: McGraw–Hill.

Johnson, N. L. and S. Kotz. 1997. Kolmogorov, Andrei Nikolayevich. In *Leading Personalities in Statistical Sciences from the Seventeenth Century to the Present*, ed. N. L. Johnson and S. Kotz, 255–256. New York: Wiley.

Kolmogorov, A. N. 1933. Sulla determinazione empirica di una legge di distribuzione. *Giornale dell' Istituto Italiano degli Attuari* 4: 83–91.

Riffenburgh, R. H. 1999. *Statistics in Medicine*. San Diego: Academic Press.

Smirnov, N. V. 1939. Estimate of deviation between empirical distribution functions in two independent samples (in Russian). *Bulletin Moscow University* 2(2): 3–16.

Also See

Related: [R] **runtest**, [R] **sktest**, [R] **swilk**

Title

> **kwallis** — Kruskal–Wallis equality-of-populations rank test

Syntax

kwallis *varname* $\left[\,if\,\right]$ $\left[\,in\,\right]$, by(*groupvar*)

Description

kwallis tests the hypothesis that several samples are from the same population. In the syntax diagram above, *varname* refers to the variable recording the outcome, and *groupvar* refers to the variable denoting the population. Note that by() is required.

Option

by(*groupvar*) is required. It specifies a binary variable that identifies the two groups.

Remarks

▷ Example 1

We have data on the 50 states. The data contain the median age of the population medage and the region of the country region for each state. We wish to test for the equality of the median age distribution across all four regions simultaneously:

```
. use http://www.stata-press.com/data/r9/census
(1980 Census data by state)
. kwallis medage, by(region)
Test: Equality of populations (Kruskal-Wallis test)
```

region	Obs	Rank Sum
NE	9	376.50
N Cntrl	12	294.00
South	16	398.00
West	13	206.50

```
chi-squared =      17.041 with 3 d.f.
probability =       0.0007
chi-squared with ties =      17.062 with 3 d.f.
probability =       0.0007
```

From the output, we see that we can reject the hypothesis that the populations are the same at any level below 0.07%.

◁

Saved Results

kwallis saves in r():

Scalars

r(df)	degrees of freedom	r(chi2)	χ^2
		r(chi2_adj)	χ^2 adjusted for ties

28

Methods and Formulas

`kwallis` is implemented as an ado-file.

The Kruskal–Wallis test (Kruskal and Wallis 1952; also see Altman 1991, 213–215, Conover 1999, 288–297, and Riffenburgh 1999, 279–283) is a multiple-sample generalization of the two-sample Wilcoxon (also called Mann–Whitney) rank sum test (Wilcoxon 1945; Mann and Whitney 1947). Samples of sizes n_j, $j = 1, \ldots, m$ are combined and ranked in ascending order of magnitude. Tied values are assigned the average ranks. Let n denote the overall sample size, and let R_j denote the sum of the ranks for the jth sample. The Kruskal–Wallis one-way analysis-of-variance test H is defined as

$$H = \frac{12}{n(n+1)} \sum_{j=1}^{m} \frac{R_j^2}{n_j} - 3(n+1)$$

The sampling distribution of H is approximately χ^2 with $m - 1$ degrees of freedom.

William Henry Kruskal was born in New York City in 1919. He studied mathematics and statistics at Antioch College, Harvard, and Columbia, and joined the University of Chicago in 1951. He has made many outstanding contributions to linear models, nonparametric statistics, government statistics, and the history and methodology of statistics.

Wilson Allen Wallis (1912–1998) was born in Philadelphia. He studied psychology and economics at the Universities of Minnesota and Chicago and at Columbia. He taught at Yale, Stanford, and Chicago, before moving as President (later Chancellor) to the University of Rochester in 1962. He also served in several Republican administrations. Wallis served as Editor of the *Journal of the American Statistical Association*, co-authored a popular introduction to statistics, and contributed to nonparametric statistics.

References

Altman, D. G. 1991. *Practical Statistics for Medical Research*. London: Chapman & Hall.

Conover, W. J. 1999. *Practical Nonparametric Statistics*. 3rd ed. New York: Wiley.

Kruskal, W. H. and W. A. Wallis. 1952a. Use of ranks in one-criterion variance analysis. *Journal of the American Statistical Association* 47: 583–621.

——. 1952b. Errata for use of ranks in one-criterion variance analysis. *Journal of the American Statistical Association* 48: 907–911.

Mann, H. B. and D. R. Whitney. 1947. On a test of whether one of two random variables is stochastically larger than the other. *Annals of Mathematical Statistics* 18: 50–60.

Olkin, I. 1991. A conversation with W. Allen Wallis. *Statistical Science* 6: 121–140.

Riffenburgh, R. H. 1999. *Statistics in Medicine*. San Diego: Academic Press.

Wilcoxon, F. 1945. Individual comparisons by ranking methods. *Biometrics* 1: 80–83.

Zabell, S. 1994. A conversation with William Kruskal. *Statistical Science* 9: 285–303.

Also See

Related: [R] **nptrend**, [R] **oneway**, [R] **runtest**, [R] **signrank**

Title

> **ladder** — Ladder of powers

Syntax

Ladder of powers

> ladder *varname* [*if*] [*in*] [, <u>g</u>enerate(*newvar*) <u>noa</u>djust]

Ladder-of-powers histograms

> gladder *varname* [*if*] [*in*] [, *histogram_options combine_options*]

Ladder-of-powers quantile–normal plots

> qladder *varname* [*if*] [*in*] [, *qnorm_options combine_options*]

by may be used with ladder; see [D] **by**.

Description

ladder searches a subset of the ladder of powers (Tukey 1977) for a transform that converts *varname* into a normally distributed variable. sktest tests for normality; see [R] **sktest**. Also see [R] **boxcox**.

gladder displays nine histograms of transforms of *varname* according to the ladder of powers. gladder is useful pedagogically, but we do not advise looking at histograms for research work; ladder or qnorm (see [R] **diagnostic plots**) is preferred.

qladder displays the quantiles of transforms of *varname* according to the ladder of powers against the quantiles of a normal distribution.

Options for ladder

___ Main ___

generate(*newvar*) saves the transformed values corresponding to the minimum chi-squared value from the table. We do not recommend using generate() because it is quite literal in interpreting the minimum, thus ignoring nearly equal but perhaps more interpretable transforms.

noadjust is the noadjust option to sktest; see [R] **sktest**.

Options for gladder

histogram_options affect the rendition of the histograms across all relevant transformations. See [R] **histogram**. Here the normal option is assumed, so you must supply the nonormal option to suppress the overlaid normal density. Also, gladder does not allow the width(#) option of histogram.

combine_options are any of the options documented in [G] **graph combine**. These include options for titling the graph (see [G] *title_options*) and options for saving the graph to disk (see [G] *saving_option*).

Options for qladder

qnorm_options affect the rendition of the quantile–normal plots across all relevant transformations. See [R] **diagnostic plots**.

combine_options are any of the options documented in [G] **graph combine**. These include options for titling the graph (see [G] *title_options*) and options for saving the graph to disk (see [G] *saving_option*).

Remarks

▷ Example 1: ladder

We have data on the mileage rating of 74 automobiles and wish to find a transform that makes the variable normally distributed:

```
. use http://www.stata-press.com/data/r9/auto
(1978 Automobile Data)

. ladder mpg
```

Transformation	formula	chi2(2)	P(chi2)
cubic	mpg^3	43.59	0.000
square	mpg^2	27.03	0.000
raw	mpg	10.95	0.004
square-root	sqrt(mpg)	4.94	0.084
log	log(mpg)	0.87	0.647
reciprocal root	1/sqrt(mpg)	0.20	0.905
reciprocal	1/mpg	2.36	0.307
reciprocal square	1/(mpg^2)	11.99	0.002
reciprocal cubic	1/(mpg^3)	24.30	0.000

If we had typed `ladder mpg, gen(mpgx)`, the variable `mpgx` containing $1/\sqrt{mpg}$ would have been automatically generated for us. This is the perfect example of why you should not, in general, specify the `generate()` option. Note that we also cannot reject the hypothesis that the reciprocal of `mpg` is normally distributed and that $1/mpg$—gallons per mile—has a better interpretation. It is a measure of energy consumption.

◁

▷ Example 2: gladder

`gladder` explores the same transforms as `ladder` but presents results graphically:

(*Continued on next page*)

```
. gladder mpg, fraction
```

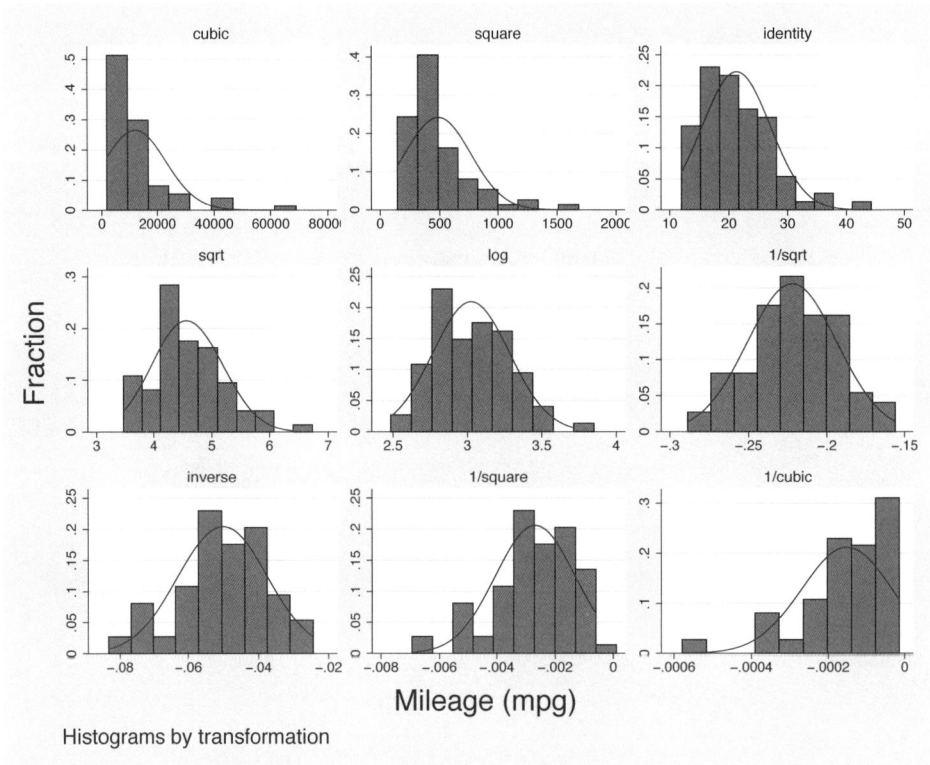

Histograms by transformation

◁

❑ Technical Note

gladder is useful pedagogically, but be careful when using it for research work, especially with large numbers of observations. For instance, consider the following data on the average July temperature in degrees Fahrenheit for 954 U.S. cities:

```
. use http://www.stata-press.com/data/r9/citytemp
(City Temperature Data)

. ladder tempjuly
```

Transformation	formula	chi2(2)	P(chi2)
cubic	tempjuly^3	47.49	0.000
square	tempjuly^2	19.70	0.000
raw	tempjuly	3.83	0.147
square-root	sqrt(tempjuly)	1.83	0.400
log	log(tempjuly)	5.40	0.067
reciprocal root	1/sqrt(tempjuly)	13.72	0.001
reciprocal	1/tempjuly	26.36	0.000
reciprocal square	1/(tempjuly^2)	64.43	0.000
reciprocal cubic	1/(tempjuly^3)	.	0.000

The period in the last line indicates that the χ^2 is very large; see [R] **sktest**.

From the table, we see that there is certainly a difference in terms of normality between the square and square-root transform. If, however, you can see the difference between the transforms in the diagram below, you have better eyes than we do:

```
. gladder tempjuly, l1title("") ylabel(none) xlabel(none)
```

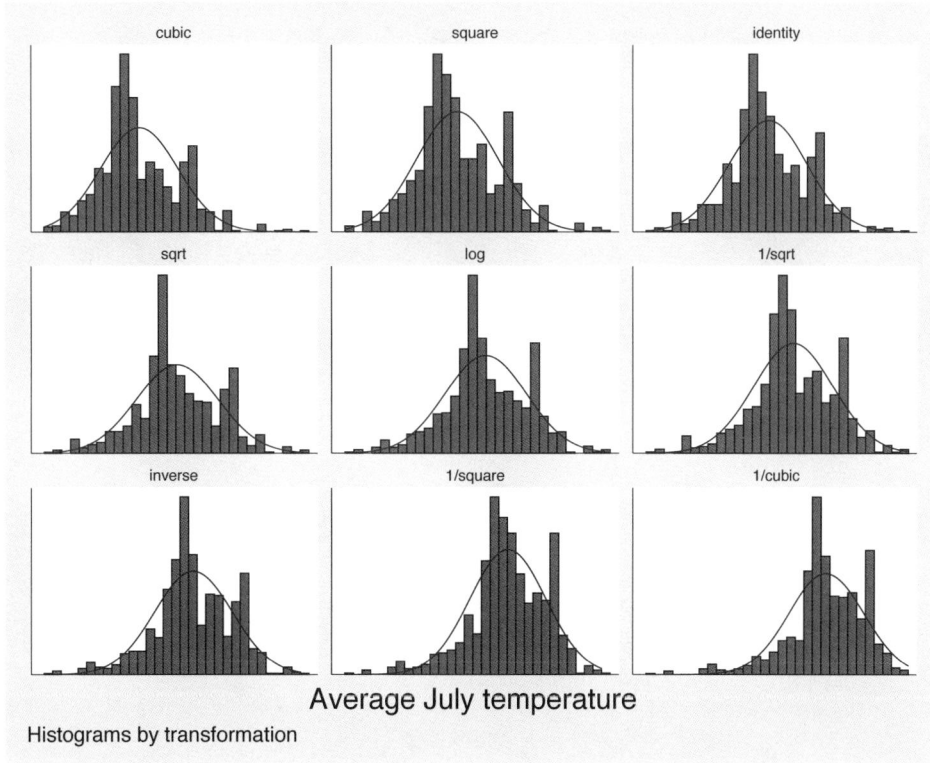

Average July temperature

Histograms by transformation

❏

▷ Example 3: qladder

A better graph for seeing normality is the quantile–normal graph, which can be produced by qladder.

(Continued on next page)

```
. qladder tempjuly, ylabel(none) xlabel(none)
```

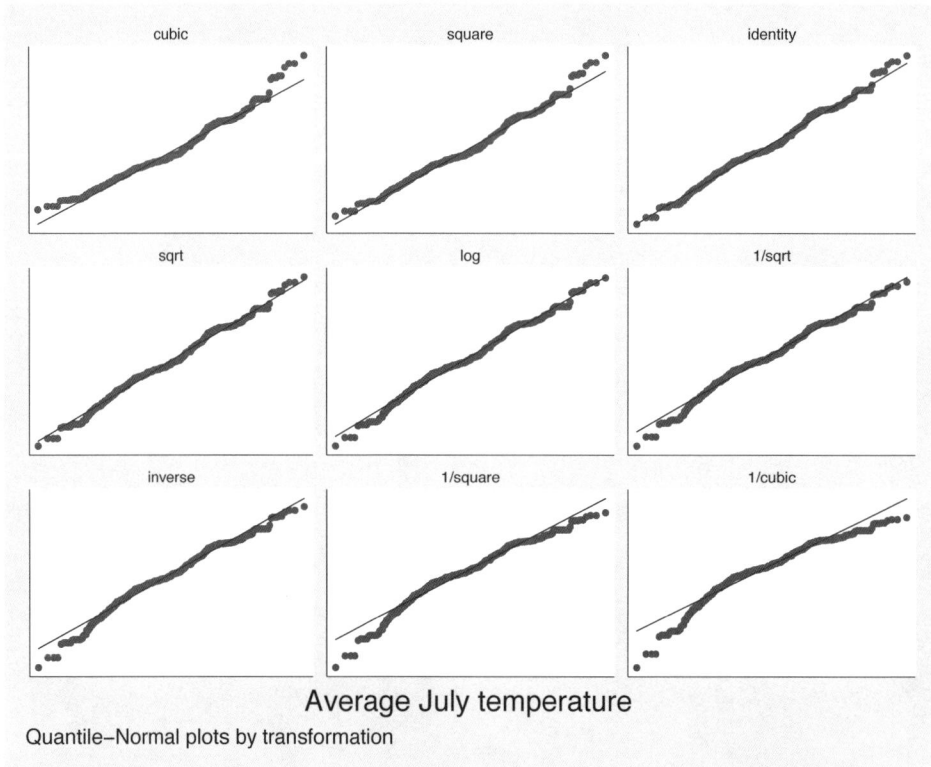

Average July temperature

Quantile–Normal plots by transformation

This graph shows that for the square transform, the upper tail—and only the upper tail—diverges from what would be expected. This is detected by `sktest` as a problem with skewness, as we would learn from using `sktest` to examine `tempjuly` squared and square-rooted.

◁

(Continued on next page)

Saved Results

ladder saves in r():

Scalars

r(N)	number of observations
r(cube)	χ^2 for cubic transformation
r(P_cube)	significance level for cubic transformation
r(square)	χ^2 for square transformation
r(P_square)	significance level for square transformation
r(raw)	χ^2 for untransformed data
r(P_raw)	significance level for untransformed data
r(sqrt)	χ^2 for square-root transformation
r(P_sqrt)	significance level for square-root transformation
r(log)	χ^2 for log transformation
r(P_log)	significance level for log transformation
r(invsqrt)	χ^2 for reciprocal-root transformation
r(P_invsqrt)	significance level for reciprocal-root transformation
r(inv)	χ^2 for reciprocal transformation
r(P_inv)	significance level for reciprocal transformation
r(invsq)	χ^2 for reciprocal-square transformation
r(P_invsq)	significance level for reciprocal-square transformation
r(invcube)	χ^2 for reciprocal-cubic transformation
r(P_invcube)	significance level for reciprocal-cubic transformation

Methods and Formulas

ladder, gladder, and qladder are implemented as ado-files.

For ladder, results are as reported by sktest; see [R] **sktest**. If generate() is specified, the transform with the minimum χ^2 value is chosen.

gladder sets the number of bins to $\min(\sqrt{n}, 10\log_{10} n)$, rounded to the closest integer, where n is the number of unique values of *varname*. See [R] **histogram** for a discussion of the optimal number of bins.

Also see Findley (1990) for a ladder-of-powers variable transformation program that produces one-way graphs with overlaid box plots, in addition to histograms with overlaid normals. Buchner and Findley (1990) discuss ladder-of-powers transformations as one aspect of preliminary data analysis. Also see Hamilton (1992, 18–23) and Hamilton (2004, 127–129).

Acknowledgment

qladder was written by Jeroen Weesie, Utrecht University, Netherlands.

References

Buchner, D. M. and T. W. Findley. 1990. Research in physical medicine and rehabilitation: VIII preliminary data analysis. *American Journal of Physical Medicine and Rehabilitation* 69: 154–169.

Findley, T. W. 1990. sed3: Variable transformation and evaluation. *Stata Technical Bulletin* 2: 15. Reprinted in *Stata Technical Bulletin Reprints*, vol. 1, pp. 85–86.

Hamilton, L. C. 1992. *Regression with Graphics*. Pacific Grove, CA: Brooks/Cole.

——. 2004. *Statistics with Stata*. Belmont, CA: Brooks/Cole.

Tukey, J. W. 1977. *Exploratory Data Analysis*. Reading, MA: Addison–Wesley.

Also See

Related:	[R] **boxcox**, [R] **diagnostic plots**, [R] **lnskew0**, [R] **lv**, [R] **sktest**
Background:	*Stata Graphics Reference Manual*

Title

level — Set default confidence level

Syntax

<u>set</u> <u>level</u> # [, <u>permanently</u>]

Description

set level specifies the default confidence level for confidence intervals for all commands that report confidence intervals. The initial value is 95, meaning 95% confidence intervals. # may be between 10.00 and 99.99, and # can have at most two digits after the decimal point.

Option

permanently specifies that, in addition to making the change right now, the level setting be remembered and become the default setting when you invoke Stata.

Remarks

To change the level of confidence intervals reported by a particular command, you need not reset the default confidence level. All commands that report confidence intervals have a level(#) option. When you do not specify the option, the confidence intervals are calculated for the default level set by set level, or 95% if you have not reset it.

▷ Example 1

We use the ci command to obtain the confidence interval for the mean of mpg:

```
. use http://www.stata-press.com/data/r9/auto
(1978 Automobile Data)
. ci mpg
```

Variable	Obs	Mean	Std. Err.	[95% Conf. Interval]	
mpg	74	21.2973	.6725511	19.9569	22.63769

To obtain 90% confidence intervals, we would type

```
. ci mpg, level(90)
```

Variable	Obs	Mean	Std. Err.	[90% Conf. Interval]	
mpg	74	21.2973	.6725511	20.17683	22.41776

or

```
. set level 90
. ci mpg
```

Variable	Obs	Mean	Std. Err.	[90% Conf. Interval]	
mpg	74	21.2973	.6725511	20.17683	22.41776

If we opt for the second alternative, the next time that we fit a model (say with `regress`), 90% confidence intervals will be reported. If we wanted 95% confidence intervals, we could specify `level(95)` on the estimation command, or we could reset the default by typing `set level 95`.

The current setting of `level()` is stored as the c-class value `c(level)`; see [P] **creturn**.

◁

Also See

Complementary:	[R] **query**,
	[P] **creturn**
Related:	[R] **ci**
Background:	[U] **20 Estimation and postestimation commands**,
	[U] **20.6 Specifying the width of confidence intervals**

Title

> **lincom** — Linear combinations of estimators

Syntax

lincom *exp* [, *options*]

options	description
ef̲orm	generic label; exp(b); the default
or	odds ratio
hr	hazard ratio
ir̲r	incidence-rate ratio
rr̲r	relative-risk ratio
le̲vel(#)	set confidence level; default is level(95)

where *exp* is any linear combination of coefficients that is a valid syntax for test; see [R] **test**. Note, however, that *exp* must not contain an equal sign.

Description

lincom computes point estimates, standard errors, t or z statistics, p-values, and confidence intervals for linear combinations of coefficients after any estimation command. Results can optionally be displayed as odds ratios, hazard ratios, incidence-rate ratios, or relative risk ratios.

To obtain linear combinations of estimators after svy, see [SVY] **svy postestimation**.

Options

level(#) specifies the confidence level, as a percentage, for confidence intervals. The default is level(95) or as set by set level; see [U] **20.6 Specifying the width of confidence intervals**.

eform, or, hr, irr, and rrr all report coefficient estimates as $\exp(\widehat{\beta})$ rather than $\widehat{\beta}$. Standard errors and confidence intervals are similarly transformed. Note that or is the default after logistic. The only difference in these options is how the output is labeled.

Option	Label	Explanation	Example commands
eform	exp(b)	Generic label	
or	Odds Ratio	Odds ratio	logistic, logit
hr	Haz. Ratio	Hazard ratio	stcox, streg
irr	IRR	Incidence-rate ratio	poisson
rrr	RRR	Relative risk ratio	mlogit

Note that *exp* may not contain any additive constants when you use the eform, or, hr, irr, or rrr options.

Remarks

Remarks are presented under the headings

Using lincom
Odds ratios and incidence-rate ratios
Multiple-equation models

Using lincom

After fitting a model and obtaining estimates for coefficients $\beta_1, \beta_2, \ldots, \beta_k$, you may want to view estimates for linear combinations of the β_i, such as $\beta_1 - \beta_2$. lincom can display estimates for any linear combination of the form $c_0 + c_1\beta_1 + c_2\beta_2 + \cdots + c_k\beta_k$.

lincom works after any estimation command for which test works. Any valid expression for test syntax 1 (see [R] **test**) is a valid expression for lincom.

lincom is useful for viewing odds ratios, hazard ratios, etc., for one group (i.e., one set of covariates) relative to another group (i.e., another set of covariates). See the examples below.

▷ Example 1

We perform a linear regression:

```
. use http://www.stata-press.com/data/r9/regress
. regress y x1 x2 x3
```

Source	SS	df	MS			
Model	3259.3561	3	1086.45203			
Residual	1627.56282	144	11.3025196			
Total	4886.91892	147	33.2443464			

Number of obs = 148
F(3, 144) = 96.12
Prob > F = 0.0000
R-squared = 0.6670
Adj R-squared = 0.6600
Root MSE = 3.3619

y	Coef.	Std. Err.	t	P>|t|	[95% Conf. Interval]	
x1	1.457113	1.07461	1.36	0.177	-.666934	3.581161
x2	2.221682	.8610358	2.58	0.011	.5197797	3.923583
x3	-.006139	.0005543	-11.08	0.000	-.0072345	-.0050435
_cons	36.10135	4.382693	8.24	0.000	27.43863	44.76407

To see the difference of the coefficients of x2 and x1, we type

```
. lincom x2 - x1
 ( 1)  - x1 + x2 = 0
```

y	Coef.	Std. Err.	t	P>|t|	[95% Conf. Interval]	
(1)	.7645682	.9950282	0.77	0.444	-1.20218	2.731316

The expression can be any linear combination.

```
. lincom 3*x1 + 500*x3
 ( 1)  3 x1 + 500 x3 = 0
```

y	Coef.	Std. Err.	t	P>|t|	[95% Conf. Interval]	
(1)	1.301825	3.396624	0.38	0.702	-5.411858	8.015507

Nonlinear expressions are not allowed.

```
. lincom x2/x1
not possible with test
r(131);
```

For information about estimating nonlinear expressions, see [R] **nlcom**.

◁

❑ Technical Note

lincom uses the same shorthands for coefficients as test (see [R] test). When you type x1, for instance, lincom knows that you mean the coefficient of x1. The formal syntax for referencing this coefficient is actually _b[x1], or alternatively, _coef[x1]. So, more formally, in the last example we could have typed

```
. lincom 3*_b[x1] + 500*_b[x3]
  (output omitted)
```

❑

Odds ratios and incidence-rate ratios

After logistic regression, the or option can be specified with lincom to display odds ratios for any effect. Incidence-rate ratios after commands such as poisson can be obtained in a similar fashion by specifying the irr option.

▷ Example 2

Consider the low birthweight dataset from Hosmer and Lemeshow (2000, 25). We fit a logistic regression model of low birthweight (variable low) on the following variables:

Variable	Description	Coding
age	age in years	
black	race black	1 if black, 0 otherwise
other	race other	1 if race other, 0 otherwise
smoke	smoking status	1 if smoker, 0 if nonsmoker
ht	history of hypertension	1 if yes, 0 if no
ui	uterine irritability	1 if yes, 0 if no
lwd	maternal weight before pregnancy	1 if weight < 110 lb., 0 otherwise
ptd	history of premature labor	1 if yes, 0 if no
agelwd	age $\times$ lwd	
smokelwd	smoke $\times$ lwd	

We first fit a model without the interaction terms agelwd and smokelwd (Hosmer and Lemeshow 1989, table 4.8) by using logit.

```
. use http://www.stata-press.com/data/r9/lbw3
(Hosmer & Lemeshow data)

. logit low age lwd black other smoke ptd ht ui

Iteration 0:  log likelihood =   -117.336
Iteration 1:  log likelihood = -99.431174
Iteration 2:  log likelihood = -98.785718
Iteration 3:  log likelihood =   -98.778
Iteration 4:  log likelihood = -98.777998
```

```
Logistic regression                          Number of obs   =        189
                                             LR chi2(8)      =      37.12
                                             Prob > chi2     =     0.0000
Log likelihood = -98.777998                  Pseudo R2       =     0.1582
```

low	Coef.	Std. Err.	z	P>|z|	[95% Conf. Interval]	
age	-.0464796	.0373888	-1.24	0.214	-.1197603	.0268011
lwd	.8420615	.4055338	2.08	0.038	.0472299	1.636893
black	1.073456	.5150752	2.08	0.037	.0639273	2.082985
other	.815367	.4452979	1.83	0.067	-.0574008	1.688135
smoke	.8071996	.404446	2.00	0.046	.0145001	1.599899
ptd	1.281678	.4621157	2.77	0.006	.3759478	2.187408
ht	1.435227	.6482699	2.21	0.027	.1646415	2.705813
ui	.6576256	.4666192	1.41	0.159	-.2569313	1.572182
_cons	-1.216781	.9556797	-1.27	0.203	-3.089878	.656317

To get the odds ratio for black smokers relative to white nonsmokers (the reference group), we type

```
. lincom black + smoke, or

 ( 1)  black + smoke = 0
```

low	Odds Ratio	Std. Err.	z	P>|z|	[95% Conf. Interval]	
(1)	6.557805	4.744692	2.60	0.009	1.588176	27.07811

lincom computed $\exp(\beta_{\text{black}} + \beta_{\text{smoke}}) = 6.56$. To see the odds ratio for white smokers relative to black nonsmokers, we type

```
. lincom smoke - black, or

 ( 1)  - black + smoke = 0
```

low	Odds Ratio	Std. Err.	z	P>|z|	[95% Conf. Interval]	
(1)	.7662425	.4430176	-0.46	0.645	.2467334	2.379603

Now let's add the interaction terms to the model (Hosmer and Lemeshow 1989, table 4.10). This time, we will use logistic rather than logit. By default, logistic displays odds ratios.

```
. generate agelwd = age*lwd

. generate smokelwd = smoke*lwd
```

```
. logistic low age black other smoke ht ui lwd ptd agelwd smokelwd
```

Logistic regression

Number of obs = 189
LR chi2(10) = 42.66
Prob > chi2 = 0.0000

Log likelihood = -96.00616

Pseudo R2 = 0.1818

low	Odds Ratio	Std. Err.	z	P>|z|	[95% Conf. Interval]	
age	.9194513	.041896	-1.84	0.065	.8408967	1.005344
black	2.95383	1.532788	2.09	0.037	1.068277	8.167462
other	2.137589	.9919132	1.64	0.102	.8608713	5.307749
smoke	3.168096	1.452377	2.52	0.012	1.289956	7.780755
ht	3.893141	2.5752	2.05	0.040	1.064768	14.2346
ui	2.071284	.9931385	1.52	0.129	.8092928	5.301191
lwd	.1772934	.3312383	-0.93	0.354	.0045539	6.902359
ptd	3.426633	1.615282	2.61	0.009	1.360252	8.632086
agelwd	1.15883	.09602	1.78	0.075	.9851216	1.36317
smokelwd	.2447849	.2003996	-1.72	0.086	.0491956	1.217988

Hosmer and Lemeshow (1989, table 4.13) consider the effects of smoking (smoke = 1) and low maternal weight before pregnancy (lwd = 1). The effect of smoking among non-low-weight mothers (lwd = 0) is given by the odds ratio 3.17 for smoke in the logistic output. The effect of smoking among low-weight mothers is given by

```
. lincom smoke + smokelwd
( 1)  smoke + smokelwd = 0
```

low	Odds Ratio	Std. Err.	z	P>|z|	[95% Conf. Interval]	
(1)	.7755022	.5749508	-0.34	0.732	.1813465	3.316322

Note that we did not have to specify the or option. After logistic, lincom assumes or by default.

The effect of low weight (lwd = 1) is more complicated since we fit an age × lwd interaction. We must specify the age of mothers for the effect. The effect among 30-year-old nonsmokers is given by

```
. lincom lwd + 30*agelwd
( 1)  lwd + 30.0 agelwd = 0
```

low	Odds Ratio	Std. Err.	z	P>|z|	[95% Conf. Interval]	
(1)	14.7669	13.56689	2.93	0.003	2.439266	89.39625

lincom computed $\exp(\beta_{\text{lwd}} + 30\beta_{\text{agelwd}}) = 14.8$. It may seem odd that we entered it as lwd + 30*agelwd, but remember that lwd and agelwd are just lincom's (and test's) shorthand for _b[lwd] and _b[agelwd]. We could have typed

```
. lincom _b[lwd] + 30*_b[agelwd]
( 1)  lwd + 30.0 agelwd = 0
```

low	Odds Ratio	Std. Err.	z	P>|z|	[95% Conf. Interval]	
(1)	14.7669	13.56689	2.93	0.003	2.439266	89.39625

◁

Multiple-equation models

lincom also works with multiple-equation models. The only difference is how you refer to the coefficients. Recall that for multiple-equation models, coefficients are referenced using the syntax

[*eqno*] *varname*

where *eqno* is the equation number or equation name and *varname* is the corresponding variable name for the coefficient; see [U] **13.5 Accessing coefficients and standard errors** and [R] **test** for details.

▷ Example 3

Let's consider the example from [R] **mlogit** (Tarlov et al. 1989; Wells et al. 1989).

```
. use http://www.stata-press.com/data/r9/sysdsn2
(Health insurance data)

. mlogit insure age male nonwhite site2 site3, nolog
```

Multinomial logistic regression

Number of obs	=	615
LR chi2(10)	=	42.99
Prob > chi2	=	0.0000

Log likelihood = -534.36165

Pseudo R2	=	0.0387

insure	Coef.	Std. Err.	z	P>|z|	[95% Conf. Interval]
Prepaid						
age	-.011745	.0061946	-1.90	0.058	-.0238862	.0003962
male	.5616934	.2027465	2.77	0.006	.1643175	.9590693
nonwhite	.9747768	.2363213	4.12	0.000	.5115955	1.437958
site2	.1130359	.2101903	0.54	0.591	-.2989296	.5250013
site3	-.5879879	.2279351	-2.58	0.010	-1.034733	-.1412433
_cons	.2697127	.3284422	0.82	0.412	-.3740222	.9134476
Uninsure						
age	-.0077961	.0114418	-0.68	0.496	-.0302217	.0146294
male	.4518496	.3674867	1.23	0.219	-.268411	1.17211
nonwhite	.2170589	.4256361	0.51	0.610	-.6171725	1.05129
site2	-1.211563	.4705127	-2.57	0.010	-2.133751	-.2893747
site3	-.2078123	.3662926	-0.57	0.570	-.9257327	.510108
_cons	-1.286943	.5923219	-2.17	0.030	-2.447872	-.1260135

(insure==Indemnity is the base outcome)

To see the estimate of the sum of the coefficient of male and the coefficient of nonwhite for the Prepaid outcome, we type

```
. lincom [Prepaid]male + [Prepaid]nonwhite

( 1)  [Prepaid]male + [Prepaid]nonwhite = 0
```

insure	Coef.	Std. Err.	z	P>|z|	[95% Conf. Interval]
(1)	1.53647	.3272489	4.70	0.000	.8950741	2.177866

To view the estimate as a ratio of relative risks (see [R] **mlogit** for the definition and interpretation), we specify the **rrr** option.

```
. lincom [Prepaid]male + [Prepaid]nonwhite, rrr

 ( 1)  [Prepaid]male + [Prepaid]nonwhite = 0
```

insure	RRR	Std. Err.	z	P>|z|	[95% Conf. Interval]	
(1)	4.648154	1.521103	4.70	0.000	2.447517	8.827451

◁

Saved Results

lincom saves in r():

Scalars
r(estimate)   point estimate
r(se)   estimate of standard error
r(df)   degrees of freedom

Methods and Formulas

lincom is implemented as an ado-file.

References

Hosmer, D. W., Jr., and S. Lemeshow. 1989. *Applied Logistic Regression.* New York: Wiley.

——. 2000. *Applied Logistic Regression.* 2nd ed. New York: Wiley.

Tarlov, A. R., J. E. Ware, Jr., S. Greenfield, E. C. Nelson, E. Perrin, and M. Zubkoff. 1989. The medical outcomes study. *Journal of the American Medical Association* 262: 925–930.

Wells, K. E., R. D. Hays, M. A. Burnam, W. H. Rogers, S. Greenfield, and J. E. Ware, Jr. 1989. Detection of depressive disorder for patients receiving prepaid or fee-for-service care. *Journal of the American Medical Association* 262: 3298–3302.

Also See

Related: [R] **nlcom**, [R] **test**, [R] **testnl**,

[SVY] **svy postestimation**

Background: [U] **13.5 Accessing coefficients and standard errors**,

[U] **20 Estimation and postestimation commands**

Title

linktest — Specification link test for single-equation models

Syntax

linktest [*if*] [*in*] [, *cmd_options*]

When *if* and *in* are not specified, the link test is performed on the same sample as the previous estimation.

Description

linktest performs a link test for model specification after any single-equation estimation command, such as logistic, regress, stcox, etc.

Option

‾‾‾‾‾‾‾‾| Main |‾‾‾

cmd_options must be the same options specified with the underlying estimation command.

Remarks

The form of the link test implemented here is based on an idea of Tukey (1949), which was further described by Pregibon (1980), elaborating on work in his unpublished thesis (Pregibon 1979). See *Methods and Formulas* below for more details.

▷ Example 1

We want to explain the mileage ratings of cars in our automobile dataset using the weight, engine displacement, and whether the car is manufactured outside the U.S.:

```
. use http://www.stata-press.com/data/r9/auto
(1978 Automobile Data)

. regress mpg weight displ foreign
```

Source	SS	df	MS		Number of obs =	74
					F(3, 70) =	45.88
Model	1619.71935	3	539.906448		Prob > F =	0.0000
Residual	823.740114	70	11.7677159		R-squared =	0.6629
					Adj R-squared =	0.6484
Total	2443.45946	73	33.4720474		Root MSE =	3.4304

mpg	Coef.	Std. Err.	t	P>|t|	[95% Conf. Interval]	
weight	-.0067745	.0011665	-5.81	0.000	-.0091011	-.0044479
displacement	.0019286	.0100701	0.19	0.849	-.0181556	.0220129
foreign	-1.600631	1.113648	-1.44	0.155	-3.821732	.6204699
_cons	41.84795	2.350704	17.80	0.000	37.15962	46.53628

46

Based on the R^2, we are reasonably pleased with this model.

If our model really is specified correctly, then if we were to regress mpg on the prediction and the prediction squared, the prediction squared would have no explanatory power. This is what linktest does:

```
. linktest
```

Source	SS	df	MS		Number of obs =	74
					F(2, 71) =	76.75
Model	1670.71514	2	835.357572		Prob > F =	0.0000
Residual	772.744316	71	10.8837228		R-squared =	0.6837
					Adj R-squared =	0.6748
Total	2443.45946	73	33.4720474		Root MSE =	3.299

mpg	Coef.	Std. Err.	t	P>|t|	[95% Conf. Interval]	
_hat	-.4127198	.6577736	-0.63	0.532	-1.724283	.8988434
_hatsq	.0338198	.015624	2.16	0.034	.0026664	.0649732
_cons	14.00705	6.713276	2.09	0.041	.6211539	27.39294

We find that the prediction squared does have explanatory power, so our specification is not as good as we thought.

Although linktest is formally a test of the specification of the dependent variable, it is often interpreted as a test that, conditional on the specification, the independent variables are specified incorrectly. We will follow that interpretation and now include weight squared in our model:

```
. generate weight2 = weight*weight
. regress mpg weight weight2 displ foreign
```

Source	SS	df	MS		Number of obs =	74
					F(4, 69) =	39.37
Model	1699.02634	4	424.756584		Prob > F =	0.0000
Residual	744.433124	69	10.7888859		R-squared =	0.6953
					Adj R-squared =	0.6777
Total	2443.45946	73	33.4720474		Root MSE =	3.2846

mpg	Coef.	Std. Err.	t	P>|t|	[95% Conf. Interval]	
weight	-.0173257	.0040488	-4.28	0.000	-.0254028	-.0092486
weight2	1.87e-06	6.89e-07	2.71	0.008	4.93e-07	3.24e-06
displacement	-.0101625	.0106236	-0.96	0.342	-.031356	.011031
foreign	-2.560016	1.123506	-2.28	0.026	-4.801349	-.3186832
_cons	58.23575	6.449882	9.03	0.000	45.36859	71.10291

Now we perform the link test on our new model:

(Continued on next page)

```
. linktest

      Source |       SS       df       MS              Number of obs =      74
-------------+------------------------------           F(  2,    71) =   81.08
       Model | 1699.39489      2  849.697445           Prob > F      =  0.0000
    Residual |  744.06457     71  10.4797827           R-squared     =  0.6955
-------------+------------------------------           Adj R-squared =  0.6869
       Total | 2443.45946     73  33.4720474           Root MSE      =  3.2372

---------------------------------------------------------------------------------
         mpg |      Coef.   Std. Err.      t    P>|t|     [95% Conf. Interval]
-------------+-------------------------------------------------------------------
        _hat |   1.141987   .7612218     1.50   0.138    -.3758456    2.659821
      _hatsq |  -.0031916   .0170194    -0.19   0.852    -.0371272    .0307441
       _cons |   -1.50305   8.196444    -0.18   0.855    -17.84629    14.84019
---------------------------------------------------------------------------------
```

We now pass the link test.

◁

▷ Example 2

Above we followed a standard misinterpretation of the link test—when we discovered a problem, we focused on the explanatory variables of our model. We might consider varying exactly what the link test tests. The link test told us that our dependent variable was misspecified. For those with an engineering background, mpg is indeed a strange measure. It would make more sense to model energy consumption—gallons per mile—in terms of weight and displacement:

```
. gen gpm = 1/mpg
. regress gpm weight displ foreign

      Source |       SS       df       MS              Number of obs =      74
-------------+------------------------------           F(  3,    70) =   76.33
       Model | .009157962      3  .003052654           Prob > F      =  0.0000
    Residual | .002799666     70  .000039995           R-squared     =  0.7659
-------------+------------------------------           Adj R-squared =  0.7558
       Total | .011957628     73  .000163803           Root MSE      =  .00632

---------------------------------------------------------------------------------
         gpm |      Coef.   Std. Err.      t    P>|t|     [95% Conf. Interval]
-------------+-------------------------------------------------------------------
      weight |   .0000144   2.15e-06     6.72   0.000     .0000102    .0000187
displacement |   .0000186   .0000186     1.00   0.319    -.0000184    .0000557
     foreign |   .0066981   .0020531     3.26   0.002     .0026034    .0107928
       _cons |   .0008917   .0043337     0.21   0.838    -.0077515     .009535
---------------------------------------------------------------------------------
```

This model looks every bit as reasonable as our original model.

```
. linktest

      Source |       SS       df       MS              Number of obs =      74
-------------+------------------------------           F(  2,    71) =  117.06
       Model | .009175219      2  .004587609           Prob > F      =  0.0000
    Residual | .002782409     71  .000039189           R-squared     =  0.7673
-------------+------------------------------           Adj R-squared =  0.7608
       Total | .011957628     73  .000163803           Root MSE      =  .00626

---------------------------------------------------------------------------------
         gpm |      Coef.   Std. Err.      t    P>|t|     [95% Conf. Interval]
-------------+-------------------------------------------------------------------
        _hat |   .6608413    .515275     1.28   0.204    -.3665877     1.68827
      _hatsq |   3.275857   4.936655     0.66   0.509    -6.567553    13.11927
       _cons |    .008365   .0130468     0.64   0.523    -.0176496    .0343795
---------------------------------------------------------------------------------
```

Specifying the model in terms of gallons-per-mile also solves the specification problem and results in a more parsimonious specification.

◁

▷ Example 3

The link test can be used with any single-equation estimation procedure, not solely regression. Let's turn our problem around and attempt to explain whether a car is manufactured outside the U.S. by its mileage rating and weight. To save paper, we will specify logit's nolog option, which suppresses the iteration log:

```
. logit foreign mpg weight, nolog
```

Logistic regression

Number of obs   =   74
LR chi2(2)   =   35.72
Prob > chi2   =   0.0000
Log likelihood = -27.175156

Pseudo R2   =   0.3966

foreign	Coef.	Std. Err.	z	P>|z|	[95% Conf. Interval]	
mpg	-.1685869	.0919174	-1.83	0.067	-.3487418	.011568
weight	-.0039067	.0010116	-3.86	0.000	-.0058894	-.001924
_cons	13.70837	4.518707	3.03	0.002	4.851864	22.56487

When we run linktest after logit, the result is another logit specification:

```
. linktest, nolog
```

Logistic regression

Number of obs   =   74
LR chi2(2)   =   36.83
Prob > chi2   =   0.0000
Log likelihood = -26.615714

Pseudo R2   =   0.4090

foreign	Coef.	Std. Err.	z	P>|z|	[95% Conf. Interval]	
_hat	.8438531	.2738759	3.08	0.002	.3070661	1.38064
_hatsq	-.1559115	.1568642	-0.99	0.320	-.4633596	.1515366
_cons	.2630557	.4299598	0.61	0.541	-.57965	1.105761

The link test reveals no problems with our specification.

If there had been a problem, we would have been virtually forced to accept the misinterpretation of the link test—we would have reconsidered our specification of the independent variables. When using logit, we have no control over the specification of the dependent variable other than to change likelihood functions.

We admit to having seen a dataset once for which the link test rejected the logit specification. We did change the likelihood function, refitting the model using probit, and satisfied the link test. Probit has thinner tails than logit. In general, however, you will not be so lucky.

◁

(*Continued on next page*)

❑ Technical Note

You should specify exactly the same options with `linktest` that you do with the estimation command, although you do not have to follow this advice as literally as we did in the preceding example. `logit`'s `nolog` option merely suppresses a part of the output, not what is estimated. We specified `nolog` both times to save paper.

If you are testing a tobit model, you must specify the censoring points just as you do with the `tobit` command.

If you are not sure which options are important, duplicate exactly what you specified on the estimation command.

If you do not specify if *exp* or in *range* with `linktest`, Stata will by default perform the link test on the same sample as the previous estimation. Suppose that you omitted some data when performing your estimation, but want to calculate the link test on all the data, which you might do if you believe the model is appropriate for all the data. You would type `linktest if e(sample) <` `.` to do this.

❑

Saved Results

`linktest` saves in `r()`:

Scalars
    `r(t)` t statistic on `_hatsq`   `r(df)`   degrees of freedom

`linktest` is *not* an estimation command in the sense that it leaves previous estimation results unchanged. For instance, after running a regression and performing the link test, typing `regress` without arguments after the link test still replays the original regression.

In terms of integrating an estimation command with `linktest`, `linktest` assumes that the name of the estimation command is stored in `e(cmd)` and that the name of the dependent variable is stored in `e(depvar)`. After estimation, it assumes that the number of degrees of freedom for the t test is given by `e(df_m)` if the macro is defined.

If the estimation command reports Z statistics instead of t statistics, `linktest` will also report Z statistics. The Z statistic, however, is still returned in `r(t)`, and `r(df)` is set to a missing value.

Methods and Formulas

`linktest` is implemented as an ado-file.

The link test is based on the idea that if a regression or regression-like equation is properly specified, you should not be able to find any additional independent variables that are significant except by chance. One kind of specification error is called a link error. In regression, this means that the dependent variable needs a transformation or "link" function to properly relate to the independent variables. The idea of a link test is to add an independent variable to the equation that is especially likely to be significant if there is a link error.

Let

$$\mathbf{y} = f(\mathbf{X}\beta)$$

be the model and $\widehat{\beta}$ be the parameter estimates. `linktest` calculates

$$\_\mathtt{hat} = \mathbf{X}\widehat{\boldsymbol{\beta}}$$

and

$$\_\mathtt{hatsq} = \_\mathtt{hat}^2$$

The model is then refitted with these two variables, and the test is based on the significance of
_hatsq. This is the form suggested by Pregibon (1979) based on an idea of Tukey (1949). Pregibon
(1980) suggests a slightly different method that has come to be known as "Pregibon's goodness-of-link
test". We prefer the older version because it is universally applicable, straightforward, and a good
second-order approximation. It can be applied to any single-equation estimation technique, whereas
Pregibon's more recent tests are estimation-technique specific.

References

Pregibon, D. 1979. *Data Analytic Methods for Generalized Linear Models.* Ph.D. Dissertation. University of Toronto.

——. 1980. Goodness of link tests for generalized linear models. *Applied Statistics* 29: 15–24.

Tukey, J. W. 1949. One degree of freedom for nonadditivity. *Biometrics* 5: 232–242.

Also See

Related: [R] **lrtest**, [R] **test**, [R] **testnl**

Title

> **lnskew0** — Find zero-skewness log or Box–Cox transform

Syntax

Zero-skewness log

>   lnskew0 *newvar* = *exp* $\left[\,if\,\right]$ $\left[\,in\,\right]$ $\left[\,,\,options\,\right]$

Zero-skewness Box–Cox transform

>   bcskew0 *newvar* = *exp* $\left[\,if\,\right]$ $\left[\,in\,\right]$ $\left[\,,\,options\,\right]$

options	description
Main	
<u>d</u>elta(*#*)	increment for derivative of skewness function; default is delta(0.02) for lnskew0 and delta(0.01) for bcskew0
<u>z</u>ero(*#*)	value for determining convergence; default is zero(0.001)
<u>l</u>evel(*#*)	set confidence level; default is level(95)

Description

lnskew0 creates *newvar* $= \ln(\pm exp - k)$, choosing k and the sign of *exp* so that the skewness of *newvar* is zero.

bcskew0 creates *newvar* $= (exp^{\lambda} - 1)/\lambda$, the Box–Cox power transformation (Box and Cox 1964), choosing λ so that the skewness of *newvar* is zero. *exp* must be strictly positive. Also see [R] **boxcox** for maximum likelihood estimation of λ.

Options

> Main

delta(*#*) specifies the increment used for calculating the derivative of the skewness function with respect to k (lnskew0) or λ (bcskew0). The default values are 0.02 for lnskew0 and 0.01 for bcskew0.

zero(*#*) specifies a value for skewness to determine convergence that is small enough to be considered zero and is, by default, 0.001.

level(*#*) specifies the confidence level for the confidence interval for k (lnskew0) or λ (bcskew0). The confidence interval is calculated only if level() is specified. *#* is specified as an integer; 95 means 95% confidence intervals. The level() option is honored only if the number of observations exceeds 7.

Remarks

▷ Example 1: lnskew0

Using our automobile dataset (see [U] **1.2.1 Sample datasets**), we want to generate a new variable equal to $\ln(\text{mpg} - k)$ to be approximately normally distributed. mpg records the miles per gallon for each of our cars. One feature of the normal distribution is that it has skewness 0.

```
. use http://www.stata-press.com/data/r9/auto
(Automobile Data)

. lnskew0 lnmpg = mpg
```

Transform	k	[95% Conf. Interval]	Skewness
ln(mpg-k)	5.383659	(not calculated)	-7.05e-06

This created the new variable $\text{lnmpg} = \ln(\text{mpg} - 5.384)$:

```
. describe lnmpg
```

variable name	storage type	display format	value label	variable label
lnmpg	float	%9.0g		ln(mpg-5.383659)

Since we did not specify the level() option, no confidence interval was calculated. At the outset, we could have typed

```
. lnskew0 lnmpg = mpg, level(95)
```

Transform	k	[95% Conf. Interval]	Skewness
ln(mpg-k)	5.383659	-17.12339 9.892416	-7.05e-06

The confidence interval is calculated under the assumption that $\ln(\text{mpg} - k)$ really does have a normal distribution. It would be perfectly reasonable to use lnskew0, even if we did not believe the transformed variable would have a normal distribution—if we literally wanted the zero-skewness transform—although, in that case the confidence interval would be an approximation of unknown quality to the true confidence interval. If we now wanted to test the believability of the confidence interval, we could also test our new variable lnmpg using swilk with the lnnormal option.

◁

❑ Technical Note

lnskew0 and bcskew0 report the resulting skewness of the variable merely to reassure you of the accuracy of its results. In our example above, lnskew0 found k such that the resulting skewness was $-7 \cdot 10^{-6}$, a number close enough to zero for all practical purposes. If we wanted to make it even smaller, we could specify the zero() option. Typing lnskew0 new=mpg, zero(1e-8) changes the estimated k to -5.383552 from -5.383659 and reduces the calculated skewness to $-2 \cdot 10^{-11}$.

When you request a confidence interval, lnskew0 may report the lower confidence interval as '.', which should be taken as indicating the lower confidence limit $k_L = -\infty$. (This cannot happen with bcskew0.)

As an example, consider a sample of size n on x and assume that the skewness of x is positive, but not significantly so, at the desired significance level—say, 5%. Then no matter how large and negative you make k_L, there is no value extreme enough to make the skewness of $\ln(x - k_L)$ equal the corresponding percentile (97.5 for a 95% confidence interval) of the distribution of skewness in a normal distribution of the same sample size. You cannot do this because the distribution of $\ln(x - k_L)$ tends to that of x—apart from location and scale shift—as $x \to \infty$. This "problem" never applies to the upper confidence limit, k_U, because the skewness of $\ln(x - k_U)$ tends to $-\infty$ as k tends upwards to the minimum value of x.

❑

▷ Example 2: bcskew0

In example 1, using lnskew0 with a variable such as mpg is probably undesirable. mpg has a natural zero, and we are shifting that zero arbitrarily. On the other hand, use of lnskew0 with a variable such as temperature measured in Fahrenheit or Celsius would be more appropriate, as the zero is indeed arbitrary.

For a variable like mpg, it makes more sense to use the Box–Cox power transform (Box and Cox 1964):

$$y^{(\lambda)} = \frac{y^\lambda - 1}{\lambda}$$

λ is free to take on any value, but note that $y^{(1)} = y - 1$, $y^{(0)} = \ln(y)$, and $y^{(-1)} = 1 - 1/y$.

bcskew0 works like lnskew0:

```
. bcskew0 bcmpg = mpg, level(95)

        Transform |        L    [95% Conf. Interval]    Skewness
    --------------+------------------------------------------------
       (mpg^L-1)/L |  -.3673283    -1.212752   .4339645    .0001898
```

Note that the 95% confidence interval includes $\lambda = -1$ (λ is labeled L in the output), which has a rather more pleasing interpretation—gallons per mile—rather than $(\text{mpg}^{-.3673} - 1)/(-.3673)$. The confidence interval, however, is calculated assuming that the power transformed variable is normally distributed. It makes perfect sense to use bcskew0, even when you do not believe that the transformed variable will be normally distributed, but in that case, the confidence interval is an approximation of unknown quality. If you believe that the transformed data are normally distributed, you can alternatively use boxcox to estimate λ; see [R] **boxcox**.

◁

Saved Results

lnskew0 and bcskew0 save in r():

Scalars
      r(gamma) k (lnskew0)
      r(lambda) λ (bcskew0)
      r(lb)   lower bound of confidence interval
      r(ub)   upper bound of confidence interval
      r(skewness) resulting skewness of transformed variable

Methods and Formulas

lnskew0 and bcskew0 are implemented as ado-files.

Skewness is as calculated by summarize; see [R] **summarize**. Newton's method with numeric, uncentered derivatives is used to estimate k (lnskew0) and λ (bcskew0). In the case of lnskew0, the initial value is chosen so that the minimum of $x - k$ is 1, and thus $\ln(x - k)$ is 0. bcskew0 starts with $\lambda = 1$.

Acknowledgment

lnskew0 and bcskew0 were written by Patrick Royston of the MRC Clinical Trials Unit, London.

References

Box, G. E. P. and D. R. Cox. 1964. An analysis of transformations. *Journal of the Royal Statistical Society, Series B* 26: 211–243.

Also See

Complementary: [R] **ladder**

Related: [R] **boxcox**, [R] **swilk**

Title

| log — Echo copy of session to file or device |

Syntax

Report status of log file

   <u>log</u>

Open log file

   <u>log</u> using *filename* $\big[$, append replace $\big[$ <u>t</u>ext | <u>s</u>mcl $\big]\big]$

Close log, temporarily suspend logging, or resume logging

   <u>log</u> $\big\{$ <u>c</u>lose | <u>of</u>f | on $\big\}$

Report status of command log file

   cmdlog

Open command log file

   cmdlog using *filename* $\big[$, append replace $\big]$

Close command log, temporarily suspend logging, or resume logging

   cmdlog $\big\{$ <u>c</u>lose | on | <u>of</u>f $\big\}$

Set default format for logs

   set logtype $\big\{$ <u>t</u>ext | <u>s</u>mcl $\big\}$ $\big[$, <u>perm</u>anently $\big]$

Specify screen width

   <u>set</u> <u>li</u>nesize #

If *filename* is specified without an extension, one of the suffixes .smcl, .log, or .txt is added.

   The extension .smcl or .log is added by log, depending on whether the file format is SMCL or ASCII text.

   The extension .txt is added by cmdlog.

In addition to using the log command, you may access the capabilities of log by opening the **File > Log** menu.

Note: If *filename* contains embedded spaces, remember to enclose it in double quotes.

56

Description

log allows you to make a full record of your Stata session. A log is a file containing what you type and Stata's output.

cmdlog allows you to make a record of what you type during your Stata session. A command log contains only what you type, so it is a subset of a full log.

You can make full logs, command logs, or both simultaneously. Neither is produced until you tell Stata to start logging.

Command logs are always ASCII text files, making them easy to convert into do-files. (In this respect, it would make more sense if the default extension of a command log file was .do because command logs are do-files. The default is .txt, not .do, however, to keep you from accidentally overwriting your important do-files.)

Full logs are recorded in one of two formats: SMCL (Stata Markup and Control Language) or text (meaning ASCII). The default is SMCL, but you can use set logtype to change that, or you can specify an option to state the format you wish. We recommend SMCL because it preserves fonts and colors. SMCL logs can be converted to ASCII text or to other formats using the translate command; see [R] **translate**. You can also use translate to produce printable versions of SMCL logs. SMCL logs can be viewed and printed from the Viewer, as can any text file; see [R] **view**.

log or cmdlog, typed without arguments, reports the status of logging.

log using and cmdlog using open a log file. log close and cmdlog close close the file. Between times, log off and cmdlog off, and log on and cmdlog on can temporarily suspend and resume logging.

set logtype specifies the default format in which full logs are to be recorded. Initially, full logs are recorded in SMCL format.

set linesize specifies the width of the screen currently being used (and so really has nothing to do with logging). Note that not all Stata commands respect linesize. Also, note that there is no permanently option allowed with set linesize.

Options for use with both log and cmdlog

append specifies that results be appended onto the end of an existing file. If the file does not already exist, a new file is created.

replace specifies that *filename*, if it already exists, be overwritten. When you do not specify either replace or append, the file is assumed to be new. If the specified file already exists, an error message is issued and logging is not started.

Options for use with log

text and smcl specify the format in which the log is to be recorded. The default is complicated to describe but is what you would expect:

If you specify the file as *filename*.smcl, the default is to write the log in SMCL format (regardless of the value of set logtype).

If you specify the file as *filename*.log, the default is to write the log in text format (regardless of the value of the set logtype).

If you type *filename* without an extension and specify neither the smcl option nor the text option, the default is to write the file according to the value of set logtype. If you have not set logtype, then that default is SMCL. In addition, the *filename* you specified will be fixed to read *filename*.smcl if a SMCL log is being created or *filename*.log if a text log is being created.

If you specify either of the options text or smcl, then what you specify determines how the log is written. If *filename* was specified without an extension, the appropriate extension is added for you.

Option for use with set logtype

permanently specifies that, in addition to making the change right now, the logtype setting be remembered and become the default setting when you invoke Stata.

Remarks

For a detailed explanation of logs, see [U] **15 Printing and preserving output**.

Note that when you open a full log, the default is to show the name of the file and a time and date stamp:

```
. log using myfile

      log:  C:\data\proj1\myfile.smcl
 log type:  smcl
opened on:  12 Jan 2005, 12:28:23

.
```

The above information will appear in the log. If you do not want this information to appear, precede the command by quietly:

```
. quietly log using myfile
```

quietly will not suppress any error messages or anything else you need to know.

Similarly, when you close a full log, the default is to show the full information,

```
. log close
      log:  C:\data\proj1\myfile.smcl
 log type:  smcl
closed on:  12 Jan 2005, 12:32:41
```

and that information will also appear in the log. If you want to suppress that, type quietly log close.

Saved Results

log and cmdlog save in r():

Macros
    r(filename)   name of file
    r(status)   on or off
    r(type)   text or smcl

Also See

Complementary:	[R] **translate**; [R] **more**, [R] **query**
Background:	[GSM] **16 Logs: Printing and saving output**,
	[GSW] **16 Logs: Printing and saving output**,
	[GSU] **16 Logs: Printing and saving output**,
	[U] **7 —more— conditions**,
	[U] **11.6 File-naming conventions**,
	[U] **15 Printing and preserving output**

Title

logistic — Logistic regression, reporting odds ratios

Syntax

logistic *depvar indepvars* $[$ *if* $]$ $[$ *in* $]$ $[$ *weight* $]$ $[$, *options* $]$

options	description
Model	
<u>off</u>set(*varname*)	include *varname* in model with coefficient constrained to 1
asis	retain perfect predictor variables
SE/Robust	
vce(*vcetype*)	*vcetype* may be <u>r</u>obust, <u>boot</u>strap, or <u>jack</u>knife
<u>r</u>obust	synonym for vce(robust)
<u>c</u>luster(*varname*)	adjust standard errors for intragroup correlation
Reporting	
<u>l</u>evel(#)	set confidence level; default is level(95)
coef	report estimated coefficients
Max options	
maximize_options	control the maximization process; seldom used

depvar and *indepvars* may contain time-series operators; see [U] **11.4.3 Time-series varlists**.
bootstrap, by, jackknife, rolling, statsby, stepwise, svy, and xi are allowed; see
  [U] **11.1.10 Prefix commands**.
fweights, iweights, and pweights are allowed; see [U] **11.1.6 weight**.
See [U] **20 Estimation and postestimation commands** for additional capabilities of estimation commands.

Description

logistic fits a logistic regression model of *depvar* on *varlist*, where *depvar* is a 0/1 variable (or, more precisely, a 0/non-0 variable). Without arguments, logistic redisplays the last logistic estimates. logistic displays estimates as odds ratios; to view coefficients, type logit after running logistic. To obtain odds ratios for any covariate pattern relative to another, see [R] **lincom**.

Options

<u>└ Model └</u>

offset(*varname*); see [R] **estimation options**.

asis forces retention of perfect predictor variables and their associated perfectly predicted observations and may produce instabilities in maximization; see [R] **probit** (*sic*).

60

⌐ SE/Robust ⌐

vce(*vcetype*); see [R] **vce_option**.

robust, cluster(*varname*); see [R] **estimation options**. cluster() can be used with pweights to
produce estimates for unstratified cluster-sampled data, but see [SVY] **svy: logistic** for a command
especially designed for survey data.

⌐ Reporting ⌐

level(*#*); see [R] **estimation options**.

coef causes logistic to report the estimated coefficients rather than the odds ratios (exponentiated
coefficients). coef may be specified when the model is fitted or may be used later to redisplay
results. coef affects only how results are displayed and not how they are estimated.

⌐ Max options ⌐

maximize_options: iterate(*#*), tolerance(*#*), ltolerance(*#*); see [R] **maximize**. These options
are seldom used.

Remarks

Remarks are presented under the headings

> *logistic and logit*
> *Robust estimate of variance*

logistic and logit

logistic provides an alternative and preferred way to fit maximum-likelihood logit models, the
other choice being logit ([R] **logit**).

First, let us dispose of some confusing terminology. We use the words logit and logistic to mean
the same thing: maximum likelihood estimation. To some, one or the other of these words connotes
transforming the dependent variable and using weighted least squares to fit the model, but that is not
how we use either word here. Thus the logit and logistic commands produce the same results.

The logistic command is generally preferred to logit because logistic presents the estimates
in terms of odds ratios rather than coefficients. To a few people, this may seem a disadvantage, but
you can type logit without arguments after logistic to see the underlying coefficients.

Nevertheless, [R] **logit** is still worth reading because logistic shares the same features as logit,
including omitting variables due to collinearity or one-way causation.

For an introduction to logistic regression, see Lemeshow and Hosmer (1998), Pagano and Gauvreau
(2000, 470–487), or Pampel (2000); for a complete but nonmathematical treatment, see Kleinbaum
and Klein (2002); and for a thorough discussion, see Hosmer and Lemeshow (2000). See Gould
(2000) for a discussion of the interpretation of logistic regression. See Dupont (2002) for a discussion
of logistic regression with examples using Stata. For a discussion using Stata with an emphasis on
model specification, see Vittinghoff et al. (2005).

Stata has a variety of commands for performing estimation when the dependent variable is
dichotomous or polychotomous. See Long and Freese (2003) for a book devoted to fitting these
models using Stata. Here is a list of some estimation commands that may be of interest. See
[I] **estimation commands** for a complete list of all of Stata's estimation commands.

asmprobit	[R] **asmprobit**	Alternative-specific multinomial probit regression
binreg	[R] **binreg**	GLM models for the binomial family
biprobit	[R] **biprobit**	Bivariate probit regression
blogit	[R] **glogit**	Logit regression for grouped data
bprobit	[R] **glogit**	Probit regression for grouped data
clogit	[R] **clogit**	Conditional (fixed-effects) logistic regression
cloglog	[R] **cloglog**	Complementary log-log regression
glm	[R] **glm**	Generalized linear models
glogit	[R] **glogit**	Weighted least-squares logistic regression for grouped data
gprobit	[R] **glogit**	Weighted least-squares probit regression for grouped data
heckprob	[R] **heckprob**	Probit model with selection
hetprob	[R] **hetprob**	Heteroskedastic probit model
ivprobit	[R] **ivprobit**	Probit model with endogenous regressors
logit	[R] **logit**	Logistic regression, reporting coefficients
mlogit	[R] **mlogit**	Multinomial (polytomous) logistic regression
mprobit	[R] **mprobit**	Multinomial probit regression
nlogit	[R] **nlogit**	Nested logit regression
ologit	[R] **ologit**	Ordered logistic regression
oprobit	[R] **oprobit**	Ordered probit regression
probit	[R] **probit**	Probit regression
rologit	[R] **rologit**	Rank-ordered logistic regression
scobit	[R] **scobit**	Skewed logistic regression
slogit	[R] **slogit**	Stereotype logistic regression
svy: heckprob	[SVY] **svy: heckprob**	Survey version of heckprob
svy: logistic	[SVY] **svy: logistic**	Survey version of logistic
svy: logit	[SVY] **svy: logit**	Survey version of logit
svy: mlogit	[SVY] **svy: mlogit**	Survey version of mlogit
svy: ologit	[SVY] **svy: ologit**	Survey version of ologit
svy: oprobit	[SVY] **svy: oprobit**	Survey version of oprobit
svy: probit	[SVY] **svy: probit**	Survey version of probit
xtcloglog	[XT] **xtcloglog**	Random-effects and population-averaged cloglog models
xtgee	[XT] **xtgee**	GEE population-averaged generalized linear models
xtlogit	[XT] **xtlogit**	Fixed-effects, random-effects, and population-averaged logit models
xtprobit	[XT] **xtprobit**	Random-effects and population-averaged probit models

▷ Example 1

Consider the following dataset from a study of risk factors associated with low birthweight described in Hosmer and Lemeshow (2000, 25).

```
. use http://www.stata-press.com/data/r9/lbw
(Hosmer & Lemeshow data)

. describe

Contains data from http://www.stata-press.com/data/r9/lbw.dta
  obs:           189                          Hosmer & Lemeshow data
  vars:           11                          15 Jan 2005 05:01
  size:        3,402 (95.1% of memory free)

              storage  display    value
variable name   type   format     label      variable label

id              int    %8.0g                 identification code
low             byte   %8.0g                 birthweight<2500g
age             byte   %8.0g                 age of mother
lwt             int    %8.0g                 weight at last menstrual period
race            byte   %8.0g      race       race
smoke           byte   %8.0g                 smoked during pregnancy
ptl             byte   %8.0g                 premature labor history (count)
ht              byte   %8.0g                 has history of hypertension
ui              byte   %8.0g                 presence, uterine irritability
ftv             byte   %8.0g                 number of visits to physician
                                               during 1st trimester
bwt             int    %8.0g                 birthweight (grams)

Sorted by:
```

We want to investigate the causes of low birthweight. In this dataset, race is a categorical variable indicating whether a person is white (race = 1), black (race = 2), or other (race = 3). We want indicator (dummy) variables for race included in the regression. (One of the dummies, of course, must be omitted.) Thus before we can fit the model, we must create the dummy variables for race.

There are a number of ways we could do this, but the easiest is to let another Stata command, xi, do it for us. We type xi: in front of our logistic command and in our *varlist* include not race but i.race to indicate that we want the indicator variables for this categorical variable; see [R] **xi** for the full details.

```
. xi: logistic low age lwt i.race smoke ptl ht ui
i.race            _Irace_1-3          (naturally coded; _Irace_1 omitted)
Logistic regression                             Number of obs   =        189
                                                LR chi2(8)      =      33.22
                                                Prob > chi2     =     0.0001
Log likelihood =   -100.724                     Pseudo R2       =     0.1416
```

low	Odds Ratio	Std. Err.	z	P>|z|	[95% Conf. Interval]	
age	.9732636	.0354759	-0.74	0.457	.9061578	1.045339
lwt	.9849634	.0068217	-2.19	0.029	.9716834	.9984249
_Irace_2	3.534767	1.860737	2.40	0.016	1.259736	9.918406
_Irace_3	2.368079	1.039949	1.96	0.050	1.001356	5.600207
smoke	2.517698	1.00916	2.30	0.021	1.147676	5.523162
ptl	1.719161	.5952579	1.56	0.118	.8721455	3.388787
ht	6.249602	4.322408	2.65	0.008	1.611152	24.24199
ui	2.1351	.9808153	1.65	0.099	.8677528	5.2534

The odds ratios are for a one-unit change in the variable. If we wanted the odds ratio for `age` to be in terms of four-year intervals, we would type

```
. gen age4 = age/4
. xi: logistic low age4 lwt i.race smoke ptl ht ui
(output omitted)
```

After `logistic`, we can type `logit` to see the model in terms of coefficients and standard errors:

```
. logit
```

Logistic regression   Number of obs =   189
                                    LR chi2(8)   =   33.22
                                      Prob > chi2   =   0.0001
Log likelihood =   -100.724   Pseudo R2   =   0.1416

low	Coef.	Std. Err.	z	P>|z|	[95% Conf. Interval]	
age	-.0271003	.0364504	-0.74	0.457	-.0985418	.0443412
lwt	-.0151508	.0069259	-2.19	0.029	-.0287253	-.0015763
_Irace_2	1.262647	.5264101	2.40	0.016	.2309024	2.294392
_Irace_3	.8620792	.4391531	1.96	0.050	.0013548	1.722804
smoke	.9233448	.4008266	2.30	0.021	.1377391	1.708951
ptl	.5418366	.346249	1.56	0.118	-.136799	1.220472
ht	1.832518	.6916292	2.65	0.008	.4769494	3.188086
ui	.7585135	.4593768	1.65	0.099	-.1418484	1.658875
_cons	.4612239	1.20459	0.38	0.702	-1.899729	2.822176

If we wanted to see the `logistic` output again, we would type `logistic` without arguments.

◁

▷ Example 2

We can specify the confidence interval for the odds ratios with the `level()` option, and we can do this either at estimation time or when replaying the model. For instance, to see our previous models with narrower, 90% confidence intervals, we might type

```
. logistic, level(90)
```

Logistic regression   Number of obs =   189
                                      LR chi2(8)   =   33.22
                                      Prob > chi2   =   0.0001
Log likelihood =   -100.724   Pseudo R2   =   0.1416

low	Odds Ratio	Std. Err.	z	P>|z|	[90% Conf. Interval]	
age	.9732636	.0354759	-0.74	0.457	.9166258	1.033401
lwt	.9849634	.0068217	-2.19	0.029	.9738063	.9962483
_Irace_2	3.534767	1.860737	2.40	0.016	1.487028	8.402379
_Irace_3	2.368079	1.039949	1.96	0.050	1.149971	4.876471
smoke	2.517698	1.00916	2.30	0.021	1.302185	4.867819
ptl	1.719161	.5952579	1.56	0.118	.9726876	3.038505
ht	6.249602	4.322408	2.65	0.008	2.003487	19.49478
ui	2.1351	.9808153	1.65	0.099	1.00291	4.545424

◁

Robust estimate of variance

If you specify `robust`, Stata reports the robust estimate of variance described in [U] **20.14 Obtaining robust variance estimates**. Here is the model previously fitted with the robust estimate of variance:

```
. xi: logistic low age lwt i.race smoke ptl ht ui, robust
i.race            _Irace_1-3         (naturally coded; _Irace_1 omitted)

Logistic regression                             Number of obs   =        189
                                                Wald chi2(8)    =      29.02
                                                Prob > chi2     =     0.0003
Log pseudolikelihood =    -100.724              Pseudo R2       =     0.1416
```

low	Odds Ratio	Robust Std. Err.	z	P>|z|	[95% Conf. Interval]	
age	.9732636	.0329376	-0.80	0.423	.9108015	1.040009
lwt	.9849634	.0070209	-2.13	0.034	.9712984	.9988206
_Irace_2	3.534767	1.793616	2.49	0.013	1.307504	9.556051
_Irace_3	2.368079	1.026563	1.99	0.047	1.012512	5.538501
smoke	2.517698	.9736416	2.39	0.017	1.179852	5.372537
ptl	1.719161	.7072902	1.32	0.188	.7675715	3.850476
ht	6.249602	4.102026	2.79	0.005	1.726445	22.6231
ui	2.1351	1.042775	1.55	0.120	.8197749	5.560858

Additionally, `robust` allows you to specify `cluster()`, and can then, within cluster, relax the assumption of independence. To illustrate this, we have made some fictional additions to the low birthweight data.

Say that these data are not a random sample of mothers but instead are a random sample of mothers from a random sample of hospitals. In fact, that may be true—we do not know the history of these data.

Hospitals specialize, and it would not be too incorrect to say that some hospitals specialize in more difficult cases. We are going to show two extremes. In one, all hospitals are alike, but we are going to estimate under the possibility that they might differ. In the other, hospitals are strikingly different. In both cases, we assume that patients are drawn from 20 hospitals.

In both examples, we will fit the same model, and we will type the same command to fit it. Below are the same data we have been using but with a new variable `hospid`, which identifies from which of the 20 hospitals each patient was drawn (and which we have made up):

(Continued on next page)

```
. use http://www.stata-press.com/data/r9/hospid1

. xi: logistic low age lwt i.race smoke ptl ht ui, robust cluster(hospid)
i.race            _Irace_1-3          (naturally coded; _Irace_1 omitted)
```

```
Logistic regression                          Number of obs   =        189
                                             Wald chi2(8)    =      49.67
                                             Prob > chi2     =     0.0000
Log pseudolikelihood =   -100.724            Pseudo R2       =     0.1416
```

(Std. Err. adjusted for 20 clusters in hospid)

low	Odds Ratio	Robust Std. Err.	z	P>|z|	[95% Conf. Interval]	
age	.9732636	.0397476	-0.66	0.507	.898396	1.05437
lwt	.9849634	.0057101	-2.61	0.009	.9738352	.9962187
_Irace_2	3.534767	2.013285	2.22	0.027	1.157563	10.79386
_Irace_3	2.368079	.8451325	2.42	0.016	1.176562	4.766257
smoke	2.517698	.8284259	2.81	0.005	1.321062	4.79826
ptl	1.719161	.6676221	1.40	0.163	.8030814	3.680219
ht	6.249602	4.066275	2.82	0.005	1.745911	22.37086
ui	2.1351	1.093144	1.48	0.138	.7827337	5.824014

The standard errors are quite similar to the standard errors we have previously obtained, whether we used the robust or conventional estimators. In this example, we invented the hospital IDs randomly.

Here are the results of the estimation with the same data but with a different set of hospital IDs:

```
. use http://www.stata-press.com/data/r9/hospid2

. xi: logistic low age lwt i.race smoke ptl ht ui, robust cluster(hospid)
i.race            _Irace_1-3          (naturally coded; _Irace_1 omitted)
```

```
Logistic regression                          Number of obs   =        189
                                             Wald chi2(8)    =       7.19
                                             Prob > chi2     =     0.5167
Log pseudolikelihood =   -100.724            Pseudo R2       =     0.1416
```

(Std. Err. adjusted for 20 clusters in hospid)

low	Odds Ratio	Robust Std. Err.	z	P>|z|	[95% Conf. Interval]	
age	.9732636	.0293064	-0.90	0.368	.9174862	1.032432
lwt	.9849634	.0106123	-1.41	0.160	.9643817	1.005984
_Irace_2	3.534767	3.120338	1.43	0.153	.6265521	19.9418
_Irace_3	2.368079	1.297738	1.57	0.116	.8089594	6.932114
smoke	2.517698	1.570287	1.48	0.139	.7414969	8.548654
ptl	1.719161	.6799153	1.37	0.171	.7919046	3.732161
ht	6.249602	7.165454	1.60	0.110	.660558	59.12808
ui	2.1351	1.411977	1.15	0.251	.5841231	7.804266

Note the strikingly larger standard errors. What happened? In these data, women most likely to have low birthweight babies are sent to certain hospitals, and the decision on likeliness is based not just on age, smoking history, etc., but on other things that doctors can see but that are not recorded in our data. Thus merely because a woman is at one of the centers identifies her to be more likely to have a low birthweight baby.

Saved Results

logistic saves in e():

Scalars

e(N)	number of observations	e(ll_0)	log likelihood, constant-only model
e(df_m)	model degrees of freedom	e(N_clust)	number of clusters
e(r2_p)	pseudo-R-squared	e(chi2)	χ^2
e(ll)	log likelihood		

Macros

e(cmd)	logistic	e(chi2type)	Wald or LR; type of model χ^2 test
e(depvar)	name of dependent variable	e(vce)	*vcetype* specified in vce()
e(wtype)	weight type	e(vcetype)	title used to label Std. Err.
e(wexp)	weight expression	e(crittype)	optimization criterion
e(title)	title in estimation output	e(properties)	b V
e(clustvar)	name of cluster variable	e(estat_cmd)	program used to implement estat
e(offset)	offset	e(predict)	program used to implement predict

Matrices

e(b)	coefficient vector	e(V)	variance–covariance matrix of the estimators

Functions

e(sample)	marks estimation sample

Methods and Formulas

logistic is implemented as an ado-file.

Define $\mathbf{x}_j$ as the (row) vector of independent variables, augmented by 1, and $\mathbf{b}$ as the corresponding estimated parameter (column) vector. The logistic regression model is fitted by logit; see [R] **logit** for details of estimation.

The odds ratio corresponding to the ith coefficient is $\psi_i = \exp(b_i)$. The standard error of the odds ratio is $s_i^{\psi} = \psi_i s_i$, where s_i is the standard error of b_i estimated by logit.

Define $I_j = \mathbf{x}_j \mathbf{b}$ as the predicted index of the jth observation. The predicted probability of a positive outcome is

$$p_j = \frac{\exp(I_j)}{1 + \exp(I_j)}$$

References

Brady, A. R. 1998. sbe21: Adjusted population attributable fractions from logistic regression. *Stata Technical Bulletin* 42: 8–12. Reprinted in *Stata Technical Bulletin Reprints*, vol. 7, pp. 137–143.

Cleves, M. and A. Tosetto. 2000. sg139: Logistic regression when binary outcome is measured with uncertainty. *Stata Technical Bulletin* 55: 20–23.

Collett, D. 1991. *Modelling Binary Data*. London: Chapman & Hall.

Dupont, W. D. 2002. *Statistical Modeling for Biomedical Researchers*. Cambridge: Cambridge University Press.

Freese, J. 2002. Least likely observations in regression models for categorical outcomes. *Stata Journal* 2: 296–300.

Garrett, J. M. 1997. sbe14: Odds ratios and confidence intervals for logistic regression models with effect modification. *Stata Technical Bulletin* 36: 15–22. Reprinted in *Stata Technical Bulletin Reprints*, vol. 6, pp. 104–114.

Gould, W. W. 2000. sg124: Interpreting logistic regression in all its forms. *Stata Technical Bulletin* 53: 19–29. Reprinted in *Stata Technical Bulletin Reprints*, vol. 9, pp. 257–270.

Hilbe, J. 1997. sg63: Logistic regression: Standardized coefficients and partial correlations. *Stata Technical Bulletin* 35: 21–22. Reprinted in *Stata Technical Bulletin Reprints*, vol. 6, pp. 162–163.

Hosmer, D. W., Jr., and S. Lemeshow. 2000. *Applied Logistic Regression.* 2nd ed. New York: Wiley.

Irala-Estévez, J. de and M. A. Martínez. 2000. sg125: Automatic estimation of interaction effects and their confidence intervals. *Stata Technical Bulletin* 53: 29–31. Reprinted in *Stata Technical Bulletin Reprints*, vol. 9, pp. 270–273.

Kleinbaum, D. G. and M. Klein. 2002. *Logistic Regression: A Self-Learning Text.* 2nd ed. New York: Springer.

Lemeshow, S. and D. W. Hosmer, Jr. 1998. Logistic regression. In *Encyclopedia of Biostatistics*, ed. P. Armitage and T. Colton, 2316–2327. New York: Wiley.

Lemeshow, S. and J.-R. Le Gall. 1994. Modeling the severity of illness of ICU patients: A systems update. *Journal of the American Medical Association* 272: 1049–1055.

Mitchell, M. N. and X. Chen. 2005. Visualizing main effects and interactions for binary logit models. *Stata Journal* 5: 64–82.

Pagano, M. and K. Gauvreau. 2000. *Principles of Biostatistics.* 2nd ed. Pacific Grove, CA: Brooks/Cole.

Pampel, F. C. 2000. *Logistic Regression: A Primer.* Thousand Oaks, CA: Sage.

Paul, C. 1998. sg92: Logistic regression for data including multiple imputations. *Stata Technical Bulletin* 45: 28–30. Reprinted in *Stata Technical Bulletin Reprints*, vol. 8, pp. 180–183.

Pearce, M. S. 2000. sg148: Profile likelihood confidence intervals for explanatory variables in logistic regression. *Stata Technical Bulletin* 56: 45–47.

Pregibon, D. 1981. Logistic regression diagnostics. *Annals of Statistics* 9: 705–724.

Reilly, M. and A. Salim. 2000. sg156: Mean score method for missing covariate data in logistic regression models. *Stata Technical Bulletin* 58: 25–27. Reprinted in *Stata Technical Bulletin Reprints*, vol. 10, pp. 256–258.

Vittinghoff, E., D. V. Glidden, S. C. Shiboski, and C. E. McCulloch. 2005. *Regression Methods in Biostatistics: Linear, Logistic, Survival, and Repeated Measures Models.* New York: Springer.

Also See

Complementary:	[R] **logistic postestimation**, [R] **roc**, [R] **rocfit**
Related:	[R] **asmprobit**, [R] **brier**, [R] **clogit**, [R] **cloglog**, [R] **glm**, [R] **glogit**, [R] **ivprobit**, [R] **logit**, [R] **mlogit**, [R] **mprobit**, [R] **nlogit**, [R] **ologit**, [R] **probit**, [R] **rologit**, [R] **scobit**, [R] **slogit**, [SVY] **svy: logistic**, [SVY] **svy: logit**, [XT] **xtlogit**
Background:	[U] **11.1.10 Prefix commands**, [U] **20 Estimation and postestimation commands**, [R] **estimation options**, [R] **maximize**, [R] *vce_option*

Title

logistic postestimation — Postestimation tools for logistic

Description

The following postestimation commands are of special interest after `logistic`:

command	description
estat clas	estat classification reports various summary statistics, including the classification table
estat gof	Pearson or Hosmer–Lemeshow goodness-of-fit test
lroc	graphs the ROC curve and calculates the area under the curve
lsens	graphs sensitivity and specificity versus probability cutoff

For information about these commands, see below.

In addition, the following standard postestimation commands are available:

command	description
adjust[1]	adjusted predictions of $\mathbf{x}\beta$ or probabilities
estat	AIC, BIC, VCE, and estimation sample summary
estimates	cataloging estimation results
lincom	point estimates, standard errors, testing, and inference for linear combinations of coefficients
linktest	link test for model specification
lrtest	likelihood-ratio test
mfx	marginal effects or elasticities
nlcom	point estimates, standard errors, testing, and inference for nonlinear combinations of coefficients
predict	predictions, residuals, influence statistics, and other diagnostic measures
predictnl	point estimates, standard errors, testing, and inference for generalized predictions
suest	seemingly unrelated estimation
test	Wald tests for simple and composite linear hypotheses
testnl	Wald tests of nonlinear hypotheses

[1] adjust does not work with time-series operators.

See the corresponding entries in the *Stata Base Reference Manual* for details.

Special-interest postestimation commands

`estat classification` reports various summary statistics, including the classification table.

`estat gof` reports the Pearson goodness-of-fit test or the Hosmer–Lemeshow goodness-of-fit test.

`lroc` graphs the ROC curve and calculates the area under the curve.

lsens graphs sensitivity and specificity versus probability cutoff and optionally creates new variables containing these data.

estat classification, estat gof, lroc, and lsens produce statistics and graphs either for the estimation sample or for any set of observations. However, they always use the estimation sample by default. When weights, if, or in are used with logistic, it is not necessary to repeat them with these commands when you want statistics computed for the estimation sample. Specify if, in, or the all option only when you want statistics computed for a set of observations other than the estimation sample. Specify weights (only fweights are allowed with these commands) only when you want to use a different set of weights.

By default, estat classification, estat gof, lroc, and lsens use the last model fitted by logistic. You may also directly specify the model to lroc and lsens by inputting a vector of coefficients with the beta() option and passing the name of the dependent variable *depvar* to these commands.

The estat classification, estat gof, lroc, and lsens commands may also be used after logit or probit.

Syntax for predict

predict $[type]$ *newvar* $[if]$ $[in]$ $[$, *statistic* <u>nooff</u>set <u>rules</u> asif $]$

statistic	description
<u>pr</u>	probability of a positive outcome; the default
xb	$\mathbf{x}_j\mathbf{b}$, linear prediction
stdp	standard error of the linear prediction
*<u>dbeta</u>	Pregibon (1981) $\Delta\widehat{\beta}$ influence statistic
*<u>deviance</u>	deviance residual
*<u>dx2</u>	Hosmer and Lemeshow (2000, 174) $\Delta\chi^2$ influence statistic
*<u>ddeviance</u>	Hosmer and Lemeshow (2000, 174) ΔD influence statistic
*<u>hat</u>	Pregibon (1981) leverage
*<u>number</u>	sequential number of the covariate pattern
*<u>residuals</u>	Pearson residuals; adjusted for number sharing covariate pattern
*<u>rstandard</u>	standardized Pearson residuals; adjusted for number sharing covariate pattern
<u>sc</u>ore	first derivative of the log likelihood with respect to $\mathbf{x}_j\beta$

Unstarred statistics are available both in and out of sample; type predict ... if e(sample) ... if wanted only for the estimation sample. Starred statistics are calculated only for the estimation sample, even when if e(sample) is not specified.

Options for predict

pr, the default, calculates the probability of a positive outcome.

xb calculates the linear prediction.

stdp calculates the standard error of the linear prediction.

dbeta calculates the Pregibon (1981) $\Delta\widehat{\beta}$ influence statistic, a standardized measure of the difference in the coefficient vector due to deletion of the observation along with all others that share the same covariate pattern. In Hosmer and Lemeshow (2000, 144–145) jargon, this statistic is M-asymptotic; that is, it is adjusted for the number of observations that share the same covariate pattern.

deviance calculates the deviance residual.

dx2 calculates the Hosmer and Lemeshow (2000, 174) $\Delta\chi^2$ influence statistic, reflecting the decrease in the Pearson χ^2 due to deletion of the observation and all others that share the same covariate pattern.

ddeviance calculates the Hosmer and Lemeshow (2000, 174) ΔD influence statistic, which is the change in the deviance residual due to deletion of the observation and all others that share the same covariate pattern.

hat calculates the Pregibon (1981) leverage or the diagonal elements of the hat matrix adjusted for the number of observations that share the same covariate pattern.

number numbers the covariate patterns—observations with the same covariate pattern have the same number. Observations not used in estimation have number set to missing. The "first" covariate pattern is numbered 1, the second 2, and so on.

residuals calculates the Pearson residual as given by Hosmer and Lemeshow (2000, 145) and adjusted for the number of observations that share the same covariate pattern.

rstandard calculates the standardized Pearson residual as given by Hosmer and Lemeshow (2000, 173) and adjusted for the number of observations that share the same covariate pattern.

score calculates the equation-level score, $\partial\ln L/\partial(\mathbf{x}_j\beta)$.

nooffset is relevant only if you specified offset(*varname*) for logistic. It modifies the calculations made by predict so that they ignore the offset variable; the linear prediction is treated as $\mathbf{x}_j\mathbf{b}$ rather than as $\mathbf{x}_j\mathbf{b} + \text{offset}_j$.

rules requests that Stata use any "rules" that were used to identify the model when making the prediction. By default, Stata calculates missing for excluded observations. See [R] **logit** for an example.

asif requests that Stata ignore the rules and the exclusion criteria and calculate predictions for all observations possible using the estimated parameter from the model. See [R] **logit** for an example.

Syntax for estat classification

> estat <u>class</u>ification $\begin{bmatrix}if\end{bmatrix}$ $\begin{bmatrix}in\end{bmatrix}$ $\begin{bmatrix}weight\end{bmatrix}$ $\begin{bmatrix}, class\_options\end{bmatrix}$

class_options	description
Main	
all	display summary statistics for all observations in the data
<u>cut</u>off(*#*)	positive outcome threshold; default is cutoff(0.5)

fweights are allowed; see [U] **11.1.6 weight**.

Options for estat classification

   Main

all requests that the statistic be computed for all observations in the data, ignoring any if or in restrictions specified by logistic.

cutoff(*#*) specifies the value for determining whether an observation has a predicted positive outcome. An observation is classified as positive if its predicted probability is $\geq$ *#*. The default is 0.5.

Syntax for estat gof

estat gof [*if*] [*in*] [*weight*] [, *gof_options*]

gof_options	description
Main	
group(*#*)	perform Hosmer–Lemeshow goodness-of-fit test using *#* quantiles
all	execute test for all observations in the data
outsample	adjust degrees of freedom for samples outside estimation sample
table	display table of groups used for test

fweights are allowed; see [U] **11.1.6 weight**.

Options for estat gof

___ Main ___

group(*#*) specifies the number of quantiles to be used to group the data for the Hosmer–Lemeshow goodness-of-fit test. group(10) is typically specified. If this option is not given, the Pearson goodness-of-fit test is computed using the covariate patterns in the data as groups.

all requests that the statistic be computed for all observations in the data, ignoring any if or in restrictions specified with logistic.

outsample adjusts the degrees of freedom for the Pearson and Hosmer–Lemeshow goodness-of-fit tests for samples outside of the estimation sample. See *Samples other than estimation sample* later in this entry.

table displays a table of the groups used for the Hosmer–Lemeshow or Pearson goodness-of-fit test with predicted probabilities, observed and expected counts for both outcomes, and totals for each group.

(Continued on next page)

Syntax for lroc

lroc [*depvar*] [*if*] [*in*] [*weight*] [, *lroc_options*]

lroc_options	description
Main	
all	compute area under ROC curve and graph curve for all observations
nograph	suppress graph
Advanced	
beta(*matname*)	row vector containing coefficients for a logistic model
Plot	
marker_options	change look of markers (color, size, etc.)
marker_label_options	add marker labels; change look or position
cline_options	change the look of the line
Reference line	
rlopts(*cline_options*)	affect rendition of the reference line
Add plot	
addplot(*plot*)	add other plots to the generated graph
Y-Axis, X-Axis, Title, Caption, Legend, Overall	
twoway_options	any options other than by() documented in [G] *twoway_options*

fweights are allowed; see [U] **11.1.6 weight**.

Options for lroc

___ Main ___

all requests that the statistic be computed for all observations in the data, ignoring any if or in restrictions specified by logistic.

nograph suppresses graphical output.

___ Advanced ___

beta(*matname*) specifies a row vector containing coefficients for a logistic model. The columns of the row vector must be labeled with the corresponding names of the independent variables in the data. The dependent variable *depvar* must be specified immediately after the command name. See *Models other than last fitted model* later in this entry.

___ Plot ___

marker_options, *marker_label_options*, and *cline_options* affect the rendition of the ROC curve—the plotted points connected by lines. These options affect the size and color or markers, whether and how the markers are labeled, and whether and how the points are connected; see [G] *marker_options*, [G] *marker_label_options*, and [G] *cline_options*.

___ Reference line ___

rlopts(*cline_options*) affect the rendition of the reference line; see [G] *connect_options*.

⌐ Add plot ⌐

addplot(*plot*) provides a way to add other plots to the generated graph. See [G] ***addplot_option***.

⌐ Y-Axis, X-Axis, Title, Caption, Legend, Overall ⌐

twoway_options are any of the options documented in [G] ***twoway_options*** excluding by(). These include options for titling the graph (see [G] ***title_options***) and options for saving the graph to disk (see [G] ***saving_option***).

Syntax for lsens

lsens [*depvar*] [*if*] [*in*] [*weight*] [, *lsens_options*]

lsens_options	description
Main	
all	graph all observations in the data
genprob(*varname*)	create variable containing probability cutoffs
gensens(*varname*)	create variable containing sensitivity
genspec(*varname*)	create variable containing specificity
replace	overwrite existing variables
nograph	suppress the graph
Advanced	
beta(*matname*)	row vector containing coefficients for the model
Plot	
connect_options	affect rendition of the plotted points connected by lines
Add plot	
addplot(*plot*)	add other plots to the generated graph
Y-Axis, X-Axis, Title, Caption, Legend, Overall	
twoway_options	any options other than by() documented in [G] ***twoway_options***

fweights are allowed; see [U] **11.1.6 weight**.

Options for lsens

⌐ Main ⌐

all requests that the statistic be computed for all observations in the data, ignoring any if or in restrictions specified with logistic.

genprob(*varname*), gensens(*varname*), and genspec(*varname*) specify the names of new variables created to contain, respectively, the probability cutoffs and the corresponding sensitivity and specificity.

replace requests that existing variables specified for genprob(), gensens(), or genspec() be overwritten.

nograph suppresses graphical output.

Advanced

beta(*matname*) specifies a row vector containing coefficients for a logistic model. The columns of the row vector must be labeled with the corresponding names of the independent variables in the data. The dependent variable *depvar* must be specified immediately after the command name. See *Models other than last fitted model* later in this entry.

Plot

connect_options affect the rendition of the plotted points connected by lines; see *connect_options* in [G] **graph twoway scatter**.

Add plot

addplot(*plot*) provides a way to add other plots to the generated graph. See [G] ***addplot_option***.

Y-Axis, X-Axis, Title, Caption, Legend, Overall

twoway_options are any of the options documented in [G] ***twoway_options*** excluding by(). These include options for titling the graph (see [G] ***title_options***) and options for saving the graph to disk (see [G] ***saving_option***).

Remarks

Remarks are presented under the headings

> *predict after logistic*
> > *predict without options*
> > *predict with the xb and stdp options*
> > *predict with the residuals option*
> > *predict with the number option*
> > *predict with the deviance option*
> > *predict with the rstandard option*
> > *predict with the hat option*
> > *predict with the dx2 option*
> > *predict with the ddeviance option*
> > *predict with the dbeta option*
> *estat classification*
> *estat gof*
> *lroc*
> *lsens*
> *Samples other than the estimation sample*
> *Models other than the last fitted model*

predict after logistic

predict is used after logistic to obtain predicted probabilities, residuals, and influence statistics for the estimation sample. The suggested diagnostic graphs below are from Hosmer and Lemeshow (2000), where they are more elaborately explained. Also see Collett (1991, 120–160) for a thorough discussion of model checking.

predict without options

Typing predict *newvar* after estimation calculates the predicted probability of a positive outcome.

We previously ran the model logistic low age lwt _Irace_2 _Irace_3 smoke ptl ht ui. We obtain the predicted probabilities of a positive outcome by typing

```
. use http://www.stata-press.com/data/r9/lbw
(Hosmer & Lemeshow data)
xi: logistic low age lwt i.race smoke ptl ht ui
  (output omitted)
. predict p
(option p assumed; Pr(low))

. summarize p low
```

Variable	Obs	Mean	Std. Dev.	Min	Max
p	189	.3121693	.1913915	.0272559	.8391283
low	189	.3121693	.4646093	0	1

predict with the xb and stdp options

predict with the xb option calculates the linear combination $x_j\mathbf{b}$, where x_j are the independent variables in the jth observation and $\mathbf{b}$ is the estimated parameter vector. This is sometimes known as the index function since the cumulative distribution function indexed at this value is the probability of a positive outcome.

With the stdp option, predict calculates the standard error of the prediction, which is *not* adjusted for replicated covariate patterns in the data. The influence statistics described below are adjusted for replicated covariate patterns in the data.

predict with the residuals option

predict can calculate more than predicted probabilities. The Pearson residual is defined as the square root of the contribution of the covariate pattern to the Pearson χ^2 goodness-of-fit statistic, signed according to whether the observed number of positive responses within the covariate pattern is less than or greater than expected. For instance,

```
. predict r, residuals
. summarize r, detail
```

```
                          Pearson residual
```

	Percentiles	Smallest		
1%	-1.750923	-2.283885		
5%	-1.129907	-1.750923		
10%	-.9581174	-1.636279	Obs	189
25%	-.6545911	-1.636279	Sum of Wgt.	189
50%	-.3806923		Mean	-.0242299
		Largest	Std. Dev.	.9970949
75%	.8162894	2.23879		
90%	1.510355	2.317558	Variance	.9941981
95%	1.747948	3.002206	Skewness	.8618271
99%	3.002206	3.126763	Kurtosis	3.038448

We notice the prevalence of a few, large positive residuals:

```
. sort r
. list id r low p age race in -5/l
```

	id	r	low	p	age	race
185.	33	2.224501	1	.1681123	19	white
186.	57	2.23879	1	.166329	15	white
187.	16	2.317558	1	.1569594	27	other
188.	77	3.002206	1	.0998678	26	white
189.	36	3.126763	1	.0927932	24	white

predict with the number option

Covariate patterns play an important role in logistic regression. Two observations are said to share the same covariate pattern if the independent variables for the two observations are identical. Although we might think of having individual observations, the statistical information in the sample can be summarized by the covariate patterns, the number of observations with that covariate pattern, and the number of positive outcomes within the pattern. Depending on the model, the number of covariate patterns can approach or be equal to the number of observations, or it can be considerably less.

Stata calculates all the residual and diagnostic statistics in terms of covariate patterns, not observations. That is, all observations with the same covariate pattern are given the same residual and diagnostic statistics. Hosmer and Lemeshow (2000, 144–145) argue that such "M-asymptotic" statistics are more useful than "N-asymptotic" statistics.

To understand the difference, think of an observed positive outcome with predicted probability of 0.8. Taking the observation in isolation, the "residual" must be positive—we expected 0.8 positive responses and observed 1. This may indeed be the "correct" residual, but not necessarily. Under the M-asymptotic definition, we ask how many successes we observed across all observations with this covariate pattern. If that number were, say, 6, and there were a total of 10 observations with this covariate pattern, then the residual is negative for the covariate pattern—we expected 8 positive outcomes but observed 6. `predict` makes this kind of calculation and then attaches the same residual to all observations in the covariate pattern.

Occasionally you might want to find all observations sharing a covariate pattern. `number` allows you to do this:

```
. predict pattern, number
. summarize pattern
```

Variable	Obs	Mean	Std. Dev.	Min	Max
pattern	189	89.2328	53.16573	1	182

We previously fitted the model `logistic low age lwt _Irace_2 _Irace_3 smoke ptl ht ui` over 189 observations. There are 182 covariate patterns in our data.

predict with the deviance option

The deviance residual is defined as the square root of the contribution to the likelihood-ratio test statistic of a saturated model versus the fitted model. It has slightly different properties from the Pearson residual (see Hosmer and Lemeshow 2000, 145–147):

```
. predict d, deviance
. summarize d, detail
```

```
                        deviance residual

        Percentiles      Smallest
 1%     -1.843472       -1.911621
 5%     -1.33477        -1.843472
10%     -1.148316       -1.843472        Obs                 189
25%     -.8445325       -1.674869        Sum of Wgt.         189

50%     -.5202702                        Mean          -.1228811
                         Largest         Std. Dev.      1.049237
75%      .9129041        1.894089
90%      1.541558        1.924457        Variance       1.100898
95%      1.673338        2.146583        Skewness       .6598857
99%      2.146583        2.180542        Kurtosis       2.036938
```

predict with the rstandard option

Pearson residuals do not have a standard deviation equal to 1. `rstandard` generates Pearson residuals normalized to have an *expected* standard deviation equal to 1.

```
. predict rs, rstandard

. summarize r rs
    Variable |        Obs        Mean    Std. Dev.        Min        Max
-------------+--------------------------------------------------------
           r |        189   -.0242299    .9970949   -2.283885   3.126763
          rs |        189   -.0279135    1.026406    -2.4478   3.149081

. correlate r rs
(obs=189)
             |        r       rs
-------------+-----------------
           r |   1.0000
          rs |   0.9998   1.0000
```

Remember that we previously created `r` containing the (unstandardized) Pearson residuals. In these data, whether we use standardized or unstandardized residuals does not matter much.

predict with the hat option

`hat` calculates the leverage of a covariate pattern—a scaled measure of distance in terms of the independent variables. Large values indicate covariate patterns "far" from the average covariate pattern that can have a large effect on the fitted model even if the corresponding residual is small. Consider the following graph:

```
. predict h, hat

. scatter h r, xline(0)
```

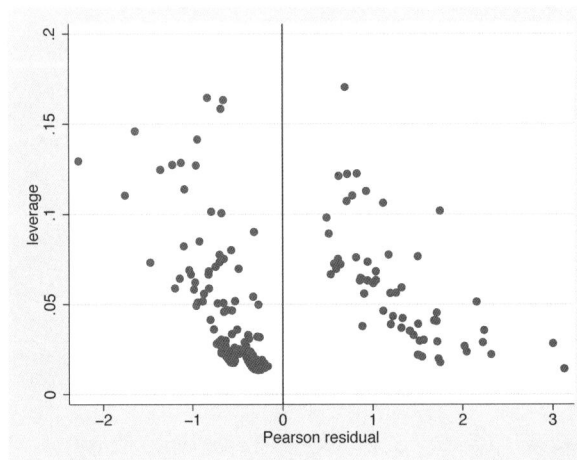

The points to the left of the vertical line are observed negative outcomes; in this case, our data contain almost as many covariate patterns as observations, so most covariate patterns are unique. In such unique patterns, we observe either 0 or 1 success and expect p, thus forcing the sign of the residual. If we had fewer covariate patterns—if we did not have continuous variables in our model—there would be no such interpretation, and we would not have drawn the vertical line at 0.

Points on the left and right edges of the graph represent large residuals—covariate patterns that are not fitted well by our model. Points at the top of our graph represent high leverage patterns. When analyzing the influence of observations on the model, we are most interested in patterns with high leverage and small residuals—patterns that might otherwise escape our attention.

predict with the dx2 option

There are many ways to measure influence, and `hat` is one example. `dx2` measures the decrease in the Pearson χ^2 goodness-of-fit statistic that would be caused by deleting an observation (and all others sharing the covariate pattern):

```
. predict dx2, dx2
. scatter dx2 p
```

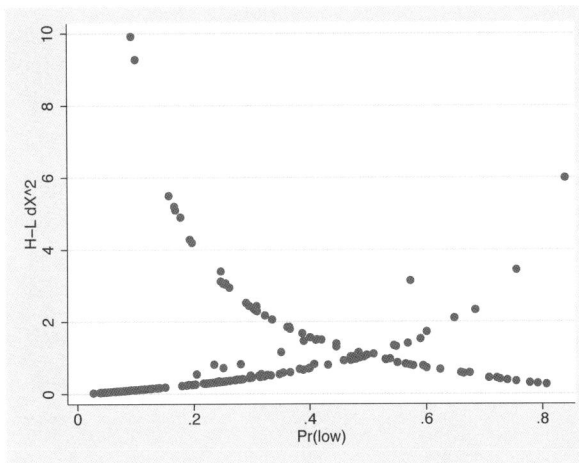

Paraphrasing Hosmer and Lemeshow (2000, 178–179), the points going from the top left to the bottom right correspond to covariate patterns with the number of positive outcomes equal to the number in the group; the points on the other curve correspond to 0 positive outcomes. In our data, most of the covariate patterns are unique, so the points tend to lie along one or the other curves; the points that are off the curves correspond to the few repeated covariate patterns in our data in which all the outcomes are not the same.

We examine this graph for large values of `dx2`—there are two at the top left.

predict with the ddeviance option

Another measure of influence is the change in the deviance residuals due to deletion of a covariate pattern:

```
. predict dd, ddeviance
```

As with `dx2`, we typically graph `ddeviance` against the probability of a positive outcome. We direct you to Hosmer and Lemeshow (2000, 178) for an example and for the interpretation of this graph.

predict with the dbeta option

One of the more direct measures of influence of interest to model fitters is the Pregibon (1981) dbeta measure, a measure of the change in the coefficient vector that would be caused by deleting an observation (and all others sharing the covariate pattern):

```
. predict db, dbeta
. scatter db p
```

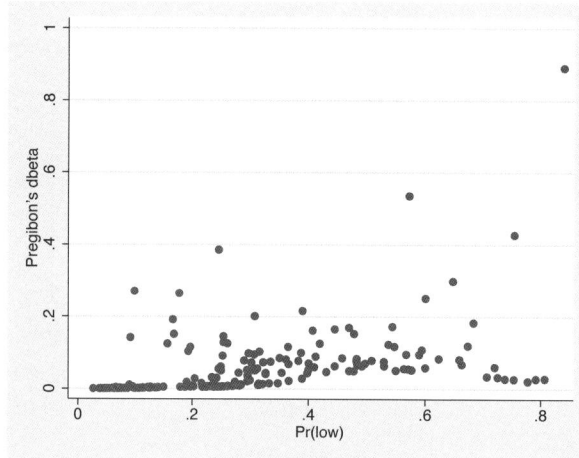

One observation has a large effect on the estimated coefficients. We can easily find this point:

```
. sort db
. list in 1
```

	id	low	age	lwt	race	smoke	ptl	ht	ui	ftv	bwt
189.	188	0	25	95	white	1	3	0	1	0	3637

_Irace_2	_Irace_3	p	r	pattern	d
0	0	.8391283	-2.283885	117	-1.911621

rs	h	dx2	dd	db
-2.4478	.1294439	5.991726	4.197658	.8909163

Hosmer and Lemeshow (2000, 180) suggest a graph that combines two of the influence measures:

```
. scatter dx2 p [w=db], title("Symbol size proportional to dBeta") mfcolor(none)
(analytic weights assumed)
```

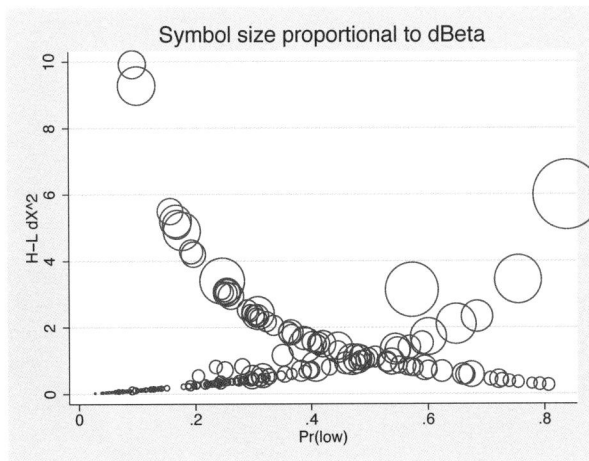

We can easily spot the most influential points by the dbeta and dx2 measures.

estat classification

▷ Example 1

estat classification presents the classification statistics and classification table after logistic.

```
. use http://www.stata-press.com/data/r9/lbw
(Hosmer & Lemeshow data)
xi: logistic low age lwt i.race smoke ptl ht ui
  (output omitted )
. estat class
Logistic model for low
```

Classified	True D	~D	Total
+	21	12	33
–	38	118	156
Total	59	130	189

```
Classified + if predicted Pr(D) >= .5
True D defined as low != 0
```

Sensitivity	Pr(+| D)	35.59%
Specificity	Pr(–|~D)	90.77%
Positive predictive value	Pr(D| +)	63.64%
Negative predictive value	Pr(~D| –)	75.64%
False + rate for true ~D	Pr(+|~D)	9.23%
False – rate for true D	Pr(–| D)	64.41%
False + rate for classified +	Pr(~D| +)	36.36%
False – rate for classified –	Pr(D| –)	24.36%
Correctly classified		73.54%

By default, estat classification uses a cutoff of 0.5, although you can vary this with the cutoff() option. You can use the lsens command to review the potential cutoffs; see lsens below.

◁

estat gof   *See also -lfit x2- (STB 44; sg87)*

estat gof computes goodness-of-fit tests; either the Pearson χ^2 test or the Hosmer–Lemeshow test.

By default, estat classification, estat gof, lroc, and lsens compute statistics for the estimation sample using the last model fitted by logistic. However, samples other than the estimation sample can be specified; see *Samples other than the estimation sample* later in this entry. Models other than the last model fitted by logistic can also be specified; see *Models other than last fitted model* later in this entry.

▷ Example 2

estat gof, typed without options, presents the Pearson χ^2 goodness-of-fit test for the fitted model. The Pearson χ^2 goodness-of-fit test is a test of the observed against expected number of responses using cells defined by the covariate patterns; see *predict with the number option* earlier in this entry for the definition of covariate patterns.

```
. estat gof
```
Logistic model for low, goodness-of-fit test

```
        number of observations =      189
number of covariate patterns =      182
          Pearson chi2(173) =      179.24
                Prob > chi2 =      0.3567
```

Our model fits reasonably well. We should note, however, that the number of covariate patterns is close to the number of observations, making the applicability of the Pearson χ^2 test questionable but not necessarily inappropriate. Hosmer and Lemeshow (2000, 147–150) suggest regrouping the data by ordering on the predicted probabilities and then forming, say, ten nearly equal-size groups. estat gof with the group() option does this:

```
. estat gof, group(10)
```
Logistic model for low, goodness-of-fit test

```
   (Table collapsed on quantiles of estimated probabilities)
        number of observations =      189
              number of groups =       10
        Hosmer-Lemeshow chi2(8) =       9.65
                  Prob > chi2 =      0.2904
```

Again we cannot reject our model. If we specify the table option, estat gof displays the groups along with the expected and observed number of positive responses (low birthweight babies):

```
. estat gof, group(10) table
```

Logistic model for low, goodness-of-fit test

(Table collapsed on quantiles of estimated probabilities)

Group	Prob	Obs_1	Exp_1	Obs_0	Exp_0	Total
1	0.0827	0	1.2	19	17.8	19
2	0.1276	2	2.0	17	17.0	19
3	0.2015	6	3.2	13	15.8	19
4	0.2432	1	4.3	18	14.7	19
5	0.2792	7	4.9	12	14.1	19
6	0.3138	7	5.6	12	13.4	19
7	0.3872	6	6.5	13	12.5	19
8	0.4828	7	8.2	12	10.8	19
9	0.5941	10	10.3	9	8.7	19
10	0.8391	13	12.8	5	5.2	18

```
        number of observations =      189
             number of groups =       10
   Hosmer-Lemeshow chi2(8) =         9.65
             Prob > chi2 =        0.2904
```

◁

❑ Technical Note

estat gof with the group() option puts all observations with the same predicted probabilities into the same group. If, as in the previous example, we request ten groups, the groups that estat gof makes are $[p_0, p_{10}]$, $(p_{10}, p_{20}]$, $(p_{20}, p_{30}]$, ..., $(p_{90}, p_{100}]$, where p_k is the kth percentile of the predicted probabilities, with p_0 the minimum and p_{100} the maximum.

If there are large numbers of ties at the quantile boundaries, as will frequently happen if all independent variables are categorical and there are only a few of them, the sizes of the groups will be uneven. If the totals in some of the groups are small, the χ^2 statistic for the Hosmer–Lemeshow test may be unreliable. In this case, fewer groups should be specified, or the Pearson goodness-of-fit test may be a better choice.

❑

▷ Example 3

The table option can be used without the group() option. We would not want to specify this for our current model because there were 182 covariate patterns in the data, caused by including the two continuous variables, age and lwt, in the model. As an aside, we fit a simpler model and specify table with estat gof:

```
. logistic low _Irace_2 _Irace_3 smoke ui
Logistic regression                             Number of obs   =        189
                                                LR chi2(4)      =      18.80
                                                Prob > chi2     =     0.0009
Log likelihood = -107.93404                     Pseudo R2       =     0.0801
```

low	Odds Ratio	Std. Err.	z	P>|z|	[95% Conf. Interval]
_Irace_2	3.052746	1.498084	2.27	0.023	1.166749 7.987368
_Irace_3	2.922593	1.189226	2.64	0.008	1.31646 6.488269
smoke	2.945742	1.101835	2.89	0.004	1.41517 6.131701
ui	2.419131	1.047358	2.04	0.041	1.03546 5.651783

```
. estat gof, tab
```
Logistic model for low, goodness-of-fit test

Group	Prob	Obs_1	Exp_1	Obs_0	Exp_0	Total
1	0.1230	3	4.9	37	35.1	40
2	0.2533	1	1.0	3	3.0	4
3	0.2907	16	13.7	31	33.3	47
4	0.2923	15	12.6	28	30.4	43
5	0.2997	3	3.9	10	9.1	13
6	0.4978	4	4.0	4	4.0	8
7	0.4998	4	4.5	5	4.5	9
8	0.5087	2	1.5	1	1.5	3
9	0.5469	2	4.4	6	3.6	8
10	0.5577	6	5.6	4	4.4	10
11	0.7449	3	3.0	1	1.0	4

Group	Prob	_Irace_2	_Irace_3	smoke	ui
1	0.1230	0	0	0	0
2	0.2533	0	0	0	1
3	0.2907	0	1	0	0
4	0.2923	0	0	1	0
5	0.2997	1	0	0	0
6	0.4978	0	1	0	1
7	0.4998	0	0	1	1
8	0.5087	1	0	0	1
9	0.5469	0	1	1	0
10	0.5577	1	0	1	0
11	0.7449	0	1	1	1

```
        number of observations =       189
number of covariate patterns =        11
            Pearson chi2(6) =         5.71
               Prob > chi2 =        0.4569
```

◁

❏ Technical Note

`logistic` and `estat gof` keep track of the estimation sample. If you type `logistic ... if` x==1, then when you type `estat gof`, the statistics will be calculated on the x==1 subsample of the data automatically.

You should specify `if` or `in` with `estat gof` only when you wish to calculate statistics for a set of observations other than the estimation sample. See *Samples other than the estimation sample* later in this entry.

If the `logistic` model was fitted with `fweights`, `estat gof` properly accounts for the weights in its calculations. (Note: `estat gof` does not allow `pweights`.) You do not have to specify the weights when you run `estat gof`. Weights should only be specified with `estat gof` when you wish to use a different set of weights.

❏

lroc

For other receiver-operating-characteristic (ROC) commands and a complete description, see [R] **roc**, [R] **rocfit**, and [R] **rocfit postestimation**.

`lroc` graphs the ROC curve—a graph of sensitivity versus one minus specificity as the cutoff c is varied—and calculates the area under it. Sensitivity is the fraction of observed positive-outcome cases that are correctly classified; specificity is the fraction of observed negative-outcome cases that are correctly classified. When the purpose of the analysis is classification, you must choose a cutoff.

The curve starts at $(0,0)$, corresponding to $c = 1$, and continues to $(1,1)$, corresponding to $c = 0$. A model with no predictive power would be a $45°$ line. The greater the predictive power, the more bowed the curve, and hence the area beneath the curve is often used as a measure of the predictive power. A model with no predictive power has area 0.5; a perfect model has area 1.

The ROC curve was first discussed in signal detection theory (Peterson, Birdsall, and Fox 1954) and then was quickly introduced into psychology (Tanner and Swets 1954). It has since been applied in other fields, particularly medicine (for instance, Metz 1978). For a classic text on ROC techniques, see Green and Swets (1974).

▷ Example 4

ROC curves are typically used when the point of the analysis is classification—which it is not in our low birthweight model. Nevertheless, the ROC curve is

```
. lroc
Logistic model for low

number of observations  =       189
area under ROC curve    =    0.7462
```

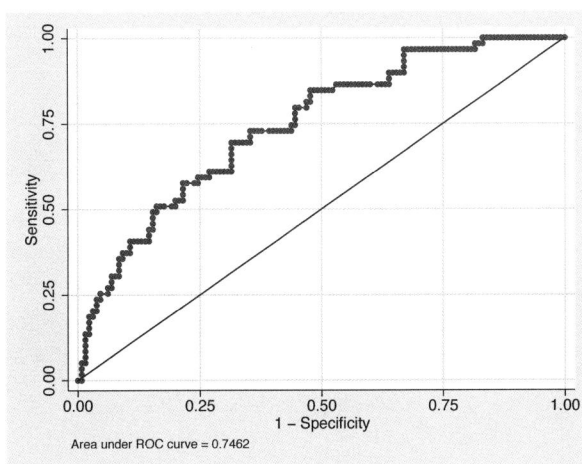

Area under ROC curve = 0.7462

We see that the area under the curve is 0.7462.

◁

lsens

lsens also plots sensitivity and specificity; it plots both sensitivity and specificity versus probability cutoff c. The graph is equivalent to what you would get from estat classification if you varied the cutoff probability from 0 to 1.

. lsens

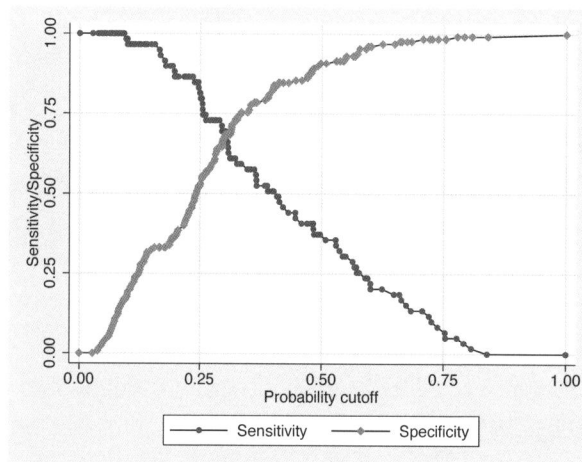

lsens optionally creates new variables containing the probability cutoff, sensitivity, and specificity.

. lsens, genprob(p) gensens(sens) genspec(spec) nograph

Note that the variables created will have $M + 2$ distinct nonmissing values: one for each of the M covariate patterns, one for $c = 0$, and another for $c = 1$.

Samples other than the estimation sample

estat gof, estat classification, lroc, and lsens can be used with samples other than the estimation sample. By default, these commands remember the estimation sample used with the last logistic command. To override this, simply use an if or in restriction to select another set of observations, or specify the all option to force the command to use all the observations in the dataset.

If you use estat gof with a sample that is completely different from the estimation sample (i.e., no overlap), you should also specify the outsample option so that the χ^2 statistic properly adjusts the degrees of freedom upward. For an overlapping sample, the conservative thing to do is to leave the degrees of freedom the same as they are for the estimation sample.

▷ Example 5

We want to develop a model for predicting low birthweight babies. One approach would be to divide our data into two groups, a developmental sample and a validation sample. See Lemeshow and Le Gall (1994) and Tilford et al. (1995) for more information on developing prediction models and severity-scoring systems.

We will do this with the low birthweight data that we considered previously. First, we randomly divide the data into two samples.

```
. use http://www.stata-press.com/data/r9/lbw
(Hosmer & Lemeshow data)

. set seed 1

. gen r = uniform()

. sort r

. gen group = 1 if _n <= _N/2
(95 missing values generated)

. replace group = 2 if group >=.
(95 real changes made)
```

Then we fit a model using the first sample (group==1), which is our developmental sample.

```
. xi: logistic low age lwt i.race smoke ptl ht ui if group==1
i.race            _Irace_1-3        (naturally coded; _Irace_1 omitted)
```

Logistic regression					Number of obs	=	94
					LR chi2(8)	=	29.14
					Prob > chi2	=	0.0003
Log likelihood = -44.293342					Pseudo R2	=	0.2475

low	Odds Ratio	Std. Err.	z	P>|z|	[95% Conf. Interval]	
age	.91542	.0553937	-1.46	0.144	.8130414	1.03069
lwt	.9744276	.0112295	-2.25	0.025	.9526649	.9966874
_Irace_2	5.063678	3.78442	2.17	0.030	1.170327	21.90913
_Irace_3	2.606209	1.657608	1.51	0.132	.7492483	9.065522
smoke	.909912	.5252898	-0.16	0.870	.2934966	2.820953
ptl	3.033543	1.507048	2.23	0.025	1.145718	8.03198
ht	21.07656	22.64788	2.84	0.005	2.565304	173.1652
ui	.988479	.6699458	-0.02	0.986	.2618557	3.731409

To test calibration in the developmental sample, we calculate the Hosmer–Lemeshow goodness-of-fit test using estat gof.

```
. estat gof, group(10)
```

Logistic model for low, goodness-of-fit test

```
    (Table collapsed on quantiles of estimated probabilities)
          number of observations =        94
             number of groups =        10
    Hosmer-Lemeshow chi2(8) =       6.67
              Prob > chi2 =     0.5721
```

Note that we did not specify an if statement with estat gof since we wanted to use the estimation sample. Since the test is not significant, we are satisfied with the fit of our model.

Running lroc gives a measure of the discrimination:

```
. lroc, nograph
Logistic model for low
number of observations =        94
area under ROC curve   =    0.8156
```

Now we test the calibration of our model by performing a goodness-of-fit test on the validation sample. We specify the outsample option so that the number of degrees of freedom is 10 rather than 8.

```
. estat gof if group==2, group(10) table outsample
```

Logistic model for low, goodness-of-fit test

(Table collapsed on quantiles of estimated probabilities)

Group	Prob	Obs_1	Exp_1	Obs_0	Exp_0	Total
1	0.0725	1	0.4	9	9.6	10
2	0.1202	4	0.8	5	8.2	9
3	0.1549	3	1.3	7	8.7	10
4	0.1888	1	1.5	8	7.5	9
5	0.2609	3	2.2	7	7.8	10
6	0.3258	4	2.7	5	6.3	9
7	0.4217	2	3.7	8	6.3	10
8	0.4915	3	4.1	6	4.9	9
9	0.6265	4	5.5	6	4.5	10
10	0.9737	4	7.1	5	1.9	9

```
        number of observations =       95
           number of groups =       10
 Hosmer-Lemeshow chi2(10) =      28.03
             Prob > chi2 =       0.0018
```

We must acknowledge that our model does not fit well on the validation sample. The model's discrimination in the validation sample is appreciably lower, as well.

```
. lroc if group==2, nograph
Logistic model for low
number of observations =      95
area under ROC curve   =   0.5839
```

◁

Models other than the last fitted model

By default, estat gof, estat classification, lroc, and lsens use the last model fitted by logistic. You can specify other models using the beta() option.

▷ Example 6

Suppose that someone publishes the following logistic model of low birthweight,

$$\Pr(\mathtt{low}=1) = F(-0.02\,\mathtt{age} - 0.01\,\mathtt{lwt} + 1.3\,\mathtt{black} + 1.1\,\mathtt{smoke} + 0.5\,\mathtt{ptl} + 1.8\,\mathtt{ht} + 0.8\,\mathtt{ui} + 0.5)$$

where F is the cumulative logistic distribution. Note that these coefficients are not odds ratios; they are the equivalent of what logit produces.

We can see whether this model fits our data. First we enter the coefficients as a row vector and label its columns with the names of the independent variables plus _cons for the constant (see [P] **matrix define** and [P] **matrix rownames**).

```
. use http://www.stata-press.com/data/r9/lbw3
(Hosmer & Lemeshow data)
. matrix input b = (-.02 -.01 1.3 1.1 .5 1.8 .8 .5)
. matrix colnames b = age lwt black smoke ptl ht ui _cons
```

We run `estat gof` using the `beta()` option to specify b. The dependent variable is entered right after the command name, and the `outsample` option gives the proper degrees of freedom.

```
. estat gof low, beta(b) group(10) outsample
```
Logistic model for low, goodness-of-fit test
```
    (Table collapsed on quantiles of estimated probabilities)
           number of observations =        189
                 number of groups =         10
        Hosmer-Lemeshow chi2(10) =       27.33
                    Prob > chi2 =        0.0023
```

Although the fit of the model is poor, `lroc` shows that it does exhibit some predictive ability.

```
. lroc low, beta(b) nograph
Logistic model for low
number of observations =        189
area under ROC curve   =     0.7275
```

Saved Results

`estat classification` saves in `r()`:

Scalars
r(P_corr)	percent correctly classified	r(P_1p)	positive predictive value
r(P_p1)	sensitivity	r(P_0n)	negative predictive value
r(P_n0)	specificity	r(P_0p)	false positive rate given classified positive
r(P_p0)	false positive rate given true negative	r(P_1n)	false negative rate given classified negative
r(P_n1)	false negative rate given true positive		

`estat gof` saves in `r()`:

Scalars
r(N)	number of observations	r(df)	degrees of freedom
r(m)	number of covariate patterns or groups	r(chi2)	χ^2

`lroc` saves in `r()`:

Scalars
r(N)	number of observations	r(area)	area under the ROC curve

`lsens` saves in `r()`:

Scalars
r(N)   number of observations

Methods and Formulas

All postestimation commands listed above are implemented as ado-files.

estat gof   See also -fit:x2- (SIB44:5937)

Let M be the total number of covariate patterns among the N observations. View the data as collapsed on covariate patterns $j = 1, 2, \ldots, M$, and define m_j as the total number of observations having covariate pattern j and y_j as the total number of positive responses among observations with covariate pattern j. Define p_j as the predicted probability of a positive outcome in covariate pattern j.

The Pearson χ^2 goodness-of-fit statistic is

$$\chi^2 = \sum_{j=1}^{M} \frac{(y_j - m_j p_j)^2}{m_j p_j (1 - p_j)}$$

This χ^2 statistic has approximately $M - k$ degrees of freedom for the estimation sample, where k is the number of independent variables, including the constant. For a sample outside of the estimation sample, the statistic has M degrees of freedom.

The Hosmer–Lemeshow goodness-of-fit χ^2 (Hosmer and Lemeshow 1980; Lemeshow and Hosmer 1982; Hosmer, Lemeshow, and Klar 1988) is calculated similarly, except that rather than using the M covariate patterns as the group definition, the quantiles of the predicted probabilities are used to form groups. Let $G = \#$ be the number of quantiles requested with group(#). The smallest index $1 \leq q(i) \leq M$, such that

$$W_{q(i)} = \sum_{j=1}^{q(i)} m_j \geq \frac{N}{G}$$

gives $p_{q(i)}$ as the upper boundary of the ith quantile for $i = 1, 2, \ldots, G$. Let $q(0) = 1$ denote the first index.

The groups are then

$$\left[p_{q(0)}, p_{q(1)} \right], \left(p_{q(1)}, p_{q(2)} \right], \ldots, \left(p_{q(G-1)}, p_{q(G)} \right]$$

If the table option is given, the upper boundaries $p_{q(1)}, \ldots, p_{q(G)}$ of the groups appear next to the group number on the output.

The resulting χ^2 statistic has approximately $G - 2$ degrees of freedom for the estimation sample. For a sample outside of the estimation sample, the statistic has G degrees of freedom.

predict after logistic

Index j will now be used to index observations, not covariate patterns. Define M_j for each observation as the total number of observations sharing j's covariate pattern. Define Y_j as the total number of positive responses among observations sharing j's covariate pattern.

The Pearson residual for the jth observation is defined as

$$r_j = \frac{Y_j - M_j p_j}{\sqrt{M_j p_j (1 - p_j)}}$$

For $M_j > 1$, the deviance residual d_j is defined as

$$d_j = \pm \left[2 \left\{ Y_j \ln \left(\frac{Y_j}{M_j p_j} \right) + (M_j - Y_j) \ln \left(\frac{M_j - Y_j}{M_j (1 - p_j)} \right) \right\} \right]^{1/2}$$

where the sign is the same as the sign of $(Y_j - M_j p_j)$. In the limiting cases, the deviance residual is given by

$$d_j = \begin{cases} -\sqrt{2M_j|\ln(1-p_j)|} & \text{if } Y_j = 0 \\ \sqrt{2M_j|\ln p_j|} & \text{if } Y_j = M_j \end{cases}$$

The *unadjusted* diagonal elements of the hat matrix h_{Uj} are given by $h_{Uj} = (\mathbf{XVX'})_{jj}$, where V is the estimated covariance matrix of parameters. The adjusted diagonal elements h_j created by hat are then $h_j = M_j p_j (1 - p_j) h_{Uj}$.

The standardized Pearson residual r_{Sj} is $r_j / \sqrt{1 - h_j}$.

The Pregibon (1981) $\Delta\widehat{\beta}_j$ influence statistic is

$$\Delta\widehat{\beta}_j = \frac{r_j^2 h_j}{(1 - h_j)^2}$$

The corresponding change in the Pearson χ^2 is r_{Sj}^2. The corresponding change in the deviance residual is $\Delta D_j = d_j^2/(1 - h_j)$.

estat classification and lsens

Again let j index observations. Define c as the cutoff() specified by the user or, if not specified, as 0.5. Let p_j be the predicted probability of a positive outcome and y_j be the actual outcome, which we will treat as 0 or 1, although Stata treats it as 0 and non-0, excluding missing observations.

A prediction is classified as *positive* if $p_j \geq c$ and otherwise is classified as *negative*. The classification is *correct* if it is *positive* and $y_j = 1$ or if it is *negative* and $y_j = 0$.

Sensitivity is the fraction of $y_j = 1$ observations that are correctly classified. *Specificity* is the percentage of $y_j = 0$ observations that are correctly classified.

lroc

The ROC curve is a graph of *specificity* against $(1 - sensitivity)$. This is guaranteed to be a monotone nondecreasing function since the number of correctly predicted successes increases and the number of correctly predicted failures decreases as the classification cutoff c decreases.

The area under the ROC curve is the area on the bottom of this graph and is determined by integrating the curve. The vertices of the curve are determined by sorting the data according to the predicted index, and the integral is computed using the trapezoidal rule.

References

Collett, D. 1991. *Modelling Binary Data*. London: Chapman & Hall.

Garrett, J. M. 2000. sg157: Predicted values calculated from linear or logistic regression models. *Stata Technical Bulletin* 58: 27–30. Reprinted in *Stata Technical Bulletin Reprints*, vol. 10, pp. 258–261.

Green, D. M. and J. A. Swets. 1974. *Signal Detection Theory and Psychophysics*. rev. ed. Huntington, NY: Krieger.

Hosmer, D. W., Jr., and S. Lemeshow. 1980. Goodness-of-fit tests for the multiple logistic regression model. *Communications in Statistics* A9: 1043–1069.

——. 2000. *Applied Logistic Regression*. 2nd ed. New York: Wiley.

Hosmer, D. W., Jr., S. Lemeshow, and J. Klar. 1988. Goodness-of-fit testing for the logistic regression model when the estimated probabilities are small. *Biometric Journal* 30: 911–924.

Lemeshow, S. and D. W. Hosmer, Jr. 1982. A review of goodness of fit statistics for use in the development of logistic regression models. *American Journal of Epidemiology* 115: 92–106.

Lemeshow, S. and J.-R. Le Gall. 1994. Modeling the severity of illness of ICU patients: A systems update. *Journal of the American Medical Association* 272: 1049–1055.

Metz, C. E. 1978. Basic principles of ROC analysis. *Seminars in Nuclear Medicine* 8: 283–298.

Mitchell, M. N. and X. Chen. 2005. Visualizing main effects and interactions for binary logit models. *Stata Journal* 5: 64–82.

Peterson, W. W., T. G. Birdsall, and W. C. Fox. 1954. The theory of signal detection. *Trans. IRE Professional Group on Information Theory*, PGIT-4: 171–212.

Pregibon, D. 1981. Logistic regression diagnostics. *Annals of Statistics* 9: 705–724.

Seed, P. T. and A. Tobias. 2001. sbe36.1: Summary statistics for diagnostic tests. *Stata Technical Bulletin* 59: 25–27. Reprinted in *Stata Technical Bulletin Reprints*, vol. 10, pp. 90–93.

Tanner, W. P., Jr., and J. A. Swets. 1954. A decision-making theory of visual detection. *Psychological Review* 61: 401–409.

Tilford, J. M., P. K. Roberson, and D. H. Fiser. 1995. sbe12: Using lfit and lroc to evaluate mortality prediction models. *Stata Technical Bulletin* 28: 14–18. Reprinted in *Stata Technical Bulletin Reprints*, vol. 5, pp. 77–81.

Tobias, A. 2000. sbe36: Summary statistics report for diagnostic tests. *Stata Technical Bulletin* 56: 16–18. Reprinted in *Stata Technical Bulletin Reprints*, vol. 10, pp. 87–90.

Tobias, A. and M. J. Campbell. 1998. sg90: Akaike's information criterion and Schwarz's criterion. *Stata Technical Bulletin* 45: 23–25. Reprinted in *Stata Technical Bulletin Reprints*, vol. 8, pp. 174–177.

Weesie, J. 1998. sg87: Windmeijer's goodness-of-fit test for logistic regression. *Stata Technical Bulletin* 44: 22–27. Reprinted in *Stata Technical Bulletin Reprints*, vol. 8, pp. 153–160.

Also See

Complementary: [R] **logistic**; [R] **adjust**, [R] **estimates**, [R] **lincom**, [R] **linktest**, [R] **lrtest**, [R] **mfx**, [R] **nlcom**, [R] **predictnl**, [R] **suest**, [R] **test**, [R] **testnl**

Related: [R] **rocfit postestimation**

Background: [U] **13.5 Accessing coefficients and standard errors**, [U] **20 Estimation and postestimation commands**, [R] **estat**, [R] **predict**

Title

> **logit** — Logistic regression, reporting coefficients

Syntax

$\underline{\text{logit}}$ *depvar* $\big[$*indepvars*$\big]$ $\big[$*if*$\big]$ $\big[$*in*$\big]$ $\big[$*weight*$\big]$ $\big[$, *options*$\big]$

options	description
Model	
<u>nocon</u>stant	suppress constant term
<u>off</u>set(*varname*)	include *varname* in model with coefficient constrained to 1
asis	retain perfect predictor variables
SE/Robust	
vce(*vcetype*)	*vcetype* may be <u>r</u>obust, <u>boot</u>strap, or jackknife
<u>r</u>obust	synonym for vce(robust)
<u>cl</u>uster(*varname*)	adjust standard errors for intragroup correlation
Reporting	
<u>l</u>evel(#)	set confidence level; default is level(95)
or	report odds ratios
Max options	
maximize_options	control the maximization process; seldom used
[†]<u>nocoef</u>	do not display coefficient table; seldom used

[†]nocoef does not appear in the dialog box.

depvar and *indepvars* may contain time-series operators; see [U] **11.4.3 Time-series varlists**.

bootstrap, by, jackknife, rolling, statsby, stepwise, svy, and xi are allowed; see
[U] **11.1.10 Prefix commands**.

fweights, iweights, and pweights are allowed; see [U] **11.1.6 weight**.

See [U] **20 Estimation and postestimation commands** for additional capabilities of estimation commands.

Description

logit fits a maximum-likelihood logit model. *depvar* = 0 indicates a negative outcome; *depvar*! = 0 & *depvar*! = . (typically *depvar* = 1) indicates a positive outcome.

Also see [R] **logistic**; logistic displays estimates as odds ratios. Many users prefer the logistic command to logit. Results are the same regardless of which you use—both are the maximum-likelihood estimator. A number of auxiliary commands that can be run after logit, probit, or logistic estimation are described in [R] **logistic postestimation**. A list of related estimation commands is given in [R] **logistic**.

If estimating on grouped data, see [R] **glogit**.

Options

noconstant, offset(*varname*); see [R] **estimation options**.

asis forces retention of perfect predictor variables and their associated perfectly predicted observations and may produce instabilities in maximization; see [R] **probit**.

vce(*vcetype*); see [R] *vce_option*.

robust, cluster(*varname*); see [R] **estimation options**. cluster() can be used with pweights to produce estimates for unstratified cluster-sampled data, but see [SVY] **svy: logit** for a command especially designed for survey data.

See [R] **logistic** for examples using robust and cluster().

level(*#*); see [R] **estimation options**.

or reports the estimated coefficients transformed to odds ratios, i.e., e^b rather than b. Standard errors and confidence intervals are similarly transformed. This option affects how results are displayed, not how they are estimated. or may be specified at estimation or when replaying previously estimated results.

maximize_options: iterate(*#*), [no]log, trace, tolerance(*#*), ltolerance(*#*); see [R] **maximize**. These options are seldom used.

The following option is available with logit but is not shown in the dialog box:

nocoef specifies that the coefficient table not be displayed. This option is sometimes used by program writers but is of no use interactively.

Remarks

Remarks are presented under the headings

> *Basic usage*
> *Model identification*

Basic usage

logit fits maximum likelihood models with dichotomous dependent (left-hand-side) variables coded as 0/1 (or, more precisely, coded as 0 and not-0).

▷ Example 1

We have data on the make, weight, and mileage rating of 22 foreign and 52 domestic automobiles. We wish to fit a logit model explaining whether a car is foreign based on its weight and mileage. Here is an overview of our data:

```
. use http://www.stata-press.com/data/r9/auto
(1978 Automobile Data)

. keep make mpg weight foreign

. describe

Contains data from http://www.stata-press.com/data/r9/auto.dta
  obs:            74                          1978 Automobile Data
  vars:            4                          13 Apr 2005 17:45
  size:         1,998 (99.7% of memory free)  (_dta has notes)
```

variable name	storage type	display format	value label	variable label
make	str18	%-18s		Make and Model
mpg	int	%8.0g		Mileage (mpg)
weight	int	%8.0gc		Weight (lbs.)
foreign	byte	%8.0g	origin	Car type

```
Sorted by:  foreign
     Note:  dataset has changed since last saved

. inspect foreign
```

```
 foreign:  Car type                      Number of Observations
```

			Total	Integers	Non-Integers
#		Negative	–	–	–
#		Zero	52	52	–
#		Positive	22	22	–
#					
#	#	Total	74	74	–
#	#	Missing	–		

```
0                       1                  74
   (2 unique values)

        foreign is labeled and all values are documented in the label.
```

The variable `foreign` takes on two unique values, 0 and 1. The value 0 denotes a domestic car, and 1 denotes a foreign car.

The model that we wish to fit is

$$\Pr(\mathtt{foreign} = 1) = F(\beta_0 + \beta_1 \mathtt{weight} + \beta_2 \mathtt{mpg})$$

where $F(z) = e^z / (1 + e^z)$ is the cumulative logistic distribution.

To fit this model, we type

(Continued on next page)

```
. logit foreign weight mpg

Iteration 0:   log likelihood =   -45.03321
Iteration 1:   log likelihood = -29.898968
Iteration 2:   log likelihood = -27.495771
Iteration 3:   log likelihood = -27.184006
Iteration 4:   log likelihood = -27.175166
Iteration 5:   log likelihood = -27.175156

Logistic regression                              Number of obs   =         74
                                                 LR chi2(2)      =      35.72
                                                 Prob > chi2     =     0.0000
Log likelihood = -27.175156                      Pseudo R2       =     0.3966
```

foreign	Coef.	Std. Err.	z	P>|z|	[95% Conf. Interval]	
weight	-.0039067	.0010116	-3.86	0.000	-.0058894	-.001924
mpg	-.1685869	.0919174	-1.83	0.067	-.3487418	.011568
_cons	13.70837	4.518707	3.03	0.002	4.851864	22.56487

We find that heavier cars are less likely to be foreign and that cars yielding better gas mileage are also less likely to be foreign, at least holding the weight of the car constant.

See [R] **maximize** for an explanation of the output.

◁

❑ Technical Note

Stata interprets a value of 0 as a negative outcome (failure) and treats all other values (except missing) as positive outcomes (successes). Thus if your dependent variable takes on the values 0 and 1, 0 is interpreted as failure and 1 as success. If your dependent variable takes on the values 0, 1, and 2, 0 is still interpreted as failure, but both 1 and 2 are treated as successes.

If you prefer a more formal mathematical statement, when you type logit y x, Stata fits the model

$$\Pr(y_j \neq 0 \mid \mathbf{x}_j) = \frac{\exp(\mathbf{x}_j \boldsymbol{\beta})}{1 + \exp(\mathbf{x}_j \boldsymbol{\beta})}$$

❑

Model identification

The logit command has one more feature, and it is probably the most useful. logit automatically checks the model for identification, and, if it is underidentified, drops whatever variables and observations are necessary for estimation to proceed. (logistic, probit, and ivprobit do this as well.)

▷ Example 2

Have you ever fitted a logit model where one or more of your independent variables perfectly predicted one or the other outcome?

For instance, consider the following data:

Outcome y	Independent Variable x
0	1
0	1
0	0
1	0

Say that we wish to predict the outcome on the basis of the independent variable. Notice that the outcome is always zero whenever the independent variable is one. In our data, $\Pr(y = 0 \mid x = 1) = 1$, which means that the logit coefficient on x must be minus infinity with a corresponding infinite standard error. At this point, you may suspect that we have a problem.

Unfortunately, not all such problems are so easily detected, especially if you have a lot of independent variables in your model. If you have ever had such difficulties, you have experienced one of the more unpleasant aspects of computer optimization. The computer has no idea that it is trying to solve for an infinite coefficient as it begins its iterative process. All it knows is that, at each step, making the coefficient a little bigger, or a little smaller, works wonders. It continues on its merry way until either (1) the whole thing comes crashing to the ground when a numerical overflow error occurs or (2) it reaches some predetermined cutoff that stops the process. In the meantime, you have been waiting. In addition, the estimates that you finally receive, if you receive any at all, may be nothing more than numerical roundoff.

Stata watches for these sorts of problems, alerts us, fixes them, and properly fits the model.

Let's return to our automobile data. Among the variables we have in the data is one called repair, which takes on three values. A value of 1 indicates that the car has a poor repair record, 2 indicates an average record, and 3 indicates a better-than-average record. Here is a tabulation of our data:

```
. use http://www.stata-press.com/data/r9/repair
(1978 Automobile Data)

. tabulate foreign repair
```

foreign	repair 1	2	3	Total
Domestic	10	27	9	46
Foreign	0	3	9	12
Total	10	30	18	58

Notice that all the cars with poor repair records (repair==1) are domestic. If we were to attempt to predict foreign on the basis of the repair records, the predicted probability for the repair==1 category would have to be zero. This in turn means that the logit coefficient must be minus infinity, and that would set most computer programs buzzing.

Let's try Stata on this problem. First, we make up two new variables, rep_is_1 and rep_is_2, which indicate the repair category.

```
. generate rep_is_1 = (repair==1)
. generate rep_is_2 = (repair==2)
```

The statement generate rep_is_1 = (repair==1) creates a new variable, rep_is_1, which takes on the value 1 when repair is 1 and zero, otherwise. Similarly, the next generate statement creates rep_is_2 that takes on the value 1 when repair is 2 and zero, otherwise. We are now ready to fit our logit model. See [R] **probit** for the corresponding probit model.

(Continued on next page)

```
. logit foreign rep_is_1 rep_is_2
Note: rep_is_1 != 0 predicts failure perfectly
      rep_is_1 dropped and 10 obs not used
Iteration 0:   log likelihood = -26.992087
Iteration 1:   log likelihood = -22.483187
Iteration 2:   log likelihood = -22.230498
Iteration 3:   log likelihood = -22.229139
Iteration 4:   log likelihood = -22.229138
```

```
Logistic regression                              Number of obs   =         48
                                                 LR chi2(1)      =       9.53
                                                 Prob > chi2     =     0.0020
Log likelihood = -22.229138                      Pseudo R2       =     0.1765
```

foreign	Coef.	Std. Err.	z	P>|z|	[95% Conf. Interval]	
rep_is_2	-2.197225	.7698003	-2.85	0.004	-3.706006	-.6884436
_cons	3.89e-16	.4714045	0.00	1.000	-.9239359	.9239359

Remember that all the cars with poor repair records (rep_is_1) are domestic, so the model cannot be fitted, or at least it cannot be fitted if we restrict ourselves to finite coefficients. Stata noted that fact: "Note: rep_is_1 != 0 predicts failure perfectly". This is Stata's mathematically precise way of saying what we said in English. When rep_is_1 is not equal to 0, the car is domestic.

Stata then went on to say, "rep_is_1 dropped and 10 obs not used". This is Stata eliminating the problem. First, the variable rep_is_1 had to be removed from the model because it would have an infinite coefficient. Then the 10 observations that led to the problem had to be eliminated, as well, so as not to bias the remaining coefficients in the model. The 10 observations that are not used are the 10 domestic cars that have poor repair records.

Finally, Stata fitted what was left of the model, using the remaining observations.

◁

❏ Technical Note

Stata is pretty smart about catching problems like this. It will catch "one-way causation by a dummy variable", as we demonstrated above.

Stata also watches for "two-way causation", that is, a variable that perfectly determines the outcome, both successes and failures. In this case, Stata says, "so-and-so predicts outcome perfectly" and stops. Statistics dictates that no model can be fitted.

Stata also checks your data for collinear variables; it will say, "so-and-so dropped due to collinearity". No observations need to be eliminated in this case, and model fitting will proceed without the offending variable.

It will also catch a subtle problem that can arise with continuous data. For instance, if we were estimating the chances of surviving the first year after an operation, and if we included in our model age, and if all the persons over 65 died within the year, Stata would say, "age > 65 predicts failure perfectly". It would then inform us about the fixup it takes, and fit what can be fitted of our model.

logit (and logistic, probit, and ivprobit) will also occasionally display messages such as

```
note: 4 failures and 0 successes completely determined.
```

There are two causes for a message like this. The first—and most unlikely—case occurs when a continuous variable (or a combination of a continuous variable with other continuous or dummy variables) is simply a great predictor of the dependent variable. Consider Stata's auto.dta dataset with six observations removed.

```
. use http://www.stata-press.com/data/r9/auto
(1978 Automobile Data)

. drop if foreign==0 & gear_ratio>3.1
(6 observations deleted)

. logit foreign mpg weight gear_ratio, nolog
```

Logistic regression					Number of obs	=	68
					LR chi2(3)	=	72.64
					Prob > chi2	=	0.0000
Log likelihood = -6.4874814					Pseudo R2	=	0.8484

foreign	Coef.	Std. Err.	z	P>|z|	[95% Conf.	Interval]
mpg	-.4944907	.2655508	-1.86	0.063	-1.014961	.0259792
weight	-.0060919	.003101	-1.96	0.049	-.0121698	-.000014
gear_ratio	15.70509	8.166234	1.92	0.054	-.3004359	31.71061
_cons	-21.39527	25.41486	-0.84	0.400	-71.20747	28.41694

```
note: 4 failures and 0 successes completely determined.
```

Note that there are no missing standard errors in the output. If you receive the "completely determined" message and have one or more missing standard errors in your output, see the second case discussed below.

Note gear_ratio's large coefficient. logit thought that the four observations with the smallest predicted probabilities were essentially predicted perfectly.

```
. predict p
(option p assumed; Pr(foreign))

. sort p

. list p in 1/4
```

	p
1.	1.34e-10
2.	6.26e-09
3.	7.84e-09
4.	1.49e-08

If this happens to you, you don't have to do anything. Computationally, the model is sound. The second case discussed below requires careful examination.

The second case occurs when the independent terms are all dummy variables or continuous ones with repeated values (e.g., age). In this case, one or more of the estimated coefficients will have missing standard errors. For example, consider this dataset consisting of five observations.

```
. list
```

	y	x1	x2
1.	0	0	0
2.	0	1	0
3.	1	1	0
4.	0	0	1
5.	1	0	1

```
. logit y x1 x2, nolog
Logistic regression                                Number of obs   =           5
                                                   LR chi2(2)      =        1.18
                                                   Prob > chi2     =      0.5530
Log likelihood = -2.7725887                        Pseudo R2       =      0.1761
```

y	Coef.	Std. Err.	z	P>|z|	[95% Conf. Interval]	
x1	18.26157	2	9.13	0.000	14.34164	22.1815
x2	18.26157	.	.	.	.	.
_cons	-18.26157	1.414214	-12.91	0.000	-21.03338	-15.48976

```
note: 1 failure and 0 successes completely determined.
. predict p
(option p assumed; Pr(y))
. list
```

	y	x1	x2	p
1.	0	0	0	1.17e-08
2.	0	1	0	.5
3.	1	1	0	.5
4.	0	0	1	.5
5.	1	0	1	.5

Two things are happening here. First, logit is able to fit the outcome ($y = 0$) for the covariate pattern $x1 = 0$ and $x2 = 0$ (i.e., the first observation) perfectly. This observation is the "1 failure ... completely determined". Second, if this observation is dropped, then x1, x2, and the constant are collinear.

This is the cause of the message "completely determined" and the missing standard errors. It happens when you have a covariate pattern (or patterns) with only one outcome and there is collinearity when the observations corresponding to this covariate pattern are dropped.

If this happens to you, confirm the causes. First, identify the covariate pattern with only one outcome. (For your data, replace x1 and x2 with the independent variables of your model.)

```
. drop p
. egen pattern = group(x1 x2)
. quietly logit y x1 x2
. predict p
(option p assumed; Pr(y))
. summarize p
```

Variable	Obs	Mean	Std. Dev.	Min	Max
p	5	.4	.2236068	1.17e-08	.5

If successes were completely determined, that means that there are predicted probabilities that are almost 1. If failures were completely determined, that means that there are predicted probabilities that are almost 0. The latter is the case here, so we locate the corresponding value of pattern:

```
. tabulate pattern if p < 1e-7
```

group(x1 x2)	Freq.	Percent	Cum.
1	1	100.00	100.00
Total	1	100.00	

Once we omit this covariate pattern from the estimation sample, `logit` can deal with the collinearity:

```
. logit y x1 x2 if pattern !=1, nolog
note: x2 dropped due to collinearity
```

Logistic regression				Number of obs	=	4
				LR chi2(1)	=	0.00
				Prob > chi2	=	1.0000
Log likelihood = -2.7725887				Pseudo R2	=	0.0000

y	Coef.	Std. Err.	z	P>|z|	[95% Conf. Interval]	
x1	0	2	0.00	1.000	-3.919928	3.919928
_cons	0	1.414214	0.00	1.000	-2.771808	2.771808

We omit the collinear variable. Then we must decide whether to include or omit the observations with pattern = 1. We could include them

```
. logit y x1, nolog
```

Logistic regression				Number of obs	=	5
				LR chi2(1)	=	0.14
				Prob > chi2	=	0.7098
Log likelihood = -3.2958369				Pseudo R2	=	0.0206

y	Coef.	Std. Err.	z	P>|z|	[95% Conf. Interval]	
x1	.6931472	1.870827	0.37	0.711	-2.973605	4.3599
_cons	-.6931472	1.224742	-0.57	0.571	-3.093597	1.707302

or exclude them:

```
. logit y x1 if pattern !=1, nolog
```

Logistic regression				Number of obs	=	4
				LR chi2(1)	=	0.00
				Prob > chi2	=	1.0000
Log likelihood = -2.7725887				Pseudo R2	=	0.0000

y	Coef.	Std. Err.	z	P>|z|	[95% Conf. Interval]	
x1	0	2	0.00	1.000	-3.919928	3.919928
_cons	0	1.414214	0.00	1.000	-2.771808	2.771808

If the covariate pattern that predicts outcome perfectly is meaningful, you may want to exclude these observations from the model. In this case, you would report that covariate pattern such and such predicted outcome perfectly and that the best model for the rest of the data is .... But, more likely, the perfect prediction was simply the result of having too many predictors in the model. In this case, you would omit the extraneous variables from further consideration and report the best model for all the data.

❏

Saved Results

logit saves in e():

Scalars
e(N)	number of observations	e(ll)	log likelihood
e(df_m)	model degrees of freedom	e(ll_0)	log likelihood, constant-only model
e(r2_p)	pseudo-R-squared	e(chi2)	χ^2
e(N_clust)	number of clusters		

Macros
e(cmd)	logit	e(chi2type)	Wald or LR; type of model χ^2 test
e(depvar)	name of dependent variable	e(vce)	*vcetype* specified in vce()
e(wtype)	weight type	e(vcetype)	title used to label Std. Err.
e(wexp)	weight expression	e(crittype)	optimization criterion
e(title)	title in estimation output	e(properties)	b V
e(clustvar)	name of cluster variable	e(estat_cmd)	program used to implement estat
e(offset)	offset	e(predict)	program used to implement predict

Matrices
e(b)	coefficient vector	e(V)	variance–covariance matrix of the estimators

Functions
e(sample)	marks estimation sample

Methods and Formulas

Cramer (2003, chapter 9) surveys the prehistory and history of the logit model. The word logit was coined by Berkson (1944) and is analogous to the word probit. For an introduction to probit and logit, see, for example, Aldrich and Nelson (1984), Johnston and DiNardo (1997), Long (1997), Long and Freese (2003), Pampel (2000), or Powers and Xie (2000).

The likelihood function for logit is

$$\ln L = \sum_{j \in S} w_j \ln F(\mathbf{x}_j \mathbf{b}) + \sum_{j \notin S} w_j \ln\left\{1 - F(\mathbf{x}_j \mathbf{b})\right\}$$

where S is the set of all observations j, such that $y_j \neq 0$, $F(z) = e^z/(1 + e^z)$, and w_j denotes the optional weights. $\ln L$ is maximized as described in [R] **maximize**.

If robust standard errors are requested, the calculation described in *Methods and Formulas* of [R] **regress** is carried forward with $\mathbf{u}_j = \{1 - F(\mathbf{x}_j \mathbf{b})\}\mathbf{x}_j$ for the positive outcomes and $-F(\mathbf{x}_j \mathbf{b})\mathbf{x}_j$ for the negative outcomes. q_c is given by its asymptotic-like formula.

Joseph Berkson (1899–1982) was born in New York City and studied at the College of the City of New York, Columbia, and Johns Hopkins, earning both an M.D. and a doctorate in statistics. He then worked at Johns Hopkins before moving to the Mayo Clinic in 1931 as a biostatistician. Among many other contributions, his most influential drew upon a long-sustained interest in the logistic function, especially his 1944 paper on bioassay, in which he introduced the term logit. Berkson was a frequent participant in controversy, sometimes humorous, sometimes bitter, on subjects such as the evidence for links between smoking and various diseases and the relative merits of probit and logit methods and of different calculation methods.

References

Aldrich, J. H. and F. D. Nelson. 1984. *Linear Probability, Logit, and Probit Models.* Newbury Park, CA: Sage.

Berkson, J. 1944. Application of the logistic function to bio-assay. *Journal of the American Statistical Association* 39: 357–365.

Cleves, M. and A. Tosetto. 2000. sg139: Logistic regression when binary outcome is measured with uncertainty. *Stata Technical Bulletin* 55: 20–23.

Cramer, J. S. 2003. *Logit Models from Economics and Other Fields.* Cambridge: Cambridge University Press.

Hosmer, D. W., Jr., and S. Lemeshow. 2000. *Applied Logistic Regression.* 2nd ed. New York: Wiley.

Johnston, J. and J. DiNardo. 1997. *Econometric Methods.* 4th ed. New York: McGraw–Hill.

Judge, G. G., W. E. Griffiths, R. C. Hill, H. Lütkepohl, and T.-C. Lee. 1985. *The Theory and Practice of Econometrics.* 2nd ed. New York: Wiley.

Long, J. S. 1997. *Regression Models for Categorical and Limited Dependent Variables.* Thousand Oaks, CA: Sage.

Long, J. S. and J. Freese. 2003. *Regression Models for Categorical Dependent Variables Using Stata.* rev. ed. College Station, TX: Stata Press.

Mitchell, M. N. and X. Chen. 2005. Visualizing main effects and interactions for binary logit models. *Stata Journal* 5: 64–82.

O'Fallon, W. M. 1998. Berkson, Joseph. In *Encyclopedia of Biostatistics*, ed. P. Armitage and T. Colton, 1: 290–295. Chichester, UK: Wiley.

Pampel, F. C. 2000. *Logistic Regression: A Primer.* Thousand Oaks, CA: Sage.

Powers, D. A. and Y. Xie. 2000. *Statistical Methods for Categorical Data Analysis.* San Diego, CA: Academic Press.

Pregibon, D. 1981. Logistic regression diagnostics. *Annals of Statistics* 9: 705–724.

Also See

Complementary:	[R] **logit postestimation**, [R] **roc**, [R] **rocfit**
Related:	[R] **asmprobit**, [R] **brier**, [R] **clogit**, [R] **cloglog**, [R] **glm**, [R] **glogit**,
	[R] **ivprobit**, [R] **logistic**, [R] **mlogit**, [R] **mprobit**, [R] **nlogit**, [R] **ologit**,
	[R] **probit**, [R] **rologit**, [R] **scobit**, [R] **slogit**,
	[SVY] **svy: logistic**, [SVY] **svy: logit**, [XT] **xtlogit**
Background:	[U] **11.1.10 Prefix commands**,
	[U] **20 Estimation and postestimation commands**,
	[R] **estimation options**, [R] **maximize**, [R] *vce_option*

Title

logit postestimation — Postestimation tools for logit

Description

The following postestimation commands are of special interest after `logit`:

command	description
estat clas	`estat classification` reports various summary statistics, including the classification table
estat gof	Pearson or Hosmer–Lemeshow goodness-of-fit test
lroc	graphs the ROC curve and calculates the area under the curve
lsens	graphs sensitivity and specificity versus probability cutoff

For information about these commands, see [R] **logistic postestimation**.

In addition, the following standard postestimation commands are available:

command	description
adjust[1]	adjusted predictions of $\mathbf{x}\beta$, probabilities, or $\exp(\mathbf{x}\beta)$
estat	AIC, BIC, VCE, and estimation sample summary
estimates	cataloging estimation results
lincom	point estimates, standard errors, testing, and inference for linear combinations of coefficients
linktest	link test for model specification
lrtest	likelihood-ratio test
mfx	marginal effects or elasticities
nlcom	point estimates, standard errors, testing, and inference for nonlinear combinations of coefficients
predict	predictions, residuals, influence statistics, and other diagnostic measures
predictnl	point estimates, standard errors, testing, and inference for generalized predictions
suest	seemingly unrelated estimation
test	Wald tests for simple and composite linear hypotheses
testnl	Wald tests of nonlinear hypotheses

[1] adjust does not work with time-series operators.

See the corresponding entries in the *Stata Base Reference Manual* for details.

Syntax for predict

predict [*type*] *newvar* [*if*] [*in*] [, *statistic* <u>nooff</u>set <u>rule</u>s asif]

statistic	description
<u>pr</u>	probability of a positive outcome; the default
xb	$\mathbf{x}_j\mathbf{b}$, fitted values
stdp	standard error of the prediction
*<u>dbeta</u>	Pregibon (1991) $\Delta\widehat{\beta}$ influence statistic
*<u>deviance</u>	deviance residual
*<u>dx2</u>	Hosmer and Lemeshow (2000) $\Delta\chi^2$ influence statistic
*<u>ddeviance</u>	Hosmer and Lemeshow (2000) ΔD influence statistic
*<u>hat</u>	Pregibon (1981) leverage
*<u>number</u>	sequential number of the covariate pattern
*<u>residuals</u>	Pearson residuals; adjusted for number sharing covariate pattern
*<u>rstandard</u>	standardized Pearson residuals; adjusted for number sharing covariate pattern
<u>score</u>	first derivative of the log likelihood with respect to $\mathbf{x}_j\beta$

Unstarred statistics are available both in and out of sample; type predict ... if e(sample) ... if wanted
only for the estimation sample. Starred statistics are calculated only for the estimation sample, even when
if e(sample) is not specified.

Options for predict

pr, the default, calculates the probability of a positive outcome.

xb calculates the linear prediction.

stdp calculates the standard error of the linear prediction.

dbeta calculates the Pregibon (1981) $\Delta\widehat{\beta}$ influence statistic, a standardized measure of the difference
in the coefficient vector due to deletion of the observation along with all others that share the
same covariate pattern. In Hosmer and Lemeshow (2000) jargon, this statistic is M-asymptotic;
that is, it is adjusted for the number of observations that share the same covariate pattern.

deviance calculates the deviance residual.

dx2 calculates the Hosmer and Lemeshow (2000) $\Delta\chi^2$ influence statistic, reflecting the decrease in
the Pearson χ^2 due to the deletion of the observation and all others that share the same covariate
pattern.

ddeviance calculates the Hosmer and Lemeshow (2000) ΔD influence statistic, which is the change
in the deviance residual due to deletion of the observation and all others that share the same
covariate pattern.

hat calculates the Pregibon (1981) leverage or the diagonal elements of the hat matrix adjusted for
the number of observations that share the same covariate pattern.

number numbers the covariate patterns—observations with the same covariate pattern have the same
number. Observations not used in estimation have number set to missing. The "first" covariate
pattern is numbered 1, the second 2, and so on.

residuals calculates the Pearson residual as given by Hosmer and Lemeshow (2000) and adjusted
for the number of observations that share the same covariate pattern.

`rstandard` calculates the standardized Pearson residual as given by Hosmer and Lemeshow (2000) and adjusted for the number of observations that share the same covariate pattern.

`score` calculates the equation-level score, $\partial \ln L / \partial(\mathbf{x}_j \boldsymbol{\beta})$.

`nooffset` is relevant only if you specified `offset(`*varname*`)` for `logit`. It modifies the calculations made by `predict` so that they ignore the offset variable; the linear prediction is treated as $\mathbf{x}_j \mathbf{b}$ rather than as $\mathbf{x}_j \mathbf{b} + \text{offset}_j$.

`rules` requests that Stata use any "rules" that were used to identify the model when making the prediction. By default, Stata calculates missing for excluded observations.

`asif` requests that Stata ignore the rules and exclusion criteria and calculate predictions for all observations possible using the estimated parameter from the model.

Remarks

Once you have fitted a logit model, you can obtain the predicted probabilities using the `predict` command for both the estimation sample and other samples; see [U] **20 Estimation and postestimation commands** and [R] **predict**. Here we will make only a few additional comments.

`predict` without arguments calculates the predicted probability of a positive outcome, i.e., $\Pr(y_j = 1) = F(\mathbf{x}_j \mathbf{b})$. With the `xb` option, `predict` calculates the linear combination $\mathbf{x}_j \mathbf{b}$, where $\mathbf{x}_j$ are the independent variables in the jth observation and $\mathbf{b}$ is the estimated parameter vector. This is sometimes known as the index function since the cumulative distribution function indexed at this value is the probability of a positive outcome.

In both cases, Stata remembers any "rules" used to identify the model and calculates missing for excluded observations, unless `rules` or `asif` is specified. For information about the other statistics available after `predict`, see [R] **logistic postestimation**.

▷ Example 1

In example 2 of [R] **logit**, we fitted the logit model `logit foreign rep_is_1 rep_is_2`. To obtain predicted probabilities, type

```
. use http://www.stata-press.com/data/r9/repair
(1978 Automobile Data)
. generate rep_is_1 = (repair==1)
. generate rep_is_2 = (repair==2)
. logit foreign rep_is_1 rep_is_2
  (output omitted )
. predict p
(option p assumed; Pr(foreign))
(10 missing values generated)
. summarize foreign p
```

Variable	Obs	Mean	Std. Dev.	Min	Max
foreign	58	.2068966	.4086186	0	1
p	48	.25	.1956984	.1	.5

Stata remembers any "rules" used to identify the model and sets predictions to missing for any excluded observations. In the previous example, `logit` dropped the variable `rep_is_1` from our model and excluded ten observations. Thus when we typed `predict p`, those same ten observations were again excluded, and their predictions were set to missing.

predict's rules option uses the rules in the prediction. During estimation, we were told "rep_is_1 != 0 predicts failure perfectly", so the rule is that when rep_is_1 is not zero, we should predict 0 probability of success or a positive outcome:

```
. predict p2, rules
. summarize foreign p p2
```

Variable	Obs	Mean	Std. Dev.	Min	Max
foreign	58	.2068966	.4086186	0	1
p	48	.25	.1956984	.1	.5
p2	58	.2068966	.2016268	0	.5

predict's asif option ignores the rules and exclusion criteria and calculates predictions for all observations possible using the estimated parameters from the model:

```
. predict p3, asif
. summarize foreign p p2 p3
```

Variable	Obs	Mean	Std. Dev.	Min	Max
foreign	58	.2068966	.4086186	0	1
p	48	.25	.1956984	.1	.5
p2	58	.2068966	.2016268	0	.5
p3	58	.2931034	.2016268	.1	.5

Which is right? What predict does by default is the most conservative approach. If a large number of observations had been excluded due to a simple rule, we could be reasonably certain that the rules prediction is correct. The asif prediction is only correct if the exclusion is a fluke, and we would be willing to exclude the variable from the analysis anyway. In that case, however, we would refit the model to include the excluded observations.

◁

Methods and Formulas

All postestimation commands listed above are implemented as ado-files.

See *Methods and Formulas* of [R] **logistic postestimation** for details.

Also See

Complementary:	[R] **logit**; [R] **logistic postestimation**; [R] **adjust**, [R] **estimates**, [R] **lincom**, [R] **linktest**, [R] **lrtest**, [R] **mfx**, [R] **nlcom**, [R] **predictnl**, [R] **suest**, [R] **test**, [R] **testnl**
Background:	[U] **13.5 Accessing coefficients and standard errors**, [U] **20 Estimation and postestimation commands**, [R] **estat**, [R] **predict**

Title

> **loneway** — Large one-way ANOVA, random effects, and reliability

Syntax

loneway *response_var group_var* $\left[\,if\,\right]$ $\left[\,in\,\right]$ $\left[\,weight\,\right]$ $\left[\,,\,options\,\right]$

options	description
Main	
mean	expected value of F distribution; default is 1
median	median of F distribution; default is 1
exact	exact confidence intervals (groups must be equal with no weights)
level(#)	set confidence level; default is level(95)

by may be used with loneway; see [D] **by**.

aweights are allowed; see [U] **11.1.6 weight**.

Description

loneway fits one-way analysis-of-variance (ANOVA) models on datasets with a large number of levels of *group_var* and presents different ancillary statistics from oneway (see [R] **oneway**):

Feature	oneway	loneway
Fit one-way model	x	x
on fewer than 376 levels	x	x
on more than 376 levels		x
Bartlett's test for equal variance	x	
Multiple-comparison tests	x	
Intragroup correlation and S.E.		x
Intragroup correlation confidence interval		x
Est. reliability of group-averaged score		x
Est. S.D. of group effect		x
Est. S.D. within group		x

Options

> Main

mean specifies that the expected value of the $F_{k-1,N-k}$ distribution be used as the reference point F_m in the estimation of ρ instead of the default value of 1.

median specifies that the median of the $F_{k-1,N-k}$ distribution be used as the reference point F_m in the estimation of ρ instead of the default value of 1.

exact requests that exact confidence intervals be computed, as opposed to the default asymptotic confidence intervals. This option is allowed only if the groups are equal in size and weights are not used.

level(#) specifies the confidence level, as a percentage, for confidence intervals of the coefficients. The default is level(95) or as set by set level; see [U] **20.6 Specifying the width of confidence intervals**.

Remarks

Remarks are presented under the headings

The one-way ANOVA model
R-squared
The random-effects ANOVA model
Intraclass correlation
Estimated reliability of the group-averaged score

The one-way ANOVA model

▷ Example 1

loneway's output looks like that of oneway, except that, at the end, additional information is presented. Using our automobile dataset, we have created a (numeric) variable called manufacturer_grp identifying the manufacturer of each car, and within each manufacturer we have retained a maximum of four models, selecting those with the lowest mpg. We can compute the intraclass correlation of mpg for all manufacturers with at least four models as follows:

```
. use http://www.stata-press.com/data/r9/auto7
(1978 Automobile Data)
. loneway mpg manufacturer_grp if nummake == 4
            One-way Analysis of Variance for mpg: Mileage (mpg)
                                          Number of obs =         36
                                          R-squared     =     0.5228

        Source             SS        df       MS          F      Prob > F

Between manufactur~p    621.88889     8    77.736111     3.70     0.0049
Within manufactur~p     567.75       27    21.027778

Total                  1189.6389     35    33.989683

        Intraclass      Asy.
        correlation     S.E.       [95% Conf. Interval]

         0.40270       0.18770      0.03481    0.77060

        Estimated SD of manufactur~p effect      3.765247
        Estimated SD within manufactur~p         4.585605
        Est. reliability of a manufactur~p mean   .72950
            (evaluated at n=4.00)
```

◁

In addition to the standard one-way ANOVA output, loneway produces the R-squared, the estimated standard deviation of the group effect, the estimated standard deviation within group, the intragroup correlation, the estimated reliability of the group-averaged mean, and, in the case of unweighted data, the asymptotic standard error and confidence interval for the intragroup correlation.

R-squared

The R-squared is, of course, simply the underlying R^2 for a regression of *response_var* on the levels of *group_var*, or mpg on the various manufacturers in this case.

The random-effects ANOVA model

`loneway` assumes that we observe a variable y_{ij} measured for n_i elements within k groups or classes such that

$$y_{ij} = \mu + \alpha_i + \epsilon_{ij}, \quad i = 1, 2, \ldots, k, \quad j = 1, 2, \ldots, n_i$$

and α_i and ϵ_{ij} are independent zero-mean random variables with variance σ_α^2 and σ_ϵ^2, respectively. This is the random-effects ANOVA model, also known as the components-of-variance model, in which it is typically assumed that the y_{ij} are normally distributed.

The interpretation with respect to our example is that the observed value of our response variable, mpg, is created in two steps. First the ith manufacturer is chosen, and a value α_i is determined—the typical mpg for that manufacturer less the overall mpg μ. Then a deviation, ϵ_{ij}, is chosen for the jth model within this manufacturer. This is how much that particular automobile differs from the typical mpg value for models from this manufacturer.

For our sample of 36 car models, the estimated standard deviations are $\sigma_\alpha = 3.8$ and $\sigma_\epsilon = 4.6$. Thus a little more than half of the variation in mpg between cars is attributable to the car model, with the rest attributable to differences between manufacturers. These standard deviations differ from those that would be produced by a (standard) fixed-effects regression in that the regression would require the sum within each manufacturer of the ϵ_{ij}, $\epsilon_{i.}$ for the ith manufacturer, to be zero, while these estimates merely impose the constraint that the sum is *expected* to be zero.

Intraclass correlation

There are various estimators of the intraclass correlation, such as the pairwise estimator, which is defined as the Pearson product-moment correlation computed over all possible pairs of observations that can be constructed within groups. For a discussion of various estimators, see Donner (1986). `loneway` computes what is termed the analysis of variance, or ANOVA, estimator. This intraclass correlation is the theoretical upper bound on the variation in *response_var* that is explainable by *group_var*, of which R-squared is an overestimate because of the serendipity of fitting. Note that this correlation is comparable to an R-squared—you do not have to square it.

In our example, the intra-manu correlation, the correlation of mpg within manufacturer, is 0.40. Since aweights were not used and the default correlation was computed (i.e., the mean and median options were not specified), `loneway` also provided the asymptotic confidence interval and standard error of the intraclass correlation estimate.

Estimated reliability of the group-averaged score

The estimated reliability of the group-averaged score or mean has an interpretation similar to that of the intragroup correlation; it is a comparable number if we average *response_var* by *group_var*, or mpg by manu in our example. It is the theoretical upper bound of a regression of manufacturer-averaged mpg on characteristics of manufacturers. Why would we want to collapse our 36-observation dataset into a 9-observation dataset of manufacturer averages? Because the 36 observations might be a mirage. When General Motors builds cars, do they sometimes put a Pontiac label and sometimes a Chevrolet label on them, so that it appears in our data as if we have two cars when we really have only one, replicated? If that were the case, and if it were the case for many other manufacturers, then we would be forced to admit that we do not have data on 36 cars; we instead have data on 9 manufacturer-averaged characteristics.

Saved Results

loneway saves in r():

Scalars

r(N)	number of observations	r(rho_t)	estimated reliability
r(rho)	intraclass correlation	r(se)	asymp. SE of intraclass correlation
r(lb)	lower bound of 95% CI for rho	r(sd_w)	estimated SD within group
r(ub)	upper bound of 95% CI for rho	r(sd_b)	estimated SD of group effect

Methods and Formulas

loneway is implemented as an ado-file.

The mean squares in the loneway's ANOVA table are computed as follows,

$$\mathrm{MS}_\alpha = \sum_i w_i . (\bar{y}_i. - \bar{y}..)^2 / (k - 1)$$

and

$$\mathrm{MS}_\epsilon = \sum_i \sum_j w_{ij}(y_{ij} - \bar{y}_i.)^2 / (N - k)$$

in which

$$w_i. = \sum_j w_{ij} \quad w.. = \sum_i w_i. \quad \bar{y}_i. = \sum_j w_{ij} y_{ij} / w_i. \quad \text{and} \quad \bar{y}.. = \sum_i w_i. \bar{y}_i. / w..$$

The corresponding expected values of these mean squares are

$$E(\mathrm{MS}_\alpha) = \sigma_\epsilon^2 + g\sigma_\alpha^2 \qquad \text{and} \qquad E(\mathrm{MS}_\epsilon) = \sigma_\epsilon^2$$

in which

$$g = \frac{w.. - \sum_i w_i.^2 / w..}{k - 1}$$

Note that in the unweighted case, we get

$$g = \frac{N - \sum_i n_i^2 / N}{k - 1}$$

As expected, $g = m$ for the case of no weights and equal group sizes in the data, i.e., $n_i = m$ for all i. Replacing the expected values with the observed values and solving yields the ANOVA estimates of σ_α^2 and σ_ϵ^2. Substituting these into the definition of the intraclass correlation

$$\rho = \frac{\sigma_\alpha^2}{\sigma_\alpha^2 + \sigma_\epsilon^2}$$

yields the ANOVA estimator of the intraclass correlation:

$$\rho_A = \frac{F_{\mathrm{obs}} - 1}{F_{\mathrm{obs}} - 1 + g}$$

Note that F_{obs} is the observed value of the F statistic from the ANOVA table. For the case of no weights and equal n_i, ρ_A = roh, which is the intragroup correlation defined by Kish (1965). Two slightly different estimators are available through the `mean` and `median` options (Gleason 1997). If either of these options is specified, the estimate of ρ becomes

$$\rho = \frac{F_{\text{obs}} - F_m}{F_{\text{obs}} + (g-1)F_m}$$

For the `mean` option, $F_m = E(F_{k-1,N-K}) = (N-k)/(N-k-2)$, i.e., the expected value of the ANOVA table's F statistic. For the `median` option, F_m is simply the median of the F statistic. Note that setting F_m to 1 gives ρ_A, so for large samples, these different point estimators are essentially the same. Also, since the intraclass correlation of the random-effects model is by definition non-negative, for any of the three possible point estimators, ρ is truncated to zero if F_{obs} is less than F_m.

For the case of no weighting, interval estimators for ρ_A are computed. If the groups are equal-sized (all n_i equal) and the `exact` option is specified, the following exact (assuming that the y_{ij} are normally distributed) $100(1-\alpha)\%$ confidence interval is computed,

$$\left\{ \frac{F_{\text{obs}} - F_m F_u}{F_{\text{obs}} + (g-1)F_m F_u}, \frac{F_{\text{obs}} - F_m F_l}{F_{\text{obs}} + (g-1)F_m F_l} \right\}$$

with $F_m = 1$, $F_l = F_{\alpha/2,k-1,N-k}$, and $F_u = F_{1-\alpha/2,k-1,N-k}$, $F_{\cdot,k-1,N-k}$ being the cumulative distribution function for the F distribution with $k-1$ and $N-k$ degrees of freedom. Note that if `mean` or `median` is specified, F_m is defined as above. If the groups are equal sized and `exact` is not specified, the following asymptotic $100(1-\alpha)\%$ confidence interval for ρ_A is computed,

$$\left[\rho_A - z_{\alpha/2}\sqrt{V(\rho_A)}, \rho_A + z_{\alpha/2}\sqrt{V(\rho_A)} \right]$$

where $z_{\alpha/2}$ is the $100(1-\alpha/2)$ percentile of the standard normal distribution and $\sqrt{V(\rho_A)}$ is the asymptotic standard error of ρ defined below. Note that this confidence interval is also available for the case of unequal groups. It is not applicable and, therefore, not computed for the estimates of ρ provided by the `mean` and `median` options. Again since the intraclass coefficient is non-negative, if the lower bound is negative for either confidence interval, it is truncated to zero. As might be expected, the coverage probability of a truncated interval is higher than its nominal value.

The asymptotic standard error of ρ_A, assuming that the y_{ij} are normally distributed, is also computed when appropriate, namely, for unweighted data and when ρ_A is computed (neither the `mean` option nor the `median` option is specified):

$$V(\rho_A) = \frac{2(1-\rho)^2}{g^2} (A + B + C)$$

with

$$A = \frac{\{1 + \rho(g-1)\}^2}{N-k}$$

$$B = \frac{(1-\rho)\{1 + \rho(2g-1)\}}{k-1}$$

$$C = \frac{\rho^2\{\sum n_i^2 - 2N^{-1}\sum n_i^3 + N^{-2}(\sum n_i^2)^2\}}{(k-1)^2}$$

and ρ_A is substituted for ρ (Donner 1986).

The estimated reliability of the group-averaged score, known as the Spearman–Brown prediction formula in the psychometric literature (Winer, Brown, and Michels 1991, 1014), is

$$\rho_t = \frac{t\rho}{1 + (t - 1)\rho}$$

for group size t. loneway computes ρ_t for $t = g$.

The estimated standard deviation of the group effect is $\sigma_\alpha = \sqrt{(\mathrm{MS}_\alpha - \mathrm{MS}_\epsilon)/g}$. This comes from the assumption that an observation is derived by adding a group effect to a within-group effect.

The estimated standard deviation within group is the square root of the mean square due to error, or $\sqrt{\mathrm{MS}_\epsilon}$.

Methods and Formulas

loneway is implemented as an ado-file.

Acknowledgment

We would like to thank John Gleason of Syracuse University for his contributions to improving loneway.

References

Donner, A. 1986. A review of inference procedures for the intraclass correlation coefficient in the one-way random effects model. *International Statistical Review* 54: 67–82.

Gleason, J. R. 1997. sg65: Computing intraclass correlations and large ANOVAs. *Stata Technical Bulletin* 35: 25–31. Reprinted in *Stata Technical Bulletin Reprints*, vol. 6, pp. 167–176.

Kish, L. 1965. *Survey Sampling*. New York: Wiley.

Winer, B. J., D. R. Brown, and K. M. Michels. 1991. *Statistical Principles in Experimental Design*. 3rd ed. New York: McGraw–Hill.

Also See

Related: [R] **anova**, [R] **oneway**

Title

> **lowess** — Lowess smoothing

Syntax

> `lowess` *yvar* *xvar* [*if*] [*in*] [*, options*]

options	description
Main	
mean	running-mean smooth; default is running-line least-squares
noweight	suppress weighted regressions; default is tricube weighting function
bwidth(*#*)	use *#* for the bandwidth; default is bwidth(0.8)
logit	transform dependent variable to logits
adjust	adjust smoothed mean to equal mean of dependent variable
nograph	suppress graph
generate(*newvar*)	create *newvar* containing smoothed values of *yvar*
Scatterplot	
marker_options	change look of markers (color, size, etc.)
marker_label_options	add marker labels; change look or position
Smoothed line	
lineopts(*cline_options*)	affect rendition of the smoothed line
Add plot	
addplot(*plot*)	add other plots to generated graph
Y-Axis, X-Axis, Title, Caption, Legend, Overall, By	
twoway_options	any of the options documented in [G] *twoway_options*

yvar and *xvar* may contain time-series operators; see [U] **11.4.3 Time-series varlists**.

Description

`lowess` carries out a locally weighted regression of *yvar* on *xvar*, displays the graph, and optionally saves the smoothed variable.

Warning: `lowess` is computationally intensive and may therefore take a long time to run on a slow computer. Lowess calculations on 1,000 observations, for instance, require performing 1,000 regressions.

Options

> Main

`mean` specifies running-mean smoothing; the default is running-line least-squares smoothing.

`noweight` prevents the use of Cleveland's (1979) tricube weighting function; the default is to use this weighting function.

`bwidth(#)` specifies the bandwidth. Centered subsets of `bwidth()` $\times N$ observations are used for calculating smoothed values for each point in the data except for the end points, where smaller, uncentered subsets are used. The greater the `bwidth()`, the greater the smoothing. The default is 0.8.

`logit` transforms the smoothed *yvar* into logits. Predicted values less than .0001 or greater than .9999 are set to $1/N$ and $1 - 1/N$, respectively, before taking logits.

`adjust` adjusts the mean of the smoothed *yvar* to equal the mean of *yvar* by multiplying by an appropriate factor. This is useful when smoothing binary (0/1) data.

`nograph` suppresses displaying the graph.

`generate(`*newvar*`)` creates *newvar* containing the smoothed values of *yvar*.

⎧ Scatterplot ⎫
---

marker_options affect the rendition of markers drawn at the plotted points, including their shape, size, color, and outline; see [G] ***marker_options***.

marker_label_options specify if and how the markers are to be labeled; see [G] ***marker_label_options***.

⎧ Smoothed line ⎫

`lineopts(`*cline_options*`)` affect the rendition of the lowess-smoothed line; see [G] ***cline_options***.

⎧ Add plot ⎫

`addplot(`*plot*`)` provides a way to add other plots to the generated graph. See [G] ***addplot_option***.

⎧ Y-Axis, X-Axis, Title, Caption, Legend, Overall, By ⎫
---

twoway_options are any of the options documented in [G] ***twoway_options***. These include options for titling the graph (see [G] ***title_options***), options for saving the graph to disk (see [G] ***saving_option***), and the `by()` option (see [G] ***by_option***).

Remarks

By default, `lowess` provides locally weighted scatterplot smoothing. The basic idea is to create a new variable (*newvar*) that, for each *yvar* y_i, contains the corresponding smoothed value. The smoothed values are obtained by running a regression of *yvar* on *xvar* using only the data (x_i, y_i) and a small amount of the data near this point. In `lowess`, the regression is weighted so that the central point (x_i, y_i) gets the highest weight and points that are farther away (based on the distance $|x_j - x_i|$) receive less weight. The estimated regression line is then used to predict the smoothed value $\widehat{y}_i$ for y_i only. The procedure is repeated to obtain the remaining smoothed values, which means that a separate weighted regression is performed for every point in the data.

Lowess is a desirable smoother because of its locality—it tends to follow the data. Polynomial smoothing methods, for instance, are global in that what happens on the extreme left of a scatterplot can affect the fitted values on the extreme right.

▷ Example 1

The amount of smoothing is affected by `bwidth(#)`. You are warned to experiment with different values. For instance,

```
. use http://www.stata-press.com/data/r9/lowess1
(example data for lowess)
. lowess h1 depth
```

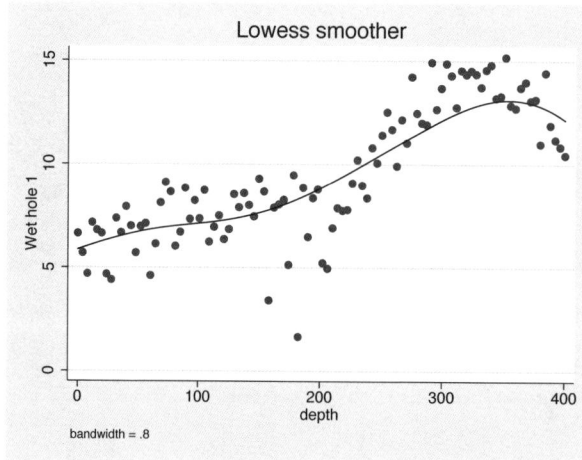

Now compare that with

```
. lowess h1 depth, bwidth(.4)
```

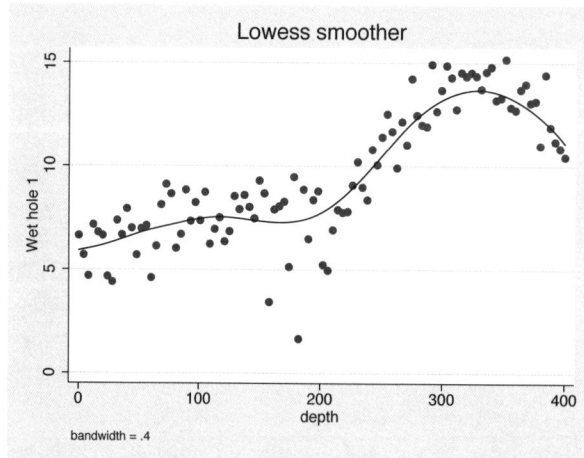

In the first case, the default bandwidth of 0.8 is used, meaning that 80% of the data are used in smoothing each point. In the second case, we explicitly specified a bandwidth of 0.4. Smaller bandwidths follow the original data more closely.

◁

▷ Example 2

Two lowess options are especially useful with binary (0/1) data: adjust and logit. adjust adjusts the resulting curve (by multiplication) so that the mean of the smoothed values is equal to the mean of the unsmoothed values. logit specifies that the smoothed curve be in terms of the log of the odds ratio:

```
. use http://www.stata-press.com/data/r9/auto
(1978 Automobile Data)
. lowess foreign mpg, ylabel(0 "Domestic" 1 "Foreign") jitter(5) adjust
```

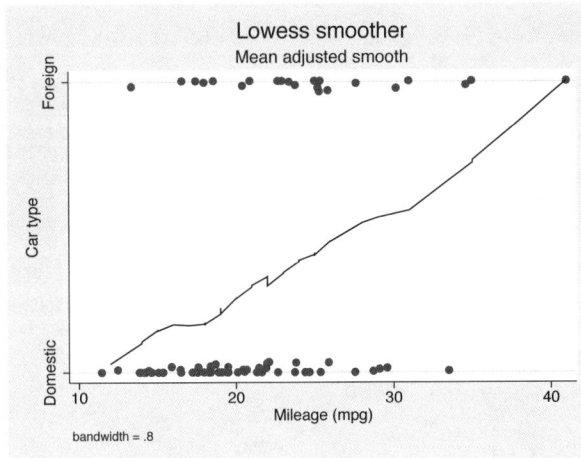

```
. lowess foreign mpg, logit yline(0) ylabel("logit of Car type")
```

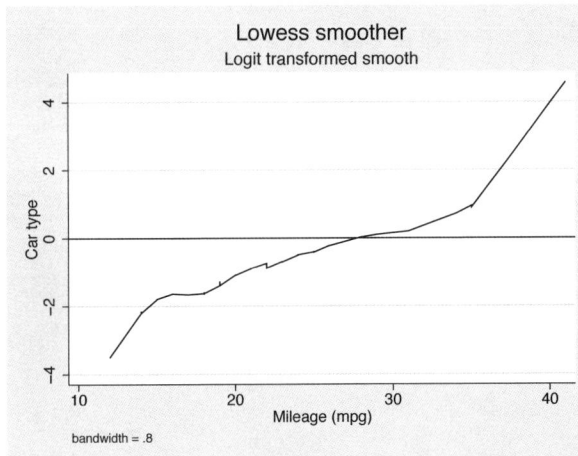

With binary data, if you do not use the logit option, it is a good idea to specify graph's jitter() option; see [G] **graph twoway scatter**. Since the underlying data (whether the car was manufactured outside the United States in this case) take on only two values, raw data points are more likely to be on top of each other, thus making it impossible to tell how many points there are. graph's jitter() option adds some noise to the data to shift the points around. This noise affects only the location of points on the graph, not the lowess curve.

When you specify the `logit` option, the display of the raw data is suppressed.

◁

❑ Technical Note

`lowess` can be used for more than just lowess smoothing. Lowess can be usefully thought of as a combination of two smoothing concepts: the use of predicted values from regression (rather than means) for imputing a smoothed value and the use of the tricube weighting function (rather than a constant weighting function). `lowess` allows you to combine these concepts freely. You can use line smoothing without weighting (specify `noweight`), mean smoothing with tricube weighting (specify `mean`), or mean smoothing without weighting (specify `mean` and `noweight`).

❑

Methods and Formulas

`lowess` is implemented as an ado-file.

Let y_i and x_i be the two variables, and assume that the data are ordered so that $x_i \leq x_{i+1}$ for $i = 1, \ldots, N - 1$. For each y_i, a smoothed value y_i^s is calculated.

The subset used in calculating y_i^s is indices $i_- = \max(1, i - k)$ through $i_+ = \min(i + k, N)$, where $k = \lfloor (N \cdot \texttt{bwidth} - 0.5)/2 \rfloor$. The weights for each of the observations between $j = i_-, \ldots, i_+$ are either 1 (`noweight`) or the tricube (default),

$$
w_j = \left\{ 1 - \left(\frac{|x_j - x_i|}{\Delta} \right)^3 \right\}^3
$$

where $\Delta = 1.0001 \max(x_{i_+} - x_i, x_i - x_{i_-})$. The smoothed value y_i^s is then the (weighted) mean or the (weighted) regression prediction at x_i.

> William S. Cleveland (1943–) studied mathematics and statistics at Princeton and Yale. He worked for several years at Bell Labs in New Jersey and now teaches statistics and computer science at Purdue. He has made key contributions in many areas of statistics, including graphics and data visualization, time series, environmental applications, and analysis of Internet traffic data.

Acknowledgment

`lowess` is a modified version of a command originally written by Patrick Royston of the MRC Clinical Trials Unit, London.

References

Chambers, J. M., W. S. Cleveland, B. Kleiner, and P. A. Tukey. 1983. *Graphical Methods for Data Analysis*. Belmont, CA: Wadsworth.

Cleveland, W. S. 1979. Robust locally weighted regression and smoothing scatterplots. *Journal of the American Statistical Association* 74: 829–836.

——. 1993. *Visualizing Data*. Summit, NJ: Hobart.

——. 1994. *The Elements of Graphing Data*. Summit, NJ: Hobart.

Goodall, C. 1990. A survey of smoothing techniques. In *Modern Methods of Data Analysis*, ed. J. Fox and J. S. Long, 126–176. Newbury Park, CA: Sage.

Royston, P. 1991. gr6: Lowess smoothing. *Stata Technical Bulletin* 3: 7–9. Reprinted in *Stata Technical Bulletin Reprints*, vol. 1, pp. 41–44.

Salgado-Ugarte, I. H. and M. Shimizu. 1995. snp8: Robust scatterplot smoothing: Enhancements to Stata's ksm. *Stata Technical Bulletin* 25: 23–26. Reprinted in *Stata Technical Bulletin Reprints*, vol. 5, pp. 190–194.

Sasieni, P. 1994. snp7: Natural cubic splines. *Stata Technical Bulletin* 22: 19–22. Reprinted in *Stata Technical Bulletin Reprints*, vol. 4, pp. 171–174.

Also See

Related:	[R] **kdensity**, [R] **smooth**,
	[D] **ipolate**
Background:	*Stata Graphics Reference Manual*

Title

> **lrtest** — Likelihood-ratio test after estimation

Syntax

> lrtest *modelspec*$_1$ [*modelspec*$_2$] [, *options*]

where *modelspec* is

$$name \mid . \mid (namelist)$$

where *name* is the name under which estimation results were saved using `estimates store`, and "." refers to the last estimation results, whether or not these were already stored.

options	description
stats	display statistical information about the two models
dir	display descriptive information about the two models
df(#)	override the automatic degrees-of-freedom calculation; seldom used
force	force testing even when apparently invalid

Description

lrtest performs a likelihood-ratio test for the null hypothesis that the parameter vector of a statistical model satisfies some smooth constraint. To conduct the test, both the unrestricted and the restricted models must be fitted using the maximum likelihood method (or some equivalent method), and the results of at least one must be stored using `estimates store`.

modelspec$_1$ and *modelspec*$_2$ specify the restricted and unrestricted model in any order. *modelspec*$_1$ and *modelspec*$_2$ cannot have names in common; for example, lrtest (A B C) (C D E) is not allowed since both model specifications include C. If *modelspec*$_2$ is not specified, the last estimation result is used; this is equivalent to specifying *modelspec*$_2$ as a period (.).

lrtest supports "composite models" specified by a parenthesized list of model names. In a composite model, we assume that the log likelihood and dimension (number of free parameters) of the full model are obtained as the sum of the log-likelihood values and dimensions of the constituting models.

lrtest provides an important alternative to test for models fitted via maximum likelihood or equivalent methods.

Options

stats displays statistical information about the unrestricted and restricted models, including the information indices of Akaike and Schwarz.

dir displays descriptive information about the unrestricted and restricted models; see [R] **estimates**.

df(#) is seldom specified; it overrides the automatic degrees-of-freedom calculation.

force forces the likelihood-ratio test calculations to take place in situations where lrtest would normally refuse to do so and issue an error. Such situations arise when one or more assumptions of the test are violated, for example, if the models were fitted with robust, cluster(), or pweights; the dependent variables in the two models differ; the null log-likelihoods differ; the samples differ; or the estimation commands differ. If you use the force option, there is no guarantee as to the validity or interpretability of the resulting test.

Remarks

The standard way to use lrtest is to

1. Fit either the restricted model or the unrestricted model using one of Stata's estimation commands and then store the results using estimates store *name*.

2. Fit the alternative model (the unrestricted or restricted model) and then type 'lrtest *name* .'. lrtest determines for itself which of the two models is the restricted model by comparing the degrees of freedom.

Often you may want to store the alternative model with estimates store *name₂*, for instance, if you plan additional tests against models yet to be fitted. The likelihood-ratio test is then obtained as lrtest *name name₂*.

Remarks are presented under the headings

> *Nested models*
> *Composite models*

Nested models

lrtest may be used with any estimation command that reports a log likelihood, including logit, poisson, streg, heckman, and stcox. It is your responsibility to check that one of the model specifications implies a statistical model that is *nested within* the model implied by the other specification. In most cases, this means that both models are fitted with the same estimation command (e.g., both are fitted by logit, with the same dependent variables) and that the set of covariates of one model is a subset of the covariates of the other model. Second, lrtest is valid only for models that are fitted by maximum likelihood or by some equivalent method, so it does not apply to cases in which models were fitted with probability weights or clusters. Specifying the robust option similarly would indicate that you are worried about the valid specification of the model, so you would not use lrtest. Third, lrtest assumes that under the null hypothesis, the test statistic is (approximately) distributed as chi-squared. This is not true for likelihood-ratio tests of "boundary conditions", such as tests for the presence of overdispersion or random effects (Gutierrez, Carter, and Drukker 2001).

▷ Example 1

We have data on infants born with low birthweights along with the characteristics of the mother (Hosmer and Lemeshow 2000; see also [R] **logistic**). We fit the following model:

(Continued on next page)

```
. use http://www.stata-press.com/data/r9/lbw2
(Hosmer & Lemeshow data)

. logistic low age lwt race2 race3 smoke ptl ht ui

Logistic regression                             Number of obs   =         189
                                                LR chi2(8)      =       33.22
                                                Prob > chi2     =      0.0001
Log likelihood =   -100.724                     Pseudo R2       =      0.1416

       low │  Odds Ratio   Std. Err.      z    P>|z|     [95% Conf. Interval]
───────────┼────────────────────────────────────────────────────────────────
       age │   .9732636   .0354759    -0.74   0.457     .9061578    1.045339
       lwt │   .9849634   .0068217    -2.19   0.029     .9716834    .9984249
     race2 │   3.534767   1.860737     2.40   0.016     1.259736    9.918406
     race3 │   2.368079   1.039949     1.96   0.050     1.001356    5.600207
     smoke │   2.517698   1.00916      2.30   0.021     1.147676    5.523162
       ptl │   1.719161   .5952579     1.56   0.118     .8721455    3.388787
        ht │   6.249602   4.322408     2.65   0.008     1.611152    24.24199
        ui │    2.1351    .9808153     1.65   0.099     .8677528      5.2534
```

We now wish to test the constraint that the coefficients on age, lwt, ptl, and ht are all zero or, equivalently in this case, that the odds ratios are all 1. One solution is to type

```
. test age lwt ptl ht

 ( 1)  age = 0
 ( 2)  lwt = 0
 ( 3)  ptl = 0
 ( 4)  ht = 0

           chi2(  4) =    12.38
         Prob > chi2 =    0.0147
```

This test is based on the inverse of the information matrix and is therefore based on a quadratic approximation to the likelihood function; see [R] **test**. A more precise test would be to refit the model, applying the proposed constraints, and then calculate the likelihood-ratio test.

We first save the current model:

```
. estimates store full
```

We then fit the constrained model, which in this case is the model omitting age, lwt, ptl, and ht:

```
. logistic low race2 race3 smoke ui

Logistic regression                             Number of obs   =         189
                                                LR chi2(4)      =       18.80
                                                Prob > chi2     =      0.0009
Log likelihood = -107.93404                     Pseudo R2       =      0.0801

       low │  Odds Ratio   Std. Err.      z    P>|z|     [95% Conf. Interval]
───────────┼────────────────────────────────────────────────────────────────
     race2 │   3.052746   1.498084     2.27   0.023     1.166749    7.987368
     race3 │   2.922593   1.189226     2.64   0.008      1.31646    6.488269
     smoke │   2.945742   1.101835     2.89   0.004      1.41517    6.131701
        ui │   2.419131   1.047358     2.04   0.041      1.03546    5.651783
```

That done, lrtest compares this model with the model we previously saved:

```
. lrtest full .

Likelihood-ratio test                           LR chi2(4)  =       14.42
(Assumption: . nested in full)                  Prob > chi2 =      0.0061
```

Comparing results, test reported that age, lwt, ptl, and ht were jointly significant at the 1.5% level; lrtest reports they are significant at the 0.6% level. Given the quadratic approximation made by test, we could argue that lrtest's results are more accurate.

Note that lrtest explicates the assumption that, based on a comparison of the degrees of freedom, it has assessed that the last fitted model (.) is nested within the model stored as full. In other words, full is the unconstrained model and . is the constrained model.

In windowed versions of Stata, the names in "(Assumption: . nested in full)" are actually links. Click on a name, and the results for that model are replayed.

◁

❏ Technical Note

lrtest determines the degrees of freedom of a model as the rank of the (co)variance matrix e(V). There are two issues here. First the *numerical* determination of the rank of a matrix is a subtle problem that can, for instance, be affected by the scaling of the variables in the model. The rank of a matrix depends on the number of (independent) linear combinations of coefficients that sum exactly to zero. In the world of numerical mathematics, it is hard to tell whether a very small number is really nonzero or is a real zero that happens to be slightly off because of round-off error from the finite precision with which computers make floating-point calculations. Whether a small number is being classified as one or the other, typically based on a threshold, affects the determined degrees of freedom. While Stata generally makes sensible choices, it is bound to make mistakes occasionally. The moral of this story is to make sure the calculated degrees of freedom are as you expect before interpreting the results.

❏

❏ Technical Note

A second issue involves regress and related commands such as anova. Mainly for historical reasons, regress does not treat the residual variance σ^2 the same way it treats the regression coefficients. Type vce after regress, and you will see the regression coefficients, not $\widehat{\sigma}^2$. Most estimation commands for models with ancillary parameters (e.g., streg and heckman) treat all parameters as equals. There is nothing technically wrong with regress here; we are usually focused on the regression coefficients, and their estimators are uncorrelated with $\widehat{\sigma}^2$. But, formally, σ^2 adds a degree of freedom to the model. This does not matter if you are comparing two regression models by a likelihood-ratio test. This test depends on the difference in the degrees of freedom, and hence being "off by 1" in each does not matter. But, if you are comparing a regression model with a larger model—e.g., a heteroskedastic regression model fitted by arch— the automatic determination of the degrees of freedom is incorrect, and you have to specify the df(#) option.

❏

▷ Example 2

Returning to the low birthweight data in the first example, we now wish to test that the coefficient on race2 is equal to that on race3. The base model is still stored under the name full, so we need only fit the constrained model and perform the test. Letting z be the index of the logit model, the base model is

$$z = \beta_0 + \beta_1 \texttt{age} + \beta_2 \texttt{lwt} + \beta_3 \texttt{race2} + \beta_4 \texttt{race3} + \cdots$$

If $\beta_3 = \beta_4$, this can be written as

$$z = \beta_0 + \beta_1 \texttt{age} + \beta_2 \texttt{lwt} + \beta_3 (\texttt{race2} + \texttt{race3}) + \cdots$$

To fit the constrained model, we create a variable equal to the sum of race2 and race3 and fit the model, which has the sum in place of the two variables.

```
. generate race23 = race2 + race3
. logistic low age lwt race23 smoke ptl ht ui
```

Logistic regression					Number of obs	=	189
					LR chi2(7)	=	32.67
					Prob > chi2	=	0.0000
Log likelihood = -100.9997					Pseudo R2	=	0.1392

low	Odds Ratio	Std. Err.	z	P>|z|	[95% Conf.	Interval]
age	.9716799	.0352638	-0.79	0.429	.9049649	1.043313
lwt	.9864971	.0064627	-2.08	0.038	.9739114	.9992453
race23	2.728186	1.080206	2.53	0.011	1.255586	5.927907
smoke	2.664498	1.052379	2.48	0.013	1.228633	5.778414
ptl	1.709129	.5924775	1.55	0.122	.8663666	3.371691
ht	6.116391	4.215585	2.63	0.009	1.58425	23.61385
ui	2.09936	.9699702	1.61	0.108	.8487997	5.192407

Comparing this model with our original model, we obtain

```
. lrtest full .
```

Likelihood-ratio test	LR chi2(1)	=	0.55
(Assumption: . nested in full)	Prob > chi2	=	0.4577

By comparison, typing `test race2=race3` after fitting our base model results in a significance level of .4572. Alternatively, we can first store the restricted model, in this case using the name `equal`. Next, `lrtest` is invoked specifying the names of the restricted and unrestricted models (we don't care about the order). This time, we also add the option `stats` requesting a table of model statistics, including the model selection indices AIC and BIC.

```
. estimates store equal
. lrtest equal full, stats
```

Likelihood-ratio test	LR chi2(1)	=	0.55
(Assumption: equal nested in full)	Prob > chi2	=	0.4577

Model	nobs	ll(null)	ll(model)	df	AIC	BIC
equal	189	-117.336	-100.9997	8	217.9994	243.9334
full	189	-117.336	-100.724	9	219.448	248.6237

◁

Composite models

lrtest supports composite models, that is, models that can be fitted by fitting a series of simpler models or by fitting models on subsets of the data. Theoretically, a composite model is one in which the likelihood function $L(\theta)$ of the parameter vector θ can be written as the product

$$L(\theta) = L_1(\theta_1) \times L_2(\theta_2) \times \cdots \times L_k(\theta_k)$$

of likelihood terms with $\theta = (\theta_1, \ldots, \theta_k)$ a partitioning of the full parameter vector. In such a case, the full-model likelihood $L(\theta)$ is maximized by maximizing the likelihood terms $L_j(\theta_j)$ in turn. Obviously, $\log L(\widehat{\theta}) = \sum_{j=1}^{k} \log L_j(\widehat{\theta}_j)$. The degrees of freedom for the composite model is obtained as the sum of the degrees of freedom of the constituting models.

▷ Example 3

As an example of the application of composite models, we consider a test of the hypothesis that the coefficients of a statistical model do not differ between different portions ("regimes") of the covariate space. Economists call a test for such an hypothesis a *Chow test*.

We continue the analysis of the data on children of low birthweight using logistic regression modeling and study whether the regression coefficients are the same between the three races: white, black, and other. A likelihood-ratio Chow test can be obtained by fitting the logistic regression model for each of the races and then comparing the "combined" results with those of the model previously stored as full. Since the full model included dummies for the three races, this version of the Chow test allows the intercept of the logistic regression model to vary between the regimes (races).

```
. logistic low age lwt smoke ptl ht ui if race==1, nolog
Logistic regression                               Number of obs   =         96
                                                  LR chi2(6)      =      13.86
                                                  Prob > chi2     =     0.0312
Log likelihood = -45.927061                       Pseudo R2       =     0.1311
```

low	Odds Ratio	Std. Err.	z	P>|z|	[95% Conf. Interval]	
age	.9869674	.0527756	-0.25	0.806	.8887649	1.096021
lwt	.9900874	.0106101	-0.93	0.353	.9695089	1.011103
smoke	4.208697	2.680132	2.26	0.024	1.20808	14.66222
ptl	1.592145	.7474264	0.99	0.322	.6344379	3.995544
ht	2.900166	3.193536	0.97	0.334	.3350554	25.10319
ui	1.229523	.9474768	0.27	0.789	.2715165	5.567715

```
. estimates store white
. logistic low age lwt smoke ptl ht ui if race==2, nolog
Logistic regression                               Number of obs   =         26
                                                  LR chi2(6)      =      10.12
                                                  Prob > chi2     =     0.1198
Log likelihood = -12.654157                       Pseudo R2       =     0.2856
```

low	Odds Ratio	Std. Err.	z	P>|z|	[95% Conf. Interval]	
age	.8735313	.1377809	-0.86	0.391	.6412385	1.189974
lwt	.9747736	.0166888	-1.49	0.136	.9426068	1.008038
smoke	16.50373	24.36988	1.90	0.058	.9134256	298.1884
ptl	4.866916	9.331296	0.83	0.409	.1135671	208.5715
ht	85.05606	214.6317	1.76	0.078	.6050219	11957.47
ui	67.61338	133.3291	2.14	0.033	1.417488	3225.12

```
. estimates store black

. logistic low age lwt smoke ptl ht ui if race==3, nolog
Logistic regression                            Number of obs   =          67
                                               LR chi2(6)      =       14.06
                                               Prob > chi2     =      0.0289
Log likelihood = -37.228444                    Pseudo R2       =      0.1589
```

low	Odds Ratio	Std. Err.	z	P>|z|	[95% Conf. Interval]	
age	.9263905	.0665385	-1.06	0.287	.8047408	1.066429
lwt	.9724499	.015762	-1.72	0.085	.9420424	1.003839
smoke	.7979034	.6340581	-0.28	0.776	.1680887	3.787582
ptl	2.845675	1.777942	1.67	0.094	.8363061	9.682898
ht	7.767504	10.00536	1.59	0.112	.6220776	96.98808
ui	2.925006	2.046473	1.53	0.125	.742311	11.52571

```
. estimates store other
```

We are now ready to perform the likelihood-ratio Chow test:

```
. lrtest (full) (white black other), stats
Likelihood-ratio test                          LR chi2(12) =          9.83
                                               Prob > chi2 =        0.6310

Assumption: (full) nested in (white, black, other)
```

Model	nobs	ll(null)	ll(model)	df	AIC	BIC
full	189	-117.336	-100.724	9	219.448	248.6237
white	96	-52.85752	-45.92706	7	105.8541	123.8046
black	26	-17.71291	-12.65416	7	39.30831	48.11499
other	67	-44.26039	-37.22844	7	88.45689	103.8897

We cannot reject the hypothesis that the logistic regression model applies to each of the races at any reasonable significance level. By specifying the option stats, we can verify the degrees of freedom of the test: $12 = 7 + 7 + 7 - 9$. We can obtain the same test by fitting an expanded model with interactions between all covariates and the variables race using the xi prefix.

(Continued on next page)

```
. xi: logistic low i.race*age i.race*lwt i.race*smoke i.race*ptl i.race*ht i.race*ui
i.race            _Irace_1-3        (naturally coded; _Irace_1 omitted)
i.race*age        _IracXage_#       (coded as above)
i.race*lwt        _IracXlwt_#       (coded as above)
i.race*smoke      _IracXsmoke_#     (coded as above)
i.race*ptl        _IracXptl_#       (coded as above)
i.race*ht         _IracXht_#        (coded as above)
i.race*ui         _IracXui_#        (coded as above)
note: _Irace_2 dropped due to collinearity
note: _Irace_3 dropped due to collinearity
  (output omitted )
note: _Irace_3 dropped due to collinearity
```

```
Logistic regression                       Number of obs   =        189
                                          LR chi2(20)     =      43.05
                                          Prob > chi2     =     0.0020
Log likelihood = -95.809661               Pseudo R2       =     0.1835
```

low	Odds Ratio	Std. Err.	z	P>|z|	[95% Conf. Interval]	
age	.9869674	.0527756	-0.25	0.806	.888765	1.09602
_IracXage_2	.885066	.1474075	-0.73	0.464	.6385697	1.226713
_IracXage_3	.9386232	.0840486	-0.71	0.479	.7875367	1.118695
_Irace_3	100.3769	309.5859	1.49	0.135	.2378648	42358.23
lwt	.9900874	.0106101	-0.93	0.353	.9695089	1.011103
_IracXlwt_2	.9845329	.0198857	-0.77	0.440	.9463191	1.02429
_IracXlwt_3	.9821859	.0190847	-0.93	0.355	.9454839	1.020313
smoke	4.208697	2.680129	2.26	0.024	1.208082	14.6622
_IracXsmok~2	3.921338	6.305976	0.85	0.395	.1677265	91.67841
_IracXsmok~3	.1895844	.19306	-1.63	0.102	.025763	1.395111
ptl	1.592145	.7474262	0.99	0.322	.6344381	3.995543
_IracXptl_2	3.05683	6.034072	0.57	0.571	.0638308	146.3903
_IracXptl_3	1.787322	1.396789	0.74	0.457	.3863583	8.268282
ht	2.900166	3.193535	0.97	0.334	.3350557	25.10318
_IracXht_2	29.328	80.74795	1.23	0.220	.1329515	6469.514
_IracXht_3	2.678297	4.538714	0.58	0.561	.0966918	74.18701
_Irace_2	99.62138	402.0819	1.14	0.254	.0365441	271573.6
ui	1.229523	.9474763	0.27	0.789	.2715167	5.567711
_IracXui_2	54.99156	116.4272	1.89	0.058	.8672537	3486.951
_IracXui_3	2.378977	2.476123	0.83	0.405	.3093352	18.29578

```
. lrtest full .
Likelihood-ratio test                     LR chi2(12) =        9.83
(Assumption: full nested in .)            Prob > chi2 =      0.6310
```

Applying `lrtest` for the full model against the model with all interactions yields exactly the same test statistic and *p*-value as for the full model against the composite model for the three regimes. In this case, the specification of the model with interactions was very convenient, and `logistic` did not have any problem computing the estimates for the expanded model. In models with more complicated likelihoods, such as Heckman's selection model (see [R] **heckman**) or complicated survival-time models (see [ST] **streg**), fitting the models with all interactions may be numerically demanding and may be much more time consuming than fitting a series of models separately for each regime.

Given the model with all interactions, we could also test the hypothesis of no differences between the regions (races) by a Wald version of the Chow test using the `testparm` command.

```
. testparm _IracX*

 ( 1)   _IracXage_2 = 0
 ( 2)   _IracXage_3 = 0
 ( 3)   _IracXlwt_2 = 0
 ( 4)   _IracXlwt_3 = 0
 ( 5)   _IracXsmoke_2 = 0
 ( 6)   _IracXsmoke_3 = 0
 ( 7)   _IracXptl_2 = 0
 ( 8)   _IracXptl_3 = 0
 ( 9)   _IracXht_2 = 0
 (10)   _IracXht_3 = 0
 (11)   _IracXui_2 = 0
 (12)   _IracXui_3 = 0

          chi2( 12) =     8.24
        Prob > chi2 =    0.7663
```

We conclude that, in this case, the Wald version of the Chow test is very similar to the likelihood-ratio version of the Chow test.

◁

Saved Results

lrtest saves in r():

Scalars

r(p)	level of significance	r(chi2)	LR-test statistic
r(df)	degrees of freedom		

Programmers wishing their estimation commands to be compatible with lrtest should note that lrtest requires that the following results be returned:

e(cmd)	name of estimation command
e(ll)	log-likelihood value
e(V)	the (co)variance matrix
e(N)	number of observations

lrtest also verifies that e(N), e(ll_0), and e(depvar) are consistent between two noncomposite models.

Methods and Formulas

lrtest is implemented as an ado-file.

Let L_0 and L_1 be the log-likelihood values associated with the full and constrained models, respectively. The test statistic of the likelihood-ratio test is $LR = -2(L_1 - L_0)$. If the constrained model is true, LR is approximately χ^2 distributed with $d_0 - d_1$ degrees of freedom, where d_0 and d_1 are the model degrees of freedom associated with the full and constrained models, respectively (Judge et al. 1985, 216–217).

lrtest determines the degrees of freedom of a model as the rank of e(V), computed as the number of nonzero diagonal elements of invsym(e(V)).

References

Gutierrez, R. G., S. L. Carter, and D. M. Drukker. 2001. On boundary-value likelihood-ratio tests. Stata Technical Bulletin 60: 15–18. Reprinted in *Stata Technical Bulletin Reprints*, vol. 10, pp. 269–273.

Hosmer, D. W., Jr., and S. Lemeshow. 2000. *Applied Logistic Regression.* 2nd ed. New York: Wiley.

Judge, G. G., W. E. Griffiths, R. C. Hill, H. Lütkepohl, and T.-C. Lee. 1985. *The Theory and Practice of Econometrics.* 2nd ed. New York: Wiley.

Kleinbaum, D. G. and M. Klein. 2002. *Logistic Regression: A Self-Learning Text.* 2nd ed. New York: Springer.

Pérez-Hoyos, S. and A. Tobias. 1999. sg111: A modified likelihood-ratio test command. *Stata Technical Bulletin* 49: 24–25. Reprinted in *Stata Technical Bulletin Reprints*, vol. 9, pp. 171–173.

Wang, Z. 2000. sg133: Sequential and drop one term likelihood-ratio tests. *Stata Technical Bulletin* 54: 46–47. Reprinted in *Stata Technical Bulletin Reprints*, vol. 9, pp. 332–334.

Also See

Related: [R] **linktest**, [R] **test**, [R] **testnl**

Title

> **lv** — Letter-value displays

Syntax

lv [*varlist*] [*if*] [*in*] [, generate <u>tail</u>(*#*)]

by may be used with lv; see [D] **by**.

Description

lv shows a letter-value display (Tukey 1977, 44–49; Hoaglin 1983) for each variable in *varlist*. If no variables are specified, letter-value displays are shown for each numeric variable in the data.

Options

Main

generate adds four new variables to the data: _mid, containing the midsummaries; _spread, containing the spreads; _psigma, containing the pseudosigmas; and _z2, containing the squared values from a standard normal distribution corresponding to the particular letter value. If the variables _mid, _spread, _psigma, and _z2 already exist, their contents are replaced. At most, only the first 11 observations of each variable are used; the remaining observations contain missing. If *varlist* specifies more than one variable, the newly created variables contain results for the last variable specified. The generate option may not be used with the by prefix.

tail(*#*) indicates the inverse of the tail density through which letter values are to be displayed. 2 corresponds to the median (meaning half in each tail), 4 to the fourths (roughly the 25th and 75th percentiles), 8 to the eighths, and so on. *#* may be specified as 4, 8, 16, 32, 64, 128, 256, 512, or 1,024 and defaults to a value of *#* that has corresponding depth just greater than 1. The default is taken as 1,024 if the calculation results in a number larger than 1,024. Given the intelligent default, this option is rarely specified.

Remarks

Letter-value displays are a collection of observations drawn systematically from the data, focusing especially on the tails rather than the middle of the distribution. The displays are called letter-value displays because letters have been (almost arbitrarily) assigned to tail densities:

Letter	Tail Area	Letter	Tail Area
M	1/2	B	1/64
F	1/4	A	1/128
E	1/8	Z	1/256
D	1/16	Y	1/512
C	1/32	X	1/1024

▷ Example 1

We have data on the mileage ratings of 74 automobiles. To obtain a letter-value display, we type

```
. use http://www.stata-press.com/data/r9/auto
(1978 Automobile Data)
. lv mpg
```

#	74		Mileage (mpg)			spread	pseudosigma
M	37.5		20				
F	19	18	21.5	25		7	5.216359
E	10	15	21.5	28		13	5.771728
D	5.5	14	22.25	30.5		16.5	5.576303
C	3	14	24.5	35		21	5.831039
B	2	12	23.5	35		23	5.732448
A	1.5	12	25	38		26	6.040635
	1	12	26.5	41		29	6.16562

					# below	# above
inner fence		7.5		35.5	0	1
outer fence		-3		46	0	0

The decimal points can be made to line up and thus the output made more readable by specifying a display format for the variable; see [U] **12.5 Formats: controlling how data are displayed**.

```
. format mpg %9.2f
. lv mpg
```

#	74		Mileage (mpg)			spread	pseudosigma
M	37.5		20.00				
F	19	18.00	21.50	25.00		7.00	5.22
E	10	15.00	21.50	28.00		13.00	5.77
D	5.5	14.00	22.25	30.50		16.50	5.58
C	3	14.00	24.50	35.00		21.00	5.83
B	2	12.00	23.50	35.00		23.00	5.73
A	1.5	12.00	25.00	38.00		26.00	6.04
	1	12.00	26.50	41.00		29.00	6.17

					# below	# above
inner fence		7.50		35.50	0	1
outer fence		-3.00		46.00	0	0

At the top, the number of observations is indicated as 74. The first line shows the statistics associated with M, the letter value that puts half the density in each tail, or the median. The median has *depth* 37.5 (that is, in the ordered data, M is 37.5 observations in from the extremes) and has value 20. The next line shows the statistics associated with F or the fourths. The fourths have depth 19 (that is, in the ordered data, the lower fourth is observation 19, and the upper fourth is observation $74 - 19 + 1$), and the values of the lower and upper fourths are 18 and 25. The number in the middle is the point halfway between the fourths—called a midsummary. If the distribution were perfectly symmetric, the midsummary would equal the median. The spread is the difference between the lower and upper summaries ($25 - 18 = 7$). For fourths, half of the data lies within a 7-mpg band. The pseudosigma is a calculation of the standard deviation using only the lower and upper summaries and assuming that the variable is normally distributed. If the data really were normally distributed, all the pseudosigmas would be roughly equal.

After the letter values, the line labeled with depth 1 reports the minimum and maximum values. In this case, the halfway point between the extremes is 26.5, which is greater than the median, indicating that 41 is more extreme than 12, at least relative to the median. Also note that, with each letter value, the midsummaries are increasing—our data are skewed. The pseudosigmas are also increasing, indicating that the data are spreading out relative to a normal distribution, although, given the evident skewness, this elongation may be an artifact of the skewness.

At the end is an attempt to identify outliers, although the points so identified are merely outside some predetermined cutoff. Points outside the inner fence are called *outside values* or *mild outliers*. Points outside the outer fence are called *severe outliers*. The inner fence is defined as $(3/2)$IQR and the outer fence as 3IQR above and below the F summaries, where the IQR is the spread of the fourths.

◁

❏ Technical Note

The form of the letter-value display has varied slightly with different authors. `lv` displays appear as described by Hoaglin (1983) but as modified by Emerson and Stoto (1983), where they included the midpoint of each of the spreads. This format was later adopted by Hoaglin (1985). If the distribution is symmetric, the midpoints will all be roughly equal. On the other hand, if the midpoints vary systematically, the distribution is skewed.

The pseudosigmas are obtained from the lower and upper summaries for each letter value. For each letter value, they are the standard deviation a normal distribution would have if its spread for the given letter value were to equal the observed spread. If the pseudosigmas are all roughly equal, the data are said to have *neutral elongation*. If the pseudosigmas increase systematically, the data are said to be more elongated than a normal, i.e., have thicker tails. If the pseudosigmas decrease systematically, the data are said to be less elongated than a normal, i.e., have thinner tails.

Interpretation of the number of mild and severe outliers is more problematic. The following discussion is drawn from Hamilton (1991):

Obviously, the presence of any such outliers does not rule out that the data have been drawn from a normal distribution; in large datasets, there will most certainly be observations outside $(3/2)$IQR and 3IQR. Severe outliers, however, comprise about two per million (.0002%) of a normal population. In samples, they lie far enough out to have substantial effects on means, standard deviations, and other classical statistics. The .0002%, however, should be interpreted carefully; outliers appear more often in small samples than one might expect from population proportions due to sampling variation in estimated quartiles. Monte Carlo simulation by Hoaglin, Iglewicz, and Tukey (1986) obtained these results on the percentages and numbers of outliers in random samples from a normal population:

| n | percentage | | number | |
	any outliers	severe	any outliers	severe
10	2.83	.362	.283	.0362
20	1.66	.074	.332	.0148
50	1.15	.011	.575	.0055
100	.95	.002	.95	.002
200	.79	.001	1.58	.002
300	.75	.001	2.25	.003
∞	.70	.0002	∞	∞

Thus the presence of any severe outliers in samples of less than 300 is sufficient to reject normality. Hoaglin, Iglewicz, and Tukey (1981) suggested the approximation $.00698 + .4/n$ for the fraction of mild outliers in a sample of size n or, equivalently, $.00698n + .4$ for the number of outliers.

❏

▷ Example 2

The generate option adds the variables _mid, _spread, _psigma, and _z2 to our data, making possible many of the diagnostic graphs suggested by Hoaglin (1985).

```
. lv mpg, generate
(output omitted)
. list _mid _spread _psigma _z2 in 1/12
```

	_mid	_spread	_psigma	_z2
1.	20	.	.	.
2.	21.5	7	5.216359	.4501955
3.	21.5	13	5.771728	1.26828
4.	22.25	16.5	5.576303	2.188846
5.	24.5	21	5.831039	3.24255
6.	23.5	23	5.732448	4.024532
7.	25	26	6.040635	4.631499
8.	.	.	.	.
9.	.	.	.	.
10.	.	.	.	.
11.	26.5	29	6.16562	5.53073
12.	.	.	.	.

Observations 12 through the end are missing for these new variables. The definition of the observations is always the same. The first observation contains the M summary, the second the F, the third the E, and so on. Observation 11 always contains the summary for depth 1. Observations 8 through 10—corresponding to letter values Z, Y, and X—contain missing because these statistics were not calculated. We have only 74 observations, and their depth would be 1.

Hoaglin (1985) suggests graphing the midsummary against z^2. If the distribution is not skewed, the points in the resulting graph will be along a horizontal line:

(*Continued on next page*)

```
. scatter _mid _z2
```

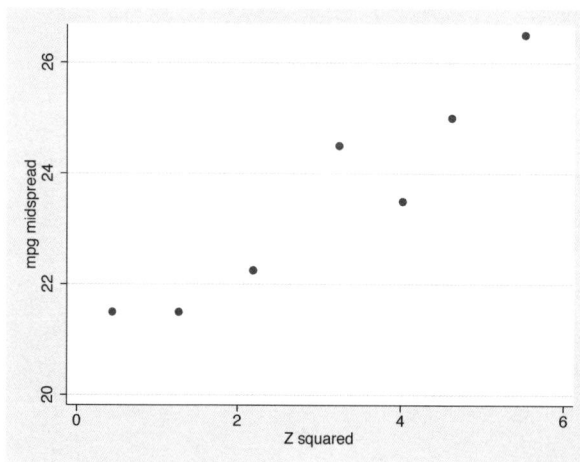

The graph clearly indicates the skewness of the distribution. We might also graph _psigma against _z2 to examine elongation.

◁

Saved Results

lv saves in r():

Scalars

r(N)	number of observations	r(u_C)	upper 32nd
r(min)	minimum	r(l_B)	lower 64th
r(max)	maximum	r(u_B)	upper 64th
r(median)	median	r(l_A)	lower 128th
r(l_F)	lower 4th	r(u_A)	upper 128th
r(u_F)	upper 4th	r(l_Z)	lower 256th
r(l_E)	lower 8th	r(u_Z)	upper 256th
r(u_E)	upper 8th	r(l_Y)	lower 512th
r(l_D)	lower 16th	r(u_Y)	upper 512th
r(u_D)	upper 16th	r(l_X)	lower 1024th
r(l_C)	lower 32nd	r(u_X)	upper 1024th

The lower/upper 8ths, 16ths, ..., 1024ths will be defined only if there are sufficient data.

Methods and Formulas

lv is implemented as an ado-file.

Let N be the number of (nonmissing) observations on x, and let $x_{(i)}$ refer to the ordered data when i is an integer. Define $x_{(i+.5)} = (x_{(i)} + x_{(i+1)})/2$; the median is defined as $x_{((N+1)/2)}$.

Define $x_{[d]}$ as the pair of numbers $x_{(d)}$ and $x_{(N+1-d)}$, where d is called the *depth*. Thus $x_{[1]}$ refers to the minimum and maximum of the data. Define $m = (N+1)/2$ as the depth of the median, $f = (\lfloor m \rfloor + 1)/2$ as the depth of the fourths, $e = (\lfloor f \rfloor + 1)/2$ as the depth of the eighths, and so on. Depths are reported on the far left of the letter-value display. The corresponding fourths of the data are $x_{[f]}$, the eighths $x_{[e]}$, and so on. These values are reported inside the display. The middle value is defined as the corresponding midpoint of $x_{[\cdot]}$. The spreads are defined as the difference in $x_{[\cdot]}$.

The corresponding point z_i on a standard normal distribution is obtained as (Hoaglin 1985, 456–457)

$$z_i = \begin{cases} F^{-1}\{(d_i - 1/3)/(N + 1/3)\} & \text{if } d_i > 1 \\ F^{-1}\{0.695/(N + 0.390)\} & \text{otherwise} \end{cases}$$

where d_i is the depth of the letter value. The corresponding pseudosigma is obtained as the ratio of the spread to $-2z_i$ (Hoaglin 1985, 431).

Define $(F_l, F_u) = x_{[f]}$. The inner fence has cutoffs $F_l - \frac{3}{2}(F_u - F_l)$ and $F_u + \frac{3}{2}(F_u - F_l)$. The outer fence has cutoffs $F_l - 3(F_u - F_l)$ and $F_u + 3(F_u - F_l)$.

The inner-fence values reported by `lv` are almost exactly equal to those used by `graph`, `box` to identify outside points. The only difference is that `graph` uses a slightly different definition of fourths, namely, the 25th and 75th percentiles as defined by `summarize`.

References

Emerson, J. D. and M. A. Stoto. 1983. Transforming data. In *Understanding Robust and Exploratory Data Analysis*, ed. D. C. Hoaglin, F. Mosteller, and J. W. Tukey, 97–128. New York: Wiley.

Fox, J. 1990. Describing univariate distributions. In *Modern Methods of Data Analysis*, ed. J. Fox and J. S. Long, 58–125. Newbury Park, CA: Sage.

Hamilton, L. C. 1991. sed4: Resistant normality check and outlier identification. *Stata Technical Bulletin* 3: 15–18. Reprinted in *Stata Technical Bulletin Reprints*, vol. 1, pp. 86–90.

Hoaglin, D. C. 1983. Letter values: A set of selected order statistics. In *Understanding Robust and Exploratory Data Analysis*, ed. D. C. Hoaglin, F. Mosteller, and J. W. Tukey, 33–57. New York: Wiley.

——. 1985. Using quantiles to study shape. In *Exploring Data Tables, Trends, and Shapes*, ed. D. C. Hoaglin, F. Mosteller, and J. W. Tukey, 417–460. New York: Wiley.

Hoaglin, D. C., B. Iglewicz, and J. W. Tukey. 1981. Small-sample performance of a resistant rule for outlier detection. *1980 Proceedings of the Statistical Computing Section*, 144–152. Washington, DC: American Statistical Association.

——. 1986. Performance of some resistant rules for outlier labeling. *Journal of the American Statistical Association* 81: 991–999.

Tukey, J. W. 1977. *Exploratory Data Analysis*. Reading, MA: Addison–Wesley.

Also See

Related: [R] **diagnostic plots**, [R] **stem**, [R] **summarize**

Title

matsize — Set the maximum number of variables in a model

Syntax

<u>set</u> <u>mat</u>size *#* [, <u>perm</u>anently]

where $10 \leq \# \leq 11000$ for Stata/SE and where $10 \leq \# \leq 800$ for Intercooled Stata.

Description

set matsize sets the maximum number of variables that can be included in any of Stata's estimation commands.

For Stata/SE, the default value is 400, but it may be changed upward or downward. The upper limit is 11,000.

For Intercooled Stata, the initial value is 200, but it may be changed upward or downward. The upper limit is 800.

This command may not be used with Small Stata; matsize is permanently frozen at 40.

Option

permanently specifies that, in addition to making the change right now, the matsize setting be remembered and become the default setting when you invoke Stata.

Remarks

set matsize controls the internal size of matrices that Stata uses. The default of 200 for Intercooled Stata, for instance, means that linear regression models are limited to 198 independent variables—198 because the constant uses one position and the dependent variable another, making a total of 200.

You may change matsize with data in memory, but increasing matsize increases the amount of memory consumed by Stata, increasing the probability of page faults and thus of making Stata run more slowly.

▷ Example 1

We wish to fit a model of y on the variables x1 through x200. Without thinking, we type

```
. regress y x1-x200
matsize too small
    You have attempted to create a matrix with more than 200 rows or columns
    or to fit a model with more than 200 variables plus ancillary parameters.
    You need to increase matsize using the set matsize command; see help
    matsize.
r(908);
```

We realize that we need to increase `matsize`, so we type

```
. set matsize 250
. regress y x1-x200
 (output omitted )
```

◁

Programmers should note that the current setting of `matsize` is stored as the c-class value `c(matsize)`; see [P] **creturn**.

Also See

Related: [D] **memory**, [P] **creturn**

Background: [U] **6 Setting the size of memory**

Title

> **maximize** — Details of iterative maximization

Syntax

Maximum likelihood optimization

  mle_cmd ... $\big[$, *options* $\big]$

Set default maximum iterations

  set maxiter # $\big[$, <u>permanently</u> $\big]$

options	description
$\big[$<u>no</u>$\big]$<u>log</u>	display an iteration log of the log likelihood; typically, the default
<u>trace</u>	display current parameter vector in iteration log
<u>gradient</u>	display current gradient vector in iteration log
<u>hessian</u>	display current negative Hessian matrix in iteration log
showstep	report steps within an iteration in iteration log
shownrtolerance	report the current value of $\mathbf{g}\mathbf{H}^{-1}\mathbf{g}'$ in iteration log
<u>techn</u>ique(*algorithm_spec*)	maximization technique
<u>iter</u>ate(#)	perform maximum of # iterations; default is iterate(16000)
<u>tol</u>erance(#)	tolerance for the coefficient vector; see *Options* for the defaults
<u>ltol</u>erance(#)	tolerance for the log likelihood; see *Options* for the defaults
<u>gtol</u>erance(#)	optional tolerance for the gradient relative to the coefficients
<u>nrtol</u>erance(#)	tolerance for the scaled gradient; see *Options* for the defaults
nonrtolerance	ignore the nrtolerance() option
<u>diff</u>icult	use a different stepping algorithm in nonconcave regions
from(*init_specs*)	initial values for the coefficients

  where *algorithm_spec* is

      algorithm $\big[$ # $\big[$ *algorithm* $\big[$ # $\big]$ $\big]$ ... $\big]$

  algorithm is $\big\{$ nr | bhhh | dfp | bfgs $\big\}$

  and *init_specs* is one of

    matname $\big[$, skip copy $\big]$

    $\big\{$ $\big[$*eqname*:$\big]$*name* = # | /*eqname* = # $\big\}$ $\big[$ ... $\big]$

    # $\big[$ # ... $\big]$, copy

Description

Stata has two maximum likelihood optimizers: One is used by internally coded commands, and the other is the `ml` command used by estimators implemented as ado-files. Both optimizers use the Newton–Raphson method with step halving (to avoid downhill steps) and special fixups when they encounter nonconcave regions of the likelihood. The two optimizers are similar but differ in the details of their implementation. For information about programming maximum likelihood estimators in ado-files, see [R] **ml** and *Maximum Likelihood Estimation with Stata*, 2nd edition (Gould, Pitblado, and Sribney 2003).

`set maxiter` specifies the default maximum number of iterations for estimation commands that iterate. The initial value is 16000, and # can be 0 to 16000. To change the maximum number of iterations performed by a particular estimation command, you need not reset `maxiter`; you can specify the `iterate(#)` option. When `iterate(#)` is not specified, the `maxiter` value is used.

Maximization options

`log` and `nolog` specify whether an iteration log showing the progress of the log likelihood is to be displayed. For most commands, the log is displayed by default, and `nolog` suppresses it. For a few commands (such as the `svy` maximum likelihood estimators), you must specify `log` to see the log.

`trace` adds to the iteration log a display of the current parameter vector.

`gradient` (`ml`-programmed estimators only) adds to the iteration log a display of the current gradient vector.

`hessian` (`ml`-programmed estimators only) adds to the iteration log a display of the current negative Hessian matrix.

`showstep` (`ml`-programmed estimators only) adds to the iteration log a report on the steps within an iteration. This option was added so that developers at StataCorp could view the stepping when they were improving the `ml` optimizer code. At this point, it mainly provides entertainment.

`shownrtolerance` (`ml`-programmed estimators only) adds to the iteration log the current value of $g\mathbf{H}^{-1}g'$, which is compared with the value of `nrtolerance()` to test for convergence. This value is only computed and reported when all other necessary stopping criteria have been met.

`technique`(*algorithm_spec*) (`ml`-programmed estimators only) specifies how the likelihood function is to be maximized. The following algorithms are currently implemented in `ml`. For details, see Gould, Pitblado, and Sribney (2003).

`technique(nr)` specifies Stata's modified Newton–Raphson (NR) algorithm.

`technique(bhhh)` specifies the Berndt–Hall–Hall–Hausman (BHHH) algorithm.

`technique(dfp)` specifies the Davidon–Fletcher–Powell (DFP) algorithm.

`technique(bfgs)` specifies the Broyden–Fletcher–Goldfarb–Shanno (BFGS) algorithm.

The default is `technique(nr)`.

You can switch between algorithms by specifying more than one in the `technique()` option. By default, `ml` will use an algorithm for five iterations before switching to the next algorithm. To specify a different number of iterations include the number after the technique in the option. For example, specifying `technique(bhhh 10 nr 1000)` requests that `ml` perform 10 iterations using the BHHH algorithm followed by 1000 iterations using the NR algorithm, and then switch back to BHHH for 10 iterations, and so on. The process continues until convergence or until the maximum number of iterations is reached.

iterate(#) specifies the maximum number of iterations. When the number of iterations equals iterate(), the optimizer stops and presents the current results. If convergence is declared before this threshold is reached, it will stop when convergence is declared. Specifying iterate(0) is useful for viewing results evaluated at the initial value of the coefficient vector. Specifying iterate(0) and from() together allows you to view results evaluated at a specified coefficient vector; note, however, that only a few commands allow the from() option. The default value of iterate(#) for both estimators programmed internally and estimators programmed with ml is the current value of set maxiter, which is iterate(16000) by default.

Below we describe the four different types of convergence tolerances employed by Stata estimators, and we describe the nonrtolerance option. After these descriptions, we explain how the various tolerances are used to determine whether the maximization algorithm has converged.

tolerance(#) specifies the tolerance for the coefficient vector. When the relative change in the coefficient vector from one iteration to the next is less than or equal to tolerance(), the tolerance() convergence criterion is satisfied.

  tolerance(1e-4) is the default for estimators programmed internally in Stata.

  tolerance(1e-6) is the default for estimators programmed with ml.

ltolerance(#) specifies the tolerance for the log likelihood. When the relative change in the log likelihood from one iteration to the next is less than or equal to ltolerance(), the ltolerance() convergence is satisfied.

  ltolerance(0) is the default for estimators programmed internally in Stata.

  ltolerance(1e-7) is the default for estimators programmed with ml.

gtolerance(#) (ml-programmed estimators only) specifies the tolerance for the gradient relative to the coefficients. When $|g_i \, b_i| \leq$ gtolerance() for all parameters b_i and the corresponding elements of the gradient g_i, the gradient tolerance criterion is met. By default, this criterion is not checked and so there is no default value for the gradient tolerance.

nrtolerance(#) (ml-programmed estimators only) specifies the tolerance for the scaled gradient. Convergence is declared when $\mathbf{g}\mathbf{H}^{-1}\mathbf{g}' <$ nrtolerance(). nrtolerance() differs from gtolerance() in that the gradient is scaled by $\mathbf{H}$. The default is nrtolerance(1e-5).

nonrtolerance (ml-programmed estimators only) specifies that the default nrtolerance criterion be turned off.

For internally-programmed Stata estimators, convergence is declared when either the tolerance() or ltolerance() criterion has first been met. No other criteria are checked.

For ml-programmed estimators, by default convergence is declared when the nrtolerance() criterion *and* either of the tolerance() or ltolerance() criteria have been met. If nonrtolerance is specified, then convergence is declared when either of the tolerance() or ltolerance() criteria has been met.

If gtolerance() is specified, then the gtolerance() criterion must be met *in addition* to any other required criteria in order for convergence to be declared.

---

difficult (ml-programmed estimators only) specifies that the likelihood function is likely to be difficult to maximize due to nonconcave regions. When the message "not concave" appears repeatedly, ml's standard stepping algorithm may not be working well. difficult specifies that a different stepping algorithm be used in nonconcave regions. There is no guarantee that difficult will work better than the default; sometimes it is better and sometimes it is worse. You should use the difficult option only when the default stepper declares convergence and the last iteration is "not concave", or when the default stepper is repeatedly issuing "not concave" messages and only producing tiny improvements in the log likelihood.

from() specifies initial values for the coefficients. Note that only a few estimators in Stata currently support this option. You can specify the initial values in one of three ways: by specifying the name of a vector containing the initial values (e.g., from(b0), where b0 is a properly labeled vector); by specifying coefficient names with the values (e.g., from(age=2.1 /sigma=7.4)); or by specifying a list of values (e.g., from(2.1 7.4, copy)). from() is intended for use when doing bootstraps (see [R] **bootstrap**) and in other special situations (e.g., with iterate(0)). Even when the values specified in from() are close to the values that maximize the likelihood, only a few iterations may be saved. Poor values in from() may lead to convergence problems.

skip specifies that any parameters found in the specified initialization vector that are not also found in the model be ignored. The default action is to issue an error message.

copy specifies that the list of values or the initialization vector be copied into the initial-value vector by position rather than by name.

Option for set maxiter

permanently specifies that, in addition to making the change right now, the maxiter setting be remembered and become the default setting when you invoke Stata.

Remarks

Only in rare circumstances would you ever need to specify any of these options, with the exception of nolog. The nolog option is useful for reducing the amount of output appearing in log files.

The following is an example of an iteration log:

```
Iteration 0:    log likelihood = -3791.0251
Iteration 1:    log likelihood =  -3761.738
Iteration 2:    log likelihood = -3758.0632   (not concave)
Iteration 3:    log likelihood = -3758.0447
Iteration 4:    log likelihood = -3757.5861
Iteration 5:    log likelihood =  -3757.474
Iteration 6:    log likelihood = -3757.4613
Iteration 7:    log likelihood = -3757.4606
Iteration 8:    log likelihood = -3757.4606
    (table of results omitted )
```

At iteration 8, the model converged. The only notable thing about this iteration log is the message "not concave" at the second iteration. This example was produced using the heckman command; its likelihood is not globally concave, so it is not surprising that this message sometimes appears. The other message that is occasionally seen is "backed up". Neither of these messages should be of any concern unless they appear at the final iteration.

If a "not concave" message appears at the last step, there are two possibilities. One is that the result is valid, but there is collinearity in the model that the command did not otherwise catch. Stata checks for obvious collinearity among the independent variables before performing the maximization, but strange collinearities or near collinearities can sometimes arise between coefficients and ancillary parameters. The second, more likely cause for a "not concave" message at the final step is that the optimizer entered a very flat region of the likelihood and prematurely declared convergence.

If a "backed up" message appears at the last step, there are also two possibilities. One is that Stata found a perfect maximum and could not step to a better point; if this is the case, all is fine, but this is a highly unlikely occurrence. The second is that the optimizer worked itself into a bad concave spot where the computed gradient and Hessian gave a bad direction for stepping.

If either of these messages appears at the last step, perform the maximization again with the gradient option. If the gradient goes to zero, the optimizer has found a maximum that may not be unique but is a maximum. From the standpoint of maximum likelihood estimation, this is a valid result. If the gradient is not zero, it is not a valid result, and you should try tightening up the convergence criterion, or try ltol(0) tol(1e-7) or gtol(0.1) (with the default ltol() tol()) to see if the optimizer can work its way out of the bad region.

If you get repeated "not concave" steps with little progress being made at each step, try specifying the difficult option. Sometimes difficult works wonderfully, reducing the number of iterations and producing convergence at a good (i.e., concave) point. Other times, difficult works poorly, taking much longer to converge than the default stepper.

(*Continued on next page*)

Saved Results

Maximum likelihood estimators save in e():

Scalars

e(N)	number of observations	always saved
e(k)	number of parameters	always saved
e(df_m)	model degrees of freedom	always saved
e(k_eq)	number of equations	usually saved
e(k_dv)	number of dependent variables	usually saved
e(ic)	number of iterations	usually saved
e(converged)	1 if converged, 0 otherwise	usually saved
e(rc)	return code	usually saved
e(rank)	rank of e(V)	always saved
e(rank0)	rank of e(V) for constant-only model	saved when constant-only model is fitted
e(N_clust)	number of clusters	saved when cluster is specified; see [U] **20.14 Obtaining robust variance estimates**
e(ll)	log likelihood	always saved
e(ll_0)	log likelihood, constant-only model	saved when constant-only model is fitted
e(chi2)	χ^2	usually saved
e(r2_p)	pseudo-R-squared	sometimes saved

Macros

e(depvar)	names of dependent variables	always saved
e(cmd)	name of command	always saved
e(user)	name of likelihood-evaluator program	always saved
e(opt)	type of optimization	always saved
e(crittype)	optimization criterion	always saved
e(properties)	estimator properties	always saved
e(predict)	program used to implement predict	usually saved
e(wtype)	weight type	saved when weights are specified or implied
e(wexp)	weight expression	saved when weights are specified or implied
e(clustvar)	name of cluster variable	saved when cluster is specified; see [U] **20.14 Obtaining robust variance estimates**
e(chi2type)	Wald or LR; type of model χ^2 test	usually saved
e(technique)	from technique() option	sometimes saved
e(vce)	*vcetype* specified in vce()	sometimes saved
e(vcetype)	title used to label Std. Err.	sometimes saved

Matrices

e(b)	coefficient vector	always saved
e(V)	variance–covariance matrix of the estimators	always saved
e(gradient)	gradient vector	usually saved
e(ilog)	iteration log (up to 20 iterations)	usually saved

Functions

e(sample)	marks estimation sample	always saved

In addition, ml saves the constraint matrix in matrix Cns, which can be obtained by typing matrix *name* = get(Cns).

See *Saved Results* in the manual entry for any maximum likelihood estimator for a complete list of returned results.

Methods and Formulas

Let L_1 be the log likelihood of the full model (i.e., the log-likelihood value shown on the output), and let L_0 be the log likelihood of the "constant-only" model. The likelihood-ratio χ^2 model test is defined as $2(L_1 - L_0)$. The pseudo-R^2 (Judge et al. 1985) is defined as $1 - L_1/L_0$. This is simply the log likelihood on a scale where 0 corresponds to the "constant-only" model and 1 corresponds to perfect prediction for a discrete model (in which case the overall log likelihood is 0).

Some maximum likelihood routines can report coefficients in an exponentiated form, e.g., odds ratios in `logistic`. Let b be the unexponentiated coefficient, s its standard error, and b_0 and b_1 the reported confidence interval for b. In exponentiated form, the point estimate is e^b, the standard error $e^b s$, and the confidence interval e^{b_0} and e^{b_1}. The displayed Z (or t) statistics and p-values are the same as those for the unexponentiated results. This is justified since $e^b = 1$ and $b = 0$ are equivalent hypotheses, and normality is more likely to hold in the b metric.

References

Gould, W., J. Pitblado, and W. Sribney. 2003. *Maximum Likelihood Estimation with Stata.* 2nd ed. College Station, TX: Stata Press.

Judge, G. G., W. E. Griffiths, R. C. Hill, H. Lütkepohl, and T.-C. Lee. 1985. *The Theory and Practice of Econometrics.* 2nd ed. New York: Wiley.

Also See

Complementary:	[R] **lrtest**, [R] **ml**, [R] **test**,
	[SVY] **ml for svy**
Background:	[U] **20 Estimation and postestimation commands**

Title

mean — Estimate means

Syntax

mean *varlist* [*if*] [*in*] [*weight*] [, *options*]

options	description
Model	
<u>std</u>ize(*varname*)	variable identifying strata for standardization
<u>stdw</u>eight(*varname*)	weight variable for standardization
<u>nostd</u>rescale	do not rescale the standard weight variable
if/in/over	
over(*varlist*[, <u>nol</u>abel])	group over subpopulations defined by *varlist*; optionally, suppress group labels
SE/Cluster	
vce(*vcetype*)	*vcetype* may be <u>boot</u>strap or <u>jack</u>knife
<u>cl</u>uster(*varname*)	adjust standard errors for intragroup correlation
Reporting	
<u>l</u>evel(*#*)	set confidence level; default is level(95)
<u>noh</u>eader	suppress the table header
<u>nol</u>egend	suppress the table legend

svy may be used with mean; see [SVY] **svy: mean**.
fweights, aweights, iweights, and pweights are allowed; see [U] **11.1.6 weight**.
See [U] **20 Estimation and postestimation commands** for additional capabilities of estimation commands.

Description

mean produces estimates of means, along with standard errors.

Options

◜ Model ◝

stdize(*varname*) specifies that the point estimates be adjusted by direct standardization across the strata identified by *varname*. This option requires the stdweight() option.

stdweight(*varname*) specifies the weight variable associated with the strata identified in the stdize() option. The standardization weights must be constant within the strata identified in the stdize() option.

nostdrescale prevents the standardization weights from being rescaled within the over() groups. This option requires stdize() but is ignored if the over() option is not specified.

| if/in/over |

over(*varlist* [, <u>nolabel</u>]) specifies that estimates be computed for multiple subpopulations, which are identified by the different values of the variables in *varlist*.

When this option is supplied with a single variable name, such as over(*varname*), the value labels of *varname* are used to identify the subpopulations. If *varname* does not have labeled values (or there are unlabeled values), the values themselves are used, provided that they are non-negative integers. Noninteger values, negative values, and labels that are not valid Stata names are substituted with a default identifier.

When over() is supplied with multiple variable names, each subpopulation is assigned a unique default identifier.

nolabel requests that value labels attached to the variables identifying the subpopulations be ignored.

| SE/Cluster |

vce(*vcetype*); see [R] **vce_option**.

cluster(*varname*); see [R] **estimation options**.

| Reporting |

level(*#*); see [R] **estimation options**.

noheader prevents the table header from being displayed. This option implies nolegend.

nolegend prevents the table legend identifying the subpopulations from being displayed.

Remarks

▷ Example 1

Using the fuel data from [R] **ttest**, we estimate the average mileage of the cars without the fuel treatment (mpg1) and those with the fuel treatment (mpg2).

```
. use http://www.stata-press.com/data/r9/fuel
. mean mpg1 mpg2
Mean estimation                      Number of obs   =        12

             |       Mean    Std. Err.     [95% Conf. Interval]
-------------+------------------------------------------------
        mpg1 |         21    .7881701      19.26525    22.73475
        mpg2 |      22.75    .9384465      20.68449    24.81551
```

Using these results, we can test the equality of the mileage between the two groups of cars.

```
. test mpg1 = mpg2
 ( 1)  mpg1 - mpg2 = 0
        F(  1,    11) =     5.04
             Prob > F =    0.0463
```

◁

▷ Example 2

In example 1, the joint observations of mpg1 and mpg2 were used to estimate a covariance between their means.

```
. matrix list e(V)
symmetric e(V)[2,2]
            mpg1        mpg2
mpg1    .62121212
mpg2     .4469697    .88068182
```

If the data were organized this way out of convenience but the two variables represent independent samples of cars (coincidentally of the same sample size), we should reshape the data and use the over() option to ensure that the covariance between the means is zero.

```
. use http://www.stata-press.com/data/r9/fuel
. stack mpg1 mpg2, into(mpg) clear
. mean mpg, over(_stack)
Mean estimation                    Number of obs    =      24
            1: _stack = 1
            2: _stack = 2
```

	Over	Mean	Std. Err.	[95% Conf. Interval]
mpg					
	1	21	.7881701	19.36955	22.63045
	2	22.75	.9384465	20.80868	24.69132

```
. matrix list e(V)
symmetric e(V)[2,2]
              mpg:        mpg:
                1           2
mpg:1    .62121212
mpg:2            0    .88068182
```

Now we can test the equality of the mileage between the two independent groups of cars.

```
. test [mpg]1 = [mpg]2
 ( 1)  [mpg]1 - [mpg]2 = 0
       F(  1,    23) =    2.04
            Prob > F =    0.1667
```

◁

▷ Example 3: standardized means

Suppose that we collected the blood-pressure data from [R] **dstdize**, and we wish to obtain standardized high blood-pressure rates for each city in the years 1990 and 1992, using, as the standard, the age, sex, and race distribution of the four cities and two years combined. Our rate is really the mean of a variable that indicates whether a sampled individual has high blood pressure. First we generate the strata and weight variables from our standard distribution, then use mean to compute the rates.

```
. use http://www.stata-press.com/data/r9/hbp, clear
. egen strata = group(age race sex) if inlist(year, 1990, 1992)
(675 missing values generated)
```

```
. by strata, sort: gen stdw = _N

. mean hbp, over(city year) stdize(strata) stdweight(stdw)

Mean estimation

N. of std strata =        24          Number of obs      =      455

              Over: city year
        _subpop_1: 1 1990
        _subpop_2: 1 1992
        _subpop_3: 2 1990
        _subpop_4: 2 1992
        _subpop_5: 3 1990
        _subpop_6: 3 1992
        _subpop_7: 5 1990
        _subpop_8: 5 1992
```

Over	Mean	Std. Err.	[95% Conf. Interval]	
hbp				
_subpop_1	.058642	.0296273	.0004182	.1168657
_subpop_2	.0117647	.0113187	-.0104789	.0340083
_subpop_3	.0488722	.0238958	.0019121	.0958322
_subpop_4	.014574	.007342	.0001455	.0290025
_subpop_5	.1011211	.0268566	.0483425	.1538998
_subpop_6	.0810577	.0227021	.0364435	.1256719
_subpop_7	.0277778	.0155121	-.0027066	.0582622
_subpop_8	.0548926	.	.	.

Note that the standard error of the high blood-pressure-rate estimate is missing for city 5 in 1992. This is because there was only one individual with high blood pressure and that individual was the only person observed in the stratum of white males 30-35 years old.

By default, mean rescales the standard weights within the over() groups. In the following, we use the nostdrescale option to prevent this, thus reproducing the results in [R] **dstdize**.

```
. mean hbp, over(city year) nolegend stdize(strata) stdweight(stdw)
> nostdrescale

Mean estimation

N. of std strata =        24          Number of obs      =      455
```

Over	Mean	Std. Err.	[95% Conf. Interval]	
hbp				
_subpop_1	.0417582	.0210973	.0002978	.0832187
_subpop_2	.0087912	.0084579	-.0078304	.0254128
_subpop_3	.044898	.0219526	.0017566	.0880393
_subpop_4	.0142857	.0071968	.0001426	.0284288
_subpop_5	.0884532	.0234921	.0422864	.1346201
_subpop_6	.0463187	.0129726	.0208249	.0718125
_subpop_7	.0223443	.0124779	-.0021772	.0468658
_subpop_8	.0505495	.	.	.

◁

Saved Results

mean saves in e():

Scalars

e(N)	number of observations	e(df_r)	sample degrees of freedom
e(N_over)	number of subpopulations	e(N_clust)	number of clusters
e(N_stdize)	number of standard strata		

Macros

e(cmd)	mean	e(over)	*varlist* from over()
e(varlist)	*varlist*	e(over_labels)	labels from over() variables
e(stdize)	*varname* from stdize()	e(over_namelist)	names from e(over_labels)
e(stdweight)	*varname* from stdweight()	e(vce)	*vcetype* specified in vce()
e(wtype)	weight type	e(vcetype)	title used to label Std. Err.
e(wexp)	weight expression	e(estat_cmd)	program used to implement estat
e(title)	title in estimation output	e(properties)	b V
e(cluster)	name of cluster variable		

Matrices

e(b)	vector of mean estimates
e(V)	(co)variance estimates
e(_N)	vector of numbers of nonmissing observations
e(_N_stdsum)	number of nonmissing observations within the standard strata
e(_p_stdize)	standardizing proportions

Functions

e(sample)	marks estimation sample

Methods and Formulas

mean is implemented as an ado-file.

Let y be the variable on which we want to calculate the mean and y_j an individual observation on y, where $j = 1, \ldots, n$ and n is the sample size. Let w_j be the weight, and if no weight is specified, define $w_j = 1$ for all j. In the case of aweights and pweights, the w_j are normalized to sum to n.

Let W be the sum of the weights

$$W = \sum_{j=1}^{n} w_j$$

The mean is defined as

$$\overline{y} = \frac{1}{W} \sum_{j=1}^{n} w_j y_j$$

The default variance estimator for the mean is

$$\widehat{V}(\overline{y}) = \frac{1}{W(W-1)} \sum_{j=1}^{n} w_j (y_j - \overline{y})^2$$

The standard error of the mean is the square root of the variance.

If x, x_j, and $\overline{x}$ are similarly defined for another variable (observed jointly with y), the covariance estimator between $\overline{x}$ and $\overline{y}$ is

$$\widehat{\mathrm{Cov}}(\overline{x}, \overline{y}) = \frac{1}{W(W-1)} \sum_{j=1}^{n} w_j (x_j - \overline{x})(y_j - \overline{y})$$

See [SVY] **direct standardization** for details regarding standardized means.

References

Cochran, W. G. 1977. *Sampling Techniques*. 3rd ed. New York: Wiley.

Stuart, A. and J. K. Ord. 1994. *Kendall's Advanced Theory of Statistics, Vol. I.* 6th ed. London: Arnold.

Also See

Complementary:	[R] **mean postestimation**
Related:	[SVY] **svy: mean**;
	[R] **ameans**, [R] **proportion**, [R] **ratio**, [R] **summarize**, [R] **total**
Background:	[U] **20 Estimation and postestimation commands**,
	[R] **estimation options**, [R] *vce_option*

Title

> **mean postestimation** — Postestimation tools for mean

Description

The following postestimation commands are available for `mean`:

command	description
estat	VCE
estimates	cataloging estimation results
lincom	point estimates, standard errors, testing, and inference for linear combinations of coefficients
nlcom	point estimates, standard errors, testing, and inference for nonlinear combinations of coefficients
test	Wald tests for simple and composite linear hypotheses
testnl	Wald tests of nonlinear hypotheses

See the corresponding entries in the *Stata Base Reference Manual* for details.

Methods and Formulas

All postestimation commands listed above are implemented as ado-files.

Also See

Complementary: [R] **mean**; [R] **estimates**, [R] **lincom**, [R] **nlcom**, [R] **test**, [R] **testnl**

Related: [SVY] **svy: mean postestimation**

Background: [U] **13.5 Accessing coefficients and standard errors**,

[U] **13.6 Accessing results from Stata commands**,

[R] **estat**

Title

meta — Meta-analysis

Remarks

Stata does not have a meta-analysis command. Stata users, however, have developed an excellent suite of commands for performing meta-analysis, many of which have been published in the *Stata Journal* (SJ) or the *Stata Technical Bulletin* (STB).

For information about meta-analysis, in addition to the articles that have appeared in the *Stata Journal* and the *Stata Technical Bulletin*, see Abramson and Abramson (2001, chapter F), Dohoo, Martin, and Stryhn (2003, chapter 24), Egger, Smith, and Altman (2001), and Sutton et al. (2000).

(Continued on next page)

Issue	insert	author(s)	command	description
STB-38	sbe16	S. Sharp, J. Sterne	`meta`	meta-analysis for an outcome of two exposures or two treatment regimens
STB-42	sbe16.1	S. Sharp, J. Sterne	`meta`	update of sbe16
STB-43	sbe16.2	S. Sharp, J. Sterne	`meta`	update; *install this version*
STB-41	sbe19	T. J. Steichen	`metabias`	performs the Begg and Mazumdar (1994) adjusted rank correlation test for publication bias and the Egger et al. (1997) regression asymmetry test for publication bias
STB-44	sbe19.1	T. J. Steichen, M. Egger, J. Sterne	`metabias`	update of sbe19
STB-57	sbe19.2	T. J. Steichen	`metabias`	update of sbe19
STB-58	sbe19.3	T. J. Steichen	`metabias`	update of sbe19
STB-61	sbe19.4	T. J. Steichen	`metabias`	update of sbe19
SJ-3-4	sbe19.4	T. J. Steichen	`metabias`	update; *install this version*
STB-41	sbe20	A. Tobias	`galbr`	performs the Galbraith plot (1988), which is useful for investigating heterogeneity in meta-analysis
STB-56	sbe20.1	A. Tobias	`galbr`	update; *install this version*
STB-42	sbe22	J. Sterne	`metacum`	performs cumulative meta-analysis, using fixed- or random-effects models, and graphs the result
STB-42	sbe23	S. Sharp	`metareg`	extends a random-effects meta-analysis to estimate the extent to which one or more covariates, with values defined for each study in the analysis, explains heterogeneity in the treatment effects
STB-44	sbe24	M. J. Bradburn, J. J. Deeks, D. G. Altman	`metan,` `funnel,` `labbe`	meta-analysis of studies with two groups funnel plot of precision versus treatment effect L'Abbé plot
STB-45	sbe24.1	M. J. Bradburn, J. J. Deeks, D. G. Altman	`funnel`	update; *install this version*
STB-47	sbe26	A. Tobias	`metainf,` `meta`	graphical technique to look for influential studies in the meta-analysis estimate
STB-56	sbe26.1	A. Tobias	`metainf`	update; *install this version*
STB-49	sbe28	A. Tobias	`metap`	combines p-values using either Fisher's method or Edgington's method
STB-56	sbe28.1	A. Tobias	`metap`	update; *install this version*
STB-57	sbe39	T. J. Steichen	`metatrim`	performs the Duval and Tweedie (2000) nonparametric "trim and fill" method of accounting for publication bias in meta-analysis
STB-58	sbe39.1	T. J. Steichen	`metatrim`	update of sbe39
STB-61	sbe39.2	T. J. Steichen	`metatrim`	update; *install this version*
SJ-4-2	st0061	J. Sterne, R. Harbord	`metafunnel`	funnel plots
SJ-4-2	pr0012	T. J. Steichen		submenu and dialogs for meta-analysis commands

Additional commands may be available; enter Stata and type `search meta analysis`.

To download and install from the Internet the Sharp and Stern `meta` command, for instance, in Stata you could

1. Select **Help > SJ and User-written Programs**.

2. Click on *STB*.

3. Click on *stb43*.

4. Click on *sbe16_2*.

5. Click on *click here to install*.

or you could instead do the following:

1. Navigate to the appropriate STB issue:
    a. Type `net from http://www.stata.com`
       Type `net cd stb`
       Type `net cd stb43`
    or
    b. Type `net from http://www.stata.com/stb/stb43`

2. Type `net describe sbe16_2`

3. Type `net install sbe16_2`

References

Abramson, J. H. and Z. H. Abramson. 2001. *Making Sense of Data: A Self-Instruction Manual on the Interpretation of Epidemiological Data.* 3rd ed. New York: Oxford University Press.

Begg, C. B. and M. Mazumdar. 1994. Operating characteristics of a rank correlation test for publication bias. *Biometrics* 50: 1088–1101.

Bradburn, M. J., J. J. Deeks, and D. G. Altman. 1998a. sbe24: metan—an alternative meta-analysis command. *Stata Technical Bulletin* 44: 4–15. Reprinted in *Stata Technical Bulletin Reprints*, vol. 8, pp. 86–100.

——. 1998b. sbe24.1: Correction to funnel plot. *Stata Technical Bulletin* 45: 21. Reprinted in *Stata Technical Bulletin Reprints*, vol. 8, pp. 100.

Dohoo, I., W. Martin, and H. Stryhn. 2003. *Veterinary Epidemiologic Research.* Charlottetown, Prince Edward Island: AVC.

Egger, M., G. D. Smith, and D. G. Altman. 2001. *Systematic Reviews in Health Care: Meta-analysis in Context.* London: BMJ.

Egger, M., G. D. Smith, M. Schneider, and C. Minder. 1997. Bias in meta-analysis detected by a simple, graphical test. *British Medical Journal* 315: 629–634.

Galbraith, R. F. 1988. A note on graphical display of estimated odds ratios from several clinical trials. *Statistics in Medicine* 7: 889–894.

L'Abbé, K. A., A. S. Detsky, and K. O'Rourke. 1987. Meta-analysis in clinical research. *Annals of Internal Medicine* 107: 224–233.

Sharp, S. 1998. sbe23: Meta-analysis regression. *Stata Technical Bulletin* 42: 16–22. Reprinted in *Stata Technical Bulletin Reprints*, vol. 7, pp. 148–155.

Sharp, S. and J. Sterne. 1997. sbe16: Meta-analysis. *Stata Technical Bulletin* 38: 9–14. Reprinted in *Stata Technical Bulletin Reprints*, vol. 7, pp. 100–106.

——. 1998a. sbe16.1: New syntax and output for the meta-analysis command. *Stata Technical Bulletin* 42: 6–8. Reprinted in *Stata Technical Bulletin Reprints*, vol. 7, pp. 106–108.

——. 1998b. sbe16.2: Corrections to the meta-analysis command. *Stata Technical Bulletin* 43: 15. Reprinted in *Stata Technical Bulletin Reprints*, vol. 8, p. 84.

Steichen, T. J. 1998. sbe19: Tests for publication bias in meta-analysis. *Stata Technical Bulletin* 41: 9–15. Reprinted in *Stata Technical Bulletin Reprints*, vol. 7, pp. 125–133.

——. 2000a. sbe19.2: Update of tests for publication bias in meta-analysis. *Stata Technical Bulletin* 57: 4. Reprinted in *Stata Technical Bulletin Reprints*, vol. 10, p. 70.

——. 2000b. sbe39: Nonparametric trim and fill analysis of publication bias in meta-analysis. *Stata Technical Bulletin* 57: 8–14. Reprinted in *Stata Technical Bulletin Reprints*, vol. 10, pp. 108–117.

——. 2000c. sbe19.3: Tests for publication bias in meta-analysis: Erratum. *Stata Technical Bulletin* 58: 8. Reprinted in *Stata Technical Bulletin Reprints*, vol. 10, p. 71.

——. 2000d. sbe39.1: Nonparametric trim and fill analysis of publication bias in meta-analysis: Erratum. *Stata Technical Bulletin* 58: 8–9. Reprinted in *Stata Technical Bulletin Reprints*, vol. 10, pp. 117–118.

——. 2001a. sbe19.4: Update to metabias to work under version 7. *Stata Technical Bulletin* 61: 11. Reprinted in *Stata Technical Bulletin Reprints*, vol. 10, pp. 71–72.

——. 2001b. sbe39.2: Update of metatrim to work under version 7. *Stata Technical Bulletin* 61: 11. Reprinted in *Stata Technical Bulletin Reprints*, vol. 10, p. 118.

——. 2004. Submenu and dialogs for meta-analysis commands. *Stata Journal* 4: 124–126.

Steichen, T. J., M. Egger, and J. Sterne. 1998. sbe19.1: Tests for publication bias in meta-analysis. *Stata Technical Bulletin* 44: 3–4. Reprinted in *Stata Technical Bulletin Reprints*, vol. 8, pp. 84–85.

Sterne, J. 1998. sbe22: Cumulative meta-analysis. *Stata Technical Bulletin* 42: 13–16. Reprinted in *Stata Technical Bulletin Reprints*, vol. 7, pp. 143–147.

Sterne, J. A. C. and R. M. Harbord. 2004. Funnel plots in meta-analysis. *Stata Journal* 4: 127–141.

Sutton, A. J., K. R. Abrams, D. R. Jones, T. A. Sheldon, and F. Song. 2000. *Methods for Meta-Analysis in Medical Research*. New York: Wiley.

Tobias, A. 1998. sbe20: Assessing heterogeneity in meta-analysis: The Galbraith plot. *Stata Technical Bulletin* 41: 15–17. Reprinted in *Stata Technical Bulletin Reprints*, vol. 7, pp. 133–136.

——. 1999a. sbe26: Assessing the influence of a single study in the meta-analysis estimate. *Stata Technical Bulletin* 47: 15–17. Reprinted in *Stata Technical Bulletin Reprints*, vol. 8, p. 108–110.

——. 1999b. sbe28: Meta-analysis of p-values. *Stata Technical Bulletin* 49: 15–17. Reprinted in *Stata Technical Bulletin Reprints*, vol. 9, pp. 138–140.

——. 2000a. sbe20.1: Update of galbr. *Stata Technical Bulletin* 56: 14. Reprinted in *Stata Technical Bulletin Reprints*, vol. 10, p. 72.

——. 2000b. sbe26.1: Update of metainf. *Stata Technical Bulletin* 56: 15. Reprinted in *Stata Technical Bulletin Reprints*, vol. 10, p. 72.

——. 2000c. sbe28.1: Update of metap. *Stata Technical Bulletin* 56: 15. Reprinted in *Stata Technical Bulletin Reprints*, vol. 10, p. 73.

Title

> **mfp** — Multivariable fractional polynomial models

Syntax

mfp *regression_cmd yvar xvarlist* $\big[$ *if* $\big]$ $\big[$ *in* $\big]$ $\big[$ *weight* $\big]$ $\big[$, *options* $\big]$

options	description
Model 2	
<u>seq</u>uential	use the Royston and Altman model-selection algorithm; default uses closed-test algorithm
cycles(*#*)	maximum number of iteration cycles; default is cycles(5)
<u>df</u>default(*#*)	default maximum degrees of freedom; default is dfdefault(4)
adjust(*adj_list*)	adjustment for each predictor
alpha(*alpha_list*)	*p*-values for testing between FP models; default is alpha(0.05)
df(*df_list*)	degrees of freedom for each predictor
<u>pow</u>ers(*numlist*)	list of fractional polynomial powers to use; default is powers(-2 -1(.5)1 2 3)
Adv. model	
<u>x</u>order(+ | − | n)	order of entry into model-selection algorithm; default is xorder(+)
<u>sel</u>ect(*select_list*)	nominal *p*-values for selection on each predictor
xpowers(*xp_list*)	fractional polynomial powers for each predictor
<u>zer</u>o(*varlist*)	treat nonpositive values of specified predictors as zero when FP transformed
<u>catz</u>ero(*varlist*)	add indicator variable for specified predictors
regression_cmd_options	other options accepted by chosen regression commands

All weight types supported by *regression_cmd* are allowed; see [U] **11.1.6 weight**.

See [U] **20 Estimation and postestimation commands** for additional capabilities of estimation commands.

fracgen may be used to create new variables containing fractional polynomial powers. See [R] **fracpoly**.

where

> *regression_cmd* may be clogit, glm, logistic, logit, poisson, probit, qreg, regress, stcox, streg, or xtgee.
>
> *yvar* is not allowed for streg and stcox. For these commands, you must first stset your data.
>
> *xvarlist* has elements of type *varlist* and/or (*varlist*), e.g., x1 x2 (x3 x4 x5)
>
> Elements enclosed in parentheses are tested jointly for inclusion in the model and are not eligible for fractional polynomial transformation.

Description

mfp selects the fractional polynomial (FP) model that best predicts the outcome variable from the RHS variables in *xvarlist*.

Options

Model 2

sequential chooses the sequential FP selection algorithm (see *Methods of FP model selection*).

cycles(#) sets the maximum number of iteration cycles permitted. cycles(5) is the default.

dfdefault(#) determines the default maximum degrees of freedom (df) for a predictor. The default is dfdefault(4) (second degree FP).

adjust(*adj_list*) defines the adjustment for the covariates *xvar1*, *xvar2*, ... of *xvarlist*. The default is adjust(mean), except for binary covariates, where it is adjust(#), with # being the lower of the two distinct values of the covariate. A typical item in *adj_list* is *varlist*:{mean | # | no}. Items are separated by commas. The first item is special in that *varlist* is optional, and, if it is omitted, the default is reset to the specified value (mean, #, or no). For example, adjust(no, age:mean) sets the default to no (i.e., no adjustment) and the adjustment for age to mean.

alpha(*alpha_list*) sets the significance levels for testing between FP models of different degree. The rules for *alpha_list* are the same as those for *df_list* in the df() option (see below). The default nominal *p*-value (significance level, selection level) is 0.05 for all variables.

Example: alpha(0.01) specifies that all variables have an FP selection level of 1 percent.

Example: alpha(0.05, weight:0.1) specifies that all variables except weight have FP selection level 5 percent; weight has level 10 percent.

df(*df_list*) sets the degrees of freedom (df) for each predictor. The df (not counting the regression constant, _cons) are twice the degree of the FP, so, for example, an *xvar* fitted as a second-degree FP (FP2) has 4 df. The first item in *df_list* may be either # or *varlist*:#. Subsequent items must be *varlist*:#. Items are separated by commas, and *varlist* is specified in the usual way for variables. With the first type of item, the df for all predictors are taken to be #. With the second type of item, all members of *varlist* (which must be a subset of *xvarlist*) have # df.

The default number of degrees of freedom for a predictor of type *varlist* specified in *xvarlist* but not in *df_list* is assigned according to the number of distinct (unique) values of the predictor, as follows:

# of distinct values	default df
1	(invalid predictor)
2–3	1
4–5	min(2, dfdefault())
≥ 6	dfdefault()

Example: df(4)
All variables have 4 df.

Example: df(2, weight displ:4)
weight and displ have 4 df; all other variables have 2 df.

Example: df(weight displ:4, mpg:2)
weight and displ have 4 df, mpg has 2 df; all other variables have default df.

powers(*numlist*) is the set of fractional polynomial powers to be used. The default set is $-2,-1,-0.5,0,0.5,1,2,3$ (0 means log).

xorder(+ | - | n) determines the order of entry of the covariates into the model selection algorithm. The default is xorder(+), which enters them in decreasing order of significance in a multiple linear regression (most significant first). xorder(-) places them in reverse significance order, whereas xorder(n) respects the original order in *xvarlist*.

select(*select_list*) sets the nominal p-values (significance levels) for variable selection by backward elimination. A variable is dropped if its removal causes a nonsignificant increase in deviance. The rules for *select_list* are the same as those for *df_list* in the df() option (see above). Using the default selection level of 1 for all variables forces them all into the model. Setting the nominal p-value to be 1 for a given variable forces it into the model, leaving others to be selected or not. The nominal p-value for elements of *xvarlist* bound by parentheses is specified by including (*varlist*) in *select_list*.

Example: select(0.05)
All variables have a nominal p-value of 5 percent.

Example: select(0.05, weight:1)
All variables except weight have a nominal p-value of 5 percent; weight is forced into the model.

Example: select(a (b c):0.05)
All variables except a, b, and c are forced into the model. b and c are tested jointly with 2 df at the 5-percent level, and a is tested singly at the 5-percent level.

xpowers(*xp_list*) sets the permitted fractional polynomial powers for covariates individually. The rules for *xp_list* are the same as for *df_list* in the df() option. The default selection is the same as those for the powers() option.

Example: xpowers(-1 0 1)
All variables have powers $-1,0,1$.

Example: xpowers(x5:-1 0 1)
All variables except x5 have default powers; x5 has powers $-1,0,1$.

zero(*varlist*) treats negative and zero values of members of *varlist* as zero when FP transformations are applied. By default, such variables are subjected to a preliminary linear transformation to avoid negative and zero values (see [R] **fracpoly**). *varlist* must be part of *xvarlist*.

catzero(*varlist*) is a variation on zero(); see *Zeroes and zero categories* below. *varlist* must be part of *xvarlist*.

regression_cmd_options may be any of the options appropriate to *regression_cmd*.

Remarks

Remarks are presented under the headings

> *Iteration report*
> *Estimation algorithm*
> *Methods of FP model selection*
> *Zeroes and zero categories*

For elements in *xvarlist* not enclosed in parentheses, mfp leaves variables in the data named Ixvar__1, Ixvar__2, ..., where *xvar* represents the first four letters of the name of *xvar1*, and so on, for *xvar2*, *xvar3*, etc. The new variables contain the best-fitting fractional polynomial powers of *xvar1*, *xvar2*, ...

Iteration report

By default, for each continuous predictor, *x*, mfp compares null, linear, and FP1 models for *x* with an FP2 model. The deviance for each of these nested submodels is given in the column headed "Deviance". The line labeled "Final" gives the deviance for the selected model and its powers. All the other predictors currently selected are included, with their transformations (if any). For models specified as having 1 df, the only choice is whether the variable enters the model or not.

Estimation algorithm

The estimation algorithm in mfp processes the *xvars* in turn. Initially, mfp silently arranges *xvarlist* in order of increasing *p*-value (i.e., of decreasing statistical significance) for omitting each predictor from the model comprising *xvarlist*, with each term linear. The aim is to model relatively important variables before unimportant ones. This may help to reduce potential model-fitting difficulties caused by collinearity or, more generally, "concurvity" among the predictors. See the xorder() option above for details on how to change the ordering.

At the initial cycle, the best-fitting FP function for *xvar1* (the first of *xvarlist*) is determined, with all the other variables assumed to be linear. Either the default or the alternative procedure is used (see *Methods of FP model selection* below). The functional form (but not the estimated regression coefficients) for *xvar1* is kept, and the process is repeated for *xvar2*, *xvar3*, etc. The first iteration concludes when all the variables have been processed in this way. The next cycle is similar, except that the functional forms from the initial cycle are retained for all variables except the one currently being processed.

A variable whose functional form is prespecified to be linear (i.e., to have 1 df) is tested for exclusion within the above procedure when its nominal *p*-value (selection level) according to select() is less than 1; otherwise, it is included.

Updating of FP functions and candidate variables continues until the functions and variables included in the overall model do not change (convergence). Convergence is usually achieved within 1 to 4 cycles.

Methods of FP model selection

mfp includes two algorithms for FP model selection, both of which are types of backward elimination. They start from a most-complex permitted FP model and attempt to simplify the model by reducing the degree. The default algorithm resembles a "closed test procedure", a sequence of tests maintaining the overall type I error rate at a prespecified nominal level, such as 5%. All significance tests are approximate; therefore, the algorithm is not precisely a closed test procedure.

The closed test algorithm for choosing an FP model with maximum permitted degree $m = 2$ (i.e., an FP2 model with 4 df) for a single continuous predictor, *x*, is as follows:

1. Inclusion: test FP2 against the null model for x on 4 df at the significance level determined by `select()`. If x is significant, continue; otherwise, drop x from the model.

2. Nonlinearity: test FP2 against a straight line in x on 3 df at the significance level determined by `alpha()`. If significant, continue; otherwise, stop, with the chosen model for x being a straight line.

3. Simplification: test FP2 against FP1 on 2 df at the significance level determined by `alpha()`. If significant, the final model is FP2; otherwise, it is FP1.

The first step is omitted if x is to be retained in the model, that is, if its nominal p-value, according to the `select()` option, is 1.

An alternative algorithm is available with the `sequential` option, as originally suggested by Royston and Altman (1994):

1. Test FP2 against FP1 on 2 df at the `alpha()` significance level. If significant, the final model is FP2; otherwise, continue.

2. Test FP1 against a straight line on 1 df at the `alpha()` level. If significant, the final model is FP1; otherwise, continue.

3. Test a straight line against omitting x on 1 df at the `select()` level. If significant, the final model is a straight line; otherwise, drop x.

The final step is omitted if x is to be retained in the model, that is, if its nominal p-value, according to the `select()` option, is 1.

If x is uninfluential, the overall type I error rate of this procedure is about double that of the "closed" test procedure, for which the rate is close to the nominal value. This inflated type I error rate confers increased apparent power to detect nonlinear relationships.

Zeroes and zero categories

The `zero()` option permits fitting an FP model to the positive values of a covariate, taking nonpositive values as zero. An application is the assessment of the effect of cigarette smoking as a risk factor in an epidemiological study. Nonsmokers may be qualitatively different from smokers, so the effect of smoking (regarded as a continuous variable) may not be continuous between one and zero cigarettes. To allow for this, the risk may be modeled as constant for the nonsmokers and as an FP function of the number of cigarettes for the smokers:

```
. generate byte nonsmokr = cond(n_cigs==0, 1, 0) if n_cigs != .
. mfp logit case n_cigs nonsmokr age, zero(n_cigs) df(4, nonsmokr:1)
```

Omission of `zero(n_cigs)` would cause `n_cigs` to be transformed before analysis by the addition of a suitable constant, probably 1.

A closely related approach involves the `catzero()` option. The command

```
. mfp logit case n_cigs age, catzero(n_cigs)
```

would achieve a similar result to the previous command but with important differences. First, `mfp` would create the equivalent of the binary variable `nonsmokr` automatically and include it in the model. Second, the two smoking variables would be treated as a single predictor in the model. With the `select()` option active, the two variables would be tested jointly for inclusion in the model.

▷ Example 1

We illustrate two of the analyses performed by Sauerbrei and Royston (1999). We use brcancer.dta, which contains prognostic factors data from the German Breast Cancer Study Group of patients with node-positive breast cancer. The response variable is recurrence-free survival time (rectime), and the censoring variable is censrec. There are 686 patients with 299 events. We use Cox regression to predict the log hazard of recurrence from prognostic factors of which 5 are continuous (x1, x3, x5, x6, x7) and 3 are binary (x2, x4a, x4b). Hormonal therapy (hormon) is known to reduce recurrence rates and is forced into the model. We use mfracpol to build a model from the initial set of eight predictors using the backfitting model selection algorithm. We set the nominal p-value for variable and FP selection to 0.05 for all variables except hormon, for which it is set to 1:

```
. use http://www.stata-press.com/data/r9/brcancer
(German breast cancer data)
. stset rectime, fail(censrec)
  (output omitted)
. mfp stcox x1 x2 x3 x4a x4b x5 x6 x7 hormon, nohr alpha(.05)
> select(.05, hormon:1)
Deviance for model with all terms untransformed =  3471.637, 686 observations
```

Variable	Model	(vs.)	Deviance	Dev diff.	P	Powers	(vs.)
x5	null	FP2	3503.610	61.366	0.000*	.	.5 3
	lin.		3471.637	29.393	0.000+	1	
	FP1		3449.203	6.959	0.031+	0	
	Final		3442.244			.5 3	
x6	null	FP2	3464.113	29.917	0.000*	.	-2 .5
	lin.		3442.244	8.048	0.045+	1	
	FP1		3435.550	1.354	0.508	.5	
	Final		3435.550			.5	

[hormon included with 1 df in model]

Variable	Model	(vs.)	Deviance	Dev diff.	P	Powers	(vs.)
x4a	null	lin.	3440.749	5.199	0.023*	1	.
	Final		3435.550			1	
x3	null	FP2	3436.832	3.560	0.469	.	-2 3
	Final		3436.832			.	
x2	null	lin.	3437.589	0.756	0.384	1	.
	Final		3437.589			.	
x4b	null	lin.	3437.848	0.259	0.611	1	.
	Final		3437.848			.	
x1	null	FP2	3437.893	18.085	0.001*	.	-2 -.5
	lin.		3437.848	18.040	0.000+	1	
	FP1		3433.628	13.820	0.001+	-2	
	Final		3419.808			-2 -.5	
x7	null	FP2	3420.805	3.715	0.446	.	-.5 3
	Final		3420.805			.	

```
Cycle 1: deviance =    3420.805
```

Variable	Model	(vs.)	Deviance	Dev diff.	P	Powers	(vs.)
x5	null	FP2	3494.867	74.143	0.000*	.	-2 -1
	lin.		3451.795	31.071	0.000+	1	
	FP1		3428.023	7.299	0.026+	0	
	Final		3420.724			-2 -1	
x6	null	FP2	3452.093	32.704	0.000*	.	0 0
	lin.		3427.703	8.313	0.040+	1	
	FP1		3420.724	1.334	0.513	.5	
	Final		3420.724			.5	

[hormon included with 1 df in model]

x4a	null	lin.	3425.310	4.586	0.032*	1		.	
	Final		3420.724			1			
x3	null	FP2	3420.724	5.305	0.257	.		-.5 0	
	Final		3420.724			.			
x2	null	lin.	3420.724	0.214	0.644	1		.	
	Final		3420.724			.			
x4b	null	lin.	3420.724	0.145	0.703	1		.	
	Final		3420.724			.			
x1	null	FP2	3440.057	19.333	0.001*	.		-2 -.5	
	lin.		3440.038	19.314	0.000+	1			
	FP1		3436.949	16.225	0.000+	-2			
	Final		3420.724			-2 -.5			
x7	null	FP2	3420.724	2.152	0.708	.		-1 3	
	Final		3420.724			.			

Fractional polynomial fitting algorithm converged after 2 cycles.

Transformations of covariates:

```
-> gen double Ix1__1 = X^-2-.0355294635 if e(sample)
-> gen double Ix1__2 = X^-.5-.4341573547 if e(sample)
   (where: X = x1/10)
-> gen double Ix2__1 = x2-1 if e(sample)
-> gen double Ix3__1 = x3-29.32944606 if e(sample)
-> gen double Ix5__1 = X-.5010204082 if e(sample)
-> gen double Ix5__2 = X^2-.2510214494 if e(sample)
   (where: X = x5/10)
-> gen double Ix6__1 = x6-109.9956268 if e(sample)
-> gen double Ix7__1 = x7-96.25218659 if e(sample)
```

Final multivariable fractional polynomial model for _t

Variable	——Initial——			——Final——		
	df	Select	Alpha	Status	df	Powers
x1	4	0.0500	0.0500	in	4	-2 -.5
x2	1	0.0500	0.0500	out	0	
x3	4	0.0500	0.0500	out	0	
x4a	1	0.0500	0.0500	in	1	1
x4b	1	0.0500	0.0500	out	0	
x5	4	0.0500	0.0500	in	4	-2 -1
x6	4	0.0500	0.0500	in	2	.5
x7	4	0.0500	0.0500	out	0	
hormon	1	1.0000	0.0500	in	1	1

Cox regression -- Breslow method for ties
Entry time _t0

Log likelihood = -1710.3619

Number of obs	=	686
LR chi2(7)	=	155.62
Prob > chi2	=	0.0000
Pseudo R2	=	0.0435

_t _d	Coef.	Std. Err.	z	P>|z|	[95% Conf. Interval]	
Ix1__1	44.73377	8.256682	5.42	0.000	28.55097	60.91657
Ix1__2	-17.92302	3.909611	-4.58	0.000	-25.58571	-10.26032
x4a	.5006982	.2496324	2.01	0.045	.0114276	.9899687
Ix5__1	.0387904	.0076972	5.04	0.000	.0237041	.0538767
Ix5__2	-.5490645	.0864255	-6.35	0.000	-.7184554	-.3796736
Ix6__1	-1.806966	.3506314	-5.15	0.000	-2.494191	-1.119741
hormon	-.4024169	.1280843	-3.14	0.002	-.6534575	-.1513763

Deviance: 3420.724.

Some explanation of the output from the model selection algorithm is desirable. Consider the first few lines of output in the iteration log:

```
1. Deviance for model with all terms untransformed = 3471.637, 686 observations

   Variable     Model (vs.)   Deviance  Dev diff.   P      Powers  (vs.)

2. x5           null  FP2      3503.610   61.366  0.000*    .        .5 3
3.              lin.           3471.637   29.393  0.000+    1
4.              FP1            3449.203    6.959  0.031+    0
5.              Final          3442.244                    .5 3
```

Line 1 gives the deviance ($-2 \times$ log partial likelihood) for the Cox model with all terms linear, the place where the algorithm starts. The model is modified variable by variable in subsequent steps. The most significant linear term turns out to be x5, which is therefore processed first. Line 2 compares the best-fitting FP2 for x5 with a model omitting x5. The FP has powers (0.5,3), and the test for inclusion of x5 is highly significant. The reported deviance of 3503.610 is for the null model, not for the FP2 model. The deviance for the FP2 model may be calculated by subtracting the deviance difference (Dev diff.) from the reported deviance, giving $3503.610 - 61.366 = 3442.244$. Line 3 shows that the FP2 model is also a significantly better fit than a straight line (lin.) and line 4 that FP2 is also somewhat better than FP1 ($p = 0.031$). Thus at this stage in the model-selection procedure, the final model for x5 (line 5) is FP2 with powers (0.5,3). The overall model with an FP2 for x5 and all other terms linear has a deviance of 3442.244.

After all the variables have been processed (cycle 1) and reprocessed (cycle 2) in this way, convergence is achieved since the functional forms (FP powers and variables included) after cycle 2 are the same as they were after cycle 1. The model finally chosen is Model II as given in tables 3 and 4 of Sauerbrei and Royston (1999). Due to scaling of variables, the regression coefficients reported there are different, but the model and its deviance are identical. The model includes x1 with powers $(-2, -0.5)$, x4a, x5 with powers $(-2, -1)$, and x6 with power 0.5. There is strong evidence of nonlinearity for x1 and for x5, the deviance differences for comparison with a straight line model (FP2 vs lin.) being respectively 19.3 and 31.1 at convergence (cycle 2). Predictors x2, x3, x4b and x7 are dropped, as may be seen from their status out in the table Final multivariable fractional polynomial model for _t (the assumed *depvar* when using stcox).

Note that all predictors except x4a and hormon, which are binary, have been adjusted to the mean of the original variable. For example, the mean of x1 (age) is 53.05 years. The first FP-transformed variable for x1 is x1^-2 and is created by the expression gen double Ix1__1 = X^-2-.0355 if e(sample). The value .0355 is obtained from $(53.05/10)^{-2}$. The division by 10 is applied automatically to improve the scaling of the regression coefficient for Ix1__1.

According to Sauerbrei and Royston (1999), medical knowledge dictates that the estimated risk function for x5 (number of positive nodes), which was based on the above FP with powers $(-2, -1)$, should be monotonic, but it was not. They improved Model II by estimating a preliminary exponential transformation, x5e $= \exp(-0.12 \cdot$ x5$)$, for x5 and fitting a degree 1 FP for x5e, thus obtaining a monotonic risk function. The value of -0.12 was estimated univariately using nonlinear Cox regression with the ado-file boxtid (Royston and Ambler 1999). To ensure a negative exponent, Sauerbrei and Royston (1999) restricted the powers for x5e to be positive. Their Model III may be fitted by using the following command:

```
. mfp stcox x1 x2 x3 x4a x4b x5e x6 x7 hormon, alpha(.05) select(.05, hormon:1)
> df(x5e:2) xpowers(x5e:0.5 1 2 3)
```

Other than the customization for x5e, the command is the same as it was before. The resulting model is as reported in table 4 of Sauerbrei and Royston (1999):

```
. use http://www.stata-press.com/data/r9/brcancer, clear
(German breast cancer data)
```

```
. stset rectime, fail(censrec)
```
(*output omitted*)
```
. mfp stcox x1 x2 x3 x4a x4b x5e x6 x7 hormon, nohr alpha(.05)
> select(.05, hormon:1) df(x5e:2) xpowers(x5e:0.5 1 2 3)
```
(*output omitted*)

Final multivariable fractional polynomial model for _t

| Variable | Initial | | | Final | | |
	df	Select	Alpha	Status	df	Powers
x1	4	0.0500	0.0500	in	4	-2 -.5
x2	1	0.0500	0.0500	out	0	
x3	4	0.0500	0.0500	out	0	
x4a	1	0.0500	0.0500	in	1	1
x4b	1	0.0500	0.0500	out	0	
x5e	2	0.0500	0.0500	in	1	1
x6	4	0.0500	0.0500	in	2	.5
x7	4	0.0500	0.0500	out	0	
hormon	1	1.0000	0.0500	in	1	1

```
Cox regression -- Breslow method for ties
Entry time _t0                              Number of obs   =        686
                                            LR chi2(6)      =     153.11
                                            Prob > chi2     =     0.0000
Log likelihood = -1711.6186                 Pseudo R2       =     0.0428
```

_t _d	Coef.	Std. Err.	z	P>|z|	[95% Conf. Interval]	
Ix1__1	43.55382	8.253433	5.28	0.000	27.37738	59.73025
Ix1__2	-17.48136	3.911882	-4.47	0.000	-25.14851	-9.814212
x4a	.5174351	.2493739	2.07	0.038	.0286713	1.006199
Ix5e__1	-1.981213	.2268903	-8.73	0.000	-2.425909	-1.536516
Ix6__1	-1.84008	.3508432	-5.24	0.000	-2.52772	-1.15244
hormon	-.3944998	.128097	-3.08	0.002	-.6455654	-.1434342

Deviance: 3423.237.

◁

(Continued on next page)

Saved Results

In addition to what *regression_cmd* saves, mfp saves in e():

Scalars

e(fp_nx)	number of predictors in *xvarlist*
e(fp_dev)	deviance of final model fitted
e(Fp_id#)	initial degrees of freedom for the #th element of *xvarlist*
e(Fp_fd#)	final degrees of freedom for the #th element of *xvarlist*
e(Fp_al#)	FP selection level for the #th element of *xvarlist*
e(Fp_se#)	Backward elimination selection level for the #th element of *xvarlist*

Macros

e(fp_cmd)	fracpoly
e(fp_cmd2)	mfp
e(fp_fvl)	variables in final model
e(fp_depv)	*yvar*
e(fp_opts)	estimation command options
e(fp_x1)	first variable in *xvarlist*
e(fp_x2)	second variable in *xvarlist*
...	
e(fp_xN)	last variable in *xvarlist*, N=e(fp_nx)
e(fp_k1)	power for first variable in *xvarlist* (*)
e(fp_k2)	power for second variable in *xvarlist* (*)
...	
e(fp_kN)	power for last var. in *xvarlist* (*), N=e(fp_nx)

Notes: (*) contains '.' if variable is not selected in final model.

Methods and Formulas

mfp is implemented as an ado-file.

Acknowledgments

mfp is an update of mfracpol by Royston and Ambler (1998).

References

Ambler, G. and P. Royston. 2001. Fractional polynomial model selection procedures: Investigation of Type I error rate. *Journal of Statistical Simulation and Computation* 69: 89–108.

Royston, P. and D. G. Altman. 1994. Regression using fractional polynomials of continuous covariates: Parsimonious parametric modelling (with discussion). *Applied Statistics* 43(3): 429–467.

Royston, P. and G. Ambler. 1998. sg81: Multivariable fractional polynomials. *Stata Technical Bulletin* 43: 24–32. Reprinted in *Stata Technical Bulletin Reprints*, vol. 8, pp. 123–132.

——. 1999a. sg81.1: Multivariable fractional polynomials: Update. *Stata Technical Bulletin* 49: 17–23. Reprinted in *Stata Technical Bulletin Reprints*, vol. 9, pp. 161–168.

——. 1999b. sg81.2: Multivariable fractional polynomials: Update. *Stata Technical Bulletin* 50: 25. Reprinted in *Stata Technical Bulletin Reprints*, vol. 9, p. 168.

——. 1999c. sg112: Nonlinear regression models involving power of exponential functions of covariates. *Stata Technical Bulletin* 49: 25–30. Reprinted in *Stata Technical Bulletin Reprints*, vol. 9, pp. 173–179.

——. 1999d. sg112.1: Nonlinear regression models involving power of exponential functions: Update. *Stata Technical Bulletin* 50: 26. Reprinted in *Stata Technical Bulletin Reprints*, vol. 9, p. 180.

Sauerbrei, W. and P. Royston. 1999. Building multivariable prognostic and diagnostic models: Transformation of the predictors by using fractional polynomials. *Journal of the Royal Statistical Society, Series A* 162: 71–94.

——. 2002. Corrigendum: Building multivariable prognostic and diagnostic models: Transformation of the predictors by using fractional polynomials. *Journal of the Royal Statistical Society, Series A* 165: 299–300.

Also See

Complementary:	[R] **mfp postestimation**; [R] **fracpoly**,
	[P] **break**
Background:	[U] **20 Estimation and postestimation commands**

Title

mfp postestimation — Postestimation tools for mfp

Description

The following postestimation commands are of special interest after `mfp`:

command	description
fracplot	plot data and fit from most recently fitted fractional polynomial model
fracpred	create variable containing prediction, deviance residuals, or SEs of fitted values

For `fracplot` and `fracpred`, see [R] **fracpoly postestimation**.

In addition, the following standard postestimation commands are available:

command	description
adjust	adjusted predictions of $\mathbf{x}\beta$, probabilities, or $\exp(\mathbf{x}\beta)$
estat	AIC, BIC, VCE, and estimation sample summary
estimates	cataloging estimation results
lincom	point estimates, standard errors, testing, and inference for linear combinations of coefficients
linktest	link test for model specification
lrtest	likelihood-ratio test
mfx	marginal effects or elasticities
nlcom	point estimates, standard errors, testing, and inference for nonlinear combinations of coefficients
test	Wald tests for simple and composite linear hypotheses
testnl	Wald tests of nonlinear hypotheses

See the corresponding entries in the *Stata Base Reference Manual* for details.

Methods and Formulas

All postestimation commands listed above are implemented as ado-files.

Also See

Complementary:	[R] **mfp**; [R] **fracpoly postestimation**; [R] **adjust**, [R] **estimates**, [R] **lincom**, [R] **linktest**, [R] **lrtest**, [R] **mfx**, [R] **nlcom**, [R] **test**, [R] **testnl**
Background:	[U] **13.5 Accessing coefficients and standard errors**, [U] **20 Estimation and postestimation commands**, [R] **estat**

Title

> **mfx** — Obtain marginal effects or elasticities after estimation

Syntax

mfx [c̲ompute] [*if*] [*in*] [, *options*]

mfx r̲eplay [, l̲evel(*#*)]

options	description
Model	
pr̲edict(*predict_option*)	calculate marginal effects (elasticities) for *predict_option*
var̲list(*varlist*)	calculate marginal effects (elasticities) for *varlist*
dydx	calculate marginal effects; the default
eyex	calculate elasticities in the form of $\partial \log y / \partial \log x$
dyex	calculate elasticities in the form of $\partial y / \partial \log x$
eydx	calculate elasticities in the form of $\partial \log y / \partial x$
no̲discrete	treat dummy (indicator) variables as continuous
no̲se	do not calculate standard errors
Model 2	
at(*atlist*)	estimate marginal effects (elasticities) at these values
noe̲sample	do not restrict calculation of means, medians, and mean offsets to the estimation sample
no̲wght	ignore weights when calculating means, medians, and mean offsets
Adv. model	
no̲nlinear	do not use the linear method
force	calculate marginal effects and standard errors in cases where it would otherwise refuse to do so
Reporting	
le̲vel(*#*)	set confidence level; default is level(95)
dia̲gnostics(beta)	report suitability of marginal-effect calculation
dia̲gnostics(vce)	report suitability of standard-error calculation
dia̲gnostics(all)	report all diagnostic information
tr̲acelvl(*#*)	report increasing levels of detail during calculations

where *atlist* is *numlist* or *matname* or

$$[\text{mean} \mid \text{median} \mid \text{zero}] \; [\textit{varname} = \# \; [, \; \textit{varname} = \#] \; [\dots]]$$

where mean is the default.

Description

mfx numerically calculates the marginal effects or the elasticities and their standard errors after estimation. Exactly what mfx can calculate is determined by the previous estimation command and the predict(*predict_option*) option. The values at which the marginal effects or elasticities are to be evaluated is determined by the at(*atlist*) option. By default, mfx calculates the marginal effects or elasticities at the means of the independent variables using the default prediction option associated with the previous estimation command.

Some disciplines use the term *partial effects*, rather than marginal effects, for what is computed by mfx.

mfx replay replays the results of the previous mfx computation.

Options

⌐ Model ⌐

predict(*predict_option*) specifies the function (that is, the form of y) for which to calculate the marginal effects or elasticities. The default is to use the default predict option of the preceding estimation command. To see which predict options are available, see help for that estimation command.

varlist(*varlist*) specifies the variables for which to calculate marginal effects (elasticities). The default is all variables.

dydx specifies that marginal effects be calculated. This is the default.

eyex specifies that elasticities be calculated in the form of $\partial \log y / \partial \log x$.

dyex specifies that elasticities be calculated in the form of $\partial y / \partial \log x$.

eydx specifies that elasticities be calculated in the form of $\partial \log y / \partial x$.

nodiscrete treats dummy variables as continuous. A dummy variable is one that takes on the value 0 or 1 in the estimation sample. If nodiscrete is not specified, the marginal effect of a dummy variable is calculated as the discrete change in y as the dummy variable changes from 0 to 1. This option is irrelevant to the computation of the elasticities because all dummy variables are treated as continuous when computing elasticities.

nose specifies that standard errors of the marginal effects (elasticities) not be computed.

⌐ Model 2 ⌐

at(*atlist*) specifies the values at which the marginal effects (elasticities) are to be estimated. The default is to evaluate at the means of the independent variables.

at(*numlist*) specifies that the marginal effects (elasticities) be evaluated at the *numlist*. The order of the values in the *numlist* is the same as the variables in the preceding estimation command, that is, from left to right, without repetition.

at(*matname*) specifies the points in a matrix format. The ordering of the variables is the same as that of *numlist*.

at(mean | median | zero [*varname* = # [, *varname* = #] [...]]) specifies that the marginal effects (elasticities) be evaluated at means, at medians of the independent variables, or at zeros. It also allows users to specify particular values for one or more independent variables, assuming that the rest are means, medians, or zeros.

at (*varname* = # $\big[$, *varname* = # $\big]$ $\big[\dots\big]$) specifies that the marginal effects or the elasticities be particular values for one or more independent variables, assuming that the rest are means.

noesample affects at (*atlist*), any offsets used in the preceding estimation, and the determination of dummy variables. It specifies that the whole dataset be considered instead of only those marked in the e(sample) defined by the previous estimation command.

nowght only affects at (*atlist*) and offsets. It specifies that weights be ignored when calculating the means or medians for the *atlist* and when calculating the means for any offsets.

Adv. model

nonlinear specifies that y, the function to be calculated for the marginal effects or the elasticities, does not meet the linear-form restriction. By default, mfx assumes that y meets the linear-form restriction, unless one or more dependent variables are shared by multiple equations. For instance, predictions after

```
. heckman mpg price, sel(foreign=rep78)
```

meet the linear-form restriction, but those after

```
. heckman mpg price, sel(foreign=rep78 price)
```

do not. If y meets the linear-form restriction, specifying nonlinear should produce the same results as not specifying it. However, the nonlinear method is generally more time consuming. Most likely, you do not need to specify nonlinear after an official Stata command. For user-written commands, if you are not sure whether y is of linear form, specifying nonlinear is a safe choice.

force specifies that marginal effects and their standard errors be calculated in cases where it would otherwise refuse to do so. Such cases arise, for instance, when the marginal effect is a function of a random quantity other than the coefficients of the model (e.g., a residual). If you specify this option, there is no guarantee that the resulting marginal effects and standard errors are correct.

Reporting

level(#) specifies the confidence level, as a percentage, for confidence intervals. The default is level(95) or as set by set level; see [U] **23.5 Specifying the width of confidence intervals**.

diagnostics(*diaglist*) asks mfx to display various diagnostic information.

diagnostics(beta) shows the information used to determine whether the prediction option is suitable for computing marginal effects.

diagnostics(vce) shows the information used to determine whether the prediction option is suitable for computing the standard errors of the marginal effects.

diagnostics(all) shows all the above diagnostic information.

tracelvl(#) shows increasing levels of detail during calculations. # may be 1, 2, 3, or 4. Level 1 shows the marginal effects and standard errors as they are computed, and which method, either linear or nonlinear, was used. Level 2 shows, in addition, the components of the matrix of partial derivatives needed for each standard error as they are computed. Level 3 shows counts of iterations in obtaining a suitable finite difference for each numerical derivative. Level 4 shows the values of these finite differences.

Remarks

Remarks are presented under the headings

Obtaining marginal effects after single-equation (SE) estimation
Obtaining marginal effects after multiple-equation (ME) estimation
Specifying the evaluation points
Obtaining three forms of elasticities

Obtaining marginal effects after single-equation (SE) estimation

Before running `mfx`, type `help` *estimation_cmd* to see what can be predicted after the estimation and to see the default prediction.

▷ Example 1

We fit a logit model using the auto dataset:

```
. use http://www.stata-press.com/data/r9/auto
(1978 Automobile Data)

. logit foreign mpg price, nolog
```

```
Logistic regression                             Number of obs   =         74
                                                LR chi2(2)      =      17.14
                                                Prob > chi2     =     0.0002
Log likelihood = -36.462189                     Pseudo R2       =     0.1903
```

foreign	Coef.	Std. Err.	z	P>|z|	[95% Conf. Interval]	
mpg	.2338353	.0671449	3.48	0.000	.1022338	.3654368
price	.000266	.0001166	2.28	0.022	.0000375	.0004945
_cons	-7.648111	2.043673	-3.74	0.000	-11.65364	-3.642586

To determine the marginal effects of `mpg` and `price` for the probability of a positive outcome at their mean values, we can issue the `mfx` command without the `predict` option because the default prediction after `logit` is the probability of a positive outcome. The calculation is requested at the mean values by default.

```
. mfx

Marginal effects after logit
      y  = Pr(foreign) (predict)
         = .26347633
```

variable	dy/dx	Std. Err.	z	P>|z|	[	95% C.I.	]	X
mpg	.0453773	.0131	3.46	0.001	.019702	.071053		21.2973
price	.0000516	.00002	2.31	0.021	7.8e-06	.000095		6165.26

The first line of the output indicates that the marginal effects were calculated after a `logit` estimation. The second line of the output describes the form of y and the `predict` command that we would type to calculate y separately. The third line of the output gives the value of y given the values of X, which are displayed in the last column of the table.

To calculate the marginal effects at particular data points, say, `mpg` $= 20$, `price` $= 6000$, specify the `at()` option:

```
. mfx, at(mpg=20, price=6000)
Marginal effects after logit
      y  = Pr(foreign) (predict)
         = .20176601
```

variable	dy/dx	Std. Err.	z	P>|z|	[95% C.I.]	X
mpg	.0376607	.00961	3.92	0.000	.018834	.056488	20
price	.0000428	.00002	2.47	0.014	8.8e-06	.000077	6000

To calculate the marginal effects for the linear prediction (xb) instead of the probability, specify predict(xb). Note that the marginal effects for the linear prediction are the coefficients themselves.

```
. mfx, predict(xb)
Marginal effects after logit
      y  = Linear prediction (predict, xb)
         =-1.0279779
```

variable	dy/dx	Std. Err.	z	P>|z|	[95% C.I.]	X
mpg	.2338353	.06714	3.48	0.000	.102234	.365437	21.2973
price	.000266	.00012	2.28	0.022	.000038	.000495	6165.26

If there is a dummy variable as an independent variable, mfx calculates the discrete change as the dummy variable changes from 0 to 1.

```
. generate goodrep = 0
. replace goodrep = 1 if rep > 3
(34 real changes made)
. logit foreign mpg goodrep, nolog
```

Logistic regression				Number of obs	=	74
				LR chi2(2)	=	26.27
				Prob > chi2	=	0.0000
Log likelihood = -31.898321				Pseudo R2	=	0.2917

foreign	Coef.	Std. Err.	z	P>|z|	[95% Conf. Interval]	
mpg	.1079219	.0565077	1.91	0.056	-.0028311	.2186749
goodrep	2.435068	.7128444	3.42	0.001	1.037918	3.832217
_cons	-4.689347	1.326547	-3.54	0.000	-7.28933	-2.089363

```
. mfx
Marginal effects after logit
      y  = Pr(foreign) (predict)
         = .21890034
```

variable	dy/dx	Std. Err.	z	P>|z|	[95% C.I.]	X
mpg	.0184528	.01017	1.81	0.070	-.001475	.038381	21.2973
goodrep*	.4271707	.10432	4.09	0.000	.222712	.63163	.459459

(*) dy/dx is for discrete change of dummy variable from 0 to 1

If nodiscrete is specified, mfx treats the dummy variable as continuous.

```
. mfx, nodiscrete
Marginal effects after logit
      y  = Pr(foreign) (predict)
         = .21890034
```

variable	dy/dx	Std. Err.	z	P>|z|	[95% C.I.]	X
mpg	.0184528	.01017	1.81	0.070	−.001475 .038381	21.2973
goodrep	.4163552	.10733	3.88	0.000	.205994 .626716	.459459

◁

❏ Technical Note

By default, mfx uses the estimation sample to determine which independent variables are dummies. A variable is declared a dummy if its only values in the estimation sample are zero or one. This determination may be affected by the option noesample. For example,

```
. replace rep78=rep78-3
(69 real changes made)
. logit foreign mpg rep78 if rep78==0|rep78==1, nolog
```

```
Logistic regression                          Number of obs   =         48
                                             LR chi2(2)      =      21.88
                                             Prob > chi2     =     0.0000
Log likelihood = -16.050909                  Pseudo R2       =     0.4053
```

foreign	Coef.	Std. Err.	z	P>|z|	[95% Conf. Interval]
mpg	.347422	.120328	2.89	0.004	.1115834 .5832605
rep78	2.105463	.9194647	2.29	0.022	.3033449 3.90758
_cons	-9.683393	2.876752	-3.37	0.001	-15.32172 -4.045063

```
. mfx
Marginal effects after logit
      y  = Pr(foreign) (predict)
         =   .1357189
```

variable	dy/dx	Std. Err.	z	P>|z|	[95% C.I.]	X
mpg	.0407523	.01531	2.66	0.008	.010755 .07075	20.2708
rep78*	.3027042	.14694	2.06	0.039	.01471 .590699	.375

(*) dy/dx is for discrete change of dummy variable from 0 to 1

When noesample is specified, the value of rep78 is considered for all observations in the dataset. Since observations with rep78 not equal to zero or one do exist, mfx will conclude it is not a dummy variable.

```
.table rep78
```

Repair Record 1978	Freq.
-2	2
-1	8
0	30
1	18
2	11

```
. mfx, noesample
Marginal effects after logit
     y  = Pr(foreign) (predict)
        = .19312144
```

variable	dy/dx	Std. Err.	z	P>|z|	[	95% C.I.	]	X
mpg	.0541372	.02043	2.65	0.008	.014089	.094185		21.2973
rep78	.3280849	.14321	2.29	0.022	.047401	.608769		.405797

The qualifiers if and in have no effect on the determination of dummy variables.

❏

Obtaining marginal effects after multiple-equation (ME) estimation

If you have not read the discussion above on using mfx after SE estimations, please do so. As a general introduction to the ME models, the following examples will demonstrate mfx after heckman and mlogit.

▷ Example 2

```
. use http://www.stata-press.com/data/r9/auto, clear
(1978 Automobile Data)

. heckman mpg weight length, sel(foreign = displacement) nolog
Heckman selection model                      Number of obs   =        74
(regression model with sample selection)     Censored obs    =        52
                                             Uncensored obs  =        22

                                             Wald chi2(2)    =      7.27
Log likelihood = -87.58426                   Prob > chi2     =    0.0264
```

	Coef.	Std. Err.	z	P>|z|	[95% Conf. Interval]	
mpg						
weight	-.0039923	.0071948	-0.55	0.579	-.0180939	.0101092
length	-.1202545	.2093074	-0.57	0.566	-.5304895	.2899805
_cons	56.72567	21.68463	2.62	0.009	14.22458	99.22676
foreign						
displacement	-.0250297	.0067241	-3.72	0.000	-.0382088	-.0118506
_cons	3.223625	.8757406	3.68	0.000	1.507205	4.940045
/athrho	-.9840858	.8112212	-1.21	0.225	-2.57405	.6058785
/lnsigma	1.724306	.2794524	6.17	0.000	1.176589	2.272022
rho	-.7548292	.349014			-.9884463	.5412193
sigma	5.608626	1.567344			3.243293	9.698997
lambda	-4.233555	3.022645			-10.15783	1.690721

```
LR test of indep. eqns. (rho = 0):   chi2(1) =     1.37   Prob > chi2 = 0.2413
```

heckman estimated two equations, mpg and foreign; see [R] **heckman**. Two of the prediction options after heckman are the expected value of the dependent variable and the probability of being observed. To obtain the marginal effects of all the independent variables for the expected value of the dependent variable, we specify predict(yexpected) with mfx.

```
. mfx, predict(yexpected)
```

Marginal effects after heckman
     y = E(mpg*|Pr(foreign)) (predict, yexpected)
       = .56522778

variable	dy/dx	Std. Err.	z	P>|z|	[	95% C.I.	]	X
weight	-.0001725	.00041	-0.42	0.675	-.000979	.000634		3019.46
length	-.0051953	.01002	-0.52	0.604	-.02483	.01444		187.932
displa~t	-.0340055	.02793	-1.22	0.223	-.088739	.020728		197.297

To calculate the marginal effects for the probability of being observed, we specify predict(psel) with mfx. Since only the independent variables in equation foreign affect the probability of being observed, some of the marginal effects will be zero. Using the option varlist(*varlist*) with mfx will restrict the calculation of marginal effects to the independent variables in the *varlist*.

```
. mfx, predict(psel)
```

Marginal effects after heckman
     y = Pr(foreign) (predict, psel)
       = .04320292

variable	dy/dx	Std. Err.	z	P>|z|	[	95% C.I.	]	X
weight	0	0	.	.	0	0		3019.46
length	0	0	.	.	0	0		187.932
displa~t	-.0022958	.00153	-1.50	0.133	-.005287	.000696		197.297

```
. mfx, predict(psel) varlist(displacement)
```

Marginal effects after heckman
     y = Pr(foreign) (predict, psel)
       = .04320292

variable	dy/dx	Std. Err.	z	P>|z|	[	95% C.I.	]	X
displa~t	-.0022958	.00153	-1.50	0.133	-.005287	.000696		197.297

◁

▷ Example 3

predict after mlogit, unlike most other estimation commands, can predict multiple new variables by issuing predict only once; see [R] **mlogit**. To calculate the marginal effects for the probability of more than one outcome, we run mfx separately for each outcome.

(Continued on next page)

```
. mlogit rep78 mpg, nolog
Multinomial logistic regression                    Number of obs   =         69
                                                   LR chi2(4)      =      15.88
                                                   Prob > chi2     =     0.0032
Log likelihood = -85.752375                        Pseudo R2       =     0.0847
```

rep78	Coef.	Std. Err.	z	P>|z|	[95% Conf. Interval]	
1						
mpg	.0708122	.1471461	0.48	0.630	-.2175888	.3592132
_cons	-4.137144	3.15707	-1.31	0.190	-10.32489	2.050599
2						
mpg	-.0164251	.0926724	-0.18	0.859	-.1980597	.1652095
_cons	-1.005118	1.822129	-0.55	0.581	-4.576426	2.566189
4						
mpg	.0958626	.0633329	1.51	0.130	-.0282676	.2199927
_cons	-2.474187	1.341131	-1.84	0.065	-5.102756	.154381
5						
mpg	.2477469	.0764076	3.24	0.001	.0979908	.397503
_cons	-6.653164	1.841793	-3.61	0.000	-10.26301	-3.043316

```
(rep78==3 is the base outcome)
. mfx, predict(outcome(1))
Marginal effects after mlogit
      y  = Pr(rep78==1) (predict, outcome(1))
         = .03233059
```

variable	dy/dx	Std. Err.	z	P>|z|	[95% C.I.]		X
mpg	.0004712	.0045	0.10	0.917	-.00835	.009292	21.2899

◁

Specifying the evaluation points

By default mfx evaluates the marginal effects at the means of the independent variables. To evaluate elsewhere, you would specify the option at with mfx. This option allows a number of different syntaxes.

▷ Example 4

Using the *numlist* and *matname* syntax, we must specify the evaluation points in the same order as the variables in the preceding estimation command, that is, from left to right, without repetition.

```
. sureg (price disp weight) (mpg foreign disp)
Seemingly unrelated regression
```

Equation	Obs	Parms	RMSE	"R-sq"	chi2	P
price	74	2	2466.937	0.2909	29.98	0.0000
mpg	74	2	4.061588	0.5004	74.03	0.0000

	Coef.	Std. Err.	z	P>|z|	[95% Conf. Interval]	
price						
displacement	2.660349	7.043538	0.38	0.706	-11.14473	16.46543
weight	1.747666	.8322723	2.10	0.036	.1164423	3.37889
_cons	363.3701	1441.681	0.25	0.801	-2462.273	3189.014
mpg						
foreign	-.6825672	1.30813	-0.52	0.602	-3.246455	1.881321
displacement	-.0465529	.0065556	-7.10	0.000	-.0594017	-.033704
_cons	30.68498	1.632312	18.80	0.000	27.4857	33.88425

```
. mfx, predict(xb) at(200 3000 0.5)
Marginal effects after sureg
      y = Linear prediction (predict, xb)
        = 6138.4383
```

variable	dy/dx	Std. Err.	z	P>|z|	[95% C.I.]	X
displa~t	2.660349	7.04354	0.38	0.706	-11.1447 16.4654	200
weight	1.747666	.83227	2.10	0.036	.116442 3.37889	3000
foreign*	0	0	.	.	0 0	.5

```
(*) dy/dx is for discrete change of dummy variable from 0 to 1
```

◁

❑ Technical Note

When using the *numlist* or *matname* syntax together with `varlist`, you must specify values for all the independent variables, not just those for which marginal effects will be calculated. These values are used in the estimation of the marginal effects, and although they don't display in the output of `mfx`, they are included in the saved results.

```
. probit foreign mpg price, nolog
```

Probit estimates				Number of obs	=	74
				LR chi2(2)	=	17.53
				Prob > chi2	=	0.0002
Log likelihood = -36.266068				Pseudo R2	=	0.1947

foreign	Coef.	Std. Err.	z	P>|z|	[95% Conf. Interval]	
mpg	.1404876	.0373595	3.76	0.000	.0672644	.2137108
price	.0001571	.0000641	2.45	0.014	.0000315	.0002827
_cons	-4.592058	1.115907	-4.12	0.000	-6.779195	-2.404921

```
. capture noisily mfx, at(6000) varlist(pri)
numlist too short in at()
```

```
. mfx, at(20 6000) varlist(pri)
Marginal effects after probit
      y  = Pr(foreign) (predict)
         = .2005512
```

variable	dy/dx	Std. Err.	z	P>|z|	[95% C.I.]	X
price	.0000441	.00002	2.62	0.009	.000011 .000077	6000

```
. matrix list e(Xmfx_X)
e(Xmfx_X)[1,2]
       mpg  price
r1      20   6000
```

❑

Obtaining three forms of elasticities

mfx can also be used to obtain all three forms of elasticities.

option	elasticity
eyex	$\partial \log y / \partial \log x$
dyex	$\partial y / \partial \log x$
eydx	$\partial \log y / \partial x$

▷ Example 5

We fit a regression model using the auto dataset. The marginal effects for the predicted value y after a regress are the same as the coefficients. To obtain the elasticities of form $\partial \log y / \partial \log x$, we specify the eyex option:

```
. regress mpg weight length
```

Source	SS	df	MS
Model	1616.08062	2	808.040312
Residual	827.378835	71	11.653223
Total	2443.45946	73	33.4720474

```
Number of obs =      74
F(  2,    71) =   69.34
Prob > F      =  0.0000
R-squared     =  0.6614
Adj R-squared =  0.6519
Root MSE      =  3.4137
```

mpg	Coef.	Std. Err.	t	P>|t|	[95% Conf. Interval]
weight	-.0038515	.001586	-2.43	0.018	-.0070138 -.0006891
length	-.0795935	.0553577	-1.44	0.155	-.1899736 .0307867
_cons	47.88487	6.08787	7.87	0.000	35.746 60.02374

```
. mfx, eyex
Elasticities after regress
      y  = Fitted values (predict)
         = 21.297297
```

variable	ey/ex	Std. Err.	z	P>|z|	[95% C.I.]	X
weight	-.5460497	.22509	-2.43	0.015	-.987208 -.104891	3019.46
length	-.7023518	.48867	-1.44	0.151	-1.66012 .255414	187.932

The first line of the output indicates that the elasticities were calculated after a `regress` estimation. The title of the second column of the table gives the form of the elasticities, $\partial \log y / \partial \log x$, the percent change in y for a 1 percent change in x.

If the independent variables have been log-transformed already, we will want the elasticities of the form $\partial \log y / \partial x$ instead.

```
. generate lnweight = ln(weight)

. generate lnlength = ln(length)

. regress mpg lnweight lnlength
      Source |       SS       df       MS              Number of obs =      74
-------------+------------------------------           F(  2,    71) =   74.00
       Model | 1651.28916      2  825.644581           Prob > F      =  0.0000
    Residual | 792.170298     71  11.1573281           R-squared     =  0.6758
-------------+------------------------------           Adj R-squared =  0.6667
       Total | 2443.45946     73  33.4720474           Root MSE      =  3.3403

------------------------------------------------------------------------------
         mpg |      Coef.   Std. Err.      t    P>|t|     [95% Conf. Interval]
-------------+----------------------------------------------------------------
    lnweight |  -13.5974    4.692504    -2.90   0.005    -22.95398   -4.240811
    lnlength | -9.816726    10.40316    -0.94   0.349    -30.56005    10.92659
       _cons |  181.1196    22.18429     8.16   0.000     136.8853    225.3538
------------------------------------------------------------------------------

. mfx, eydx

Elasticities after regress
      y  = Fitted values (predict)
         = 21.297297

------------------------------------------------------------------------------
variable |      ey/dx   Std. Err.      z    P>|z|  [    95% C.I.   ]        X
---------+--------------------------------------------------------------------
lnweight | -.6384565      .22064   -2.89   0.004   -1.0709 -.206009   7.97875
lnlength | -.4609376      .48855   -0.94   0.345   -1.41847  .496594   5.22904
------------------------------------------------------------------------------
```

Note that, although the interpretation is the same, the results for `eyex` and `eydx` differ since we are fitting different models.

If the dependent variable were log-transformed, we would specify `dyex` instead.   ◁

Saved Results

In addition to the `e()` results from the preceding estimation, `mfx` saves in `e()`:

Scalars
    e(Xmfx_y)   value of y given X
    e(Xmfx_off)   value of mean of the offset variable or `log` of the exposure variable
    e(Xmfx_off#)   value of mean of the offset variable for equation #

Macros
    e(Xmfx_type)   dydx, eyex, eydx or dyex
    e(Xmfx_discrete)   discrete or nodiscrete
    e(Xmfx_cmd)   mfx
    e(Xmfx_label_p)   label of the predict option
    e(Xmfx_predict)   the predict option
    e(Xmfx_dummy)   corresponding to independent variables; 1 means dummy, 0 means continuous
    e(Xmfx_variables)   corresponding to independent variables; 1 means marginal effect calculated, 0 otherwise
    e(Xmfx_method)   linear or nonlinear

Matrices

e(Xmfx_dydx)	marginal effects
e(Xmfx_se_dydx)	standard errors of the marginal effects
e(Xmfx_eyex)	elasticities of form eyex
e(Xmfx_se_eyex)	standard errors of elasticities of form eyex
e(Xmfx_eydx)	elasticities of form eydx
e(Xmfx_se_eydx)	standard errors of elasticities of form eydx
e(Xmfx_dyex)	elasticities of form dyex
e(Xmfx_se_dyex)	standard errors of elasticities of form dyex
e(Xmfx_X)	values around which marginal effects (elasticities) were estimated

Methods and Formulas

mfx is implemented as an ado-file.

After an estimation, mfx calculates marginal effects (elasticities) and their standard errors. A marginal effect of a continuous independent variable x is the partial derivative, with respect to x, of the prediction function f specified in mfx's predict option; see Greene (2003, 668) for more information about marginal effects. If no prediction function is specified, the default prediction for the preceding estimation command is used. This derivative is evaluated at the values of the independent variables specified in the at option of mfx or, if none is specified, at the default values that are the means of the independent variables. If there were any offsets in the preceding estimation, the derivative is evaluated at the means of the offset variables.

For a dummy variable—one that only takes the values zero or one in the estimation sample—a difference rather than a derivative is computed. The difference is the value of the prediction function at one, minus its value at zero.

For a continuous variable, the derivative is calculated numerically, which means that it approximates the derivative by using the following formula with an appropriately small h,

$$\frac{\partial y}{\partial x_j} = \lim_{h \to 0} \frac{f(x_1, \ldots, x_j + h, \ldots, x_p, \beta_0, \ldots, \beta_p) - f(x_1, \ldots, x_j, \ldots, x_p, \beta_0, \ldots, \beta_p)}{h}. \quad (1)$$

The delta method is used to estimate the variance of the marginal effect (Greene 2003, 70). The marginal effect is a function of only the coefficients of the model since all other variables are held constant at the values at which the marginal effect is sought.

$$\text{Var}\left(\frac{\partial y}{\partial x_j}\right) = \mathbf{D}_j' \mathbf{V} \mathbf{D}_j,$$

where $\mathbf{V}$ is the variance–covariance matrix from the estimation and $\mathbf{D}_j$ is the column vector whose kth entry is the partial derivative of the marginal effect of x_j, with respect to the coefficient of the kth independent variable:

$$(\mathbf{D}_j)_k = \frac{\partial}{\partial \beta_k} \frac{\partial y}{\partial x_j}$$

Thus to compute a single standard error, the derivative of the marginal effect is computed with respect to each coefficient in the model.

Computing the derivative of a function f with respect to a variable x can be time consuming because an iterative algorithm must be used to find an appropriately small change in x for use in (1). The command `mfx` avoids this type of iteration as much as possible. If the independent variables and coefficients appear in the formula for the prediction function f only in the sum

$$\mathbf{x}\boldsymbol{\beta} = \beta_0 + \mathbf{x}\boldsymbol{\beta}_x = \beta_0 + \sum_{j=1}^{p} x_j \beta_j,$$

a marked simplification in the computation of the marginal effects and their standard errors can be made.

An example of a prediction that satisfies this condition is the predicted probability of success following `logistic`:

$$f(x_1, \ldots x_p, \beta_0, \ldots, \beta_p) = \frac{\exp(\mathbf{x}\boldsymbol{\beta})}{1 + \exp(\mathbf{x}\boldsymbol{\beta})}$$

An example when this condition is not satisfied is the predicted hazard ratio following `streg` without the option `noconstant`:

$$f(x_1, \ldots x_p, \beta_0, \ldots, \beta_p) = \exp(\mathbf{x}\boldsymbol{\beta}_x)$$

The constant β_0 is missing from the sum.

For this condition to be satisfied after multiple-equation estimation, say for example, an estimation with two equations, the variables and coefficients can only appear as part of the two sums, that of the first equation $\mathbf{x}\boldsymbol{\beta}$ and that of the second equation $\mathbf{z}\boldsymbol{\gamma}$. If the same variable appears in both equations, even the linear predictor for the first equation does not satisfy this condition: if $x_{j_0} = z_{k_0}$ for some j_0 and k_0, then

$$\mathbf{x}\boldsymbol{\beta} = \sum_{j=0}^{p} x_j \beta_j = z_{k_0}\beta_{j_0} + \sum_{j=0}^{j_0-1} x_j \beta_j + \sum_{j=j_0+1}^{p} x_j \beta_j$$

so z_{k_0} is appearing, but not as part of the sum $\mathbf{z}\boldsymbol{\gamma} = \sum_{k=0}^{q} z_k \gamma_k$.

If this condition is satisfied, the linear-form restriction has been met, and the linear method, to be described below, is used to estimate the marginal effects and their standard errors. If not, the usual method, described above, is used and is called the nonlinear method. Following a multiple-equation estimation with any independent variables common to more than one equation, the nonlinear method will always be used.

We begin our description of the linear method with the easiest case, a single-equation estimation. Using the chain rule, we can write the marginal effect as

$$\frac{\partial y}{\partial x_j} = \frac{dy}{d(\mathbf{x}\boldsymbol{\beta})} \frac{\partial(\mathbf{x}\boldsymbol{\beta})}{\partial x_j} = \frac{dy}{d(\mathbf{x}\boldsymbol{\beta})} \beta_j$$

Note that the same derivative, $dy/d(\mathbf{x}\boldsymbol{\beta})$, is used for every x_j. To calculate it, we use the same formula in reverse. Since it doesn't matter which variable is used, we use the first one x_1:

$$\frac{dy}{d(\mathbf{x}\boldsymbol{\beta})} = \frac{1}{\beta_1} \frac{\partial y}{\partial x_1}$$

Therefore only one derivative $\partial y/\partial x_1$ needs to be calculated by the usual, nonlinear, method, and all marginal effects are then obtained by multiplying the derivative by the appropriate coefficient. Thus the linear method is generally much faster than the nonlinear method.

To compute the standard errors, we need the second derivatives. If j is not equal to k, using the chain rule we have

$$\frac{\partial}{\partial \beta_k} \frac{\partial y}{\partial x_j} = \beta_j \ x_k \ \frac{d}{d(\mathbf{x}\boldsymbol{\beta})} \ \frac{dy}{d(\mathbf{x}\boldsymbol{\beta})} \tag{2}$$

If j is equal to k, using the product rule we have

$$\frac{\partial}{\partial \beta_j} \frac{\partial y}{\partial x_j} = \beta_j \ x_j \ \frac{d}{d(\mathbf{x}\boldsymbol{\beta})} \ \frac{dy}{d(\mathbf{x}\boldsymbol{\beta})} \ + \ \frac{dy}{d(\mathbf{x}\boldsymbol{\beta})} \tag{3}$$

We obtain the second derivative, again using the chain rule

$$\frac{d}{d(\mathbf{x}\boldsymbol{\beta})} \ \frac{dy}{d(\mathbf{x}\boldsymbol{\beta})} = \frac{1}{\beta_1^2} \ \frac{\partial^2 y}{\partial x_1^2}$$

Now we turn to multiple equations. The linear method will only be used when there are no variables in common between the equations. For marginal effects, the formulas developed above apply in each equation separately. For standard errors, we will consider the case of two equations. Suppose that x_j and x_k are both in the first equation and j is not equal to k. Then $\partial/\partial \beta_k (\partial y/\partial x_j)$ is calculated as in (2) above, since x_j does not appear in the second equation, making $\partial(\mathbf{z}\boldsymbol{\gamma})/\partial x_j = 0$, and β_k is in equation one, making $\partial(\mathbf{z}\boldsymbol{\gamma})/\partial \beta_k = 0$. If x_j and x_k are both in the first equation and j is equal to k, we use the product rule and obtain the same as (3) above. Now suppose that x_j is in equation one and z_k is in equation two. Then

$$\frac{\partial}{\partial \gamma_k} \frac{\partial y}{\partial x_j} = \beta_j \ z_k \ \frac{d}{d(\mathbf{z}\boldsymbol{\gamma})} \ \frac{dy}{d(\mathbf{x}\boldsymbol{\beta})}$$

The second derivative is calculated by $d/d(\mathbf{z}\boldsymbol{\gamma})\{dy/d(\mathbf{x}\boldsymbol{\beta})\} = 1/(\gamma_1 \beta_1)\{\partial^2 y/(\partial z_1 \partial x_1)\}$.

In the case of multiple equations, it is possible to have an equation that is a constant only, such as an ancillary parameter. Then it is not possible to obtain $d^2y/d(\mathbf{z}\boldsymbol{\gamma})d(\mathbf{x}\boldsymbol{\beta})$ by converting to derivatives with respect to the independent variables, so it is evaluated directly. For example, if equation two had only a constant term so that $\mathbf{z}\boldsymbol{\gamma} = \gamma_0$, then $d^2y/d(\mathbf{z}\boldsymbol{\gamma})d(\mathbf{x}\boldsymbol{\beta}) = 1/\beta_1 \ \partial/\partial x_1(dy/d\gamma_0)$.

References

Greene, W. H. 2003. *Econometric Analysis*. 5th ed. Upper Saddle River, NJ: Prentice Hall.

Also See

Background: [U] **20 Estimation and postestimation commands**,

[R] **predict**,

[P] **_predict**

Title

mkspline — Linear spline construction

Syntax

Linear spline with knots at specified points

> mkspline *newvar₁* #₁ [*newvar₂* #₂ [...]] *newvarₖ* = *oldvar* [*if*] [*in*] [, marginal]

Linear spline with knots equally spaced or at percentiles of data

> mkspline *stubname* # = *oldvar* [*if*] [*in*] [, marginal pctile]

Description

mkspline creates variables containing a linear spline of *oldvar*.

In the first syntax, mkspline creates *newvar₁*, ..., *newvarₖ* containing a linear spline of *oldvar* with knots at the specified #₁, ..., #ₖ₋₁.

In the second syntax, mkspline creates # variables named *stubname*1, ..., *stubname*# containing a linear spline of *oldvar*. The knots are equally spaced over the range of *oldvar* or are placed at the percentiles of *oldvar*.

Options

___| Options |_____

marginal specifies that the new variables be constructed so that, when used in estimation, the coefficients represent the change in the slope from the preceding interval. The default is to construct the variables so that, when used in estimation, the coefficients measure the slopes for the interval.

pctile is allowed only with the second syntax. It specifies that the knots be placed at percentiles of the data rather than being equally spaced based on the range.

Remarks

Linear splines allow estimating the relationship between y and x as a piecewise linear function, which is a function composed of linear segments—straight lines. One linear segment represents the function for values of x below x_0, another linear segment handles values between x_0 and x_1, and so on. The linear segments are arranged so that they join at x_0, x_1, ..., which are called the knots. An example of a piecewise linear function is shown below.

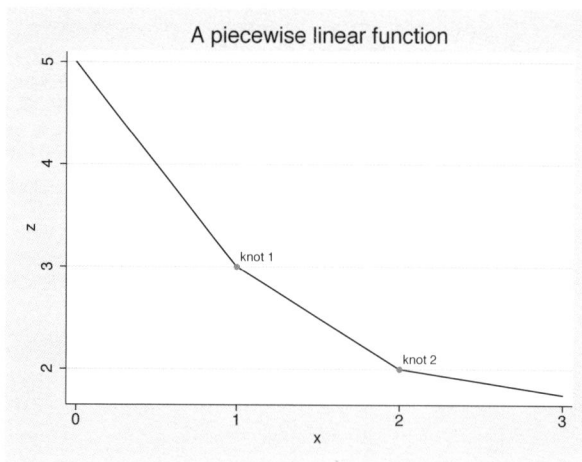

A piecewise linear function

▷ Example 1

We wish to fit a model of log income on education and age using a piecewise linear function for age:

$$\text{lninc} = b_0 + b_1\,\text{educ} + f(\text{age}) + u$$

The knots are to be placed at ten-year intervals: 20, 30, 40, 50, and 60.

```
. mkspline age1 20 age2 30 age3 40 age4 50 age5 60 age6 = age, marginal
. regress lninc educ age1-age6
  (output omitted )
```

Since we specified the `marginal` option, we could test whether the age effect is the same in the 30–40 and 40–50 intervals by asking whether the `age4` coefficient is zero. With the `marginal` option, coefficients measure the change in slope from the preceding group. Specifying `marginal` changes only the interpretation of the coefficients; the same model is fitted in either case. Without the `marginal` option, the interpretation of the coefficients would have been

$$\frac{dy}{d\text{age}} = \begin{cases} a_1 & \text{if age} < 20 \\ a_2 & \text{if } 20 \leq \text{age} < 30 \\ a_3 & \text{if } 30 \leq \text{age} < 40 \\ a_4 & \text{if } 40 \leq \text{age} < 50 \\ a_5 & \text{if } 50 \leq \text{age} < 60 \\ a_6 & \text{otherwise.} \end{cases}$$

With the `marginal` option specified, the interpretation is

$$\frac{dy}{d\text{age}} = \begin{cases} a_1 & \text{if age} < 20 \\ a_1 + a_2 & \text{if } 20 \leq \text{age} < 30 \\ a_1 + a_2 + a_3 & \text{if } 30 \leq \text{age} < 40 \\ a_1 + a_2 + a_3 + a_4 & \text{if } 40 \leq \text{age} < 50 \\ a_1 + a_2 + a_3 + a_4 + a_5 & \text{if } 50 \leq \text{age} < 60 \\ a_1 + a_2 + a_3 + a_4 + a_5 + a_6 & \text{otherwise.} \end{cases}$$

◁

▷ Example 2

As a second example, say that we have a binary outcome variable called outcome. We are beginning an analysis and wish to parameterize the effect of dosage on outcome. We wish to divide the data into five equal-width groups of dosage for the piecewise linear function.

```
. mkspline dose 5 = dosage
. logistic outcome dose1-dose5
  (output omitted )
```

mkspline dose 5 = dosage creates five variables—dose1, dose2, ..., dose5—equally spacing the knots over the range of dosage. If dosage varied between 0 and 100, mkspline dose 5 = dosage has the same effect as typing

```
. mkspline dose1 20 dose2 40 dose3 60 dose4 80 dose5 = dosage
```

The pctile option sets the knots to divide the data into five equal sample-size groups rather than five equal-width ranges. Typing

```
. mkspline dose 5 = dosage, pctile
```

places the knots at the 20th, 40th, 60th, and 80th percentiles of the data.

◁

Methods and Formulas

mkspline is implemented as an ado-file.

Let V_i, $i = 1, \ldots, n$, be the variables to be created, k_i, $i = 1, \ldots, n-1$, be the corresponding knots, and $\mathcal{V}$ be the original variable (the command is mkspline V_1 k_1 V_2 k_2 ... V_n = $\mathcal{V}$). Then

$$V_1 = \min(\mathcal{V}, k_1)$$

$$V_i = \max\left\{\min(\mathcal{V}, k_i), k_{i-1}\right\} - k_{i-1} \quad i = 2, \ldots, n$$

If the marginal option is specified, the definitions are

$$V_1 = \mathcal{V}$$

$$V_i = \max(0, \mathcal{V} - k_{i-1}) \quad i = 2, \ldots, n$$

In the second syntax, mkspline stubname # = $\mathcal{V}$, so let m and M be the minimum and maximum of $\mathcal{V}$. Without the pctile option, knots are set at $m + (M - m)(i/n)$ for $i = 1, \ldots, n-1$. If pctile is specified, knots are set at the $100(i/n)$ percentiles, for $i = 1, \ldots, n-1$. Percentiles are calculated by egen's pctile() function.

References

Gould, W. W. 1993. sg19: Linear splines and piecewise linear functions. *Stata Technical Bulletin* 15: 13–17. Reprinted in *Stata Technical Bulletin Reprints*, vol. 3, pp. 98–104.

Greene, W. H. 2003. *Econometric Analysis*. 5th ed. Upper Saddle River, NJ: Prentice Hall.

Newson, R. 2000. *B*-splines parameterized by their values at reference points on the *x*-axis. *Stata Technical Bulletin* 57: 20–27.

Panis, C. 1994. sg24: The piecewise linear spline transformation. *Stata Technical Bulletin* 18: 27–29. Reprinted in *Stata Technical Bulletin Reprints*, vol. 3, pp. 146–149.

Also See

Related: [R] **fracpoly**

Title

ml — Maximum likelihood estimation

Syntax

ml model in interactive mode

> ml mo̲del   *method progname eq* [*eq* ...] [*if*] [*in*] [*weight*]
>
> [, *model_options* sv̲y *diparm_options*]

ml model in noninteractive mode

> ml mo̲del   *method progname eq* [*eq* ...] [*if*] [*in*] [*weight*] , ma̲ximize
>
> [*model_options* sv̲y *diparm_options* *noninteractive_options*]

Noninteractive mode is invoked by specifying option maximize. Use maximize when you want to use ml as a subroutine of another ado-file or program and you want to carry forth the problem, from definition to posting of final results, in one command.

> ml clear

> ml qu̲ery

> ml check

> ml sea̲rch   [[/]*eqname*[:] $#_{lb}$ $#_{ub}$] [...] [, *search_options*]

> ml pl̲ot   [*eqname*:]*name* [# [# [#]]] [, sa̲ving(*filename*[, replace])]

> ml init   { [*eqname*:]*name=#* | /*eqname=#* } [...]

> ml init   # [# ...] , copy

> ml init   *matname* [, copy skip]

> ml rep̲ort

> ml trace   { on | off }

> ml count   [clear | on | off]

> ml ma̲ximize   [, *ml_maximize_options* *display_options* *eform_option*]

> ml gr̲aph   [#] [, sa̲ving(*filename*[, replace])]

> ml di̲splay   [, *display_options* *eform_option*]

> ml fo̲otnote

where *method* is { lf | d0 | d1 | d1debug | d2 | d2debug },

186

eq is the equation to be estimated, enclosed in parentheses, and optionally with a name to be given to the equation, preceded by a colon,

$$\left(\left[\,eqname:\,\right]\,\left[\,varnames\,=\,\right]\,\left[\,varnames\,\right]\,\left[\,,\,eq\_options\,\right]\right)$$

or *eq* is the name of a parameter, such as sigma with a slash in front

/*eqname*   which is equivalent to   (*eqname*:)

and *diparm_options* is one or more diparm(*diparm_args*) options where *diparm_args* is either __sep__ or anything accepted by the _diparm command; see help _diparm.

eq_options	description
noconstant	do not include an intercept in the equation
offset(*varname_o*)	include *varname_o* in model with coefficient constrained to 1
exposure(*varname_e*)	include ln(*varname_e*) in model with coefficient constrained to 1

model_options	description
robust	compute standard errors using the robust/sandwich estimator
cluster(*varname*)	adjust standard errors for intragroups correlation; implies robust
constraints(*numlist*)	constraints by number to be applied
constraints(*matname*)	matrix that contains the constraints to be applied
nocnsnotes	do not display notes when constraints are dropped
title(*string*)	place a title on the estimation output
nopreserve	do not preserve the estimation subsample in memory
collinear	keep collinear variables within equations
missing	keep observations containing variables with missing values
lf0($\#_k \#_{ll}$)	number of parameters and log-likelihood value of the "constant-only" model
continue	specifies that a model has been fitted and sets the initial values b_0 for the model to be fitted based on those results
waldtest(#)	perform a Wald test; see *Options for use with ml model in interactive or noninteractive mode* below
obs(#)	number of observations
noscvars	do not create and pass score variables to likelihood-evaluator program; seldom used
crittype(*string*)	describe the criterion optimized by ml
subpop(*varname*)	compute estimates for the single subpopulation
srssubpop	compute deff and deft using an estimate of simple-random-sampling variance
nosvyadjust	carry out Wald test as $W/k \sim F(k, d)$
technique(nr)	Stata's modified Newton–Raphson (NR) algorithm
technique(bhhh)	Berndt–Hall–Hall–Hausman (BHHH) algorithm
technique(dfp)	Davidon–Fletcher–Powell (DFP) algorithm
technique(bfgs)	Broyden–Fletcher–Goldfarb–Shanno (BFGS) algorithm
vce(oim)	estimate using the observed information matrix
vce(opg)	estimate using the outer product of the coefficient gradients
vce(robust)	synonym for robust
vce(native)	see *Options for use with ml model in interactive or noninteractive mode* below

noninteractive_options	description
init(*ml_init_args*)	set the initial values $\mathbf{b}_0$
search(on)	equivalent to ml search, repeat(0); the default
search(norescale)	equivalent to ml search, repeat(0) norescale
search(quietly)	same as search(on), except that output is suppressed
search(off)	prevents calling ml search
repeat(#)	ml search's repeat() option; see below
bounds(*ml_search_bounds*)	specify bounds for ml search
nowarning	suppress "convergence not achieved" message of iterate(0)
novce	substitute the zero matrix for the variance matrix
score(*newvars*)	new variables containing the contribution to the score
maximize_options	control the maximization process; seldom used

search_options	description
repeat(#)	number of random attempts to find better initial-value vector; default is repeat(10) in interactive mode and repeat(0) in noninteractive mode
restart	use random actions to find starting values; not recommended
norescale	do not rescale to improve parameter vector; not recommended
maximize_options	control the maximization process; seldom used

ml_maximize_options	description
nowarning	suppress "convergence not achieved" message of iterate(0)
novce	substitute the zero matrix for the variance matrix
score(*newvars* | *stub**)	new variables containing the contribution to the score
nooutput	suppress the display of the final results
noclear	do not clear ml problem definition after model has converged
maximize_options	control the maximization process; seldom used

display_options	description
noheader	suppress display of the header above the coefficient table
nofootnote	suppress display of the footnote below the coefficient table
level(#)	set confidence level; default is level(95)
first	display coefficient table reporting results for first equation only
neq(#)	display coefficient table reporting first # equations
plus	display coefficient table ending in dashes–plus-sign–dashes

eform_option	description
<u>ef</u>orm(*string*)	display exponentiated coefficients; column title is "*string*"
<u>ef</u>orm	display exponentiated coefficients; column title is "exp(b)"
hr	report hazard ratios
<u>irr</u>	report incidence-rate ratios
or	report odds ratios
<u>rrr</u>	report relative-risk ratios

fweights, aweights, iweights, and pweights are allowed; see [U] **11.1.6 weight**. With all but method lf,
  you must write your likelihood-evaluation program carefully if pweights are to be specified, and
  pweights may not be specified with method d0. See Gould, Pitblado, Sribney (2003, chapter 4) for details.
See [U] **20 Estimation and postestimation commands** for additional capabilities of estimation commands.
  To redisplay results, type ml display.

Syntax of subroutines for use by method d0, d1, and d2 evaluators

mleval   *newvar* = *vecname* $\left[\,, \texttt{eq}(\#)\,\right]$

mleval   *scalarname* = *vecname*, scalar $\left[\texttt{eq}(\#)\right]$

mlsum   *scalarname*$_{\mathrm{lnf}}$ = *exp* $\left[\textit{if}\right]$ $\left[\,, \underline{\texttt{noweight}}\right]$

mlvecsum   *scalarname*$_{\mathrm{lnf}}$ *rowvecname* = *exp* $\left[\textit{if}\right]$ $\left[\,, \texttt{eq}(\#)\right]$

mlmatsum   *scalarname*$_{\mathrm{lnf}}$ *matrixname* = *exp* $\left[\textit{if}\right]$ $\left[\,, \texttt{eq}(\#\left[\,,\#\right])\right]$

mlmatbysum *scalarname*$_{\mathrm{lnf}}$ *matrixname* *varname*$_a$ *varname*$_b$ $\left[\textit{varname}_c\right]$ $\left[\textit{if}\right]$,

            by(*varname*) $\left[\,\texttt{eq}(\#\left[\,,\#\right])\right]$

Syntax of user-written evaluator

Summary of notation

The log-likelihood function is $\ln L(\theta_{1j}, \theta_{2j}, \ldots, \theta_{Ej})$, where $\theta_{ij} = \mathbf{x}_{ij}\mathbf{b}_i$, $j = 1, \ldots, N$ indexes
observations, and $i = 1, \ldots, E$ indexes the linear equations defined by ml model. If the likelihood
satisfies the linear-form restrictions, it can be decomposed as $\ln L = \sum_{j=1}^{N} \ln \ell(\theta_{1j}, \theta_{2j}, \ldots, \theta_{Ej})$.

Method lf evaluators:

```
program progname
        version 9
        args lnf theta1 [theta2 ... ]
        /* if you need to create any intermediate results: */
        tempvar tmp1 tmp2 ...
        quietly gen double 'tmp1' = ...
        ...
        quietly replace 'lnf' = ...
end
```

```
where
'lnf'       variable to be filled in with observation-by-observation values of ln ℓ_j
'theta1'    variable containing evaluation of 1st equation θ_1j=x_1j b_1
'theta2'    variable containing evaluation of 2nd equation θ_2j=x_2j b_2
```
where
'lnf'   variable to be filled in with observation-by-observation values of $\ln \ell_j$
'theta1'   variable containing evaluation of 1st equation $\theta_{1j}=\mathbf{x}_{1j}\mathbf{b}_1$
'theta2'   variable containing evaluation of 2nd equation $\theta_{2j}=\mathbf{x}_{2j}\mathbf{b}_2$

Method d0 evaluators:

```
program progname
        version 9
        args todo b lnf

        tempvar theta1 theta2 ...
        mleval 'theta1' = 'b', eq(1)
        mleval 'theta2' = 'b', eq(2) // if there is a θ₂
        ...

        // if you need to create any intermediate results:
        tempvar tmp1 tmp2 ...
        gen double 'tmp1' = ...
        ...

        mlsum 'lnf' = ...
end
```

where
'todo'	always contains 1 (may be ignored)
'b'	full parameter row vector $\mathbf{b} = (\mathbf{b}_1, \mathbf{b}_2, ..., \mathbf{b}_E)$
'lnf'	scalar to be filled in with overall $\ln L$

Method d1 evaluators:

```
program progname
        version 9
        args todo b lnf g [negH g1 [g2 ... ] ]

        tempvar theta1 theta2 ...
        mleval 'theta1' = 'b', eq(1)
        mleval 'theta2' = 'b', eq(2) // if there is a θ₂
        ...

        // if you need to create any intermediate results:
        tempvar tmp1 tmp2 ...
        gen double 'tmp1' = ...
        ...

        mlsum 'lnf' = ...
        if ('todo'==0 | 'lnf'>=.) exit

        tempname d1 d2 ...
        mlvecsum 'lnf' 'd1' = formula for ∂ ln ℓⱼ/∂θ₁ⱼ, eq(1)
        mlvecsum 'lnf' 'd2' = formula for ∂ ln ℓⱼ/∂θ₂ⱼ, eq(2)
        ...

        matrix 'g' = ('d1','d2', ... )
end
```

where
'todo'	contains 0 or 1
	$0 \Rightarrow$ 'lnf' to be filled in; $1 \Rightarrow$ 'lnf' and 'g' to be filled in
'b'	full parameter row vector $\mathbf{b} = (\mathbf{b}_1, \mathbf{b}_2, ..., \mathbf{b}_E)$
'lnf'	scalar to be filled in with overall $\ln L$
'g'	row vector to be filled in with overall $\mathbf{g} = \partial \ln L / \partial \mathbf{b}$
'negH'	argument to be ignored
'g1'	variable optionally to be filled in with $\partial \ln \ell_j / \partial \mathbf{b}_1$
'g2'	variable optionally to be filled in with $\partial \ln \ell_j / \partial \mathbf{b}_2$
...	

Method d2 evaluators:

```
program progname
        version 9
        args todo b lnf g negH [g1 [g2 ... ] ]

        tempvar theta1 theta2 ...
        mleval 'theta1' = 'b', eq(1)
        mleval 'theta2' = 'b', eq(2) // if there is a θ2
        ...

        // if you need to create any intermediate results:
        tempvar tmp1 tmp2 ...
        gen double 'tmp1' = ...
        ...

        mlsum 'lnf' = ...
        if ('todo'==0 | 'lnf'>=.) exit

        tempname d1 d2 ...
        mlvecsum 'lnf' 'd1' = formula for ∂ln ℓ_j/∂θ_{1j}, eq(1)
        mlvecsum 'lnf' 'd2' = formula for ∂ln ℓ_j/∂θ_{2j}, eq(2)
        ...
        matrix 'g' = ('d1','d2', ... )
        if ('todo'==1 | 'lnf'>=.) exit

        tempname d11 d12 d22 ...
        mlmatsum 'lnf' 'd11' = formula for -∂² ln ℓ_j/∂θ²_{1j}, eq(1)
        mlmatsum 'lnf' 'd12' = formula for -∂² ln ℓ_j/∂θ_{1j}∂θ_{2j}, eq(1,2)
        mlmatsum 'lnf' 'd22' = formula for -∂² ln ℓ_j/∂θ²_{2j}, eq(2)
        ...
        matrix 'negH' = ('d11','d12', ... \ 'd12','d22', ... )
end
```

where

'todo'	contains 0, 1, or 2
	$0 \Rightarrow$ 'lnf' to be filled in; $1 \Rightarrow$ 'lnf' and 'g' to be filled in;
	$2 \Rightarrow$ 'lnf', 'g', and 'negH' to be filled in
'b'	full parameter row vector $\mathbf{b}=(\mathbf{b}_1, \mathbf{b}_2, ..., \mathbf{b}_E)$
'lnf'	scalar to be filled in with overall $\ln L$
'g'	row vector to be filled in with overall $\mathbf{g} = \partial \ln L/\partial \mathbf{b}$
'negH'	matrix to be filled in with overall negative Hessian $-\mathbf{H} = -\partial^2 \ln L/\partial\mathbf{b}\partial\mathbf{b}'$
'g1'	variable optionally to be filled in with $\partial \ln \ell_j/\partial \mathbf{b}_1$
'g2'	variable optionally to be filled in with $\partial \ln \ell_j/\partial \mathbf{b}_2$
...	

Global macros for use by all evaluators

$ML_y1	name of first dependent variable
$ML_y2	name of second dependent variable, if any
...	
$ML_samp	variable containing 1 if observation to be used; 0 otherwise
$ML_w	variable containing weight associated with observation or 1 if no weights specified

Method lf evaluators can ignore $ML_samp, but restricting calculations to the $ML_samp==1 subsample will speed execution. Method lf evaluators must ignore $ML_w; application of weights is handled by the method itself.

Methods d0, d1, and d2 can ignore $ML_samp as long as ml model's nopreserve option is not specified. Methods d0, d1, and d2 will run more quickly if nopreserve is specified. Methods d0, d1, and d2 evaluators can ignore $ML_w only if they use mlsum, mlvecsum, mlmatsum, and mlmatbysum to produce all final results.

Description

ml model defines the current problem.

ml clear clears the current problem definition. This command is rarely, if ever, used because, when you type ml model, any previous problem is automatically cleared.

ml query displays a description of the current problem.

ml check verifies that the log-likelihood evaluator you have written seems to work. We strongly recommend using this command.

ml search searches for (better) initial values. We recommend using this command.

ml plot provides a graphical way of searching for (better) initial values.

ml init provides a way to specify initial values.

ml report reports the values of $\ln L$, its gradient, and its negative Hessian at the initial values or current parameter estimates $\mathbf{b}_0$.

ml trace traces the execution of the user-defined log-likelihood evaluation program.

ml count counts the number of times the user-defined log-likelihood evaluation program is called; this command is seldom used. ml count clear clears the counter. ml count on turns on the counter. ml count without arguments reports the current values of the counters. ml count off stops counting calls.

ml maximize maximizes the likelihood function and reports final results. Once ml maximize has successfully completed, the previously mentioned ml commands may no longer be used unless noclear is specified. ml graph and ml display may be used whether or not noclear is specified.

ml graph graphs the log-likelihood values against the iteration number.

ml display redisplays final results.

ml footnote displays a warning message when the model did not converge within the specified number of iterations.

progname is the name of a program you write to evaluate the log-likelihood function. In this documentation, it is referred to as the user-written evaluator or sometimes simply as the evaluator. The program you write is written in the style required by the method you choose. The methods are lf, d0, d1, and d2. Thus if you choose to use method lf, your program is called a method lf evaluator. Method lf evaluators are required to evaluate the observation-by-observation log likelihood $\ln \ell_j$, $j = 1, \ldots, N$. Method d0 evaluators are required to evaluate the overall log likelihood $\ln L$. Method d1 evaluators are required to evaluate the overall log likelihood and its gradient vector $\mathbf{g} = \partial \ln L / \partial \mathbf{b}$. Method d2 evaluators are required to evaluate the overall log likelihood, its gradient, and its negative Hessian matrix $-H = -\partial^2 \ln L / \partial \mathbf{b} \partial \mathbf{b}'$.

mleval is a subroutine used by method d0, d1, and d2 evaluators to evaluate the coefficient vector $\mathbf{b}$ that they are passed.

mlsum is a subroutine used by method d0, d1, and d2 evaluators to define the value $\ln L$ that is to be returned.

mlvecsum is a subroutine used by method d1 and d2 evaluators to define the gradient vector $\mathbf{g}$ that is to be returned. It is suitable for use only when the likelihood function meets the linear-form restrictions.

mlmatsum is a subroutine used by method d2 evaluators to define the negative Hessian matrix, $-\mathbf{H}$, that is to be returned. It is suitable for use only when the likelihood function meets the linear-form restrictions.

`mlmatbysum` is a subroutine used by method d2 evaluators to help define the negative Hessian matrix, $-\mathbf{H}$, that is to be returned. It is suitable for use when the likelihood function contains terms made up of grouped sums, such as in panel-data models. For such models, use `mlmatsum` to compute the observation-level outer products and `mlmatbysum` to compute the group-level outer products. `mlmatbysum` requires that the data be sorted by the variable identified in the `by()` option.

Options for use with ml model in interactive or noninteractive mode

`robust` and `cluster(`*varname*`)` specify the robust variance estimator, as does specifying `pweights` or the `svy` option.

These options will work with a method lf evaluator; all you need do is specify them.

These options will not work with a method d0 evaluator, and specifying these options will result in an error message.

With method d1 or d2 evaluators in which the likelihood function satisfies the linear-form restrictions, these options will work only if you fill in the equation scores; otherwise, specifying these options will result in an error message.

`constraints(`*numlist* | *matname*`)` specifies the linear constraints to be applied during estimation. `constraints(`*numlist*`)` specifies the constraints by number. Constraints are defined using the `constraint` command; see [R] **constraint**. `constraint(`*matname*`)` specifies a matrix that contains the constraints.

`nocnsnotes` prevents notes from being displayed when constraints are dropped. A constraint will be dropped if it is inconsistent, contradicts other constraints, or causes some other error when the constraint matrix is being built. Constraints are checked in the order they are specified.

`title(`*string*`)` specifies the title to be placed on the estimation output when results are complete.

`nopreserve` specifies that it is not necessary for `ml` to ensure that only the estimation subsample is in memory when the user-written likelihood evaluator is called. `nopreserve` is irrelevant when you use method lf.

For the other methods, if `nopreserve` is not specified, `ml` saves the data in a file (preserves the original dataset) and drops the irrelevant observations before calling the user-written evaluator. This way, even if the evaluator does not restrict its attentions to the `$ML_samp==1` subsample, results will still be correct. Later, `ml` automatically restores the original dataset.

`ml` need not go through these machinations in the case of method lf because the user-written evaluator calculates observation-by-observation values, and `ml` itself sums the components.

`ml` goes through these machinations if and only if the estimation sample is a subsample of the data in memory. If the estimation sample includes every observation in memory, `ml` does not preserve the original dataset. Thus programmers must not alter the original dataset unless they `preserve` the data themselves.

We recommend that interactive users of `ml` not specify `nopreserve`; the speed gain is not worth the possibility of getting incorrect results.

We recommend that programmers specify `nopreserve`, but only after verifying that their evaluator really does restrict its attentions solely to the `$ML_samp==1` subsample.

`collinear` specifies that `ml` not remove the collinear variables within equations. There is no reason you would want to leave collinear variables in place, but this option is of interest to programmers who, in their code, have already removed collinear variables and do not want `ml` to waste computer time checking again.

missing specifies that observations containing variables with missing values not be eliminated from the estimation sample. There are two reasons you might want to specify missing:

Programmers may wish to specify missing because, in other parts of their code, they have already eliminated observations with missing values and do not want ml to waste computer time looking again.

You may wish to specify missing if your model explicitly deals with missing values. Stata's heckman command is a good example of this. In such cases, there will be observations where missing values are allowed and other observations where they are not—where their presence should cause the observation to be eliminated. If you specify missing, it is your responsibility to specify an if *exp* that eliminates the irrelevant observations.

lf0($\#_k$ $\#_{ll}$) is typically used by programmers. It specifies the number of parameters and log-likelihood value of the "constant-only" model so that ml can report a likelihood-ratio test rather than a Wald test. These values may have been analytically determined, or they may have been determined by a previous fitting of the constant-only model on the estimation sample.

Also see the continue option directly below.

If you specify lf0(), it must be safe for you to specify the missing option, too, else how did you calculate the log likelihood for the constant-only model on the same sample? You must have identified the estimation sample, and done so correctly, so there is no reason for ml to waste time rechecking your results. All of which is to say, do not specify lf0() unless you are certain your code identifies the estimation sample correctly.

lf0(), even if specified, is ignored if robust, cluster(), pweights, or svy is specified because, in that case, a likelihood-ratio test would be inappropriate.

continue is typically specified by programmers. It does two things:

First, it specifies that a model has just been fitted by either ml or some other estimation command, such as logit, and that the likelihood value stored in e(ll) and the number of parameters stored in e(b) as of this instant are the relevant values of the constant-only model. The current value of the log likelihood is used to present a likelihood-ratio test unless robust, cluster(), pweights, or svy is specified because, in that case, a likelihood-ratio test would be inappropriate.

Second, continue sets the initial values, b_0, for the model about to be fitted according to the e(b) currently stored.

The comments made about specifying missing with lf0() apply equally well in this case.

(*Continued on next page*)

waldtest(#) is typically specified by programmers. By default, ml presents a Wald test, but that is overridden if options lf0() or continue are specified. A Wald test is performed if robust, cluster(), or pweights are specified.

waldtest(0) prevents even the Wald test from being reported.

waldtest(-1) is the default. It specifies that a Wald test be performed by constraining all coefficients except for the intercept to 0 in the first equation. Remaining equations are to be unconstrained. A Wald test is performed if neither lf0() nor continue was specified, and a Wald test is forced if robust, cluster(), or pweights were specified.

waldtest(k) for $k \leq -1$ specifies that a Wald test be performed by constraining all coefficients except for intercepts to 0 in the first $|k|$ equations; remaining equations are to be unconstrained. A Wald test is performed if neither lf0() nor continue was specified, and a Wald test is forced if robust, cluster(), or pweights were specified.

waldtest(k) for $k \geq 1$ works like the options above, except that it forces a Wald test to be reported even if the information to perform the likelihood-ratio test is available and even if none of robust, cluster(), or pweights were specified. waldtest(k), $k \geq 1$, may not be specified with lf0().

obs(#) is used mostly by programmers. It specifies that the number of observations reported and ultimately stored in e(N) be #. Ordinarily, ml works that out for itself. Programmers may want to specify this option when, in order for the likelihood-evaluator to work for N observations, they first had to modify the dataset so that it contained a different number of observations.

noscvars is used mostly by programmers. It specifies that method d0, d1, or d2 is being used but that the likelihood-evaluation program does not calculate or use arguments 'g1', 'g2', etc., which are the score vectors. Thus ml can save a little time by not generating and passing those arguments.

crittype(string) is used mostly by programmers. It allows programmers to supply a string (up to 32 characters long) that describes the criterion that is being optimized by ml. The default is "log likelihood" for nonrobust and "log pseudolikelihood" for robust estimation.

svy indicates that ml is to pick up the svy settings set by svyset and use the robust variance estimator. This option requires the data to be svyset. svy may not be supplied with *weights* or the cluster() options.

subpop(*varname*) specifies that estimates be computed for the single subpopulation defined by the observations for which *varname* $\neq 0$. Typically, *varname* $= 1$ defines the subpopulation, and *varname* $= 0$ indicates observations not belonging to the subpopulation. For observations whose subpopulation status is uncertain, *varname* should be set to missing ('.'). This option requires the svy option.

srssubpop can be specified only if subpop() is specified. srssubpop requests that deff and deft be computed using an estimate of simple-random-sampling variance for sampling within a subpopulation. If srssubpop is not specified, deff and deft are computed using an estimate of simple-random-sampling variance for sampling from the entire population. Typically, srssubpop would be given when computing subpopulation estimates by strata or by groups of strata.

nosvyadjust specifies that the model Wald test be carried out as $W/k \sim F(k, d)$, where W is the Wald test statistic, k is the number of terms in the model excluding the constant term, d = total number of sampled PSUs minus the total number of strata, and $F(k, d)$ is an F distribution with k numerator degrees of freedom and d denominator degrees of freedom. By default, an adjusted Wald test is conducted: $(d - k + 1)W/(kd) \sim F(k, d - k + 1)$. See Korn and Graubard (1990) for a discussion of the Wald test and the adjustments thereof. This option requires the svy option.

technique(*algorithm_spec*) specifies how the likelihood function is to be maximized. The following algorithms are currently implemented in ml. For details, see Gould, Pitblado, and Sribney (2003).

technique(nr) specifies Stata's modified Newton–Raphson (NR) algorithm.

technique(bhhh) specifies the Berndt–Hall–Hall–Hausman (BHHH) algorithm.

technique(dfp) specifies the Davidon–Fletcher–Powell (DFP) algorithm.

technique(bfgs) specifies the Broyden–Fletcher–Goldfarb–Shanno (BFGS) algorithm.

The default is technique(nr).

You can switch between algorithms by specifying more than one in the technique() option. By default, ml will use an algorithm for five iterations before switching to the next algorithm. To specify a different number of iterations include the number after the technique in the option. For example, technique(bhhh 10 nr 1000) requests that ml perform 10 iterations using the BHHH algorithm followed by 1000 iterations using the NR algorithm, and then switch back to BHHH for 10 iterations, and so on. The process continues until convergence or until the maximum number of iterations is reached.

vce(oim | opg | robust | native) specifies the type of variance–covariance matrix. This option may not be combined with the svy option.

vce(oim) specifies that the standard errors and coefficient covariance matrix be estimated using the observed information matrix (that is, the inverse of the negative Hessian matrix). This is the default for all optimization algorithms except technique(bhhh).

vce(opg) specifies that the standard errors and coefficient covariance matrix be estimated using the outer product of the coefficient gradients with respect to the observation likelihoods. This is the default for technique(bhhh), except when the robust option is supplied.

vce(robust) specifies that the standard errors and coefficient covariance matrix be estimated using the Huber/White/sandwich estimator.

vce(native) specifies that the standard errors and coefficient covariance matrix be estimated using the information matrix defined by the algorithm used to maximize the likelihood function. For technique(nr), vce(native) is a synonym for vce(oim); and for technique(bhhh), vce(native) is a synonym for vce(opg). vce(native) is not allowed when switching optimization techniques.

Options for use with ml model in noninteractive mode

The following additional options are for use with ml model in noninteractive mode. Noninteractive mode is for programmers who use ml as a subroutine and want to issue a single command that will carry forth the estimation from start to finish.

maximize is required. It specifies noninteractive mode.

init(*ml_init_args*) sets the initial values $\mathbf{b}_0$. *ml_init_args* are whatever you would type after the ml init command.

search(on | norescale | quietly | off) specifies whether ml search is to be used to improve the initial values. search(on) is the default and is equivalent to running separately ml search, repeat(0). search(norescale) is equivalent to running separately ml search, repeat(0) norescale. search(quietly) is equivalent to search(on), except that it suppresses ml search's output. search(off) prevents the calling of ml search altogether.

repeat(#) is ml search's repeat() option. repeat(0) is the default.

bounds(*ml_search_bounds*) specifies the search bounds. The command ml model issues is ml search *ml_search_bounds*, repeat(#). Specifying search bounds is optional.

nowarning, novce, and score() are ml maximize's equivalent options.

maximize_options: <u>diff</u>icult, <u>tech</u>nique(*algorithm_spec*), <u>iter</u>ate(#), [<u>no</u>]log, <u>tra</u>ce, <u>grad</u>ient, showstep, <u>hess</u>ian, <u>shownrtol</u>erance, <u>tol</u>erance(#), <u>ltol</u>erance(#), <u>gtol</u>erance(#), <u>nrtol</u>erance(#), <u>nonrtol</u>erance, from(*init_specs*); see [R] **maximize**. These options are seldom used.

Options for use when specifying equations

noconstant specifies that the equation not include an intercept.

offset(*varname_o*) specifies that the equation be $\mathbf{xb} + varname_o$—that it include $varname_o$ with coefficient constrained to be 1.

exposure(*varname_e*) is an alternative to offset(*varname_o*); it specifies that the equation be $\mathbf{xb} + \ln(varname_e)$. The equation is to include $\ln(varname_e)$ with coefficient constrained to be 1.

Options for use with ml search

repeat(#) specifies the number of random attempts that are to be made to find a better initial-value vector. The default is repeat(10).

repeat(0) specifies that no random attempts be made. More correctly, repeat(0) specifies that no random attempts be made if the initial initial-value vector is a feasible starting point. If it is not, ml search will make random attempts, even if you specify repeat(0), because it has no alternative. The repeat() option refers to the number of random attempts to be made to improve the initial values. When the initial starting value vector is not feasible, ml search will make up to 1,000 random attempts to find starting values. It stops the instant it finds one set of values that works and then moves into its improve-initial-values logic.

repeat(*k*), $k > 0$, specifies the number of random attempts to be made to improve the initial values.

restart specifies that random actions be taken to obtain starting values and that the resulting starting values not be a deterministic function of the current values. Generally, you should not specify this option because, with restart, ml search intentionally does not produce as good a set of starting values as it could. restart is included for use by the optimizer when it gets into serious trouble. The random actions ensure that the actions of the optimizer and ml search, working together, do not result in a long, endless loop.

restart implies norescale, which is why we recommend that you do not specify restart. In testing, sometimes rescale worked so well that, even after randomization, the rescaler would bring the starting values right back to where they had been the first time and thus defeat the intended randomization.

norescale specifies that ml search not engage in its rescaling actions to improve the parameter vector. We do not recommend specifying this option because rescaling tends to work so well.

maximize_options: [<u>no</u>]log, <u>tra</u>ce; see [R] **maximize**. These options are seldom used.

Option for use with ml plot

saving(*filename*[, replace]) specifies that the graph be saved in *filename*.gph.
See [G] *saving_option*.

Options for use with ml init

copy specifies that the list of numbers or the initialization vector be copied into the initial-value vector by position rather than by name.

skip specifies that any parameters found in the specified initialization vector that are not also found in the model be ignored. The default action is to issue an error message.

Options for use with ml maximize

nowarning is allowed only with iterate(0). nowarning suppresses the "convergence not achieved" message. Programmers might specify iterate(0) nowarning when they have a vector **b** already containing the final estimates and want ml to calculate the variance matrix and post final estimation results. In that case, specify init(b) search(off) iterate(0) nowarning nolog.

novce is allowed only with iterate(0). novce substitutes the zero matrix for the variance matrix, which in effect posts estimation results as fixed constants.

score(*newvars* | *stub**) creates new variables containing the contributions to the score for each equation and ancillary parameter in the model; see [U] **20.15 Obtaining scores**.

If score(*newvars*) is specified, the *newvars* must contain k new variables, one for each equation in the model. If score(*stub**) is specified, variables named *stub*1, *stub*2, ... *stub*k are created.

The first variable contains $\partial \ln L_j / \partial (\mathbf{x}_j \mathbf{b}_1)$; the second variable contains $\partial \ln L_j / \partial (\mathbf{x}_j \mathbf{b}_2)$; and so on.

nooutput suppresses display of the final results. This is different from prefixing ml maximize with quietly in that the iteration log is still displayed (assuming that nolog is not specified).

noclear specifies that after the model has converged, the ml problem definition not be cleared. Perhaps you are having convergence problems and intend to run the model to convergence. If so, use ml search to see if those values can be improved, and then start the estimation again.

maximize_options: difficult, iterate(*#*), [no]log, trace, gradient, showstep, hessian, shownrtolerance, tolerance(*#*), ltolerance(*#*), gtolerance(*#*), nrtolerance(*#*), nonrtolerance; see [R] **maximize**. These options are seldom used.

display_options; see *Options for use with ml display* below.

eform_option; see *Options for use with ml display* below.

Option for use with ml graph

saving(*filename*[, replace]) specifies that the graph be saved in *filename*.gph.
See [G] *saving_option*.

Options for use with ml display

noheader suppresses the display of the header above the coefficient table that displays the final log-likelihood value, the number of observations, and the model significance test.

nofootnote suppresses the display of the footnote below the coefficient table, which displays a warning if the model fitted did not converge within the specified number of iterations. Use ml footnote to display the warning (1) if you add to the coefficient table using the plus option or (2) you have your own footnotes and want the warning to be last.

level(#) is the standard confidence-level option. It specifies the confidence level, as a percentage, for confidence intervals of the coefficients. The default is level(95) or as set by set level; see [U] **20.6 Specifying the width of confidence intervals**.

first displays a coefficient table reporting results for the first equation only, and the report makes it appear that the first equation is the only equation. This is used by programmers who estimate ancillary parameters in the second and subsequent equations and who wish to report the values of such parameters themselves.

neq(#) is an alternative to first. neq(#) displays a coefficient table reporting results for the first # equations. This is used by programmers who estimate ancillary parameters in the # + 1 and subsequent equations and who wish to report the values of such parameters themselves.

plus displays the coefficient table, but rather than ending the table in a line of dashes, ends it in dashes–plus-sign–dashes. This is so that programmers can write additional display code to add more results to the table and make it appear as if the combined result is one table. Programmers typically specify plus with options first or neq(). This option implies nofootnote.

eform_option: eform(*string*), eform, hr, irr, or, and rrr display the coefficient table in exponentiated form: for each coefficient, $\exp(b)$ rather than b is displayed, and standard errors and confidence intervals are transformed. Display of the intercept, if any, is suppressed. *string* is the table header that will be displayed above the transformed coefficients and must be 11 characters or fewer in length, for example, eform("Odds ratio"). The options eform, hr, irr, or, and rrr provide a default *string* equivalent to "exp(b)", "Haz. Ratio", "IRR", "Odds Ratio", and "RRR", respectively. These options may not be combined.

Options for use with mleval

eq(#) specifies the equation number i for which $\theta_{ij} = \mathbf{x}_{ij}\mathbf{b}_i$ is to be evaluated. eq(1) is assumed if eq() is not specified.

scalar asserts that the ith equation is known to evaluate to a constant, meaning that the equation was specified as (), (*name*:), or /*name* on the ml model statement. If you specify this option, the new variable created is created as a scalar. If the ith equation does not evaluate to a scalar, an error message is issued.

Option for use with mlsum

noweight specifies that weights ($ML_w) be ignored when summing the likelihood function.

Option for use with mlvecsum

eq(#) specifies the equation for which a gradient vector $\partial \ln L / \partial \mathbf{b}_i$ is to be constructed. The default is eq(1).

Option for use with mlmatsum

eq(#[,#]) specifies the equations for which the negative Hessian matrix is to be constructed. The default is eq(1), which is the same as eq(1,1), which means $-\partial^2 \ln L / \partial \mathbf{b}_1 \partial \mathbf{b}_1'$. Specifying eq($i,j$) results in $-\partial^2 \ln L / \partial \mathbf{b}_i \partial \mathbf{b}_j'$.

Options for use with mlmatbysum

by(*varname*) is required and specifies the group variable.

eq(#[,#]) specifies the equations for which the negative Hessian matrix is to be constructed. The default is eq(1), which is the same as eq(1,1), which means $-\partial^2 \ln L / \partial \mathbf{b}_1 \partial \mathbf{b}_1'$. Specifying eq($i,j$) results in $-\partial^2 \ln L / \partial \mathbf{b}_i \partial \mathbf{b}_j'$.

Remarks

For a thorough discussion of ml, see *Maximum Likelihood Estimation with Stata*, 2nd edition (Gould, Pitblado, and Sribney 2003). The book provides a tutorial introduction to ml, notes on advanced programming issues, and a discourse on maximum likelihood estimation from both theoretical and practical standpoints. See *Survey options and ml* at the end of *Remarks* for examples of the new svy options. For more information about survey estimation, see [SVY] **survey** and [SVY] **variance estimation**.

ml requires that you write a program that evaluates the log-likelihood function and, possibly, its first and second derivatives. The style of the program you write depends upon the method you choose: methods lf and d0 require that your program evaluate the log likelihood only, method d1 requires that your program evaluate the log likelihood and gradient, and method d2 requires that your program evaluate the log likelihood, gradient, and negative Hessian. Methods lf and d0 differ from each other in that, with method lf, your program is required to produce observation-by-observation log-likelihood values $\ln \ell_j$ and it is assumed that $\ln L = \sum_j \ln \ell_j$; with method d0, your program is only required to produce the overall value $\ln L$.

Once you have written the program—called an evaluator—you define a model to be fitted using ml model and obtain estimates using ml maximize. You might type

```
        . ml model ...
        . ml maximize
```

but we recommend that you type

```
        . ml model ...
        . ml check
        . ml search
        . ml maximize
```

ml check verifies your evaluator has no obvious errors, and ml search finds better initial values.

You fill in the `ml model` statement with (1) the method you are using, (2) the name of your program, and (3) the "equations". You write your evaluator in terms of θ_1, θ_2, ..., each of which has a linear equation associated with it. That linear equation might be as simple as $\theta_i = b_0$, it might be $\theta_i = b_1\text{mpg} + b_2\text{weight} + b_3$, or it might omit the intercept b_3. The equations are specified in parentheses on the `ml model` line.

Suppose that you are using method lf and the name of your evaluator program is `myprog`. The statement

```
. ml model lf myprog (mpg weight)
```

would specify a single equation with $\theta_i = b_1\text{mpg} + b_2\text{weight} + b_3$. If you wanted to omit b_3, you would type

```
. ml model lf myprog (mpg weight, nocons)
```

and if all you wanted was $\theta_i = b_0$, you would type

```
. ml model lf myprog ()
```

With multiple equations, you list the equations one after the other; so, if you typed

```
. ml model lf myprog (mpg weight) ()
```

you would be specifying $\theta_1 = b_1\text{mpg} + b_2\text{weight} + b_3$ and $\theta_2 = b_4$. You would write your likelihood in terms of θ_1 and θ_2. If the model was linear regression, θ_1 might be the **xb** part and θ_2 the variance of the residuals.

When you specify the equations, you also specify any dependent variables. If you typed

```
. ml model lf myprog (price = mpg weight) ()
```

`price` would be the one and only dependent variable, and that would be passed to your program in `$ML_y1`. If your model had two dependent variables, you could type

```
. ml model lf myprog (price displ = mpg weight) ()
```

Then `$ML_y1` would be `price`, and `$ML_y2` would be `displ`. You can specify however many dependent variables are necessary and specify them on any equation. It does not matter on which equation you specify them; the first one specified is placed in `$ML_y1`, the second in `$ML_y2`, and so on.

▷ Example 1: Method lf

Using method lf, we want to produce observation-by-observation values of the log likelihood. The probit log-likelihood function is

$$\ln \ell_j = \begin{cases} \ln \Phi(\theta_{1j}) & \text{if } y_j = 1 \\ \ln \Phi(-\theta_{1j}) & \text{if } y_j = 0 \end{cases}$$

$$\theta_{1j} = \mathbf{x}_j \mathbf{b}_1$$

The following is the method lf evaluator for this likelihood function:

```
program myprobit
        version 9
        args lnf theta1
        quietly replace 'lnf' = ln(normal('theta1')) if $ML_y1==1
        quietly replace 'lnf' = ln(normal(-'theta1')) if $ML_y1==0
end
```

If we wanted to fit a model of foreign on mpg and weight, we would type

```
. use http://www.stata-press.com/data/r9/auto
(1978 Automobile Data)
. ml model lf myprobit (foreign = mpg weight)
. ml maximize
```

The 'foreign =' part specifies that y is foreign. The 'mpg weight' part specifies that $\theta_{1j} = b_1 \text{mpg}_j + b_2 \text{weight}_j + b_3$. The result of running this is

```
. ml model lf myprobit (foreign = mpg weight)
. ml maximize
initial:       log likelihood = -51.292891
alternative:   log likelihood = -45.055272
rescale:       log likelihood = -45.055272
Iteration 0:   log likelihood = -45.055272
Iteration 1:   log likelihood = -27.904114
Iteration 2:   log likelihood = -26.858048
Iteration 3:   log likelihood = -26.844198
Iteration 4:   log likelihood = -26.844189
Iteration 5:   log likelihood = -26.844189
```

Number of obs	=		74
Wald chi2(2)	=		20.75
Log likelihood = -26.844189	Prob > chi2	=	0.0000

foreign	Coef.	Std. Err.	z	P>|z|	[95% Conf. Interval]	
mpg	-.1039503	.0515689	-2.02	0.044	-.2050235	-.0028772
weight	-.0023355	.0005661	-4.13	0.000	-.003445	-.0012261
_cons	8.275464	2.554142	3.24	0.001	3.269438	13.28149

◁

▷ Example 2: Method lf for two-equation, two-dependent-variable model

A two-equation, two-dependent-variable model is a little different. Rather than receiving one θ, our program will receive two. Rather than there being one dependent variable in \$ML_y1, there will be dependent variables in \$ML_y1 and \$ML_y2. For instance, the Weibull regression log-likelihood function is

$$\ln \ell_j = -(t_j e^{-\theta_{1j}})^{\exp(\theta_{2j})} + d_j\{\theta_{2j} - \theta_{1j} + (e^{\theta_{2j}} - 1)(\ln t_j - \theta_{1j})\}$$
$$\theta_{1j} = \mathbf{x}_j \mathbf{b}_1$$
$$\theta_{2j} = s$$

where t_j is the time of failure or censoring and $d_j = 1$ if failure and 0 if censored. We can make the log likelihood a little easier to program by introducing some extra variables:

$$p_j = \exp(\theta_{2j})$$
$$M_j = \{t_j \exp(-\theta_{1j})\}^{p_j}$$
$$R_j = \ln t_j - \theta_{1j}$$
$$\ln \ell_j = -M_j + d_j\{\theta_{2j} - \theta_{1j} + (p_j - 1)R_j\}$$

The method lf evaluator for this is

```
program myweib
        version 9
        args lnf theta1 theta2

        tempvar p M R
        quietly gen double 'p' = exp('theta2')
        quietly gen double 'M' = ($ML_y1*exp(-'theta1'))^'p'
        quietly gen double 'R' = ln($ML_y1)-'theta1'

        quietly replace 'lnf' = -'M' + $ML_y2*('theta2'-'theta1' + ('p'-1)*'R')
end
```

We can fit a model by typing

```
. ml model lf myweib (studytime died = drug2 drug3 age) ()
. ml maximize
```

Note that we specified '()' for the second equation. The second equation corresponds to the Weibull shape parameter s, and the linear combination we want for s contains just an intercept. Alternatively, we could type

```
. ml model lf myweib (studytime died = drug2 drug3 age) /s
```

Typing /s means the same thing as typing (s:), and both really mean the same thing as (). The s, either after a slash or in parentheses before a colon, labels the equation. It makes the output look prettier, and that is all:

```
. ml model lf myweib (studytime died = drug2 drug3 age) /s

. ml max

initial:      log likelihood =        -744
alternative:  log likelihood = -356.14276
rescale:      log likelihood = -200.80201
rescale eq:   log likelihood = -136.69232
Iteration 0:  log likelihood = -136.69232  (not concave)
Iteration 1:  log likelihood = -124.11726
Iteration 2:  log likelihood = -113.94953
Iteration 3:  log likelihood = -110.30691
Iteration 4:  log likelihood = -110.26748
Iteration 5:  log likelihood = -110.26736
Iteration 6:  log likelihood = -110.26736
```

				Number of obs	=	48
				Wald chi2(3)	=	35.25
Log likelihood = -110.26736				Prob > chi2	=	0.0000

	Coef.	Std. Err.	z	P>|z|	[95% Conf. Interval]	
eq1						
drug2	1.012966	.2903917	3.49	0.000	.4438086	1.582123
drug3	1.45917	.2821195	5.17	0.000	.9062261	2.012114
age	-.0671728	.0205688	-3.27	0.001	-.1074868	-.0268587
_cons	6.060723	1.152845	5.26	0.000	3.801188	8.320259
s						
_cons	.5573333	.1402154	3.97	0.000	.2825162	.8321504

◁

▷ Example 3: Method d0

Method d0 evaluators receive $\mathbf{b} = (\mathbf{b}_1, \mathbf{b}_2, \ldots, \mathbf{b}_E)$, the coefficient vector, rather than the already evaluated $\theta_1, \theta_2, \ldots, \theta_E$, and they are required to evaluate the overall log-likelihood $\ln L$ rather than $\ln \ell_j$, $j = 1, \ldots, N$.

Use `mleval` to produce the thetas from the coefficient vector.

Use `mlsum` to sum the components that enter into $\ln L$.

In the case of Weibull, $\ln L = \sum \ln \ell_j$, and our method d0 evaluator is

```
program weib0
        version 9
        args todo b lnf

        tempvar theta1 theta2
        mleval 'theta1' = 'b', eq(1)
        mleval 'theta2' = 'b', eq(2)

        local t "$ML_y1"            // this is just for readability
        local d "$ML_y2"

        tempvar p M R
        quietly gen double 'p' = exp('theta2')
        quietly gen double 'M' = ('t'*exp(-'theta1'))^'p'
        quietly gen double 'R' = ln('t')-'theta1'

        mlsum 'lnf' = -'M' + 'd'*('theta2'-'theta1' + ('p'-1)*'R')
end
```

To fit our model using this evaluator, we would type

```
. ml model d0 weib0 (studytime died = drug2 drug3 age) /s
. ml maximize                                              ◁
```

❑ Technical Note

Method d0 does not require $\ln L = \sum_j \ln \ell_j$, $j = 1, \ldots, N$, as method lf does. Your likelihood function might have independent components only for groups of observations. Panel-data estimators have a log-likelihood value $\ln L = \sum_i \ln L_i$, where i indexes the panels, each of which contains multiple observations. Conditional logistic regression has $\ln L = \sum_k \ln L_k$, where k indexes the risk pools. Cox regression has $\ln L = \sum_{(t)} \ln L_{(t)}$, where (t) denotes the ordered failure times.

To evaluate such likelihood functions, first calculate the within-group log-likelihood contributions. This usually involves `generate` and `replace` statements prefixed with by, as in

```
                tempvar sumd
                by group: gen double 'sumd' = sum($ML_y1)
```

Structure your code so that the log-likelihood contributions are recorded in the last observation of each group. Say that a variable is named `'cont'`. To sum the contributions, code

```
                tempvar last
                quietly by group: gen byte 'last' = (_n==_N)
                mlsum 'lnf' = 'cont' if 'last'
```

You must inform `mlsum` which observations contain log-likelihood values to be summed. First, you do not want to include intermediate results in the sum. Second, `mlsum` does not skip missing values. Rather, if `mlsum` sees a missing value among the contributions, it sets the overall result, `'lnf'`, to missing. That is how `ml maximize` is informed that the likelihood function could not be evaluated at the particular value of **b**. `ml maximize` will then take action to escape from what it thinks is an infeasible area of the likelihood function.

When the likelihood function violates the linear-form restriction $\ln L = \sum_j \ln \ell_j$, $j = 1, \ldots, N$, with $\ln \ell_j$ being a function solely of values within the jth observation, use method d0. In the following examples, we will demonstrate methods d1 and d2 with likelihood functions that meet this linear-form restriction. The d1 and d2 methods themselves do not require the linear-form restriction, but the utility routines `mlvecsum` and `mlmatsum` do. Using method d1 or d2 when the restriction is violated is a difficult; however, `mlmatbysum` may be of some help for method d2 evaluators.   ❑

▷ Example 4: Method d1

Method d1 evaluators are required to produce the gradient vector $\mathbf{g} = \partial \ln L / \partial \mathbf{b}$, as well as the overall log-likelihood value. Using `mlvecsum`, we can obtain $\partial \ln L / \partial \mathbf{b}$ from $\partial \ln L / \partial \theta_i$, $i = 1, \ldots, E$. The derivatives of the Weibull log-likelihood function are

$$\frac{\partial \ln \ell_j}{\partial \theta_{1j}} = p_j (M_j - d_j)$$

$$\frac{\partial \ln \ell_j}{\partial \theta_{2j}} = d_j - R_j p_j (M_j - d_j)$$

The method d1 evaluator for this is

```
program weib1
        version 9
        args todo b lnf g                              // g is new

        tempvar t1 t2
        mleval 't1' = 'b', eq(1)
        mleval 't2' = 'b', eq(2)

        local t "$ML_y1"
        local d "$ML_y2"

        tempvar p M R
        quietly gen double 'p' = exp('t2')
        quietly gen double 'M' = ('t'*exp(-'t1'))^'p'
        quietly gen double 'R' = ln('t')-'t1'

        mlsum 'lnf' = -'M' + 'd'*('t2'-'t1' + ('p'-1)*'R')
        if ('todo'==0 | 'lnf'>=.) exit                 /* <-- new */

        tempname d1 d2                                 /* <-- new */
        mlvecsum 'lnf' 'd1' = 'p'*('M'-'d'), eq(1)     /* <-- new */
        mlvecsum 'lnf' 'd2' = 'd' - 'R'*'p'*('M'-'d'), eq(2)  /* <-- new */
        matrix 'g' = ('d1','d2')                       /* <-- new */
end
```

We obtained this code by starting with our method d0 evaluator and then adding the extra lines method d1 requires. To fit our model using this evaluator, we could type

```
. ml model d1 weib1 (studytime died = drug2 drug3 age) /s
. ml maximize
```

but we recommend substituting method d1debug for method d1 and typing

```
. ml model d1debug weib1 (studytime died = drug2 drug3 age) /s
. ml maximize
```

Method d1debug will compare the derivatives we calculate with numerical derivatives and thus verify that our program is correct. Once we are certain the program is correct, then we would switch from method d1debug to method d1.

◁

▷ Example 5: Method d2

Method d2 evaluators are required to produce $-\mathbf{H} = -\partial^2 \ln L / \partial \mathbf{b} \partial \mathbf{b}'$, the negative Hessian matrix, as well as the gradient and log-likelihood value. `mlmatsum` will help calculate $\partial^2 \ln L / \partial \mathbf{b} \partial \mathbf{b}'$ from the negative second derivatives with respect to θ. For the Weibull model, these negative second derivatives are

$$-\frac{\partial^2 \ln \ell_j}{\partial \theta_{1j}^2} = p_j^2 M_j$$

$$-\frac{\partial^2 \ln \ell_j}{\partial \theta_{1j} \partial \theta_{2j}} = -p_j(M_j - d_j + R_j p_j M_j)$$

$$-\frac{\partial^2 \ln \ell_j}{\partial \theta_{2j}^2} = p_j R_j(R_j p_j M_j + M_j - d_j)$$

The method d2 evaluator is

```
program weib2
        version 9
        args todo b lnf g negH                          // negH added
        tempvar t1 t2
        mleval 't1' = 'b', eq(1)
        mleval 't2' = 'b', eq(2)
        local t "$ML_y1"
        local d "$ML_y2"
        tempvar p M R
        quietly gen double 'p' = exp('t2')
        quietly gen double 'M' = ('t'*exp(-'t1'))^'p'
        quietly gen double 'R' = ln('t')-'t1'
        mlsum 'lnf' = -'M' + 'd'*('t2'-'t1' + ('p'-1)*'R')
        if ('todo'==0 | 'lnf'>=.) exit
        tempname d1 d2
        mlvecsum 'lnf' 'd1' = 'p'*('M'-'d'), eq(1)
        mlvecsum 'lnf' 'd2' = 'd' - 'R'*'p'*('M'-'d'), eq(2)
        matrix 'g' = ('d1','d2')
        if ('todo'==1 | 'lnf'>=.) exit                  // new from here down
        tempname d11 d12 d22
        mlmatsum 'lnf' 'd11' = 'p'^2 * 'M', eq(1)
        mlmatsum 'lnf' 'd12' = -'p'*('M'-'d' + 'R'*'p'*'M'), eq(1,2)
        mlmatsum 'lnf' 'd22' = 'p'*'R'*('R'*'p'*'M' + 'M' - 'd') , eq(2)
        matrix 'negH' = ('d11','d12' \ 'd12','d22')
end
```

We started with our previous method d1 evaluator and added the lines that method d2 requires. We could now fit a model by typing

```
. ml model d2 weib2 (studytime died = drug2 drug3 age) /s
. ml maximize
```

but we would recommend substituting method d2debug for method d2 and typing

```
. ml model d2debug weib2 (studytime died = drug2 drug3 age) /s
. ml maximize
```

Method d2debug will compare the first and second derivatives we calculate with numerical derivatives and thus verify that our program is correct. Once we are certain the program is correct, then we would switch from method d2debug to method d2.

◁

As we stated earlier, to produce the robust variance estimator with method lf, there is nothing to do except specify `robust`, `cluster()`, and/or `pweights`. For method d0, these options do not work. For methods d1 and d2, these options will work if your likelihood function meets the linear-form restrictions and you fill in the equation scores. The equation scores are defined as

$$\frac{\partial \ln \ell_j}{\partial \theta_{1j}}, \quad \frac{\partial \ln \ell_j}{\partial \theta_{2j}}, \quad \cdots$$

Your evaluator will be passed variables, one for each equation, which you fill in with the equation scores. For *both* method d1 and d2, these variables are passed in the sixth and subsequent positions of the argument list. That is, you must process the arguments as

```
args todo b lnf g negH g1 g2 ...
```

Note that for method d1, the 'negH' argument is not used; it is merely a placeholder.

▷ Example 6: Robust variance estimates

If you have used `mlvecsum` in your method d1 or d2 evaluator, it is easy to turn it into a program that allows the computation of the robust variance estimator. The expression that you specified on the right-hand side of `mlvecsum` is the equation score.

Here we turn the program that we gave earlier in the method d1 example into one that allows robust, `cluster()`, and/or `pweights`.

```
program weib1
        version 9
        args todo b lnf g negH g1 g2          // negH, g1, and g2 are new

        tempvar t1 t2
        mleval 't1' = 'b', eq(1)
        mleval 't2' = 'b', eq(2)

        local t "$ML_y1"
        local d "$ML_y2"

        tempvar p M R
        quietly gen double 'p' = exp('t2')
        quietly gen double 'M' = ('t'*exp(-'t1'))^'p'
        quietly gen double 'R' = ln('t')-'t1'

        mlsum 'lnf' = -'M' + 'd'*('t2'-'t1' + ('p'-1)*'R')
        if ('todo'==0 | 'lnf'>=.) exit

        tempname d1 d2
        quietly replace 'g1' = 'p'*('M'-'d')               /* <-- new     */
        quietly replace 'g2' = 'd' - 'R'*'p'*('M'-'d')     /* <-- new     */
        mlvecsum 'lnf' 'd1'  = 'g1', eq(1)                 /* <-- changed */
        mlvecsum 'lnf' 'd2'  = 'g2', eq(2)                 /* <-- changed */
        matrix 'g' = ('d1','d2')
end
```

To fit our model and get the robust variance estimates, we type

```
. ml model d1 weib1 (studytime died = drug2 drug3 age) /s, robust
. ml maximize
```

◁

Survey options and ml

`ml` can handle stratification, poststratification, multiple stages of clustering, and finite population corrections. Specifying the `svy` option implies that the data come from a survey design and also implies that the survey linearized variance estimator is to be used; see [SVY] **variance estimation**.

▷ Example 7

Suppose that we are interested in a probit analysis of data from a survey in which q1 is the answer to a yes/no question and d1, d2, d3 are demographic responses. The following is a d2 evaluator for the probit model that meets the requirements for `robust` (linear form and computes the scores).

```
program myd2probit
        version 9
        args todo b lnf g negH g1
        tempvar z Fz lnfj
        mleval 'z' = 'b'
        quietly gen double 'Fz'   = normal( 'z')  if $ML_y1 == 1
        quietly replace    'Fz'   = normal(-'z')  if $ML_y1 == 0
        quietly gen double 'lnfj' = log('Fz')
        mlsum 'lnf' = 'lnfj'
        if ('todo'==0 | 'lnf' >= .) exit
        quietly replace 'g1' =  normalden('z')/'Fz'  if $ML_y1 == 1
        quietly replace 'g1' = -normalden('z')/'Fz'  if $ML_y1 == 0
        mlvecsum 'lnf' 'g' = 'g1', eq(1)
        if ('todo'==1 | 'lnf' >= .) exit
        mlmatsum 'lnf' 'negH' = 'g1'*('g1'+'z'), eq(1,1)
end
```

To fit a model, we `svyset` the data, then use `svy` with `ml`.

```
. svyset psuid [pw=w], strata(strid)
. ml model d2 myd2probit (q1 = d1 d2 d3), svy
. ml maximize
```

We could also use the `subpop()` option to make inferences about the subpopulation identified by the variable `sub`:

```
. svyset psuid [pw=w], strata(strid)
. ml model d2 myd2probit (q1 = d1 d2 d3), svy subpop(sub)
. ml maximize
```

◁

Saved Results

For results saved by `ml` without the `svy` option, see [R] **maximize**.

For results saved by `ml` with the `svy` option, see [SVY] **svy**.

References

Gould, W., J. Pitblado, and W. Sribney. 2003. *Maximum Likelihood Estimation with Stata*. 2nd ed. College Station, TX: Stata Press.

Also See

Complementary: [R] **maximize**, [R] **nl**,

[P] **_estimates**, [P] **matrix**,

Mata Reference Manual

Title

> **mlogit** — Multinomial (polytomous) logistic regression

Syntax

<u>mlog</u>it *depvar* $\left[\,indepvars\,\right]$ $\left[\,if\,\right]$ $\left[\,in\,\right]$ $\left[\,weight\,\right]$ $\left[\,,\ options\,\right]$

options	description
Model	
<u>noc</u>onstant	suppress constant term
<u>b</u>aseoutcome(#)	value of *depvar* that will be the base outcome
<u>c</u>onstraints(*clist*)	apply specified linear constraints
SE/Robust	
vce(*vcetype*)	*vcetype* may be <u>r</u>obust, <u>boot</u>strap, or <u>jack</u>knife
robust	synonym for vce(robust)
<u>cl</u>uster(*varname*)	adjust standard errors for intragroup correlation
Reporting	
<u>l</u>evel(#)	set confidence level; default is level(95)
<u>rrr</u>	report relative-risk ratios
Max options	
maximize_options	control the maximization process; seldom used

where *clist* has the form $\#\left[\,-\#\,\right]\ \left[\,,\ \#\left[\,-\#\,\right]\ \dots\,\right]$

bootstrap, by, jackknife, rolling, statsby, svy, and xi are allowed; see [U] **11.1.10 Prefix commands**.
fweights, iweights, and pweights are allowed; see [U] **11.1.6 weight**.
See [U] **20 Estimation and postestimation commands** for additional capabilities of estimation commands.

Description

mlogit fits maximum-likelihood multinomial logit models, also known as polytomous logistic regression. You can define constraints to perform constrained estimation. Some people refer to conditional logistic regression as multinomial logit. If you are one of them, see [R] **clogit**.

See [R] **logistic** for a list of related estimation commands.

The model can have a maximum of 50 outcomes with Stata/SE or Intercooled Stata and 20 outcomes with Small Stata.

Options

> **Model**

noconstant; see [R] **estimation options**.

baseoutcome(#) specifies the value of *depvar* to be treated as the base outcome. The default is to choose the most frequent outcome.

constraints(*clist*); see [R] **estimation options**.

⌐ SE/Robust ⌐

vce(*vcetype*); see [R] *vce_option*.

robust, cluster(*varname*); see [R] **estimation options**. cluster() can be used with pweights to produce estimates for unstratified cluster-sampled data, but see [SVY] **svy: mlogit** for a command especially designed for survey data.

⌐ Reporting ⌐

level(#); see [R] **estimation options**.

rrr reports the estimated coefficients transformed to relative risk ratios, i.e., e^b rather than b; see *Description of the model* below for an explanation of this concept. Standard errors and confidence intervals are similarly transformed. This option affects how results are displayed, not how they are estimated. rrr may be specified at estimation or when replaying previously estimated results.

⌐ Max options ⌐

maximize_options: iterate(#), [no]log, trace, tolerance(#), ltolerance(#); see [R] **maximize**. These options are seldom used.

Remarks

Remarks are presented under the headings

  Description of the model
  Fitting unconstrained models
  Fitting constrained models

mlogit fits maximum likelihood models with discrete dependent (left-hand-side) variables when the dependent variable takes on more than two outcomes and the outcomes have no natural ordering. If the dependent variable takes on only two outcomes, estimates are identical to those produced by logistic or logit; see [R] **logistic** or [R] **logit**. If the outcomes are ordered, see [R] **ologit**.

Description of the model

For an introduction to multinomial logit models, see Aldrich and Nelson (1984, 73–77), Greene (2003, chapter 21), Hosmer and Lemeshow (2000, 260–287), Long (1997, chapter 6), and Long and Freese (2003, chapter 6). For a description emphasizing the difference in assumptions and data requirements for conditional and multinomial logit, see Judge et al. (1985, 768–772).

Consider the outcomes 1, 2, 3, ..., m recorded in y, and the explanatory variables X. Assume that there are $m = 3$ outcomes: "buy an American car", "buy a Japanese car", and "buy a European car". The values of y are then said to be "unordered". Even though the outcomes are coded 1, 2, and 3, the numerical values are arbitrary because $1 < 2 < 3$ does not imply that outcome 1 (buy American) is less than outcome 2 (buy Japanese) is less than outcome 3 (buy European). This unordered categorical property of y distinguishes the use of mlogit from regress (which is appropriate for a continuous dependent variable), from ologit (which is appropriate for ordered categorical data), and from logit (which is appropriate for two outcomes, which can be thought of as ordered).

In the multinomial logit model, you estimate a set of coefficients, $\beta^{(1)}$, $\beta^{(2)}$, and $\beta^{(3)}$, corresponding to each outcome:

$$\Pr(y=1) = \frac{e^{X\beta^{(1)}}}{e^{X\beta^{(1)}} + e^{X\beta^{(2)}} + e^{X\beta^{(3)}}}$$

$$\Pr(y=2) = \frac{e^{X\beta^{(2)}}}{e^{X\beta^{(1)}} + e^{X\beta^{(2)}} + e^{X\beta^{(3)}}}$$

$$\Pr(y=3) = \frac{e^{X\beta^{(3)}}}{e^{X\beta^{(1)}} + e^{X\beta^{(2)}} + e^{X\beta^{(3)}}}$$

The model, however, is unidentified in the sense that there is more than one solution to $\beta^{(1)}$, $\beta^{(2)}$, and $\beta^{(3)}$ that leads to the same probabilities for $y=1$, $y=2$, and $y=3$. To identify the model, you arbitrarily set one of $\beta^{(1)}$, $\beta^{(2)}$, or $\beta^{(3)}$ to 0—it does not matter which. That is, if you arbitrarily set $\beta^{(1)} = 0$, the remaining coefficients $\beta^{(2)}$ and $\beta^{(3)}$ will measure the change relative to the $y=1$ group. If you instead set $\beta^{(2)} = 0$, the remaining coefficients $\beta^{(1)}$ and $\beta^{(3)}$ will measure the change relative to the $y=2$ group. The coefficients will differ because they have different interpretations, but the predicted probabilities for $y=1$, 2, and 3 will still be the same. Thus either parameterization will be a solution to the same underlying model.

Setting $\beta^{(1)} = 0$, the equations become

$$\Pr(y=1) = \frac{1}{1 + e^{X\beta^{(2)}} + e^{X\beta^{(3)}}}$$

$$\Pr(y=2) = \frac{e^{X\beta^{(2)}}}{1 + e^{X\beta^{(2)}} + e^{X\beta^{(3)}}}$$

$$\Pr(y=3) = \frac{e^{X\beta^{(3)}}}{1 + e^{X\beta^{(2)}} + e^{X\beta^{(3)}}}$$

The relative probability of $y=2$ to the base outcome is

$$\frac{\Pr(y=2)}{\Pr(y=1)} = e^{X\beta^{(2)}}$$

Let us call this ratio the relative risk, and let us further assume that X and $\beta_k^{(2)}$ are vectors equal to $(x_1, x_2, \ldots, x_k)$ and $(\beta_1^{(2)}, \beta_2^{(2)}, \ldots, \beta_k^{(2)})'$, respectively. The ratio of the relative risk for a one-unit change in x_i is then

$$\frac{e^{\beta_1^{(2)} x_1 + \cdots + \beta_i^{(2)}(x_i + 1) + \cdots + \beta_k^{(2)} x_k}}{e^{\beta_1^{(2)} x_1 + \cdots + \beta_i^{(2)} x_i + \cdots + \beta_k^{(2)} x_k}} = e^{\beta_i^{(2)}}$$

Thus the exponentiated value of a coefficient is the relative-risk ratio for a one-unit change in the corresponding variable (risk is measured as the risk of the outcome relative to the base outcome).

Fitting unconstrained models

▷ Example 1

We have data on the type of health insurance available to 616 psychologically depressed subjects in the U.S. (Tarlov et al. 1989; Wells et al. 1989). The insurance is categorized as either an indemnity plan (i.e., regular fee-for-service insurance, which may have a deductible or coinsurance rate) or a prepaid plan (a fixed up-front payment allowing subsequent unlimited use as provided, for instance, by an HMO). The third possibility is that the subject has no insurance whatsoever. We wish to explore the demographic factors associated with each subject's insurance choice. One of the demographic factors in our data is the race of the participant, coded as white or nonwhite:

```
. use http://www.stata-press.com/data/r9/sysdsn3
(Health insurance data)

. tabulate insure nonwhite, chi2 col
```

Key
frequency
column percentage

| | nonwhite | | |
insure	0	1	Total
Indemnity	251	43	294
	50.71	35.54	47.73
Prepaid	208	69	277
	42.02	57.02	44.97
Uninsure	36	9	45
	7.27	7.44	7.31
Total	495	121	616
	100.00	100.00	100.00

Pearson chi2(2) =   9.5599   Pr = 0.008

Although `insure` appears to take on the values Indemnity, Prepaid, and Uninsure, it actually takes on the values 1, 2, and 3. The words appear because we have associated a value label with the numeric variable `insure`; see [U] **12.6.3 Value labels**.

When we fit a multinomial logit model, we can tell `mlogit` which outcome to use as the base outcome, or we can let `mlogit` choose. To fit a model of `insure` on `nonwhite`, letting `mlogit` choose the base outcome, we type

```
. mlogit insure nonwhite

Iteration 0:   log likelihood = -556.59502
Iteration 1:   log likelihood = -551.78935
Iteration 2:   log likelihood = -551.78348
Iteration 3:   log likelihood = -551.78348
```

Multinomial logistic regression

				Number of obs	=	616
				LR chi2(2)	=	9.62
				Prob > chi2	=	0.0081
Log likelihood = -551.78348				Pseudo R2	=	0.0086

insure	Coef.	Std. Err.	z	P>|z|	[95% Conf. Interval]	
Prepaid						
nonwhite	.6608212	.2157321	3.06	0.002	.2379942	1.083648
_cons	-.1879149	.0937644	-2.00	0.045	-.3716896	-.0041401
Uninsure						
nonwhite	.3779585	.407589	0.93	0.354	-.4209012	1.176818
_cons	-1.941934	.1782185	-10.90	0.000	-2.291236	-1.592632

(insure==Indemnity is the base outcome)

mlogit chose the indemnity outcome as the base outcome and presented coefficients for the outcomes prepaid and uninsured. According to the model, the probability of prepaid for whites (nonwhite = 0) is

$$\Pr(\texttt{insure} = \texttt{Prepaid}) = \frac{e^{-.188}}{1 + e^{-.188} + e^{-1.942}} = 0.420$$

Similarly, for nonwhites, the probability of prepaid is

$$\Pr(\texttt{insure} = \texttt{Prepaid}) = \frac{e^{-.188+.661}}{1 + e^{-.188+.661} + e^{-1.942+.378}} = 0.570$$

These results agree with the column percentages presented by tabulate since the mlogit model is fully saturated. That is, there are enough terms in the model to fully explain the column percentage in each cell. Note that the model chi-squared and the tabulate chi-squared are in almost perfect agreement; both test that the column percentages of insure are the same for both values of nonwhite.

◁

▷ Example 2

By specifying the baseoutcome() option, we can control which outcome of the dependent variable is treated as the base. Left to its own, mlogit chose to make outcome 1, indemnity, the base outcome. To make outcome 2, prepaid, the base, we would type

(Continued on next page)

```
. mlogit insure nonwhite, base(2)
Iteration 0:   log likelihood = -556.59502
Iteration 1:   log likelihood = -551.78935
Iteration 2:   log likelihood = -551.78348
Iteration 3:   log likelihood = -551.78348
Multinomial logistic regression                  Number of obs   =        616
                                                 LR chi2(2)      =       9.62
                                                 Prob > chi2     =     0.0081
Log likelihood = -551.78348                      Pseudo R2       =     0.0086
```

insure	Coef.	Std. Err.	z	P>|z|	[95% Conf. Interval]	
Indemnity						
nonwhite	-.6608212	.2157321	-3.06	0.002	-1.083648	-.2379942
_cons	.1879149	.0937644	2.00	0.045	.0041401	.3716896
Uninsure						
nonwhite	-.2828628	.3977302	-0.71	0.477	-1.0624	.4966741
_cons	-1.754019	.1805145	-9.72	0.000	-2.107821	-1.400217

(insure==Prepaid is the base outcome)

The baseoutcome() option requires that we specify the numeric value of the outcome, so we could not type base(Prepaid).

Although the coefficients now appear to be different, note that the summary statistics reported at the top are identical. With this parameterization, the probability of prepaid insurance for whites is

$$\Pr(\text{insure} = \text{Prepaid}) = \frac{1}{1 + e^{.188} + e^{-1.754}} = 0.420$$

This is the same answer we obtained previously.

◁

▷ Example 3

By specifying rrr, which we can do at estimation time or when we redisplay results, we see the model in terms of relative-risk ratios:

```
. mlogit, rrr
Multinomial logistic regression                  Number of obs   =        616
                                                 LR chi2(2)      =       9.62
                                                 Prob > chi2     =     0.0081
Log likelihood = -551.78348                      Pseudo R2       =     0.0086
```

insure	RRR	Std. Err.	z	P>|z|	[95% Conf. Interval]	
Indemnity						
nonwhite	.516427	.1114099	-3.06	0.002	.3383588	.7882073
Uninsure						
nonwhite	.7536232	.2997387	-0.71	0.477	.3456254	1.643247

(insure==Prepaid is the base outcome)

Looked at this way, the relative risk of choosing an indemnity over a prepaid plan is 0.52 for nonwhites relative to whites.

◁

▷ Example 4

One of the advantages of mlogit over tabulate is that we can include continuous variables and multiple categorical variables in the model. In examining the data on insurance choice, we decide that we want to control for age, gender, and site of study (the study was conducted in three sites):

```
. mlogit insure age male nonwhite site2 site3
Iteration 0:   log likelihood = -555.85446
Iteration 1:   log likelihood = -534.72983
Iteration 2:   log likelihood = -534.36536
Iteration 3:   log likelihood = -534.36165
Iteration 4:   log likelihood = -534.36165
```

Multinomial logistic regression			Number of obs	=	615
			LR chi2(10)	=	42.99
			Prob > chi2	=	0.0000
Log likelihood = -534.36165			Pseudo R2	=	0.0387

insure	Coef.	Std. Err.	z	P>|z|	[95% Conf. Interval]	
Prepaid						
age	-.011745	.0061946	-1.90	0.058	-.0238862	.0003962
male	.5616934	.2027465	2.77	0.006	.1643175	.9590693
nonwhite	.9747768	.2363213	4.12	0.000	.5115955	1.437958
site2	.1130359	.2101903	0.54	0.591	-.2989296	.5250013
site3	-.5879879	.2279351	-2.58	0.010	-1.034733	-.1412433
_cons	.2697127	.3284422	0.82	0.412	-.3740222	.9134476
Uninsure						
age	-.0077961	.0114418	-0.68	0.496	-.0302217	.0146294
male	.4518496	.3674867	1.23	0.219	-.268411	1.17211
nonwhite	.2170589	.4256361	0.51	0.610	-.6171725	1.05129
site2	-1.211563	.4705127	-2.57	0.010	-2.133751	-.2893747
site3	-.2078123	.3662926	-0.57	0.570	-.9257327	.510108
_cons	-1.286943	.5923219	-2.17	0.030	-2.447872	-.1260135

(insure==Indemnity is the base outcome)

These results suggest that the inclination of nonwhites to choose prepaid care is even stronger than it was without controlling. We also see that subjects in site 2 are less likely to be uninsured.

◁

Fitting constrained models

mlogit can fit models with subsets of coefficients constrained to be zero, with subsets of coefficients constrained to be equal both within and across equations, and with subsets of coefficients arbitrarily constrained to equal linear combinations of other estimated coefficients.

Before fitting a constrained model, you define the constraints using the constraint command; see [R] **constraint**. Once the constraints are defined, you estimate using mlogit, specifying the constraint() option. Typing constraint(4) would use the constraint you previously saved as 4. Typing constraint(1,4,6) would use the previously stored constraints 1, 4, and 6. Typing constraint(1-4,6) would use the previously stored constraints 1, 2, 3, 4, and 6.

Sometimes you will not be able to specify the constraints without knowing the omitted outcome. In such cases, assume that the omitted outcome is whatever outcome is convenient for you, and include the baseoutcome() option when you type the mlogit command.

▷ Example 5

We can use constraints to test hypotheses, among other things. In our insurance-choice model, let's test the hypothesis that there is no distinction between having indemnity insurance and being uninsured. Indemnity-style insurance was the omitted outcome, so we type

```
. test [Uninsure]
 ( 1)   [Uninsure]age = 0
 ( 2)   [Uninsure]male = 0
 ( 3)   [Uninsure]nonwhite = 0
 ( 4)   [Uninsure]site2 = 0
 ( 5)   [Uninsure]site3 = 0

          chi2(  5) =      9.31
        Prob > chi2 =     0.0973
```

If indemnity had not been the omitted outcome, we would have typed `test [Uninsure=Indemnity]`.

The results produced by `test` are an approximation based on the estimated covariance matrix of the coefficients. Since the probability of being uninsured is quite low, the log likelihood may be nonlinear for the uninsured. Conventional statistical wisdom is not to trust the asymptotic answer under these circumstances, but to perform a likelihood-ratio test instead.

To use Stata's `lrtest` likelihood-ratio test command, we must fit both the unconstrained and constrained models. The unconstrained model is the one we have previously fitted. Following the instruction in [R] **lrtest**, we first save the unconstrained model results:

```
. estimates store unconstrained
```

To fit the constrained model, we must refit our model with all the coefficients except the constant set to 0 in the `Uninsure` equation. We define the constraint and then refit:

(Continued on next page)

```
. constraint define 1 [Uninsure]

. mlogit insure age male nonwhite site2 site3, constr(1)

Iteration 0:    log likelihood = -555.85446
Iteration 1:    log likelihood = -539.80523
Iteration 2:    log likelihood = -539.75644
Iteration 3:    log likelihood = -539.75643
```

```
Multinomial logistic regression              Number of obs   =       615
                                             LR chi2(5)      =     32.20
                                             Prob > chi2     =    0.0000
Log likelihood = -539.75643                  Pseudo R2       =    0.0290
```

```
 ( 1)   [Uninsure]age = 0
 ( 2)   [Uninsure]male = 0
 ( 3)   [Uninsure]nonwhite = 0
 ( 4)   [Uninsure]site2 = 0
 ( 5)   [Uninsure]site3 = 0
```

insure	Coef.	Std. Err.	z	P>|z|	[95% Conf. Interval]	
Prepaid						
age	-.0107025	.0060039	-1.78	0.075	-.0224699	.0010649
male	.4963616	.1939683	2.56	0.010	.1161908	.8765324
nonwhite	.942137	.2252094	4.18	0.000	.5007347	1.383539
site2	.2530912	.2029465	1.25	0.212	-.1446767	.6508591
site3	-.5521774	.2187237	-2.52	0.012	-.9808678	-.1234869
_cons	.1792752	.3171372	0.57	0.572	-.4423023	.8008527
Uninsure						
age	(dropped)					
male	(dropped)					
nonwhite	(dropped)					
site2	(dropped)					
site3	(dropped)					
_cons	-1.87351	.1601099	-11.70	0.000	-2.18732	-1.5597

```
(insure==Indemnity is the base outcome)
```

We can now perform the likelihood-ratio test:

```
. lrtest unconstrained .

Likelihood-ratio test                        LR chi2(5)   =     10.79
(Assumption: . nested in unconstrained)      Prob > chi2  =    0.0557
```

The likelihood-ratio chi-squared is 10.79 with 5 degrees of freedom—just slightly greater than the magic $p = .05$ level—so we should not call this difference significant.

◁

❏ Technical Note

In certain circumstances, you should fit a multinomial logit model with conditional logit; see [R] **clogit**. With substantial data manipulation, clogit can handle the same class of models with some interesting additions. For example, if we had available the price and deductible of the most competitive insurance plan of each type, mlogit could not use this information, but clogit could.

❏

Saved Results

mlogit saves in e():

Scalars

e(N)	number of observations	e(ll_0)	log likelihood, constant-only model
e(k_out)	number of outcomes	e(N_clust)	number of clusters
e(df_m)	model degrees of freedom	e(chi2)	χ^2
e(r2_p)	pseudo-R-squared	e(ibaseout)	base outcome number
e(ll)	log likelihood	e(baseout)	the value of *depvar* to be treated as the base outcome

Macros

e(cmd)	mlogit	e(eqnames)	names of equations
e(depvar)	name of dependent variable	e(chi2type)	Wald or LR; type of model χ^2 test
e(wtype)	weight type	e(vce)	*vcetype* specified in vce()
e(wexp)	weight expression	e(vcetype)	title used to label Std. Err.
e(title)	title in estimation output	e(properties)	b V
e(clustvar)	name of cluster variable	e(predict)	program used to implement predict

Matrices

e(b)	coefficient vector	e(V)	variance–covariance matrix of the estimators
e(out)	outcome values		

Functions

e(sample)	marks estimation sample

Methods and Formulas

The multinomial logit model is described in Greene (2003, chapter 21).

Suppose that there are k categorical outcomes and—without loss of generality—let the base outcome be 1. The probability that the response for the jth observation is equal to the ith outcome is

$$p_{ij} = \Pr(y_j = i) = \begin{cases} \dfrac{1}{1 + \sum\limits_{m=2}^{k} \exp(\mathbf{x}_j \boldsymbol{\beta}_m)}, & \text{if } i = 1 \\[3ex] \dfrac{\exp(\mathbf{x}_j \boldsymbol{\beta}_i)}{1 + \sum\limits_{m=2}^{k} \exp(\mathbf{x}_j \boldsymbol{\beta}_m)}, & \text{if } i > 1 \end{cases}$$

where $\mathbf{x}_j$ is the row vector of observed values of the independent variables for the jth observation and $\boldsymbol{\beta}_m$ is the coefficient vector for outcome m. The log pseudolikelihood is

$$\ln L = \sum_j w_j \sum_{i=1}^{k} I_i(y_j) \ln p_{ik}$$

where w_j is an optional weight and

$$I_i(y_j) = \begin{cases} 1, & \text{if } y_j = i \\ 0, & \text{otherwise} \end{cases}$$

Newton–Raphson maximum likelihood is used; see [R] **maximize**.

For constrained equations, the set of constraints is orthogonalized, and a subset of maximizable parameters is selected. For example, a parameter that is constrained to zero is not a maximizable parameter. If two parameters are constrained to be equal to each other, only one is a maximizable parameter.

Let $\mathbf{r}$ be the vector of maximizable parameters. Note that $\mathbf{r}$ is physically a subset of the solution parameters, $\mathbf{b}$. A matrix, $\mathbf{T}$, and a vector, $\mathbf{m}$, are defined as

$$\mathbf{b} = \mathbf{Tr} + \mathbf{m}$$

so that

$$\frac{\partial f}{\partial \mathbf{b}} = \frac{\partial f}{\partial \mathbf{r}} \mathbf{T}'$$

$$\frac{\partial^2 f}{\partial \mathbf{b}^2} = \mathbf{T} \frac{\partial^2 f}{\partial \mathbf{r}^2} \mathbf{T}'$$

$\mathbf{T}$ consists of a block form in which one part is a permutation of the identity matrix, and the other part describes how to calculate the constrained parameters from the maximizable parameters.

References

Aldrich, J. H. and F. D. Nelson. 1984. *Linear Probability, Logit, and Probit Models.* Newbury Park, CA: Sage.

Freese, J. and J. S. Long. 2000. sg155: Tests for the multinomial logit model. *Stata Technical Bulletin* 58: 19–25. Reprinted in *Stata Technical Bulletin Reprints*, vol. 10, pp. 247–255.

Greene, W. H. 2003. *Econometric Analysis.* 5th ed. Upper Saddle River, NJ: Prentice Hall.

Hamilton, L. C. 1993. sqv8: Interpreting multinomial logistic regression. *Stata Technical Bulletin* 13: 24–28. Reprinted in *Stata Technical Bulletin Reprints*, vol. 3, pp. 176–181.

——. 2004. *Statistics with Stata.* Belmont, CA: Brooks/Cole.

Hendrickx, J. 2000. sbe37: Special restrictions in multinomial logistic regression. *Stata Technical Bulletin* 56: 18–26.

Hosmer, D. W., Jr., and S. Lemeshow. 2000. *Applied Logistic Regression.* 2nd ed. New York: Wiley.

Judge, G. G., W. E. Griffiths, R. C. Hill, H. Lütkepohl, and T.-C. Lee. 1985. *The Theory and Practice of Econometrics.* 2nd ed. New York: Wiley.

Kleinbaum, D. G. and M. Klein. 2002. *Logistic Regression: A Self-Learning Text.* 2nd ed. New York: Springer.

Long, J. S. 1997. *Regression Models for Categorical and Limited Dependent Variables.* Thousand Oaks, CA: Sage.

Long, J. S. and J. Freese. 2003. *Regression Models for Categorical Dependent Variables Using Stata.* rev. ed. College Station, TX: Stata Press.

Tarlov, A. R., J. E. Ware, Jr., S. Greenfield, E. C. Nelson, E. Perrin, and M. Zubkoff. 1989. The medical outcomes study. *Journal of the American Medical Association* 262: 925–930.

Wells, K. E., R. D. Hays, M. A. Burnam, W. H. Rogers, S. Greenfield, and J. E. Ware, Jr. 1989. Detection of depressive disorder for patients receiving prepaid or fee-for-service care. *Journal of the American Medical Association* 262: 3298–3302.

Also See

Complementary:	[R] **mlogit postestimation**; [R] **constraint**
Related:	[R] **clogit**, [R] **logistic**, [R] **logit**, [R] **nlogit**, [R] **ologit**, [R] **slogit**, [SVY] **svy: mlogit**
Background:	[U] **11.1.10 Prefix commands**, [U] **20 Estimation and postestimation commands**, [R] **estimation options**, [R] **maximize**, [R] *vce_option*

Title

| mlogit postestimation — Postestimation tools for mlogit |

Description

The following postestimation commands are available for `mlogit`:

command	description
estat	AIC, BIC, VCE, and estimation sample summary
estimates	cataloging estimation results
hausman	Hausman's specification test
lincom	point estimates, standard errors, testing, and inference for linear combinations of coefficients
lrtest	likelihood-ratio test
mfx	marginal effects or elasticities
nlcom	point estimates, standard errors, testing, and inference for nonlinear combinations of coefficients
predict	predictions, residuals, influence statistics, and other diagnostic measures
predictnl	point estimates, standard errors, testing, and inference for generalized predictions
suest	seemingly unrelated estimation
test	Wald tests for simple and composite linear hypotheses
testnl	Wald tests of nonlinear hypotheses

See the corresponding entries in the *Stata Base Reference Manual* for details.

Syntax for predict

predict $\left[\,type\,\right]$ *newvar* $\left[\,if\,\right]$ $\left[\,in\,\right]$ $\left[\,,\ statistic\ \underline{o}utcome(outcome)\,\right]$

predict $\left[\,type\,\right]$ $\left\{\ stub*\,|\,newvar_1\ \ldots newvar_{k-1}\ \right\}$ $\left[\,if\,\right]$ $\left[\,in\,\right]$, $\underline{sc}ores$

where k is the number of outcomes in the model.

statistic	description
$\underline{p}r$	probability of a positive outcome; the default
xb	linear prediction $\mathbf{x}_j\mathbf{b}$
stdp	standard error of the linear prediction
stddp	standard error of the difference in two linear predictions

Note that you specify one new variable with `xb`, `stdp`, and `stddp` and specify either one or k new variables with `pr`.

These statistics are available both in and out of sample; type `predict ... if e(sample) ...` if wanted only for the estimation sample.

Options for predict

pr, the default, calculates the probability of each of the alternatives in the model or the probability of the outcome specified in outcome(). If you specify outcome(), you only need to specify one new variable; otherwise, you must specify k new variables, one for each alternative in the model.

xb calculates the linear prediction. You must also specify the outcome(*outcome*) option.

stdp calculates the standard error of the linear prediction. You must also specify the out-come(*outcome*) option.

stddp calculates the standard error of the difference in two linear predictions. You must specify option outcome(*outcome*), and in this case, you specify the two particular outcomes of interest inside the parentheses, for example, predict sed, stddp outcome(1,3).

outcome(*outcome*) specifies the outcome for which the statistic is to be calculated. equation() is a synonym for outcome(): it does not matter which you use, and the standard rules for specifying an equation() apply.

scores calculates equation-level score variables. The number of score variables created will be one less than the number of outcomes in the model. If the number of outcomes in the model were k, then

the first new variable will contain $\partial \ln L/\partial(\mathbf{x}_j \boldsymbol{\beta}_1)$;

the second new variable will contain $\partial \ln L/\partial(\mathbf{x}_j \boldsymbol{\beta}_2)$;

. . .

the $(k-1)$st new variable will contain $\partial \ln L/\partial(\mathbf{x}_j \boldsymbol{\beta}_{k-1})$.

Remarks

Remarks are presented under the headings

> *Obtaining predicted values*
> *Testing hypotheses about coefficients*

Obtaining predicted values

▷ Example 1

After estimation, we can use predict to obtain predicted probabilities, index values, and standard errors of the index, or differences in the index. For instance, in example 4 of [R] **mlogit**, we fitted a model of insurance choice on various characteristics. We can obtain the predicted probabilities for outcome 1 by typing

```
. predict p1 if e(sample), outcome(1)
(option p assumed; predicted probability)
(29 missing values generated)
. summarize p1
```

Variable	Obs	Mean	Std. Dev.	Min	Max
p1	615	.4764228	.1032279	.1698142	.71939

Note that we included `if e(sample)` to restrict the calculation to the estimation sample. In example 4 of [R] **mlogit**, the multinomial logit model was fitted on 615 observations, so there must be missing values in our dataset.

Although we typed `outcome(1)`, specifying 1 for the indemnity outcome, we could have typed `outcome(Indemnity)`. For instance, to obtain the probabilities for prepaid, we could type

```
. predict p2 if e(sample), outcome(prepaid)
(option p assumed; predicted probability)
equation prepaid not found
r(303);

. predict p2 if e(sample), outcome(Prepaid)
(option p assumed; predicted probability)
(29 missing values generated)

. summarize p2
```

Variable	Obs	Mean	Std. Dev.	Min	Max
p2	615	.4504065	.1125962	.1964103	.7885724

We must specify the label exactly as it appears in the underlying value label (or how it appears in the `mlogit` output), including capitalization.

Here we have used `predict` to obtain probabilities for the same sample on which we estimated. That is not necessary. We could use another dataset that had the independent variables defined (in our example, age, `male`, `nonwhite`, `site2`, and `site3`) and use `predict` to obtain predicted probabilities; in this case, we would not specify `if e(sample)`.

◁

▷ Example 2

`predict` can also be used to obtain the "index" values—the $\sum x_i \widehat{\beta}_i^{(k)}$—as well as the probabilities:

```
. predict idx1, outcome(Indemnity) xb
(1 missing value generated)

. summarize idx1
```

Variable	Obs	Mean	Std. Dev.	Min	Max
idx1	643	0	0	0	0

The indemnity outcome was our base outcome—the outcome for which all the coefficients were set to 0—so the index is always 0. For the prepaid and uninsured outcomes, we type

```
. predict idx2, outcome(Prepaid) xb
(1 missing value generated)

. predict idx3, outcome(Uninsure) xb
(1 missing value generated)

. summarize idx2 idx3
```

Variable	Obs	Mean	Std. Dev.	Min	Max
idx2	643	-.0566113	.4962973	-1.298198	1.700719
idx3	643	-1.980747	.6018139	-3.112741	-.8258458

We can obtain the standard error of the index by specifying the `stdp` option:

```
. predict se2, outcome(Prepaid) stdp
(1 missing value generated)
```

```
. list p2 idx2 se2 in 1/5
```

	p2	idx2	se2
1.	.3709022	-.4831167	.2437772
2.	.4977667	.055111	.1694686
3.	.4113073	-.1712106	.1793498
4.	.5424927	.3788345	.2513701
5.	.	-.0925817	.1452616

We obtained the probability p2 in the previous example.

Finally, predict can calculate the standard error of the difference in the index values between two outcomes with the stddp option:

```
. predict se_2_3, outcome(Prepaid,Uninsure) stddp
(1 missing value generated)
. list idx2 idx3 se_2_3 in 1/5
```

	idx2	idx3	se_2_3
1.	-.4831167	-3.073253	.5469354
2.	.055111	-2.715986	.4331917
3.	-.1712106	-1.579621	.3053815
4.	.3788345	-1.462007	.4492552
5.	-.0925817	-2.814022	.4024784

In the first observation, the difference in the indexes is $-.483 - (-3.073) = 2.59$. The standard error of that difference is .547.

◁

▷ Example 3

It is more difficult to interpret the results from mlogit than those from clogit or logit since there are multiple equations. For example, suppose that one of the independent variables in our model takes on the values 0 and 1 and we are attempting to understand the effect of this variable. Assume that the coefficient on this variable for the second outcome, $\beta^{(2)}$, is positive. We might then be tempted to reason that the probability of the second outcome is higher if the variable is 1 rather than 0. Most of the time, that will be true, but occasionally we will be surprised. The probability of some other outcome could increase even more (say, $\beta^{(3)} > \beta^{(2)}$), and thus the probability of outcome 2 would actually fall relative to that outcome. We can use predict to help interpret such results.

Continuing with our previously fitted insurance-choice model, we wish to describe the model's predictions by race. For this purpose, we can use the "method of recycled predictions", in which we vary characteristics of interest across the whole dataset and average the predictions. That is, we have data on both whites and nonwhites, and our individuals have other characteristics as well. We will first pretend that all the people in our data are white, but hold their other characteristics constant. We then calculate the probabilities of each outcome. Next we will pretend that all the people in our data are nonwhite, still holding their other characteristics constant. Again we calculate the probabilities of each outcome. The difference in those two sets of calculated probabilities, then, is the difference due to race, holding other characteristics constant.

```
. gen byte nonwhold = nonwhite          // save real race
. replace nonwhite = 0                  // make everyone white
(126 real changes made)
```

```
. predict wpind, outcome(Indemnity)              // predict probabilities
(option p assumed; predicted probability)
(1 missing value generated)

. predict wpp, outcome(Prepaid)
(option p assumed; predicted probability)
(1 missing value generated)

. predict wpnoi, outcome(Uninsure)
(option p assumed; predicted probability)
(1 missing value generated)

. replace nonwhite=1                             // make everyone nonwhite
(644 real changes made)

. predict nwpind, outcome(Indemnity)
(option p assumed; predicted probability)
(1 missing value generated)

. predict nwpp, outcome(Prepaid)
(option p assumed; predicted probability)
(1 missing value generated)

. predict nwpnoi, outcome(Uninsure)
(option p assumed; predicted probability)
(1 missing value generated)

. replace nonwhite=nonwhold                      // restore real race
(518 real changes made)

. summarize wp* nwp*, sep(0)
```

Variable	Obs	Mean	Std. Dev.	Min	Max
wpind	643	.5141673	.0872679	.3092903	.71939
wpp	643	.4082052	.0993286	.1964103	.6502247
wpnoi	643	.0776275	.0360283	.0273596	.1302816
nwpind	643	.3112809	.0817693	.1511329	.535021
nwpp	643	.630078	.0979976	.3871782	.8278881
nwpnoi	643	.0586411	.0287185	.0209648	.0933874

In [R] **mlogit**, we presented a cross-tabulation of insurance type and race. Those values were unadjusted. The means reported above are the values adjusted for age, sex, and site. Combining the results gives

	Unadjusted		Adjusted	
	white	nonwhite	white	nonwhite
Indemnity	.51	.36	.52	.31
Prepaid	.42	.57	.41	.63
Uninsured	.07	.07	.08	.06

We find, for instance, after adjusting for age, sex, and site, that while 57% of nonwhites in our data had prepaid plans, 63% of nonwhites choose prepaid plans.

◁

You can also compute marginal effects to interpret the results from multinomial logit models effectively. The marginal effects show how the probabilities of each outcome change with respect to changes in the regressors. See example 3 in [R] **mfx** for an example of computing marginal effects after mlogit.

❑ Technical Note

You can use predict to classify predicted values and compare them with the observed outcomes to interpret a multinomial logit model. This is a variation on the notions of sensitivity and specificity for logistic regression. Here we will classify indemnity and prepaid as definitely predicting indemnity, definitely predicting prepaid, and ambiguous.

```
. predict indem, outcome(Indemnity) index          // obtain indexes
(1 missing value generated)
. predict prepaid, outcome(Prepaid) index
(1 missing value generated)
. gen diff = prepaid-indem                          // obtain difference
(1 missing value generated)
. predict sediff, outcome(Indemnity,Prepaid) stddp  // & its standard error
(1 missing value generated)
. gen type = 1 if diff/sediff < -1.96               // definitely indemnity
(504 missing values generated)
. replace type = 3 if diff/sediff > 1.96            // definitely prepaid
(100 real changes made)
. replace type = 2 if type>=. & diff/sediff < .     // ambiguous
(404 real changes made)
. label def type 1 "Def Ind" 2 "Ambiguous" 3 "Def Prep"
. label values type type                            // label results
. tabulate insure type
```

insure	type Def Ind	Ambiguous	Def Prep	Total
Indemnity	78	183	33	294
Prepaid	44	177	56	277
Uninsure	12	28	5	45
Total	134	388	94	616

We can see that the predictive power of this model is modest. There are a substantial number of misclassifications in both directions, though there are more correctly classified observations than misclassified observations.

Also the uninsured look overwhelmingly as though they might have come from the indemnity system rather than from the prepaid system.

❑

(Continued on next page)

Testing hypotheses about coefficients

▷ Example 4

    `test` tests hypotheses about the coefficients just as after any estimation command; see [R] **test**. Note, however, `test`'s syntax for dealing with multiple equation models. Because `test` bases its results on the estimated covariance matrix, we might prefer a likelihood-ratio test; see example 5 in [R] **mlogit** for an example of `lrtest`.

    If we simply list variables after the `test` command, we are testing that the corresponding coefficients are zero across all equations:

```
. test site2 site3
 ( 1)  [Prepaid]site2 = 0
 ( 2)  [Uninsure]site2 = 0
 ( 3)  [Prepaid]site3 = 0
 ( 4)  [Uninsure]site3 = 0
           chi2(  4) =   19.74
         Prob > chi2 =  0.0006
```

We can test that all the coefficients (except the constant) in a single equation are zero by simply typing the outcome in square brackets:

```
. test [Uninsure]
 ( 1)  [Uninsure]age = 0
 ( 2)  [Uninsure]male = 0
 ( 3)  [Uninsure]nonwhite = 0
 ( 4)  [Uninsure]site2 = 0
 ( 5)  [Uninsure]site3 = 0
           chi2(  5) =    9.31
         Prob > chi2 =  0.0973
```

We specify the outcome just as we do with `predict`; we can specify the label if the outcome variable is labeled, or we can specify the numeric value of the outcome. We would have obtained the same test as above if we had typed `test [3]` since 3 is the value of `insure` for the outcome uninsured.

    We can combine the two syntaxes. To test that the coefficients on the site variables are 0 in the equation corresponding to the outcome prepaid, we can type

```
. test [Prepaid]: site2 site3
 ( 1)  [Prepaid]site2 = 0
 ( 2)  [Prepaid]site3 = 0
           chi2(  2) =   10.78
         Prob > chi2 =  0.0046
```

We specified the outcome and then followed that with a colon and the variables we wanted to test.

We can also test that coefficients are equal across equations. To test that all coefficients except the constant are equal for the prepaid and uninsured outcomes, we can type

```
. test [Prepaid=Uninsure]
 ( 1)  [Prepaid]age - [Uninsure]age = 0
 ( 2)  [Prepaid]male - [Uninsure]male = 0
 ( 3)  [Prepaid]nonwhite - [Uninsure]nonwhite = 0
 ( 4)  [Prepaid]site2 - [Uninsure]site2 = 0
 ( 5)  [Prepaid]site3 - [Uninsure]site3 = 0
           chi2(  5) =   13.80
         Prob > chi2 =  0.0169
```

To test that only the site variables are equal, we can type

```
. test [Prepaid=Uninsure]: site2 site3
 ( 1)  [Prepaid]site2 - [Uninsure]site2 = 0
 ( 2)  [Prepaid]site3 - [Uninsure]site3 = 0
          chi2(  2) =   12.68
        Prob > chi2 =  0.0018
```

Finally we can test any arbitrary constraint by simply entering the equation and specifying the coefficients as described in [U] **13.5 Accessing coefficients and standard errors**. The following hypothesis is senseless but illustrates the point:

```
. test (([Prepaid]age+[Uninsure]site2)/2 = 2-[Uninsure]nonwhite)
 ( 1)  .5 [Prepaid]age + [Uninsure]nonwhite + .5 [Uninsure]site2 = 2
          chi2(  1) =   22.45
        Prob > chi2 =  0.0000
```

See [R] **test** for more information about test. The information there about combining hypotheses across test commands (the accum option) also applies after mlogit.

◁

Methods and Formulas

All postestimation commands listed above are implemented as ado-files.

Also See

Complementary:	[R] **mlogit**; [R] **estimates**, [R] **hausman**, [R] **lincom**, [R] **lrtest**, [R] **mfx**, [R] **nlcom**, [R] **predictnl**, [R] **suest**, [R] **test**, [R] **testnl**
Background:	[U] **13.5 Accessing coefficients and standard errors**, [U] **20 Estimation and postestimation commands**, [R] **estat**, [R] **predict**

Title

> **more** — The —more— message

Syntax

Tell Stata to pause or not pause for –more– messages

> set <u>mo</u>re { on | off } [, <u>permanently</u>]

Set number of lines between –more– messages

> set <u>pa</u>gesize #

Description

set more on, which is the default, tells Stata to wait until you press a key before continuing when a —more— message is displayed.

set more off tells Stata not to pause or display the —more— message.

set pagesize # sets the number of lines between —more— messages. The permanently option is not allowed with set pagesize.

Option

permanently specifies that, in addition to making the change right now, the more setting be remembered and become the default setting when you invoke Stata.

Remarks

When you see —more— at the bottom of the screen,

Press ...	and Stata...
letter *l* or *Enter*	displays the next line
letter *q*	acts as if you pressed *Break*
space bar or any other key	displays the next screen

In addition, you can click the **More** button or click on —more— to display the next screen.

—more— is Stata's way of telling you that it has something more to show you but that showing it to you will cause the information on the screen to scroll off.

If you type set more off, —more— conditions will never arise, and Stata's output will scroll by at full speed.

If you type set more on, —more— conditions will be restored at the appropriate places.

Programmers should see [P] **more** for information on the more programming command.

Also See

Complementary:	[R] **query**,
	[P] **creturn**, [P] **more**
Background:	[U] **7** —more— **conditions**

Title

mprobit — Multinomial probit regression

Syntax

mprobit *depvar* [*indepvars*] [*if*] [*in*] [*weight*] [, *options*]

options	description
Model	
<u>nocon</u>stant	suppress constant terms
<u>baseout</u>come(*#*)	alternative used to normalize location
<u>constr</u>aints(*constraints*)	apply specified linear constraints
SE/Robust	
vce(*vcetype*)	*vcetype* may be oim, <u>r</u>obust, opg, <u>boot</u>strap, or <u>jack</u>knife
<u>r</u>obust	synonym for vce(robust)
<u>cl</u>uster(*varname*)	adjust standard errors for intragroup correlation
Reporting	
<u>l</u>evel(*#*)	set confidence level; default is level(95)
Int options	
<u>intp</u>oints(*#*)	number of quadrature points
Max options	
maximize_options	control the maximization process; seldom used

bootstrap, by, jackknife, rolling, statsby, and xi are allowed; see [U] **11.1.10 Prefix commands**.
fweights, iweights, and pweights are allowed; see [U] **11.1.6 weight**.
See [U] **20 Estimation and postestimation commands** for additional capabilities of estimation commands.

Description

mprobit fits multinomial probit (MNP) models via maximum likelihood. *depvar* contains the choice made for each observation, and *indepvars* are the associated covariates. The error terms are assumed to be independent, standard normal, random variables. See [R] **asmprobit** for the case where the latent-variable errors are correlated or heteroskedastic and you have alternative-specific variables.

Options

⌐ Model ⌐

noconstant suppresses the $J - 1$ constant terms.

baseoutcome(*#*) specifies the alternative to use to normalize the location of the latent variable. The default is to use the most frequent outcome.

constraints(*constraints*); see [R] **estimation options**.

SE/Robust

vce(*vcetype*); see [R] **vce_option**.

robust, cluster(*varname*); see [R] **estimation options**.

Reporting

level(*#*); see [R] **estimation options**.

Int options

intpoints(*#*) specifies the number of quadrature points to use in the Gaussian quadrature approximation of the likelihood. The default is 15.

Max options

maximize_options: difficult, technique(*algorithm_spec*), iterate(*#*), [no]log, trace, gradient, showstep, hessian, shownrtolerance, tolerance(*#*), ltolerance(*#*), gtolerance(*#*), nrtolerance(*#*), nonrtolerance, from(*init_specs*); see [R] **maximize**. These options are seldom used.

Remarks

The multinomial probit (MNP) model is used with discrete dependent variables that take on more than two outcomes that do not have a natural ordering. The stochastic error terms for this implementation of the model are assumed to have independent, standard normal distributions. To use mprobit, you must have a single observation for each decision maker in the sample. See [R] **asmprobit** for another implementation of the MNP model that permits correlated and heteroskedastic errors and is suitable when you have data for each alternative a decision maker faced.

The MNP model is frequently motivated using a latent-variable framework. The latent variable for the jth alternative, $j = 1, \ldots, J$, is

$$\eta_{ij} = \mathbf{z}_i \boldsymbol{\alpha}_j + \xi_{ij}$$

where the $1 \times q$ row vector $\mathbf{z}_i$ contains the observed independent variables for the ith decision maker. Associated with $\mathbf{z}_i$ are the J vectors of regression coefficients $\boldsymbol{\alpha}_j$. The $\xi_{i,1}, \ldots, \xi_{i,J}$ are distributed independently and identically standard normal. The decision maker chooses the alternative k such that $\eta_{ik} \geq \eta_{im}$ for $m \neq k$.

Suppose that case i chooses alternative k, and take the difference between latent variable η_{ik} and the $J - 1$ others:

$$
\begin{aligned}
v_{ijk} &= \eta_{ij} - \eta_{ik} \\
&= \mathbf{z}_i(\boldsymbol{\alpha}_j - \boldsymbol{\alpha}_k) + \xi_{ij} - \xi_{ik} \\
&= \mathbf{z}_i \boldsymbol{\gamma}_{j'} + \epsilon_{ij'}
\end{aligned}
\tag{1}
$$

where $j' = j$ if $j < k$ and $j' = j - 1$ if $j > k$ so that $j' = 1, \ldots, J - 1$. Notice that $\mathrm{Var}(\epsilon_{ij'}) = \mathrm{Var}(\xi_{ij} - \xi_{ik}) = 2$ and that $\mathrm{Cov}(\epsilon_{ij'}, \epsilon_{il'}) = 1$ for $j' \neq l'$. The probability that alternative k is chosen is

$$
\begin{aligned}
\Pr(i \text{ chooses } k) &= \Pr(v_{i1k} \leq 0, \ldots, v_{i,J-1,k} \leq 0) \\
&= \Pr(\epsilon_{i1} \leq -\mathbf{z}_i \boldsymbol{\gamma}_1, \ldots, \epsilon_{i,J-1} \leq -\mathbf{z}_i \boldsymbol{\gamma}_{J-1})
\end{aligned}
$$

Hence, evaluating the likelihood function involves computing probabilities from the multivariate normal distribution. That all the covariances are equal simplifies the problem somewhat; see *Methods and Formulas* for details.

In (1), not all J of the α_j are identifiable. To remove the indeterminacy, α_l is set to the zero vector, where l is the base outcome as specified in the baseoutcome() option. That fixes the lth latent variable to zero so that the remaining variables measure the attractiveness of the other alternatives relative to the base.

▷ Example 1

As discussed in [R] **mlogit**, we have data on the type of health insurance available to 616 psychologically depressed subjects in the U.S. (Tarlov et al. 1989; Wells et al. 1989). Patients may have either an indemnity (fee-for-service) plan or a prepaid plan such as an HMO, or the patient may be uninsured. Demographic variables include age, gender, race, and site. Indemnity insurance is the most popular alternative, so mprobit will choose it as the base outcome by default.

```
. use http://www.stata-press.com/data/r9/sysdsn3, clear
(Health insurance data)

. mprobit insure age male nonwhite site2 site3
Iteration 0:   log likelihood = -548.55788
Iteration 1:   log likelihood = -539.38091
Iteration 2:   log likelihood =  -534.5748
Iteration 3:   log likelihood = -534.52836
Iteration 4:   log likelihood = -534.52833
```

Multinomial probit regression				Number of obs	=	615
				Wald chi2(10)	=	40.18
Log likelihood = -534.52833				Prob > chi2	=	0.0000

insure	Coef.	Std. Err.	z	P>|z|	[95% Conf. Interval]	
Prepaid						
age	-.0098536	.0052688	-1.87	0.061	-.0201802	.000473
male	.4774678	.1718316	2.78	0.005	.1406841	.8142515
nonwhite	.8245003	.1977582	4.17	0.000	.4369013	1.212099
site2	.0973956	.1794546	0.54	0.587	-.2543289	.4491201
site3	-.495892	.1904984	-2.60	0.009	-.869262	-.1225221
_cons	.22315	.2792424	0.80	0.424	-.324155	.7704549
Uninsure						
age	-.0050815	.0075327	-0.67	0.500	-.0198452	.0096823
male	.3332637	.2432986	1.37	0.171	-.1435928	.8101203
nonwhite	.2485859	.2767734	0.90	0.369	-.29388	.7910518
site2	-.6899478	.2804497	-2.46	0.014	-1.239619	-.1402765
site3	-.1788447	.2479898	-0.72	0.471	-.6648957	.3072063
_cons	-.9855916	.3891873	-2.53	0.011	-1.748385	-.2227985

(insure=Indemnity is the base outcome)

◁

The likelihood function for mprobit is derived under the assumption that all decision-making units face the same choice set, which is the union of all outcomes observed in the dataset. If that is not true for your model, then an alternative is to use the asmprobit command, which does not require this assumption. In order to do that, you will need to expand the dataset so that each decision maker has k_i observations, where k_i is the number of alternatives in the choice set faced by decision maker i. You will also need to create a binary variable to indicate the choice made by each decision maker. Moreover, you will need to use the correlation(independent) and stddev(homoskedastic) options with asmprobit unless you have alternative-specific variables.

Saved Results

mprobit saves in e():

Scalars

e(N)	number of observations	e(df_m)	model degrees of freedom
e(N_clust)	number of clusters	e(ll)	log simulated-likelihood
e(k_indvars)	number of independent variables	e(chi2)	χ^2
e(k)	number of parameters	e(p)	p-value for χ^2
e(k_eq)	number of equations	e(const)	flag for the constant
e(k_out)	number of outcomes	e(rank)	rank of e(V)
e(k_points)	number of quadrature points	e(rc)	return code
e(i_base)	base outcome index	e(converged)	1 if converged, 0 otherwise

Macros

e(cmd)	mprobit	e(vce)	*vcetype* specified in vce()
e(depvar)	name of dependent variable	e(vcetype)	title used to label Std. Err.
e(indvars)	independent variables	e(opt)	type of optimization
e(wtype)	weight type	e(ml_method)	type of ml method
e(wexp)	weight expression	e(user)	name of likelihood-evaluator program
e(title)	title in estimation output	e(technique)	maximization technique
e(clustvar)	name of cluster variable	e(crittype)	optimization criterion
e(chi2type)	Wald, type of model χ^2	e(properties)	b V
e(outeqs)	outcome equations	e(predict)	program used to implement predict
e(out*i*)	outcome labels, $i=1,\ldots,$e(k_out)		

Matrices

e(b)	coefficient vector	e(V)	variance–covariance matrix of the
e(gradient)	gradient vector		estimators
e(outcomes)	outcome values		

Functions

e(sample)	marks estimation sample

Methods and Formulas

As discussed in *Remarks*, the latent variables for a J-alternative model are $\eta_{ij} = \mathbf{z}_i\boldsymbol{\alpha}_j + \xi_{ij}$, for $j = 1,\ldots,J$, $i = 1,\ldots,n$, and $\{\xi_{i,1},\ldots,\xi_{i,J}\} \sim$ i.i.d $N(0,1)$. The experimenter observes alternative k for the ith observation if $\eta_{ik} > \eta_{il}$ for $l \neq k$. For $j \neq k$, let

$$
\begin{aligned}
v_{ij'} &= \eta_{ij} - \eta_{ik} \\
&= \mathbf{z}_i(\boldsymbol{\alpha}_j - \boldsymbol{\alpha}_k) + \xi_{ij} - \xi_{ik} \\
&= \mathbf{z}_i\boldsymbol{\gamma}_{j'} + \epsilon_{ij'},
\end{aligned}
$$

where $j' = j$ if $j < k$ and $j' = j - 1$ if $j > k$ so that $j' = 1,\ldots,J - 1$. Notice that $\epsilon_i = (\epsilon_{i1},\ldots,\epsilon_{i,J-1}) \sim \text{MVN}(\mathbf{0}, \boldsymbol{\Sigma})$, where

$$\boldsymbol{\Sigma} = \begin{pmatrix} 2 & 1 & 1 & \dots & 1 \\ 1 & 2 & 1 & \dots & 1 \\ 1 & 1 & 2 & \dots & 1 \\ \vdots & \vdots & \vdots & \ddots & \vdots \\ 1 & 1 & 1 & \dots & 2 \end{pmatrix}$$

Denote the deterministic part of the model as $\lambda_{ij'} = \mathbf{z}_i \boldsymbol{\gamma}_{j'}$; the probability that subject i chooses outcome k is

$$\begin{aligned} \Pr(y_i = k) &= \Pr(v_{i1} \leq 0, \dots, v_{i,J-1} \leq 0) \\ &= \Pr(\epsilon_{i1} \leq -\lambda_{i1}, \dots, \epsilon_{i,J-1} \leq -\lambda_{i,J-1}) \\ &= \frac{1}{(2\pi)^{(J-1)/2} |\boldsymbol{\Sigma}|^{1/2}} \int_{-\infty}^{-\lambda_{i1}} \cdots \int_{-\infty}^{-\lambda_{i,J-1}} \exp\left(-\tfrac{1}{2} \mathbf{z}' \boldsymbol{\Sigma}^{-1} \mathbf{z}\right) d\mathbf{z} \end{aligned}$$

Because of the exchangeable correlation structure of $\boldsymbol{\Sigma}$ (notice that $\rho_{ij} = 1/2$ for all $i \neq j$), we can utilize Dunnett's (1989) result to reduce the multidimensional integral to a single dimension:

$$\Pr(y_i = k) = \frac{1}{\sqrt{\pi}} \int_0^{\infty} \left\{ \prod_{j=1}^{J-1} \Phi\left(-z\sqrt{2} - \lambda_{ij}\right) + \prod_{j=1}^{J-1} \Phi\left(z\sqrt{2} - \lambda_{ij}\right) \right\} e^{-z^2} dz$$

Gaussian quadrature is used to approximate this integral, resulting in the K-point quadrature formula

$$\Pr(y_i = k) \approx \frac{1}{2} \sum_{k=1}^{K} w_k \left\{ \prod_{j=1}^{J-1} \Phi\left(-x_k\sqrt{2} - \lambda_{ij}\right) + \prod_{j=1}^{J-1} \Phi\left(x_k\sqrt{2} - \lambda_{ij}\right) \right\}$$

where w_k and x_k are the weights and roots of the Laguerre polynomial of order K. In mprobit, K is specified by option intpoints().

References

Dunnett, C. W. 1989. Algorithm AS 251: Multivariate normal probability integrals with product correlation structure. *Journal of the Royal Statistical Society, Series C* 38: 564–579.

Also See

Complementary: [R] **mprobit postestimation**

Related: [R] **asmprobit**, [R] **mlogit**; [R] **clogit**, [R] **nlogit**,
[R] **ologit**, [R] **oprobit**

Background: [U] **11.1.10 Prefix commands**,
[U] **20 Estimation and postestimation commands**,
[R] **estimation options**, [R] **maximize**, [R] *vce_option*

Title

> **mprobit postestimation** — Postestimation tools for mprobit

Description

The following postestimation commands are available for `mprobit`:

command	description
estat	AIC, BIC, VCE, and estimation sample summary
estimates	cataloging estimation results
lincom	point estimates, standard errors, testing, and inference for linear combinations of coefficients
lrtest	likelihood-ratio test
mfx	marginal effects or elasticities
nlcom	point estimates, standard errors, testing, and inference for nonlinear combinations of coefficients
predict	predicted probabilities, linear predictions, and standard errors
predictnl	point estimates, standard errors, testing, and inference for generalized predictions
suest	seemingly unrelated estimation
test	Wald tests for simple and composite linear hypotheses
testnl	Wald tests of nonlinear hypotheses

See the corresponding entries in the *Stata Base Reference Manual* for details.

Syntax for predict

predict [*type*] *newvar* [*if*] [*in*] [, *statistic* <u>ou</u>tcome(*outcome*)]

predict [*type*] { *stub** | *newvar*$_1$ ... *newvar*$_{k-1}$ } [*if*] [*in*] , <u>sc</u>ores

where k is the number of outcomes in the model.

statistic	description
<u>p</u>r	probability of a positive outcome; the default
xb	linear prediction
stdp	standard error of the linear prediction

Note that you specify one new variable with xb and stdp and specify either one or k new variables with pr.

Statistics are available both in and out of sample; type predict ... if e(sample) ... if wanted only for the estimation sample.

Options for predict

pr, the default, calculates the probability of each of the alternatives in the model or the probability of the outcome specified in outcome(). If you specify outcome(), you only need to specify one new variable; otherwise, you must specify k new variables, one for each alternative in the model.

xb calculates the linear prediction $\mathbf{x}_i\boldsymbol{\alpha}_j$ for alternative j and individual i. The index j corresponds to the outcome specified in outcome().

stdp calculates the standard error of the linear prediction.

outcome(*outcome*) specifies the outcome for which the statistic is to be calculated.

scores calculates the equation-level score variables. The jth new variable will contain the scores for the jth fitted equation.

Remarks

Once you have fitted a multinomial probit model, you can use predict to obtain probabilities that an individual will choose each of the alternatives for the estimation sample, as well as other samples; see [U] **20 Estimation and postestimation commands** and [R] **predict**.

▷ Example 1

In [R] **mprobit** we fit the multinomial probit model to a dataset containing the type of health insurance available to 616 psychologically depressed subjects in the U.S. (Tarlov et al. 1989; Wells et al. 1989). We can obtain the predicted probabilities by typing

```
. predict p1-p3, pr
. list p1-p3 insure in 1/10
```

	p1	p2	p3	insure
1.	.5961306	.3741823	.029687	Indemnity
2.	.4719296	.4972288	.0308416	Prepaid
3.	.4896085	.4121961	.0981953	Indemnity
4.	.3730529	.5416623	.0852848	Prepaid
5.	.5063069	.4629773	.0307158	.
6.	.4768125	.4923547	.0308327	Prepaid
7.	.5035672	.4657016	.0307312	Prepaid
8.	.3326361	.5580404	.1093235	.
9.	.4758165	.4384811	.0857024	Uninsure
10.	.5734057	.3316601	.0949342	Prepaid

Notice that insure contains a missing value for observations 5 and 8. Because of that, those two observations were not used in the estimation. However, because none of the independent variables is missing, predict can still calculate the probabilities. Had we typed

```
. predict p1-p3 if e(sample), pr
```

predict would have filled in missing values for p1, p2, and p3 for those observations since they were not used in the estimation.

◁

Methods and Formulas

All postestimation commands listed above are implemented as ado-files.

Also See

Complementary: [R] **mprobit**; [R] **estimates**, [R] **lincom**, [R] **lrtest**, [R] **mfx**,
                   [R] **nlcom**, [R] **predictnl**, [R] **suest**, [R] **test**, [R] **testnl**

Background: [U] **13.5 Accessing coefficients and standard errors**,
                   [U] **20 Estimation and postestimation commands**,
                   [R] **estat**, [R] **predict**

Title

mvreg — Multivariate regression

Syntax

mvreg *depvars* = *indepvars* $\big[$ *if* $\big]$ $\big[$ *in* $\big]$ $\big[$ *weight* $\big]$ $\big[$, *options* $\big]$

options	description
Model	
<u>noco</u>nstant	suppress constant term
Reporting	
<u>level</u>(#)	set confidence level; default is level(95)
<u>corr</u>	report correlation matrix
† <u>noh</u>eader	suppress header table from above coefficient table
† <u>nota</u>ble	suppress coefficient table

† noheader and notable are not shown in the dialog box.

depvars and *indepvars* may contain time-series operators; see [U] **11.4.3 Time-series varlists**.

bootstrap, by, jackknife, rolling, statsby, and xi may be used with mvreg; see
   [U] **11.1.10 Prefix commands**.

aweights and fweights are allowed; see [U] **11.1.6 weight**.

See [U] **20 Estimation and postestimation commands** for additional capabilities of estimation commands.

Description

mvreg fits multivariate regression models.

Options

$\overline{\quad\boxed{\text{Model}}\quad}$

noconstant suppresses the constant term (intercept) in the model.

$\overline{\quad\boxed{\text{Reporting}}\quad}$

level(#) specifies the confidence level, as a percentage, for confidence intervals. The default is
   level(95) or as set by set level; see [U] **20.6 Specifying the width of confidence intervals**.

corr displays the correlation matrix of the residuals between the equations.

The following options are available with mvreg but are not shown in the dialog box:

noheader suppresses display of the table reporting F statistics, R-squared, and root mean squared
   error above the coefficient table.

notable suppresses display of the coefficient table.

Remarks

Multivariate regression differs from multiple regression in that *several* dependent variables are jointly regressed on the same independent variables. Multivariate regression is related to Zellner's seemingly unrelated regression (see [R] **sureg**), but since the same set of independent variables is used for each dependent variable, the syntax is simpler, and the calculations are faster.

The individual coefficients and standard errors produced by `mvreg` are identical to those that would be produced by `regress` estimating each equation separately. The difference is that `mvreg`, being a joint estimator, also estimates the between-equation covariances, so you can test coefficients across equations and, in fact, the `test` syntax makes such tests more convenient.

▷ Example 1

Using the automobile data, we fit a multivariate regression for "space" variables (`headroom`, `trunk`, and `turn`) in terms of a set of other variables, including three "performance variables" (`displacement`, `gear_ratio`, and `mpg`):

```
. use http://www.stata-press.com/data/r9/auto
(1978 Automobile Data)
. mvreg headroom trunk turn = price mpg displ gear_ratio length weight
```

Equation	Obs	Parms	RMSE	"R-sq"	F	P
headroom	74	7	.7390205	0.2996	4.777213	0.0004
trunk	74	7	3.052314	0.5326	12.7265	0.0000
turn	74	7	2.132377	0.7844	40.62042	0.0000

	Coef.	Std. Err.	t	P>|t|	[95% Conf. Interval]	
headroom						
price	-.0000528	.000038	-1.39	0.168	-.0001286	.0000229
mpg	-.0093774	.0260463	-0.36	0.720	-.061366	.0426112
displacement	.0031025	.0024999	1.24	0.219	-.0018873	.0080922
gear_ratio	.2108071	.3539588	0.60	0.553	-.4956976	.9173118
length	.015886	.012944	1.23	0.224	-.0099504	.0417223
weight	-.0000868	.0004724	-0.18	0.855	-.0010296	.0008561
_cons	-.4525117	2.170073	-0.21	0.835	-4.783995	3.878972
trunk						
price	.0000445	.0001567	0.28	0.778	-.0002684	.0003573
mpg	-.0220919	.1075767	-0.21	0.838	-.2368159	.1926322
displacement	.0032118	.0103251	0.31	0.757	-.0173971	.0238207
gear_ratio	-.2271321	1.461926	-0.16	0.877	-3.145149	2.690885
length	.170811	.0534615	3.20	0.002	.0641014	.2775206
weight	-.0015944	.001951	-0.82	0.417	-.0054885	.0022997
_cons	-13.28253	8.962868	-1.48	0.143	-31.17249	4.607429
turn						
price	-.0002647	.0001095	-2.42	0.018	-.0004833	-.0000462
mpg	-.0492948	.0751542	-0.66	0.514	-.1993031	.1007136
displacement	.0036977	.0072132	0.51	0.610	-.0106999	.0180953
gear_ratio	-.1048432	1.021316	-0.10	0.919	-2.143399	1.933712
length	.072128	.0373487	1.93	0.058	-.0024204	.1466764
weight	.0027059	.001363	1.99	0.051	-.0000145	.0054264
_cons	20.19157	6.261549	3.22	0.002	7.693467	32.68967

We should have specified the `corr` option so that we would also see the correlations between the residuals of the equations. We can correct our omission because `mvreg`—like all estimation commands—typed without arguments redisplays results. The `noheader` and `notable` (read no-table) options suppress redisplaying the output we have already seen:

```
. mvreg, notable noheader corr
Correlation matrix of residuals:
            headroom      trunk       turn
headroom     1.0000
   trunk     0.4986     1.0000
    turn    -0.1090    -0.0628     1.0000
Breusch-Pagan test of independence: chi2(3) =    19.566, Pr = 0.0002
```

The Breusch–Pagan test is significant, so the residuals of these three space variables are not independent of each other.

The three performance variables among our independent variables are `mpg`, `displacement`, and `gear_ratio`. We can jointly test the significance of these three variables in all the equations by typing

```
. test mpg displacement gear_ratio
 ( 1)  [headroom]mpg = 0
 ( 2)  [trunk]mpg = 0
 ( 3)  [turn]mpg = 0
 ( 4)  [headroom]displacement = 0
 ( 5)  [trunk]displacement = 0
 ( 6)  [turn]displacement = 0
 ( 7)  [headroom]gear_ratio = 0
 ( 8)  [trunk]gear_ratio = 0
 ( 9)  [turn]gear_ratio = 0
       F(  9,    67) =    0.33
           Prob > F =    0.9622
```

These three variables are not, as a group, significant. We might have suspected this from their individual significance in the individual regressions, but this multivariate test provides an overall assessment with a single p-value.

We can also perform a test for the joint significance of all three equations:

```
. test [headroom]
 (output omitted )
. test [trunk], accum
 (output omitted )
. test [turn], accum
 ( 1)  [headroom]price = 0
 ( 2)  [headroom]mpg = 0
 ( 3)  [headroom]displacement = 0
 ( 4)  [headroom]gear_ratio = 0
 ( 5)  [headroom]length = 0
 ( 6)  [headroom]weight = 0
 ( 7)  [trunk]price = 0
 ( 8)  [trunk]mpg = 0
 ( 9)  [trunk]displacement = 0
 (10)  [trunk]gear_ratio = 0
 (11)  [trunk]length = 0
 (12)  [trunk]weight = 0
 (13)  [turn]price = 0
 (14)  [turn]mpg = 0
 (15)  [turn]displacement = 0
 (16)  [turn]gear_ratio = 0
 (17)  [turn]length = 0
```

```
(18)  [turn]weight = 0
      F( 18,   67) =   19.34
           Prob > F =    0.0000
```

The set of variables as a whole is strongly significant. We might have suspected this, too, from the individual equations.

◁

❑ Technical Note

The mvreg command provides a good way to deal with multiple comparisons. If we wanted to assess the effect of length, we might be dissuaded from interpreting any of its coefficients except that in the trunk equation. [trunk]length—the coefficient on length in the trunk equation—has a p-value of .002, but in the other two equations, it has p-values of only .224 and .058.

A conservative statistician might argue that there are 18 tests of significance in mvreg's output (not counting those for the intercept), so p-values above $.05/18 = .0028$ should be declared insignificant at the 5% level. A more aggressive but, in our opinion, reasonable approach would be to first note that the three equations are jointly significant, so we are justified in making some interpretation. Then we would work through the individual variables using test, possibly using $.05/6 = .0083$ (6 because there are 6 independent variables) for the 5% significance level. For instance, examining length:

```
. test length
 ( 1)  [headroom]length = 0
 ( 2)  [trunk]length = 0
 ( 3)  [turn]length = 0
       F( 3,   67) =    4.94
            Prob > F =    0.0037
```

The reported significance level of .0037 is less than .0083, so we will declare this variable significant. [trunk]length is certainly significant with its p-value of .002, but what about in the remaining two equations with p-values .224 and .058? We perform a joint test:

```
. test [headroom]length [turn]length
 ( 1)  [headroom]length = 0
 ( 2)  [turn]length = 0
       F( 2,   67) =    2.91
            Prob > F =    0.0613
```

At this point, reasonable statisticians could disagree. The .06 significance value suggests no interpretation, but these were the two least-significant values out of three, so we would expect the p-value to be a little high. Perhaps an equivocal statement is warranted: there seems to be an effect, but chance cannot be excluded.

❑

(Continued on next page)

Saved Results

mvreg saves in e():

Scalars
e(N)	number of observations
e(k)	number of parameters (including constant)
e(k_eq)	number of equations
e(df_r)	residual degrees of freedom
e(chi2)	Breusch–Pagan χ^2 (corr only)
e(df_chi2)	degrees of freedom for Breusch–Pagan χ^2 (corr only)

Macros
e(cmd)	mvreg
e(eqnames)	names of equations
e(wtype)	weight type
e(wexp)	weight expression
e(r2)	R-squared for each equation
e(rmse)	RMSE for each equation
e(F)	F statistic for each equation
e(p_F)	significance of F for each equation
e(properties)	b V
e(predict)	program used to implement predict

Matrices
e(b)	coefficient vector
e(V)	variance–covariance matrix of the estimators
e(Sigma)	$\widehat{\Sigma}$ matrix

Functions
e(sample)	marks estimation sample

Methods and Formulas

mvreg is implemented as an ado-file.

Given q equations and p independent variables (including the constant), the parameter estimates are given by the $p \times q$ matrix

$$\mathbf{B} = (\mathbf{X}'\mathbf{W}\mathbf{X})^{-1}\mathbf{X}'\mathbf{W}\mathbf{Y}$$

where $\mathbf{Y}$ is an $n \times q$ matrix of dependent variables and $\mathbf{X}$ is a $n \times p$ matrix of independent variables. $\mathbf{W}$ is a weighting matrix equal to $\mathbf{I}$ if no weights are specified. If weights are specified, let $\mathbf{v}: 1 \times n$ be the specified weights. If fweight frequency weights are specified, $\mathbf{W} = \text{diag}(\mathbf{v})$. If aweight analytic weights are specified, $\mathbf{W} = \text{diag}\{\mathbf{v}/(\mathbf{1}'\mathbf{v})(\mathbf{1}'\mathbf{1})\}$, meaning that the weights are normalized to sum to the number of observations.

The residual covariance matrix is

$$\mathbf{R} = \{\mathbf{Y}'\mathbf{W}\mathbf{Y} - \mathbf{B}'(\mathbf{X}'\mathbf{W}\mathbf{X})\mathbf{B}\}/(n-p)$$

The estimated covariance matrix of the estimates is $\mathbf{R} \otimes (\mathbf{X}'\mathbf{W}\mathbf{X})^{-1}$. These results are identical to those produced by sureg when the same list of independent variables is specified repeatedly; see [R] **sureg**.

The Breusch and Pagan (1980) χ^2 statistic—a Lagrange multiplier statistic—is given by

$$\lambda = n \sum_{i=1}^{q} \sum_{j=1}^{i-1} r_{ij}^2$$

where r_{ij} is the estimated correlation between the residuals of the equations and n is the number of observations. It is distributed as χ^2 with $q(q-1)/2$ degrees of freedom.

References

Breusch, T. and A. Pagan. 1980. The LM test and its applications to model specification in econometrics. *Review of Economic Studies* 47: 239–254.

Also See

Complementary:	[R] **mvreg postestimation**
Related:	[MV] **manova**;
	[R] **reg3**, [R] **regress**, [R] **regress postestimation**, [R] **sureg**
Background:	[U] **11.1.10 Prefix commands**,
	[U] **20 Estimation and postestimation commands**,
	[R] **estimation options**

Title

mvreg postestimation — Postestimation tools for mvreg

Description

The following postestimation commands are available for `mvreg`:

command	description
estat	VCE and estimation sample summary
estimates	cataloging estimation results
lincom	point estimates, standard errors, testing, and inference for linear combinations of coefficients
nlcom	point estimates, standard errors, testing, and inference for nonlinear combinations of coefficients
predict	predictions, residuals, influence statistics, and other diagnostic measures
predictnl	point estimates, standard errors, testing, and inference for generalized predictions
test	Wald tests for simple and composite linear hypotheses
testnl	Wald tests of nonlinear hypotheses

See the corresponding entries in the *Stata Base Reference Manual* for details.

Syntax for predict

predict $\begin{bmatrix} type \end{bmatrix}$ *newvar* $\begin{bmatrix} if \end{bmatrix}$ $\begin{bmatrix} in \end{bmatrix}$ $\begin{bmatrix} , \underline{eq}uation(eqno \begin{bmatrix} , eqno \end{bmatrix}) \ statistic \end{bmatrix}$

statistic	description
xb	linear prediction $\mathbf{xb}_j$; the default
stdp	standard error of the linear prediction
<u>r</u>esiduals	residuals
<u>d</u>ifference	difference between the linear predictions of two equations
<u>std</u>dp	standard error of the difference in linear predictions

These statistics are available both in and out of sample; type `predict ... if e(sample) ...` if wanted only for the estimation sample.

Options for predict

equation($eqno$ [, $eqno$]) specifies the equation to which you are referring.

> equation() is filled in with one *eqno* for options xb, stdp, and residuals. equation(#1) would mean the calculation is to be made for the first equation, equation(#2) would mean the second, and so on. Alternatively, you could refer to the equations by their names. equation(income) would refer to the equation named "income" and equation(hours) to the equation named "hours".

> If you do not specify equation(), results are the same as if you specified equation(#1).

> difference and stddp refer to between-equation concepts. To use these options, you must specify two equations, e.g., equation(#1,#2) or equation(income,hours). When two equations must be specified, equation() is required. With equation(#1,#2), difference computes the prediction of equation(#1) minus the prediction of equation(#2).

xb, the default, calculates the fitted values—the prediction of $x_j b$ for the specified equation.

stdp calculates the standard error of the prediction for the specified equation (the standard error of the predicted expected value or mean for the observation's covariate pattern). This is also referred to as the standard error of the fitted value.

residuals calculates the residuals.

difference calculates the difference between the linear predictions of two equations in the system.

stddp is allowed only after you have previously fitted a multiple-equation model. The standard error of the difference in linear predictions $(x_{1j}b - x_{2j}b)$ between equations 1 and 2 is calculated.

For more information on using predict after multiple-equation estimation commands, see [R] **predict**.

Methods and Formulas

All postestimation commands listed above are implemented as ado-files.

Also See

Complementary:	[R] **mvreg**;
	[R] **estimates**, [R] **lincom**, [R] **nlcom**, [R] **predictnl**,
	[R] **test**, [R] **testnl**
Background:	[U] **13.5 Accessing coefficients and standard errors**,
	[U] **20 Estimation and postestimation commands**,
	[R] **estat**, [R] **predict**

Title

> **nbreg** — Negative binomial regression

Syntax

Negative binomial regression model

> nbreg *depvar* [*indepvars*] [*if*] [*in*] [*weight*] [, *nbreg_options*]

Generalized negative binomial model

> gnbreg *depvar* [*indepvars*] [*if*] [*in*] [*weight*] [, *gnbreg_options*]

nbreg_options	description
Model	
<u>nocon</u>stant	suppress constant term
dispersion(<u>mean</u>)	parameterization of dispersion; dispersion(mean) is the default
dispersion(<u>constant</u>)	constant dispersion for all observations
<u>exp</u>osure(*varname_e*)	include ln(*varname_e*) in model with coefficient constrained to 1
<u>off</u>set(*varname_o*)	include *varname_o* in model with coefficient constrained to 1
<u>constr</u>aints(*constraints*)	apply specified linear constraints
SE/Robust	
vce(*vcetype*)	*vcetype* may be oim, <u>r</u>obust, opg, <u>boot</u>strap, or <u>jack</u>knife
<u>r</u>obust	synonym for vce(robust)
<u>cl</u>uster(*varname*)	adjust standard errors for intragroup correlation
Reporting	
<u>l</u>evel(#)	set confidence level; default is level(95)
<u>nolr</u>test	suppress likelihood-ratio test
<u>irr</u>	report incidence-rate ratios
Max options	
maximize_options	control the maximization process; seldom used

depvar, *indepvars*, *varname_e*, and *varname_o* may contain time-series operators; see [U] **11.4.3 Time-series varlists**.

(*Continued on next page*)

gnbreg_options	description
Model	
<u>nocon</u>stant	suppress constant term
<u>lnal</u>pha(*varlist*)	dispersion model variables
<u>exp</u>osure(*varname_e*)	include ln(*varname_e*) in model with coefficient constrained to 1
<u>off</u>set(*varname_o*)	include *varname_o* in model with coefficient constrained to 1
<u>constr</u>aints(*constraints*)	apply specified linear constraints
SE/Robust	
vce(*vcetype*)	*vcetype* may be oim, <u>r</u>obust, opg, <u>boot</u>strap, or <u>jack</u>knife
robust	synonym for vce(robust)
<u>clu</u>ster(*varname*)	adjust standard errors for intragroup correlation
Reporting	
<u>l</u>evel(#)	set confidence level; default is level(95)
<u>irr</u>	report incidence-rate ratios
Max options	
maximize_options	control the maximization process; seldom used

bootstrap, by, jackknife, rolling, statsby, stepwise, svy, and xi are allowed; see
    [U] **11.1.10 Prefix commands**.
fweights, iweights, and pweights are allowed; see [U] **11.1.6 weight**.
See [U] **20 Estimation and postestimation commands** for additional capabilities of estimation commands.

Description

nbreg fits a negative binomial regression model of *depvar* on *indepvars*, where *depvar* is a non-negative count variable. In this model, the count variable is believed to be generated by a Poisson-like process, except that the variation is greater than that of a true Poisson. This extra variation is referred to as overdispersion. See [R] **poisson** before reading this entry.

gnbreg fits a generalization of the negative binomial mean-dispersion model; the shape parameter α may also be parameterized.

If you have panel data, see [XT] **xtnbreg**.

Options for nbreg

> **Model**

noconstant; see [R] **estimation options**.

dispersion(mean | constant) specifies the parameterization of the model. dispersion(mean), the default, yields a model with dispersion equal to $1 + \alpha \exp(\mathbf{x}_j \boldsymbol{\beta} + \text{offset}_j)$; that is, the dispersion is a function of the expected mean: $\exp(\mathbf{x}_j \boldsymbol{\beta} + \text{offset}_j)$. dispersion(constant) has dispersion equal to $1 + \delta$; that is, it is a constant for all observations.

exposure(*varname_e*), offset(*varname_o*), constraints(*constraints*); see [R] **estimation options**.

SE/Robust

vce(*vcetype*); see [R] **vce_option**.

robust, cluster(*varname*); see [R] **estimation options**. cluster() can be used with pweights to produce estimates for unstratified cluster-sampled data, but see [SVY] **svy: nbreg** for a command especially designed for survey data.

Reporting

level(*#*); see [R] **estimation options**.

nolrtest suppresses fitting the Poisson model. Without this option, a comparison Poisson model is fitted, and the likelihood is used in a likelihood-ratio test of the null hypothesis that the dispersion parameter is zero.

irr reports estimated coefficients transformed to incidence-rate ratios, that is, e^{β_i} rather than β_i. Standard errors and confidence intervals are similarly transformed. This option affects how results are displayed, not how they are estimated or stored. irr may be specified at estimation or when replaying previously estimated results.

Max options

maximize_options: difficult, technique(*algorithm_spec*), iterate(*#*), [no]log, trace, gradient, showstep, hessian, shownrtolerance, tolerance(*#*), ltolerance(*#*), gtolerance(*#*), nrtolerance(*#*), nonrtolerance, from(*init_specs*); see [R] **maximize**. These options are seldom used.

Options for gnbreg

Model

noconstant; see [R] **estimation options**.

lnalpha(*varlist*) allows you to specify a linear equation for $\ln \alpha$. Specifying lnalpha(male old) means that $\ln \alpha = \gamma_0 + \gamma_1 \text{male} + \gamma_2 \text{old}$, where γ_0, γ_1, and γ_2 are parameters to be estimated along with the other model coefficients. If this option is not specified, gnbreg and nbreg will produce the same results because the shape parameter will be parameterized as a constant.

exposure(*varname_e*), offset(*varname_o*), constraints(*constraints*); see [R] **estimation options**.

SE/Robust

vce(*vcetype*); see [R] **vce_option**.

robust, cluster(*varname*); see [R] **estimation options**. cluster() can be used with pweights to produce estimates for unstratified cluster-sampled data, but see the svy: gnbreg command in [SVY] **svy: nbreg** for a command especially designed for survey data.

Reporting

level(*#*); see [R] **estimation options**.

irr reports estimated coefficients transformed to incidence-rate ratios, that is, e^{β_i} rather than β_i. Standard errors and confidence intervals are similarly transformed. This option affects how results are displayed, not how they are estimated or stored. irr may be specified at estimation or when replaying previously estimated results.

maximize_options: <u>difficult</u>, <u>technique</u>(*algorithm_spec*), <u>iterate</u>(#), [<u>no</u>]<u>log</u>, <u>trace</u>, <u>gradient</u>, <u>showstep</u>, <u>hessian</u>, <u>shownrtolerance</u>, <u>tolerance</u>(#), <u>ltolerance</u>(#), <u>gtolerance</u>(#), <u>nrtolerance</u>(#), <u>nonrtolerance</u>, <u>from</u>(*init_specs*); see [R] **maximize**. These options are seldom used.

Remarks

Remarks are presented under the headings

> *Introduction to negative binomial regression*
> *nbreg*
> *gnbreg*

Introduction to negative binomial regression

Negative binomial regression models the number of occurrences (counts) of an event when the event has extra-Poisson variation, that is, when it has overdispersion. The Poisson regression model is

$$y_j \sim \text{Poisson}(\mu_j)$$

where

$$\mu_j = \exp(\mathbf{x}_j\boldsymbol{\beta} + \text{offset}_j)$$

for observed counts y_j with covariates $\mathbf{x}_j$ for the jth observation. One derivation of the negative binomial mean-dispersion model is that individual units follow a Poisson regression model, but there is an omitted variable ν_j, such that e^{ν_j} follows a gamma distribution with mean 1 and variance α:

$$y_j \sim \text{Poisson}(\mu_j^*)$$

where

$$\mu_j^* = \exp(\mathbf{x}_j\boldsymbol{\beta} + \text{offset}_j + \nu_j)$$

and

$$e^{\nu_j} \sim \text{Gamma}(1/\alpha, \alpha)$$

Note that with this parameterization, a Gamma(a, b) distribution will have expectation ab and variance ab^2.

We refer to α as the overdispersion parameter. The larger α is, the greater the overdispersion. The Poisson model corresponds to $\alpha = 0$. nbreg parameterizes α as $\ln\alpha$. gnbreg allows $\ln\alpha$ to be modeled as $\ln\alpha_j = \mathbf{z}_j\boldsymbol{\gamma}$, a linear combination of covariates $\mathbf{z}_j$.

nbreg will fit two different parameterizations of the negative binomial model. The default, described above and also given by the option dispersion(mean), has dispersion for the jth observation equal to $1 + \alpha\exp(\mathbf{x}_j\boldsymbol{\beta} + \text{offset}_j)$. This is seen by noting that the above implies that

$$\mu_j^* \sim \text{Gamma}(1/\alpha, \alpha\mu_j)$$

and thus

$$\begin{aligned}
\text{Var}(y_j) &= E\left\{\text{Var}(y_j|\mu_j^*)\right\} + \text{Var}\left\{E(y_j|\mu_j^*)\right\} \\
&= E(\mu_j^*) + \text{Var}(\mu_j^*) \\
&= \mu_j(1 + \alpha\mu_j)
\end{aligned}$$

The alternative parameterization, given by the option `dispersion(constant)`, has dispersion equal to $1 + \delta$; that is, it is constant for all observations. This is so because the constant-dispersion model assumes instead that

$$\mu_j^* \sim \text{Gamma}(\mu_j/\delta, \delta)$$

and thus $\text{Var}(y_j) = \mu_j(1 + \delta)$. The Poisson model corresponds to $\delta = 0$.

For detailed derivations of both models, see Cameron and Trivedi (1998, 70–77). In particular, note that the mean-dispersion model is known as the NB2 model in their terminology, while the constant-dispersion model is referred to as the NB1 model.

See Long and Freese (2003) for a discussion of the negative binomial regression model with Stata examples and for a discussion of other regression models for count data.

nbreg

It is not uncommon to posit a Poisson regression model and observe a lack of model fit. The following data appeared in Rodríguez (1993):

```
. use http://www.stata-press.com/data/r9/rod93
. list, sepby(cohort)
```

	cohort	age_mos	deaths	exposure
1.	1	0.5	168	278.4
2.	1	2.0	48	538.8
3.	1	4.5	63	794.4
4.	1	9.0	89	1,550.8
5.	1	18.0	102	3,006.0
6.	1	42.0	81	8,743.5
7.	1	90.0	40	14,270.0
8.	2	0.5	197	403.2
9.	2	2.0	48	786.0
10.	2	4.5	62	1,165.3
11.	2	9.0	81	2,294.8
12.	2	18.0	97	4,500.5
13.	2	42.0	103	13,201.5
14.	2	90.0	39	19,525.0
15.	3	0.5	195	495.3
16.	3	2.0	55	956.7
17.	3	4.5	58	1,381.4
18.	3	9.0	85	2,604.5
19.	3	18.0	87	4,618.5
20.	3	42.0	70	9,814.5
21.	3	90.0	10	5,802.5

(Continued on next page)

```
. generate logexp = ln(exposure)

. quietly tab cohort, gen(coh)

. poisson deaths coh2 coh3, offset(logexp)
Iteration 0:    log likelihood = -2160.0544
Iteration 1:    log likelihood = -2159.5162
Iteration 2:    log likelihood = -2159.5159
Iteration 3:    log likelihood = -2159.5159
Poisson regression                              Number of obs    =          21
                                                LR chi2(2)       =       49.16
                                                Prob > chi2      =      0.0000
Log likelihood = -2159.5159                     Pseudo R2        =      0.0113
```

deaths	Coef.	Std. Err.	z	P>|z|	[95% Conf. Interval]	
coh2	-.3020405	.0573319	-5.27	0.000	-.4144089	-.1896721
coh3	.0742143	.0589726	1.26	0.208	-.0413698	.1897983
_cons	-3.899488	.0411345	-94.80	0.000	-3.98011	-3.818866
logexp	(offset)					

```
. estat gof
        Goodness-of-fit chi2  =  4190.689
        Prob > chi2(18)       =    0.0000
```

The extreme significance of the goodness-of-fit χ^2 indicates that the Poisson regression model is inappropriate, suggesting to us that we should try a negative binomial model:

```
. nbreg deaths coh2 coh3, offset(logexp) nolog
Negative binomial regression                    Number of obs    =          21
                                                LR chi2(2)       =        0.40
Dispersion      = mean                          Prob > chi2      =      0.8171
Log likelihood = -131.3799                      Pseudo R2        =      0.0015
```

deaths	Coef.	Std. Err.	z	P>|z|	[95% Conf. Interval]	
coh2	-.2676187	.7237203	-0.37	0.712	-1.686084	1.150847
coh3	-.4573957	.7236651	-0.63	0.527	-1.875753	.9609618
_cons	-2.086731	.511856	-4.08	0.000	-3.08995	-1.083511
logexp	(offset)					
/lnalpha	.5939963	.2583615			.0876171	1.100376
alpha	1.811212	.4679475			1.09157	3.005295

Likelihood-ratio test of alpha=0: chibar2(01) = 4056.27 Prob>=chibar2 = 0.000

Our original Poisson model is a special case of the negative binomial—it corresponds to $\alpha = 0$. nbreg, however, estimates α indirectly, estimating instead $\ln \alpha$. In our model, $\ln \alpha = 0.594$, meaning that $\alpha = 1.81$ (nbreg undoes the transformation for us at the bottom of the output).

To test $\alpha = 0$ (equivalent to $\ln \alpha = -\infty$), nbreg performs a likelihood-ratio test. The staggering χ^2 value of 4,056 asserts that the probability that we would observe these data conditional on $\alpha = 0$ is virtually zero, that is, conditional on the process being Poisson. The data are not Poisson. It is not accidental that this χ^2 value is quite close to the goodness-of-fit statistic from the Poisson regression itself.

❏ Technical Note

The usual Gaussian test of $\alpha = 0$ is omitted since this test occurs on the boundary, invalidating the usual theory associated with such tests. However, the likelihood-ratio test of $\alpha = 0$ has been modified to be valid on the boundary. In particular, the null distribution of the likelihood-ratio test statistic is not the usual χ_1^2, but rather a $50:50$ mixture of a χ_0^2 (point mass at zero) and a χ_1^2, denoted as $\overline{\chi}_{01}^2$. See Gutierrez et al. (2001) for more details.

❏

❏ Technical Note

The negative binomial model deals with cases in which there is more variation than would be expected if the process were Poisson. The negative binomial model is not helpful if there is less than Poisson variation—if the variance of the count variable is less than its mean. However, underdispersion is uncommon. Poisson models arise because of independently generated events. Overdispersion comes about if some of the parameters (causes) of the Poisson processes are unknown. To obtain underdispersion, the sequence of events somehow would have to be regulated; that is, events would not be independent, but controlled based on past occurrences.

❏

gnbreg

gnbreg is a generalization of nbreg, dispersion(mean). Whereas in nbreg a single $\ln\alpha$ is estimated, gnbreg allows $\ln\alpha$ to vary, observation by observation, as a linear combination of another set of covariates: $\ln\alpha_j = \mathbf{z}_j\boldsymbol{\gamma}$.

We will assume that the number of deaths is a function of age, whereas the $\ln\alpha$ parameter is a function of cohort. To fit the model, we type

```
. gnbreg deaths age_mos, lnalpha(coh2 coh3) offset(logexp)

Fitting constant-only model:

Iteration 0:   log likelihood =   -187.067  (not concave)
Iteration 1:   log likelihood = -137.43798
Iteration 2:   log likelihood = -132.47158
Iteration 3:   log likelihood = -131.57982
Iteration 4:   log likelihood = -131.57948
Iteration 5:   log likelihood = -131.57948

Fitting full model:

Iteration 0:   log likelihood = -124.34327
Iteration 1:   log likelihood = -117.68002
Iteration 2:   log likelihood = -117.56307
Iteration 3:   log likelihood = -117.56164
Iteration 4:   log likelihood = -117.56164
```

(Continued on next page)

```
Generalized negative binomial regression        Number of obs   =         21
                                                 LR chi2(1)      =      28.04
                                                 Prob > chi2     =     0.0000
Log likelihood = -117.56164                      Pseudo R2       =     0.1065
```

	Coef.	Std. Err.	z	P>|z|	[95% Conf. Interval]	
deaths						
age_mos	-.0516657	.0051747	-9.98	0.000	-.061808	-.0415233
_cons	-1.867225	.2227944	-8.38	0.000	-2.303894	-1.430556
logexp	(offset)					
lnalpha						
coh2	.0939546	.7187747	0.13	0.896	-1.314818	1.502727
coh3	.0815279	.7365476	0.11	0.912	-1.362079	1.525135
_cons	-.4759581	.5156502	-0.92	0.356	-1.486614	.5346978

We find that age is a significant determinant of the number of deaths. The standard errors for the variables in the $\ln \alpha$ equation suggest that the overdispersion parameter does not vary across cohorts. We can test this by typing

```
. test coh2 coh3
 ( 1)  [lnalpha]coh2 = 0
 ( 2)  [lnalpha]coh3 = 0

        chi2(  2) =     0.02
      Prob > chi2 =   0.9904
```

There is no evidence of variation by cohort in these data.

❏ Technical Note

Note the intentional absence of a likelihood-ratio test for $\alpha = 0$ in gnbreg. The test is affected by the same boundary condition that affects the comparison test in nbreg; however, when α is parameterized by more than a constant term, the null distribution becomes intractable. For this reason, we recommend using nbreg to test for overdispersion and, if you have reason to believe that overdispersion exists, only then modeling the overdispersion using gnbreg.

❏

(Continued on next page)

Saved Results

nbreg and gnbreg save in e():

Scalars

e(N)	number of observations	e(N_clust)	number of clusters
e(k)	number of parameters	e(chi2)	χ^2
e(k_eq)	number of equations	e(chi2_c)	χ^2 for comparison test
e(k_dv)	number of dependent variables	e(p)	significance
e(df_m)	model degrees of freedom	e(rank)	rank of e(V)
e(r2_p)	pseudo-R-squared	e(rank0)	rank of e(V) for constant-only
e(ll)	log likelihood		model
e(ll_0)	log likelihood, constant-only model	e(ic)	number of iterations
e(ll_c)	log likelihood, comparison model	e(rc)	return code
e(alpha)	the value of alpha	e(converged)	1 if converged, 0 otherwise

Macros

e(cmd)	nbreg or gnbreg	e(dispers)	mean or constant
e(depvar)	name of dependent variable	e(vce)	*vcetype* specified in vce()
e(wtype)	weight type	e(vcetype)	title used to label Std. Err.
e(wexp)	weight expression	e(opt)	type of optimization
e(title)	title in estimation output	e(ml_method)	type of ml method
e(clustvar)	name of cluster variable	e(user)	name of likelihood-evaluator program
e(offset#)	offset for equation #	e(technique)	maximization technique
e(chi2type)	Wald or LR; type of model χ^2 test	e(crittype)	optimization criterion
e(chi2_ct)	Wald or LR; type of model χ^2 test	e(properties)	b V
	corresponding to e(chi2_c)	e(predict)	program used to implement predict

Matrices

e(b)	coefficient vector	e(V)	variance–covariance matrix of
e(ilog)	iteration log (up to 20 iterations)		the estimators
e(gradient)	gradient vector		

Functions

e(sample)	marks estimation sample

Methods and Formulas

nbreg and gnbreg are implemented as ado-files.

See [R] **poisson** and Feller (1968, 156–164) for an introduction to the Poisson distribution.

Mean-dispersion model

A negative binomial distribution can be regarded as a gamma mixture of Poisson random variables. The number of times something occurs, y_j, is distributed as Poisson$(\nu_j \mu_j)$. That is, its conditional likelihood is

$$f(y_j \mid \nu_j) = \frac{(\nu_j \mu_j)^{y_j} e^{-\nu_j \mu_j}}{\Gamma(y_j + 1)}$$

where $\mu_j = \exp(\mathbf{x}_j \boldsymbol{\beta} + \text{offset}_j)$ and ν_j is an unobserved parameter with a Gamma$(1/\alpha, \alpha)$ density:

$$g(\nu) = \frac{\nu^{(1-\alpha)/\alpha} e^{-\nu/\alpha}}{\alpha^{1/\alpha} \Gamma(1/\alpha)}$$

This gamma distribution has mean 1 and variance α, where α is our ancillary parameter.

The unconditional likelihood for the jth observation is therefore

$$f(y_j) = \int_0^\infty f(y_j \mid \nu)g(\nu)\,d\nu = \frac{\Gamma(m + y_j)}{\Gamma(y_j + 1)\Gamma(m)}\, p_j^m (1 - p_j)^{y_j}$$

where $p_j = 1/(1 + \alpha\mu_j)$ and $m = 1/\alpha$. Solutions for α are handled by searching for $\ln\alpha$ since α is required to be greater than zero.

The log likelihood (with weights w_j and offsets) is given by

$$m = 1/\alpha \qquad p_j = 1/(1 + \alpha\mu_j) \qquad \mu_j = \exp(\mathbf{x}_j\boldsymbol{\beta} + \text{offset}_j)$$

$$\ln L = \sum_{j=1}^n w_j \left[\ln\{\Gamma(m + y_j)\} - \ln\{\Gamma(y_j + 1)\} \right.$$
$$\left. - \ln\{\Gamma(m)\} + m\ln(p_j) + y_j\ln(1 - p_j) \right]$$

In the case of gnbreg, α can vary across the observations according to the parameterization $\ln\alpha_j = \mathbf{z}_j\boldsymbol{\gamma}$.

Constant-dispersion model

The constant-dispersion model assumes that y_j is conditionally distributed as $\text{Poisson}(\mu_j^*)$, where $\mu_j^* \sim \text{Gamma}(\mu_j/\delta, \delta)$ for some dispersion parameter δ (by contrast, the mean-dispersion model assumes that $\mu_j^* \sim \text{Gamma}(1/\alpha, \alpha\mu_j)$). The log likelihood is given by

$$m_j = \mu_j/\delta \qquad p = 1/(1 + \delta)$$

$$\ln L = \sum_{j=1}^n w_j \left[\ln\{\Gamma(m_j + y_j)\} - \ln\{\Gamma(y_j + 1)\} \right.$$
$$\left. - \ln\{\Gamma(m_j)\} + m_j\ln(p) + y_j\ln(1 - p) \right]$$

with everything else defined as before in the calculations for the mean-dispersion model.

Maximization for gnbreg is done via the lf linear-form method, and for nbreg it is done via the d2 method described in [R] **ml**.

References

Cameron, A. C. and P. K. Trivedi. 1998. *Regression Analysis of Count Data*. Cambridge: Cambridge University Press.

Feller, W. 1968. *An Introduction to Probability Theory and Its Applications*, vol. 1. 3rd ed. New York: Wiley.

Gutierrez, R. G., S. L. Carter, and D. M. Drukker. 2001. On boundary-value likelihood-ratio tests. *Stata Technical Bulletin* 60: 15–18. Reprinted in *Stata Technical Bulletin Reprints*, vol. 10, pp. 269–273.

Hilbe, J. 1998. sg91: Robust variance estimators for MLE Poisson and negative binomial regression. *Stata Technical Bulletin* 45: 26–28. Reprinted in *Stata Technical Bulletin Reprints*, vol. 8, pp. 177–180.

——. 1999. sg102: Zero-truncated Poisson and negative binomial regression. *Stata Technical Bulletin* 47: 37–40. Reprinted in *Stata Technical Bulletin Reprints*, vol. 8, pp. 233–236.

Long, J. S. 1997. *Regression Models for Categorical and Limited Dependent Variables*. Thousand Oaks, CA: Sage.

Long, J. S. and J. Freese. 2001. Predicted probabilities for count models. *Stata Journal* 1: 51–57.

——. 2003. *Regression Models for Categorical Dependent Variables Using Stata*. rev. ed. College Station, TX: Stata Press.

Rodríguez, G. 1993. sbe10: An improvement to poisson. *Stata Technical Bulletin* 11: 11–14. Reprinted in *Stata Technical Bulletin Reprints*, vol. 2, pp. 94–98.

Rogers, W. H. 1991. sbe1: Poisson regression with rates. *Stata Technical Bulletin* 1: 11–12. Reprinted in *Stata Technical Bulletin Reprints*, vol. 1, pp. 62–64.

——. 1993. sg16.4: Comparison of nbreg and glm for negative binomial. *Stata Technical Bulletin* 16: 7. Reprinted in *Stata Technical Bulletin Reprints*, vol. 3, pp. 82–84.

Also See

Complementary:	[R] **nbreg postestimation**; [R] **constraint**
Related:	[R] **glm**, [R] **poisson**, [R] **zip**,
	[SVY] **svy: nbreg**, [XT] **xtnbreg**
Background:	[U] **11.1.10 Prefix commands**,
	[U] **20 Estimation and postestimation commands**,
	[R] **estimation options**, [R] **maximize**, [R] *vce_option*

Title

Description

The following postestimation commands are available for `nbreg` and `gnbreg`:

command	description
adjust[1]	adjusted predictions of $\mathbf{x}\beta$ or $\exp(\mathbf{x}\beta)$
estat	AIC, BIC, VCE, and estimation sample summary
estimates	cataloging estimation results
lincom	point estimates, standard errors, testing, and inference for linear combinations of coefficients
linktest	link test for model specification
lrtest	likelihood-ratio test
mfx	marginal effects or elasticities
nlcom	point estimates, standard errors, testing, and inference for nonlinear combinations of coefficients
predict	predictions, residuals, influence statistics, and other diagnostic measures
predictnl	point estimates, standard errors, testing, and inference for generalized predictions
suest	seemingly unrelated estimation
test	Wald tests for simple and composite linear hypotheses
testnl	Wald tests of nonlinear hypotheses

[1] `adjust` does not work with time-series operators.

See the corresponding entries in the *Stata Base Reference Manual* for details.

Syntax for predict

> predict [*type*] *newvar* [*if*] [*in*] [, *statistic* <u>nooff</u>set]

> predict [*type*] { *stub** | *newvar*_{reg} *newvar*_{disp} } [*if*] [*in*], <u>sc</u>ores

statistic	description
n	predicted number of events; the default
ir	incidence rate (equivalent to predict ..., n nooffset)
xb	linear prediction
stdp	standard error of the linear prediction

In addition, relevant only after `gnbreg` are

statistic	description
alpha	predicted values of α_j
lnalpha	predicted values of $\ln \alpha_j$
stdplna	standard error of predicted $\ln \alpha_j$

These statistics are available both in and out of sample; type predict ... if e(sample) ... if wanted only for the estimation sample.

Options for predict

n, the default, calculates the predicted number of events, which is $\exp(\mathbf{x}_j \boldsymbol{\beta})$ if neither offset(*varname_o*) nor exposure(*varname_e*) was specified when the model was fitted; $\exp(\mathbf{x}_j \boldsymbol{\beta} + \text{offset}_j)$ if offset() was specified; or $\exp(\mathbf{x}_j \boldsymbol{\beta}) \times \text{exposure}_j$ if exposure() was specified.

ir calculates the incidence rate $\exp(\mathbf{x}_j \boldsymbol{\beta})$, which is the predicted number of events when exposure is 1. This is equivalent to specifying both n and nooffset options.

xb calculates the linear prediction, which is $\mathbf{x}_j \boldsymbol{\beta}$ if neither offset() nor exposure() was specified; $\mathbf{x}_j \boldsymbol{\beta} + \text{offset}_j$ if offset() was specified; or $\mathbf{x}_j \boldsymbol{\beta} + \ln(\text{exposure}_j)$ if exposure() was specified; see nooffset below.

stdp calculates the standard error of the linear prediction.

alpha, lnalpha, and stdplna are relevant after gnbreg estimation only; they produce the predicted values of α_j, $\ln \alpha_j$, and the standard error of the predicted $\ln \alpha_j$, respectively.

nooffset is relevant only if you specified offset() or exposure() when you fitted the model. It modifies the calculations made by predict so that they ignore the offset or exposure variable; the linear prediction is treated as $\mathbf{x}_j \boldsymbol{\beta}$ rather than as $\mathbf{x}_j \boldsymbol{\beta} + \text{offset}_j$ or $\mathbf{x}_j \boldsymbol{\beta} + \ln(\text{exposure}_j)$. Specifying predict ..., nooffset is equivalent to specifying predict ..., ir.

scores calculates equation-level score variables.

The first new variable will contain $\partial \ln L / \partial(\mathbf{x}_j \boldsymbol{\beta})$.

The second new variable will contain $\partial \ln L / \partial(\ln \alpha_j)$ for dispersion(mean) and gnbreg.

The second new variable will contain $\partial \ln L / \partial(\ln \delta)$ for dispersion(constant).

(*Continued on next page*)

Remarks

After `nbreg` and `gnbreg`, `predict` returns the predicted number of events:

```
. nbreg deaths coh2 coh3, nolog
Negative binomial regression                    Number of obs   =         21
                                                LR chi2(2)      =       0.14
Dispersion     = mean                           Prob > chi2     =     0.9307
Log likelihood = -108.48841                     Pseudo R2       =     0.0007
```

deaths	Coef.	Std. Err.	z	P>|z|	[95% Conf. Interval]	
coh2	.0591305	.2978419	0.20	0.843	-.5246289	.64289
coh3	-.0538792	.2981621	-0.18	0.857	-.6382662	.5305077
_cons	4.435906	.2107213	21.05	0.000	4.0229	4.848912
/lnalpha	-1.207379	.3108622			-1.816657	-.5980999
alpha	.29898	.0929416			.1625683	.5498555

```
Likelihood-ratio test of alpha=0:  chibar2(01) =  434.62 Prob>=chibar2 = 0.000
. predict count
(option n assumed; predicted number of events)
. summarize deaths count
```

Variable	Obs	Mean	Std. Dev.	Min	Max
deaths	21	84.66667	48.84192	10	197
count	21	84.66667	4.00773	80	89.57143

Methods and Formulas

All postestimation commands listed above are implemented as ado-files.

In the following, we use the same notation as in [R] **nbreg**.

Mean-dispersion model

The equation-level scores are given by

$$\text{score}(\mathbf{x}\boldsymbol{\beta})_j = p_j(y_j - \mu_j)$$
$$\text{score}(\tau)_j = -m\left\{\frac{\alpha_j(\mu_j - y_j)}{1 + \alpha_j\mu_j} - \ln(1 + \alpha_j\mu_j) + \psi(y_j + m) - \psi(m)\right\}$$

where $\tau_j = \ln \alpha_j$, and $\psi(z)$ is the digamma function.

Constant-dispersion model

The equation-level scores are given by

$$\text{score}(\mathbf{x}\boldsymbol{\beta})_j = m_j\left\{\psi(y_j + m_j) - \psi(m_j) + \ln(p)\right\}$$
$$\text{score}(\tau)_j = y_j - (y_j + m_j)(1 - p) - \text{score}(\mathbf{x}\boldsymbol{\beta})_j$$

where $\tau_j = \ln \delta_j$.

Also See

Complementary: [R] **nbreg**; [R] **adjust**, [R] **estimates**, [R] **lincom**, [R] **linktest**, [R] **lrtest**,
[R] **mfx**, [R] **nlcom**, [R] **predictnl**, [R] **suest**, [R] **test**, [R] **testnl**

Background: [U] **13.5 Accessing coefficients and standard errors**,
[U] **20 Estimation and postestimation commands**,
[R] **estat**, [R] **predict**

Title

net — Install and manage user-written additions from the net

Syntax

Set current location for net

    net from *directory_or_url*

Change into a different net directory

    net cd *path_or_url*

Change to a different net site

    net link *linkname*

Search for installed packages

    net search   (see [R] **net search**)

Report current net location   ⁃

    net

Describe a package

    net <u>d</u>escribe *pkgname* $\left[\, , \, \underline{\text{fr}}\text{om}(\textit{directory_or_url}) \right]$

Set location where packages will be installed

    net set ado *dirname*

Set location where ancillary files will be installed

    net set other *dirname*

Report net 'from', 'ado', and 'other' settings

    net <u>q</u>uery

Install ado- and help files from a package

    net <u>ins</u>tall *pkgname* $\left[\, , \, \text{all replace } \underline{\text{fr}}\text{om}(\textit{directory_or_url}) \right]$

Install ancillary files from a package

    net get *pkgname* $\left[\, , \, \text{all replace } \underline{\text{fr}}\text{om}(\textit{directory_or_url}) \right]$

Shortcut to access Stata Journal net site

   net sj *vol-issue* [*insert*]

Shortcut to access STB net site

   net stb *issue* [*insert*]

List installed packages

   ado [, <u>f</u>ind(*string*) <u>fro</u>m(*dirname*)]

   ado dir [*pkgid*] [, <u>f</u>ind(*string*) <u>fro</u>m(*dirname*)]

Describe installed packages

   ado <u>d</u>escribe [*pkgid*] [, <u>f</u>ind(*string*) <u>fro</u>m(*dirname*)]

Uninstall an installed package

   ado uninstall *pkgid* [, <u>fro</u>m(*dirname*)]

where

pkgname is	name of a package	
pkgid is	name of a package	
	or	a number in square brackets: [#]
dirname is	a directory name	
	or	PLUS (default)
	or	PERSONAL
	or	SITE

Description

   net fetches and installs additions to Stata. The additions can be obtained from the Internet or from media. The additions can be ado-files (new commands), help files, or even datasets. Collections of files are bound together into *packages*. For instance, the package named zz49 might add the xyz command to Stata. At a minimum, such a package would contain xyz.ado, the code to implement the new command, and xyz.hlp, the online help to describe it. That the package contains two files is a detail: You use net to fetch the package zz49, regardless of the number of files.

   ado manages the packages you have installed using net. The ado command allows you to list packages you have previously installed and to uninstall them.

   You can also access the net and ado features by selecting **Help > SJ and User-written Programs**; this is the recommended method to find and install additions to Stata.

Options

   all is used with net install and net get. Typing it with either one makes the command equivalent to typing net install followed by net get.

replace is for use with net install and net get. It specifies that the fetched files replace existing files if any of the files already exist.

find(*string*) is for use with ado, ado dir, and ado describe. It specifies that the descriptions of the packages installed on your computer be searched, and that the package descriptions containing *string* be listed.

from(*dirname*) when used with ado specifies where the packages are installed. The default is from(PLUS). PLUS is a codeword that Stata understands to correspond to a particular directory on your computer that was set at installation time. On Windows computers, PLUS probably means the directory c:\ado\plus, but it might mean something else. You can find out what it means by typing sysdir, but this is irrelevant if you use the defaults.

from(*directory_or_url*) when used with net specifies the directory or URL where installable packages may be found. The directory or URL is the same as the one that would have been specified with net from.

Remarks

For an introduction to using net and ado, see [U] **28 Using the Internet to keep up to date**. The purpose of this documentation is

1. to briefly, but accurately, describe net and ado and all their features and

2. to provide documentation to those who wish to set up their own sites to distribute additions to Stata.

Remarks are presented under the headings

> *Definition of a package*
> *The purpose of the net and ado commands*
> *Content pages*
> *Package-description pages*
> *Where packages are installed*
> *A summary of the net command*
> *A summary of the ado command*
> *Relationship of net and ado to the point-and-click interface*
> *Creating your own site*
> *Format of content and package-description files*
> *Example 1*
> *Example 2*
> *Additional package directives*
> *SMCL in content and package-description files*
> *Error-free file delivery*

Definition of a package

A *package* is a collection of files—typically .ado and .hlp files—that together provide a new feature in Stata. Packages contain additions that you wish had been part of Stata at the outset. We write such additions, and so do other users.

One source of these additions is the *Stata Journal*, a printed and electronic journal with corresponding software. If you want the journal, you must subscribe, but the software is available for free from our web site.

The purpose of the net and ado commands

The net command makes it easy to distribute and install packages. The goal is to get you quickly to a package description page that summarizes the addition, for example,

```
. net describe rte_stat, from(http://www.wemakeitupaswego.edu/faculty/sgazer/)
```

---

package **rte_stat** from http://www.wemakeitupaswego.edu/faculty/sgazer/

TITLE
      rte_stat. The robust-to-everything statistic; update.

DESCRIPTION/AUTHOR(S)
      S. Gazer, Dept. of Applied Theoretical Mathematics, WMIUAWG Univ.
      Aleph-0 100% confidence intervals proved too conservative for some
      applications; Aleph-1 confidence intervals have been substituted.
      The new robust-to-everything supplants the previous robust-to-
      everything-conceivable statistic. See "Inference in the absence
      of data" (forthcoming). After installation, see help **rte**.

INSTALLATION FILES   (type **net install rte_stat**)
      rte.ado
      rte.hlp
      nullset.ado
      random.ado

If you decide that the addition might prove useful, net makes the installation easy:

```
. net install rte_stat
checking rte_stat consistency and verifying not already installed...
installing into c:\ado\plus\ ...
installation complete.
```

The ado command helps you manage packages installed with net. Perhaps you remember that you installed a package that calculates the robust-to-everything statistic, but you cannot remember the command's name. You could use ado to search what you have previously installed for the rte command,

```
. ado
[1] package sg145 from http://www.stata.com/stb/stb56
      STB-56 sg145. Scalar measures of fit for regression models.
  (output omitted)
[15] package rte_stat from http://www.wemakeitupaswego.edu/faculty/sgazer
      rte_stat.  The robust-to-everything statistic; update.
  (output omitted)
[21] package st0001 from http://www.stata-journal.com/software/sj1-1
      SJ1-1 st0001.  Flexible parametric alt. to the Cox model, and more
```

or, you might type

```
. ado, find("robust-to-everything")
[15] package rte_stat from http://www.wemakeitupaswego.edu/faculty/sgazer
      rte_stat.  The robust-to-everything statistic; update.
```

Perhaps you decide that rte, despite the author's claims, is not worth the disk space it occupies. You can use ado to erase it:

```
. ado uninstall rte_stat
package rte_stat from http://www.wemakeitupaswego.edu/faculty/sgazer
      rte_stat.  The robust-to-everything statistic; update.
(package uninstalled)
```

ado uninstall is easier than erasing the files by hand because ado uninstall erases every file associated with the package, and, moreover, ado knows where on your computer rte_stat is installed; you would have to hunt for these files.

Content pages

There are two types of pages displayed by net: content pages and package-description pages. When you type net from, net cd, net link, or net without arguments, Stata goes to the specified place and displays the content page:

```
. net from http://www.stata.com
```

```
http://www.stata.com/
StataCorp
```

```
Welcome to StataCorp.

Below we provide links to sites providing additions to Stata, including
the Stata Journal, STB, and Statalist.  These are NOT THE OFFICIAL UPDATES;
you fetch and install the official updates by typing -update-.

PLACES you could -net link- to:
    sj                  The Stata Journal

DIRECTORIES you could -net cd- to:
    stb                 materials published in the Stata Technical Bulletin
    users               materials by various people including StataCorp employees
    meetings            Stata user group meetings
    links               other locations providing additions to Stata
```

A content page tells you about other content pages and package-description pages. The example above lists other content pages only. Below we follow one of the links for the *Stata Journal*:

```
. net link sj
```

```
http://www.stata-journal.com/
The Stata Journal
```

```
The Stata Journal is a refereed, quarterly journal containing articles
of interest to Stata users.  For more details and subscription information,
visit the Stata Journal web site at http://www.stata-journal.com/.

PLACES you could -net link- to:
    stata               StataCorp web site

DIRECTORIES you could -net cd- to:
    production          Files for authors of the Stata Journal
    software            Software associated with Stata Journal articles
```

```
. net cd software
```

```
http://www.stata-journal.com/software/
The Stata Journal
```

```
PLACES you could -net link- to:
    stata               StataCorp web site
    stb                 Stata Technical Bulletin (STB) software archive

DIRECTORIES you could -net cd- to:
    sj1-1               volume 1, issue 1
  (output omitted)
```

```
. net cd sj1-1
```

```
http://www.stata-journal.com/software/sj1-1/
```
Stata Journal volume 1, issue 1

```
DIRECTORIES you could -net cd- to:
    ..              Other Stata Journals
PACKAGES you could -net describe-:
    dm0001          Sort a list of items
    gr0001          Generalized Lorenz curves and related graphs: an update
    st0001          Flexible parametric alternatives to the Cox model, and
                    more
    st0002          Predicted probabilities for count models
    st0003          Haplotype analysis in population-based association
                    studies
    st0004          Residual diagnostics for cross-section time series
                    regression models
```

---

dm0001, gr0001, ..., st0004 are links to package-description pages.

The links for the *Stata Technical Bulletin* (STB) follow steps similar to those in the *Stata Journal* example above.

```
. net cd stb
```

```
http://www.stata.com/stb/
```
The Stata Technical Bulletin

```
PLACES you could -net link- to:
    stata           StataCorp web site
    portugal        STB mirror site
DIRECTORIES you could -net cd- to:
  (output omitted )
    stb54           STB-54, March    2000
  (output omitted )
```

```
. net cd stb54
```

```
http://www.stata.com/stb/stb54/
```
STB-54 March 2000

```
DIRECTORIES you could -net cd- to:
    ..              Other STBs
PACKAGES you could -net describe-:
  (output omitted )
```

---

1. When you type net from, you follow that with a location to display the location's content page.

   a. The location could be a URL, such as *http://www.stata.com*. The content page at that location would then be listed.

   b. The location could be a: on a Windows computer or a mounted volume on a Macintosh computer. The content page on that source would be listed. That would work if you had special media obtained from StataCorp or special media prepared by another user.

   c. The location could even be a directory on your computer, but that would work only if that directory contained the right kind of files.

2. Once you have specified a location, typing `net cd` will take you into subdirectories of that location, if there are any. Typing

   ```
   . net from http://www.stata-journal.com
   . net cd software
   ```

   is equivalent to typing

   ```
   . net from http://www.stata-journal.com/software
   ```

   Typing `net cd` displays the content page from that location.

3. Typing `net` without arguments redisplays the current content page, which is the content page last displayed.

4. `net link` is similar to `net cd` in that the result is to change the location, but rather than changing to subdirectories of the current location, `net link` jumps to another location:

   ```
   . net from http://www.stata-journal.com
   ```

   ```
   http://www.stata-journal.com/
   The Stata Journal
   ```

   ```
   The Stata Journal is a refereed, quarterly journal containing articles
   of interest to Stata users.  For more details and subscription information,
   visit the Stata Journal web site at http://www.stata-journal.com/.

   PLACES you could -net link- to:
       stata           StataCorp web site
   DIRECTORIES you could -net cd- to:
       production      Files for authors of the Stata Journal
       software        Software associated with Stata Journal articles
   ```

   Typing `net link stata` would jump to *http://www.stata.com*:

   ```
   . net link stata
   ```

   ```
   http://www.stata.com/
   StataCorp
   ```

   ```
   Welcome to StataCorp.
       (output omitted )
   ```

Package-description pages

Package-description pages describe what could be installed:

```
. net from http://www.stata-journal.com/software/sj1-1
```

```
http://www.stata-journal.com/software/sj1-1/
    (output omitted )
```

```
. net describe st0001
```

---

package **st0001** from http://www.stata-journal.com/software/sj1-1

TITLE
    SJ1-1 st0001. Flexible parametric alt. to the Cox model, and more
DESCRIPTION/AUTHOR(S)
    by Patrick Royston, UK Medical Research Council
    Support: patrick.royston@ctu.mrc.ac.uk
    After installation, see help **stpm**
INSTALLATION FILES   (type **net install st0001**)
    st0001/bhcalc.ado
    st0001/bhcalc.hlp
    st0001/frac_s3b.ado
    st0001/frac_spl.ado
    st0001/mlsurvlf.ado
    st0001/stpm.ado
    st0001/stpm.hlp
    st0001/stpm_p.ado
ANCILLARY FILES   (type **net get st0001**)
    st0001/bc.dta

A package-description page describes the package and tells you how to install the component files. Package-description pages potentially describe two types of files:

1. Installation files: files that you type `net install` to install and that are required to make the addition work.

2. Ancillary files: additional files you might want to install—you type `net get` to install them—but that you can ignore. Ancillary files are typically datasets that are useful for demonstration purposes. Ancillary files are not really installed in the sense of being copied to an official place for use by Stata itself. They are merely copied into the current directory so that you may use them if you wish.

You install the official files by typing `net install` followed by the package name. For example, to install st0001, you would type

```
. net install st0001
checking st0001 consistency and verifying not already installed...
installing into c:\ado\plus\ ...
installation complete.
```

You get the ancillary files—if there are any and if you want them—by typing `net get` followed by the package name:

```
. net get st0001
checking st0001 consistency and verifying not already installed...
copying into current directory...
        copying bc.dta
ancillary files successfully copied.
```

Most users ignore the ancillary files.

Once you have installed a package—typed `net install`—use `ado` to redisplay the package-description page whenever you wish:

```
. ado describe st0001
```

---

[1] package **st0001** from http://www.stata-journal.com/software/sj1-1

TITLE
       SJ1-1 st0001. Flexible parametric alt. to the Cox model, and more

DESCRIPTION/AUTHOR(S)
       by Patrick Royston, UK Medical Research Council
       Support: patrick.royston@ctu.mrc.ac.uk
       After installation, see help **stpm**

INSTALLATION FILES
       b/bhcalc.ado
       b/bhcalc.hlp
       f/frac_s3b.ado
       f/frac_spl.ado
       m/mlsurvlf.ado
       s/stpm.ado
       s/stpm.hlp
       s/stpm_p.ado

INSTALLED ON
       11 Jun 2002

Note that the package-description page shown by `ado` includes the location from which we got the package and when we installed it. Also note that it does not mention the ancillary files that were originally part of this package because they are not tracked by `ado`.

Where packages are installed

Packages should be installed in PLUS or SITE, which are codewords that Stata understands and that correspond to some real directories on your computer. Typing `sysdir` will tell you where these are, if you care.

```
. sysdir
   STATA:  C:\STATA\
 UPDATES:  C:\STATA\ado\updates\
    BASE:  C:\STATA\ado\base\
    SITE:  C:\STATA\ado\site\
    PLUS:  c:\ado\plus\
PERSONAL:  c:\ado\personal\
OLDPLACE:  c:\ado\
```

If you type `sysdir`, you may obtain different results.

By default, `net` installs in the PLUS directory, and `ado` tells you about what is installed there. If you are on a multiple-user system, you may wish to install some packages in the SITE directory. This way, they will be available to other Stata users. To do that, before using `net install`, type

```
. net set ado SITE
```

and when reviewing what is installed or removing packages, redirect `ado` to that directory:

```
. ado ..., from(SITE)
```

In both cases, you literally type SITE because Stata will understand that SITE means the site ado-directory as defined by sysdir. To install into SITE, you must have write access to that directory.

If you reset where net installs and then, in the same session, wish to install into your private ado-directory, type

```
. net set ado PLUS
```

That is how things were originally. If you are confused as to where you are, type net query.

A summary of the net command

The net command displays content pages and package-description pages. Such pages are provided over the Internet, and most users get them there. We recommend that you start at *http://www.stata.com* and work out from there. We also recommend using net search to find packages of interest to you; see [R] **net search**.

net from moves you to a location and displays the content page.

net cd and net link change from there to other locations. net cd enters subdirectories of the original location. net link jumps from one location to another, depending on the code on the content page.

net describe lists a package-description page. Packages are named, and you type net describe *pkgname*.

net install installs a package into your copy of Stata. net get copies any additional files (ancillary files) to your current directory.

net sj and net stb make loading files from the *Stata Journal* and its predecessor, the *Stata Technical Bulletin*, easier.

```
net sj vol-issue
```

is a synonym for typing

```
net from http://www.stata-journal.com/software/sjvol-issue
```

while

```
net sj vol-issue insert
```

is a synonym for typing

```
net from http://www.stata-journal.com/software/sjvol-issue
net describe insert
```

net set controls where files are installed by net. By default, net installs in the PLUS directory; see [P] **sysdir**. net set ado SITE would cause subsequent net commands to install in the SITE directory. net set other sets where ancillary files, such as .dta files, are installed. The default is the current directory.

net query displays the current net from, net set ado, and net set other settings.

A summary of the ado command

The `ado` command lists the package descriptions of previously installed packages.

Typing `ado` without arguments is the same as typing `ado dir`. Both list the names and titles of the packages you have installed.

`ado describe` lists full package-description pages.

`ado uninstall` removes packages from your computer.

Since you can install packages from a variety of sources, the package names may not always be unique. Thus the packages installed on your computer are numbered sequentially, and you may refer to them by name or by number. For instance, say that you wanted to get rid of the robust-to-everything statistic command you installed. Type

```
. ado, find("robust-to-everything")
[15] package rte_stat from http://www.wemakeitupaswego.edu/faculty/sgazer
      rte_stat.  The robust-to-everything statistic; update.
```

You could then type

```
. ado uninstall rte_stat
```

or

```
. ado uninstall [15]
```

Typing `ado uninstall rte_stat` would work only if the name `rte_stat` were unique; otherwise, `ado` would refuse, and you would have to type the number.

The `find()` option is allowed with `ado dir` and `ado describe`. It searches the package description for the word or phrase you specify, ignoring case (`alpha` matches `Alpha`). The complete package description is searched, including the author's name and the name of the files. Thus if `rte` was the name of a command that you wanted to eliminate, but you could not remember the name of the package, you could type

```
. ado, find(rte)
[15] package rte_stat from http://www.wemakeitupaswego.edu/faculty/sgazer
      rte_stat.  The robust-to-everything statistic; update.
```

Relationship of net and ado to the point-and-click interface

Users may instead select **Help > SJ and User-written Programs**. There are advantages and disadvantages:

1. Flipping through content and package-description pages is easier; it is much like a browser. See chapter 19 in the *Getting Started* manual.

2. When browsing a product-description page, note that the `.hlp` files are highlighted. You may click on `.hlp` files to review them before installing the package.

3. You may not redirect from where `ado` searches for files.

Creating your own site

The rest of this entry concerns how to create your own site to distribute additions to Stata. The idea is that you have written additions for use with Stata—say, xyz.ado and xyz.hlp—and you wish to put them out so that coworkers or researchers at other institutions can easily install them. Or, perhaps you just have a dataset that you and others want to share.

In any case, all you need is a homepage. You place the files that you want to distribute on your homepage (or in a subdirectory), and you add two more files—a content file and a package description file—and you are done.

Format of content and package-description files

The content file describes the content page. It must be named stata.toc:

```
──────────────────────────────────────────────────────────── top of stata.toc ──────────
OFF                                    (to make site unavailable temporarily)
* lines starting with * are comments; they are ignored

* blank lines are ignored, too

* v indicates version—specify v 3; old-style toc files do not have this
v 3

* d lines display description text
* the first d line is the title, and the remaining ones are text
* blank d lines display a blank line
d title
d text
d text
d
. . .

* l lines display links
l word-to-show path-or-url [description]
l word-to-show path-or-url [description]
. . .

* t lines display other directories within the site
t path [description]
t path [description]
. . .

* p lines display packages
p pkgname [description]
p pkgname [description]
. . .
──────────────────────────────────────────────────────────── end of stata.toc ──────────
```

Package files describe packages and are named *pkgname*.pkg:

─────────────────────────────────────── top of *pkgname*.pkg ───────────

    * lines starting with * are comments; they are ignored

    * blank lines are ignored, too

    * v indicates version—specify v 3; old-style pkg files do not have this
    v 3

    * d lines display package description text
    * the first d line is the title, and the remaining ones are text
    * blank d lines display a blank line
    d *title*
    d *text*
    d *text*
    d
    . . .

    * f identifies the component files
    f [*path/*]*filename* [*description*]
    f [*path/*]*filename* [*description*]
    . . .

    * e line is optional; it means stop reading
    e

───────────────────────────────────── end of *pkgname*.pkg ───────────

Example 1

Say that we want the user to see the following:

. net from http://www.university.edu/~me

───

http://www.university.edu/~me
Chris Farrar, Uni University

───

PACKAGES you could -**net describe**-:
    xyz   interval-truncated survival

. net describe xyz

───

package **xyz** from http://www.university.edu/~me

───

TITLE
      xyz. interval-truncated survival.
DESCRIPTION/AUTHOR(S)
      C. Farrar, Uni University.
INSTALLATION FILES   (type **net install xyz**)
      xyz.ado
      xyz.hlp
ANCILLARY FILES   (type **net get xyz**)
      sample.dta

───

The files needed to do this would be

─────────────────────────────────────── top of stata.toc ───────────

    v 3
    d Chris Farrar, Uni University
    p xyz interval-truncated survival

───────────────────────────────────── end of stata.toc ───────────

```
────────────────────────────────────────── top of xyz.pkg ──────────
v 3
d xyz.  interval-truncated survival.
d C. Farrar, Uni University.
f xyz.ado
f xyz.hlp
f sample.dta
────────────────────────────────────────── end of xyz.pkg ──────────
```

On his homepage, Chris would place the following files:

stata.toc	(shown above)
xyz.pkg	(shown above)
xyz.ado	file to be delivered (for use by net install)
xyz.hlp	file to be delivered (for use by net install)
sample.dta	file to be delivered (for use by net get)

Note that Chris does nothing to distinguish ancillary files from installation files.

Example 2

S. Gazer wants to create a more complex site:

```
. net from http://www.wemakeitupaswego.edu/faculty/sgazer
```

```
http://www.wemakeitupaswego.edu/faculty/sgazer
Data-free inference materials
```

```
S. Gazer, Department of Applied Theoretical Mathematics
Also see my homepage for the preprint of "Irrefutable inference".
PLACES you could -net link- to:
    stata            StataCorp web site
DIRECTORIES you could -net cd- to:
    ir               irrefutable inference programs (work in progress)
PACKAGES you could -net describe-:
    rtec             Robust-to-everything-conceivable statistic
    rte              Robust-to-everything statistic

. net describe rte
```

```
package rte from http://www.wemakeitupaswego.edu/faculty/sgazer/
```

```
TITLE
      rte.  The robust-to-everything statistic; update.
DESCRIPTION/AUTHOR(S)
      S. Gazer, Dept. of Applied Theoretical Mathematics, WMIUAWG Univ.
      Aleph-0 100% confidence intervals proved too conservative for some
      applications; Aleph-1 confidence intervals have been substituted.
      The new robust-to-everything supplants the previous robust-to-
      everything-conceivable statistic.  See "Inference in the absence
      of data" (forthcoming).  After installation, see help rte.

      Support:  email sgazer@wemakeitupaswego.edu
INSTALLATION FILES                                      (type net install rte_stat)
      rte.ado
      rte.hlp
      nullset.ado
      random.ado
ANCILLARY FILES                                         (type net get rte_stat)
      empty.dta
```

The files needed to do this would be

```
─────────────────────────────────────────────── top of stata.toc ───────────
v 3
d Data-free inference materials
d S. Gazer, Department of Applied Theoretical Mathematics
d
d Also see my homepage for the preprint of "Irrefutable inference".
l stata http://www.stata.com
t ir irrefutable inference programs (work in progress)
p rtec Robust-to-everything-conceivable statistic
p rte  Robust-to-everything statistic
─────────────────────────────────────────────── end of stata.toc ───────────
```

```
─────────────────────────────────────────────── top of rte.pkg ───────────
v 3
d rte.  The robust-to-everything statistic; update.
d {bf:S. Gazer, Dept. of Applied Theoretical Mathematics, WMIUAWG Univ.}
d Aleph-0 100% confidence intervals proved too conservative for some
d applications; Aleph-1 confidence intervals have been substituted.
d The new robust-to-everything supplants the previous robust-to-
d everything-conceivable statistic.  See "Inference in the absence
d of data" (forthcoming).  After installation, see help {bf:rte}.
d
d Support:  email sgazer@wemakeitupaswego.edu
f rte.ado
f rte.hlp
f nullset.ado
f random.ado
f empty.dta
─────────────────────────────────────────────── end of rte.pkg ───────────
```

On his homepage, Mr. Gazer would place the following files:

stata.toc	(shown above)
rte.pkg	(shown above)
rte.ado	(file to be delivered)
rte.hlp	(file to be delivered)
nullset.ado	(file to be delivered)
random.ado	(file to be delivered)
empty.dta	(file to be delivered)
rtec.pkg	the other package referred to in stata.toc
rtec.ado	the corresponding files to be delivered
rtec.hlp	
ir/stata.toc	the contents file for when the user types net cd ir
ir/...	whatever other .pkg files are referred to
ir/...	whatever other files are to be delivered

For complex sites, a different structure may prove more convenient:

(Continued on next page)

stata.toc	(shown above)
rte.pkg	(shown above)
rtec.pkg	the other package referred to in stata.toc
rte/	directory containing rte files to be delivered:
rte/rte.ado	(file to be delivered)
rte/rte.hlp	(file to be delivered)
rte/nullset.ado	(file to be delivered)
rte/random.ado	(file to be delivered)
rte/empty.dta	(file to be delivered)
rtec/	directory containing rtec files to be delivered:
rtec/...	(files to be delivered)
ir/stata.toc	the contents file for when the user types net cd ir
ir/*.pkg	whatever other package files are referred to
ir/*/...	whatever other files are to be delivered

If you prefer this structure, it is simply a matter of changing the bottom of the rte.pkg from

```
f rte.ado
f rte.hlp
f nullset.ado
f random.ado
f empty.dta
```

to

```
f rte/rte.ado
f rte/rte.hlp
f rte/nullset.ado
f rte/random.ado
f rte/empty.dta
```

Note that, in writing paths and files, the directory separator forward slash (/) is used, regardless of operating system, because this is what the Internet uses.

Also note that it does not matter whether the files you put out are in DOS/Windows, Macintosh, or Unix format (how lines end is recorded differently). When Stata reads the files over the Internet, it will figure out the file format on its own and will automatically translate the files to what is appropriate for the receiver.

Additional package directives

F *filename* is similar to f *filename*, except that, when the file is installed, it will be copied to the system directories (and not the current directory) in all cases.

With f *filename*, the file is installed into a directory according to the file's suffix. For instance, xyz.ado would be installed in the system directories, whereas xyz.dta would be installed in the current directory.

Coding F xyz.ado would have the same result as coding f xyz.ado.

Coding F xyz.dta, however, would state that xyz.dta is to be installed in the system directories.

g *platformname* *filename* is also a variation on f *filename*. It specifies that the file be installed only if the user's operating system is of type *platformname*; otherwise, the file is ignored. The platform names are WIN, WIN64A (64-bit x86-64), and WIN64I (64-bit Itanium) for Windows; MAC for Macintosh; and AIX, AIX64, DECALPHA, HP, IRIX, LINUX, LINUX64 (64-bit x86-64), LINUX64I (64-bit Itanium), SOLARIS, and SOL64 for Unix.

G *platformname filename* is a variation on F *filename*. The file, if not ignored, is to be installed in the system directories.

g *platformname filename1 filename2* is a more detailed version of g *platformname filename*. In this case, *filename1* is the name of the file on the server (the file to be copied), and *filename2* is to be the name of the file on the user's system; e.g., you might code

```
g WIN mydll.forwin mydll.plugin
g LINUX mydll.forlinux mydll.plugin
```

When you specify one *filename*, the result is the same as specifying two identical *filenames*.

G *platformname filename1 filename2* is the install-in-system-directories version of g *platformname filename1 filename2*

h *filename* asserts that *filename* must be loaded, or this package is not to be installed; e.g., you might code

```
g WIN mydll.forwin mydll.plugin
g LINUX mydll.forlinux mydll.plugin
h mydll.plugin
```

if you were offering the plugin `mydll.plugin` for Windows and Linux only.

SMCL in content and package-description files

The text listed on the second and subsequent d lines in both `stata.toc` and *pkgname*`.pkg` may contain SMCL as long as you include v 3; see [P] **smcl**.

Thus in `rte.pkg`, note that S. Gazer coded the third line as

```
d {bf:S. Gazer, Dept. of Applied Theoretical Mathematics, WMIUAWG Univ.}
```

Error-free file delivery

Most people transport files over the Internet and never worry about the file being corrupted in the process because corruption rarely occurs. If, however, it is of great importance to you that the files be delivered perfectly or not at all, you can include checksum files in the directory.

For instance, say that `big.dta` is included in your package, and that it is of great importance that `big.dta` be sent perfectly. First use Stata to make the checksum file for `big.dta`

```
. checksum big.dta, save
```

That creates a small file called `big.sum`; see [D] **checksum**. Then copy both `big.dta` and `big.sum` to your homepage. If `set checksum` is on (the default is `off`), whenever Stata reads *filename*.*whatever* over the net, it also looks for *filename*`.sum`. If it finds such a file, it uses the information recorded in it to verify that what was copied was error free.

If you do this, be cautious. If you put `big.dta` and `big.sum` on your homepage and then later change `big.dta` without changing `big.sum`, people will think that there are transmission errors when they try to download `big.dta`.

References

Baum, C. F. and N. J. Cox. 1999. ip29: Metadata for user-written contributions to the Stata programming language. *Stata Technical Bulletin* 52: 10–12. Reprinted in *Stata Technical Bulletin Reprints*, vol. 9, pp. 121–124.

Cox, N. J. and C. F. Baum. 2000. ip29.1: Metadata for user-written contributions to the Stata programming language: Extensions. *Stata Technical Bulletin* 54: 21–22. Reprinted in *Stata Technical Bulletin Reprints*, vol. 9, pp. 124–126.

Also See

Complementary:	[R] **net search**, [R] **search**, [R] **sj**, [R] **ssc**,
	[D] **checksum**, [P] **smcl**
Related:	[R] **update**
Background:	[GSM] **19 Using the Internet**,
	[GSU] **19 Using the Internet**,
	[GSW] **19 Using the Internet**,
	[U] **28 Using the Internet to keep up to date**

Title

| net search — Search Internet for installable packages |

Syntax

net search *word* [*word* ...] [, *options*]

options	description
or	list packages that contain any of the keywords; default is all
nosj	search non-SJ and non-STB sources
tocpkg	search both tables of contents and packages (default)
toc	search table of contents only
pkg	search packages only
<u>e</u>verywhere	search packages for match
<u>f</u>ilenames	search filenames associated with package for match
errnone	make return code 111 instead of 0 when no matches found

Description

net search searches the Internet for user-written additions to Stata, including, but not limited to, user-written additions published in the *Stata Journal* (SJ) and in the *Stata Technical Bulletin* (STB). net search lists the available additions that contain the specified keywords.

The user-written materials found are available for immediate download by using the net command or by clicking on the link.

In addition to typing net search, you may select **Help > Search...** and choose **Search net resources**.

Options

or is relevant only when multiple keywords are specified. By default, only packages that include all the keywords are listed. or changes this to list packages that contain any of the keywords.

nosj specifies that net search not list matches that were published in the SJ or in the STB.

tocpkg, toc, and pkg determine what is searched. tocpkg is the default, meaning that both tables of contents (tocs) and packages (pkgs) are searched. toc restricts the search to tables of contents only. pkg restricts the search to packages only.

everywhere and filenames determine where in packages net search looks for *keywords*. The default is everywhere. filenames restricts net search to search for matches only in the filenames associated with a package. Specifying everywhere implies pkg.

errnone is a programmer's option that causes the return code to be 111 instead of 0 when no matches are found.

Remarks

net search searches the Internet for user-written additions to Stata. If you want to search the Stata documentation for a particular topic, command, or author, see [R] **search**. net search *word* [*word* ...] (without options) is equivalent to typing search *word* [*word* ...], net.

Remarks are presented under the headings

> *Topic searches*
> *Author searches*
> *Command searches*
> *Where does net search look?*
> *How does net search work?*

Topic searches

Example: find what is available about random effects

    . net search random effect

Comments:

1. It is best to search using the singular form of a word. net search random effect will find both "random effect" and "random effects".

2. net search random effect will also find "random-effect" because net search performs a string search and not a word search.

3. net search random effect lists all packages containing the words "random" and "effect", not necessarily used together.

4. If you wanted all packages containing the word "random" or the word "effect", you would type net search random effect, or.

Author searches

Example: find what is available by author Jeroen Weesie

    . net search weesie

Comments:

1. You could type net search jeroen weesie, but that might list less because sometimes the last name is used without the first.

2. You could type net search Weesie, but it would not matter. Capitalization is ignored in the search.

Example: find what is available by Jeroen Weesie excluding SJ and STB materials

    . net search weesie, nosj

1. The SJ and the STB tend to dominate search results because so much has been published in them. If you know that what you are looking for is not in the SJ or in the STB, specifying the nosj option will narrow the search.

2. net search weesie lists everything that net search weesie, nosj lists, and more. If you just type net search weesie, look down the list. SJ and STB materials are listed first, and non-SJ and non-STB materials are listed last.

Command searches

Example: find the user-written command kursus

```
. net search kursus, file
```

1. You could just type net search kursus, and that will list everything net search kursus, file lists, and more. Since you know kursus is a command, however, there must be a kursus.ado file associated with the package. Typing net search kursus, file narrows the search.

2. You could also type net search kursus.ado, file to narrow the search even more.

Where does net search look?

net search looks everywhere, not just at *http://www.stata.com*.

net search begins by looking at *http://www.stata.com*, but then follows every link, which takes it to other places, and then follows every link again, which takes it to even more places, and so on.

Authors: Please let us know if you have a site that we should include in our search by sending an email to *netsearch@stata.com*. We will then link to your site from ours to ensure that net search finds your materials. That is not strictly necessary, however, as long as your site is directly or indirectly linked from some site that is linked to ours.

How does net search work?

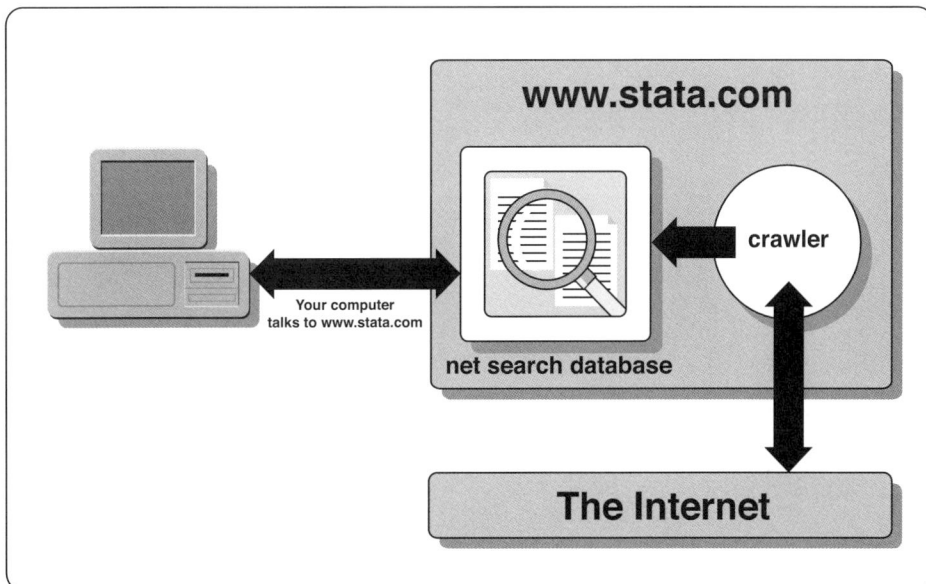

http://www.stata.com maintains a database of Stata resources. When you use `net search`, it contacts *http://www.stata.com* with your request, *http://www.stata.com* searches its database, and Stata returns the results to you.

Another part of the system is called the crawler, which searches the web for new Stata resources to add to the `net search` database and verifies that the resources already found are still available. When a new resource becomes available, the crawler takes about two days to add it to the database, and, similarly, if a resource disappears, the crawler takes roughly two days to remove it from the database.

References

Baum, C. F. and N. J. Cox. 1999. ip29: Metadata for user-written contributions to the Stata programming language. *Stata Technical Bulletin* 52: 10–12. Reprinted in *Stata Technical Bulletin Reprints*, vol. 9, pp. 121–124.

Cox, N. J. and C. F. Baum. 2000. ip29.1: Metadata for user-written contributions to the Stata programming language: Extensions. *Stata Technical Bulletin* 54: 21–22. Reprinted in *Stata Technical Bulletin Reprints*, vol. 9, pp. 124–126.

Gould, W. W. and A. R. Riley. 2000. stata55: Search web for installable packages. *Stata Technical Bulletin* 54: 4–6. Reprinted in *Stata Technical Bulletin Reprints*, vol. 9, pp. 10–13.

Also See

Complementary:	[R] **net**, [R] **sj**
Related:	[R] **search**, [R] **update**

Title

---

netio — Control Internet connections

Syntax

Turn on or off the use of a proxy server

    <u>se</u>t httpproxy {on | off} [, init]

Set proxy host name

    <u>se</u>t httpproxyhost ["]*name*["]

Set the proxy port number

    <u>se</u>t httpproxyport #

Turn on or off proxy authorization

    <u>se</u>t httpproxyauth {on | off}

Set proxy authorization user ID

    <u>se</u>t httpproxyuser ["]*name*["]

Set proxy authorization password

    <u>se</u>t httpproxypw ["]*password*["]

Set time limit for establishing initial connection

    <u>se</u>t timeout1 *#seconds* [, <u>perma</u>nently]

Set time limit for data transfer

    <u>se</u>t timeout2 *#seconds* [, <u>perma</u>nently]

Description

Several commands (e.g., net, news, and update) are designed specifically for use over the Internet. Many other Stata commands that read a file (e.g., copy, type, and use) can also read directly from a URL. In most cases, all these commands will work without your ever needing to concern yourself with the set commands discussed here. These set commands provide control over network system parameters.

If you experience problems using Stata's network features, ask your system administrator if your site uses a proxy. A proxy is a server between your computer and the rest of the Internet, and your computer may need to communicate with other computers on the Internet through this proxy. If your site uses a proxy, your system administrator can provide you with its host name and the port your computer can use to communicate with it. If your site's proxy requires you to log in to it before it will respond, your system administrator will provide you with a user ID and password.

set httpproxyhost sets the name of the host to be used as a proxy server. set httpproxyport sets the port number. set httpproxy turns on or off the use of a proxy server, leaving the proxy host name and port intact, even when not in use.

Under the Windows and Macintosh operating systems, when you set httpproxy on, Stata will attempt to obtain the values of httpproxyhost and httpproxyport from the operating system if they have not been previously set. set httpproxy on, init attempts to obtain these values from the operating system, even if they have been previously set.

If the proxy requires authorization (user ID and password), set authorization on via set httpproxyauth on. The proxy user and proxy password must also be set to the appropriate user ID and password by using set httpproxyuser and set httpproxypw.

Note that Stata remembers the various proxy settings between sessions and does not need a permanently option.

set timeout1 changes the time limit in seconds that Stata imposes for establishing the initial connection with a remote host. set timeout2 changes the time limit in seconds that Stata imposes for subsequent data transfer with the host. If these time limits are exceeded, a "connection timed out" message and error code 2 are produced. You should seldom need to change these settings.

Options

init specifies that set httpproxy on attempts to initialize httpproxyhost and httpproxyport from the operating system (Windows and Macintosh only).

permanently specifies that, in addition to making the change right now, the timeout1 and timeout2 settings be remembered and become the default setting when you invoke Stata.

The various httpproxy settings do not have a permanently option because permanently is implied.

Remarks

If you receive an error message, see *http://www.stata.com/support/faqs/web/* for the latest information.

1. remote connection failed r(677);

If you see remote connection failed r(677);

then you asked for something to be done over the web, and Stata tried but could not contact the specified host. Stata was able to talk over the network and look up the host but was not able to establish a connection to that host.

Perhaps the host is down; try again later.

If 100% of your web accesses result in this message, then perhaps your network connection is through a proxy server. If it is, then you must tell Stata.

Contact your system administrator. Ask for the name and port of the "http proxy server". Say that you are told

> http proxy server: jupiter.myuni.edu
> port number: 8080

In Stata, type

```
. set httpproxyhost jupiter.myuni.edu
. set httpproxyport 8080
. set httpproxy on
```

Your web accesses should then work.

2. connection timed out r(2);

If you see

```
connection timed out r(2);
```

an Internet connection has timed out. This can happen when

a. the connection between you and the host is slow, or

b. the connection between you and the host has disappeared, and so it eventually "timed out".

In the case of (b), wait a while (say, 5 minutes) and try again (sometimes pieces of the Internet can break for up to a day, but that is rare). In the case of (a), you can reset the limits for what constitutes "timed out". There are two numbers to set.

The time to establish the initial connection is **timeout1**. By default, Stata waits 120 seconds before declaring a timeout. You can change the limit:

```
. set timeout1 #seconds
```

You might try doubling the usual limit and specify 240; *#seconds* must be between 1 and 32,000.

The time to retrieve data from an open connection is **timeout2**. By default, Stata waits 300 seconds (5 minutes) before declaring a timeout. To change the limit, type

```
. set timeout2 #seconds
```

You might try doubling the usual limit and specify 600; *#seconds* must be between 1 and 32,000.

Also See

Related: [U] **28 Using the Internet to keep up to date**

Title

news — Report Stata news

Syntax

```
news
```

Description

news displays a brief listing of recent Stata news and information, which it obtains from Stata's web site. news requires that your computer be connected to the Internet.

You may also execute news by selecting **Help > News**.

Remarks

news provides an easy way of displaying a brief list of the latest Stata news:

```
. news
    ---  ----  ----  ----  ----
  /__    /   ____/  /   ____/
 ___/   /  /___/   /  /___/  News          The latest from http://www.stata.com
```
---

4 May 2005. Official update available for download

    Click here, or pull down **Help** and selecting **Official Updates**, or
    type "update from http://www.stata.com" to obtain the free update.

28 April 2005. NetCourse schedule updated

    See http://www.stata.com/netcourse/ for more information.

1 April 2005. Stata 9 available

    Stata 9 -- mixed models, new matrix programming language Mata, etc. --
    is now available. Click here or visit http://www.stata.com for
    more information.

21 May 2002. New book available from Stata Press

    (output omitted)

```
<end>
```

Also See

Related: [R] **net**

Background: [U] **28 Using the Internet to keep up to date**

Title

nl — Nonlinear least-squares estimation

Syntax

Interactive version

nl (*depvar* = <*sexp*>) [*if*] [*in*] [*weight*] [, *options*]

Programmed substitutable expression version

nl *sexp_prog* : *depvar* [*varlist*] [*if*] [*in*] [*weight*] [, *options*]

Function evaluator program version

nl *func_prog* @ *depvar* [*varlist*] [*if*] [*in*] [*weight*] ,

{ <u>parameters</u>(*namelist*) | <u>nparameters</u>(#) } [*options*]

where

depvar is the dependent variable;

<*sexp*> is a substitutable expression;

sexp_prog is a substitutable expression program; and

func_prog is a function evaluator program.

(*Continued on next page*)

287

options	description
Model	
<u>var</u>iables(*varlist*)	variables in model
<u>ini</u>tial(*initial_values*)	initial values for parameters
* <u>par</u>ameters(*namelist*)	parameters in model (function evaluator program version only)
* <u>np</u>arameters(*#*)	number of parameters in model (function evaluator program version only)
sexp_options	options for substitutable expression program (programmed substitutable expression version only)
func_options	options for function evaluator program (function evaluator program version only)
Model 2	
<u>lnl</u>sq(*#*)	use log least-squares where $\ln(depvar - \#)$ is assumed to be normally distributed
<u>noc</u>onstant	indicate that the model has no constant term; seldom used
<u>has</u>constant(*name*)	identify constant term; seldom used
SE/Robust	
<u>r</u>obust	use Huber/White/sandwich estimator of variance
hc2	use a bias correction for **robust** covariance matrix
hc3	use an alternative bias correction for **robust** covariance matrix
<u>cl</u>uster(*varname*)	observations are independent across groups specified by *varname*
Reporting	
<u>l</u>evel(*#*)	set confidence level; default is **level(95)**
<u>lea</u>ve	generate variables containing derivative of $E(y)$
<u>t</u>itle(*string*)	display *string* as title above the table of parameter estimates
<u>t</u>itle2(*string*)	display *string* as subtitle
Opt options	
optimization_options	control the optimization process; seldom used
eps(*#*)	specify *#* for convergence criterion; default is **eps(1e-5)**
<u>de</u>lta(*#*)	specify *#* for computing derivatives; default is **delta(4e-7)**

* For function evaluator program version, you must specify parameters(*namelist*) or nparameters(*#*), or both.
bootstrap, by, jackknife, rolling, and statsby are allowed; see [U] **11.1.10 Prefix commands**.
aweights, fweights, and iweights are allowed; see [U] **11.1.6 weight**.
See [U] **20 Estimation and postestimation commands** for additional capabilities of estimation commands.

Description

nl fits an arbitrary nonlinear regression function by least squares. With the interactive version of the command, you enter the function directly on the command line or dialog box using a *substitutable expression*. If you have a function that you use regularly, you can write a *substitutable expression program* and use the second syntax to avoid having to re-enter the function every time. The function evaluator program version gives you the most flexibility in exchange for increased complexity; with this version, your program is given a vector of parameters and a variable list, and your program computes the regression function.

When you write a substitutable expression program or function evaluator program, the first two letters of the name must be nl. *sexp_prog* and *func_prog* refer to the name of the program without the first two letters. For example, if you wrote a function evaluator program named nlregss, you would type nl regss @ ... to estimate the parameters.

Options

Model
---

variables(*varlist*) specifies the variables in the model. nl ignores observations for which any of these variables have missing values. If you do not specify variables(), nl issues an error message with return code 480 if the estimation sample contains any missing values.

initial(*initial_values*) specifies the initial values to begin the estimation. You can specify a $1 \times k$ matrix, where k is the number of parameters in the model, or you can specify a parameter name, its initial value, another parameter name, its initial value, and so on. For example, to initialize alpha to 1.23 and delta to 4.57, you would type

```
nl ... , initial(alpha 1.23 delta 4.57) ...
```

Initial values declared using this option override any that are declared within substitutable expressions. If you specify a parameter that does not appear in your model, nl exits with error code 480. If you specify a matrix, the values must be in the same order that the parameters are declared in your model. nl ignores the row and column names of the matrix.

parameters(*namelist*) specifies the names of the parameters in the model. The names of the parameters must adhere to the naming conventions of Stata's variables; see [U] **11.3 Naming conventions**. If you specify both parameters() and nparameters(), the number of names in the former must match the number specified in the latter; if not, nl issues an error message with return code 198.

nparameters(*#*) specifies the number of parameters in the model. If you do not specify names with the parameters() option, nl names them b1, b2, ..., b#. If you specify both parameters() and nparameters(), the number of names in the former must match the number specified in the latter; if not, nl issues an error message with return code 198.

sexp_options refer to any options allowed by your *sexp_prog*.

func_options refer to any options allowed by your *func_prog*.

Model 2
---

lnlsq(*#*) fits the model using log least-squares, which we define as least squares with shifted lognormal errors. In other words, $\ln(depvar - \#)$ is assumed to be normally distributed. Sums of squares and deviance are adjusted to the same scale as *depvar*.

noconstant indicates that the function does not include a constant term. This option is generally not needed, even if there is no constant term in the model, unless the coefficient of variation (over observations) of the partial derivative of the function with respect to a parameter is less than eps() and that parameter is not a constant term.

hasconstant(*name*) indicates that parameter *name* be treated as the constant term in the model and that nl should not use its default algorithm to find a constant term. As with noconstant, this option is seldom used.

SE/Robust

robust, cluster(*varname*); see [R] **estimation options**.

hc2 and hc3 specify alternative bias corrections for the robust variance calculation. hc2 and hc3 may not be specified with cluster(). In the unclustered case, robust uses $\widehat{\sigma}_j^2 = \{n/(n-k)\}u_j^2$ as an estimate of the variance of the jth observation, where u_j is the calculated residual and $n/(n-k)$ is included to improve the overall estimate's small-sample properties.

hc2 instead uses $u_j^2/(1-h_{jj})$ as the observation's variance estimate, where h_{jj} is the jth diagonal element of the hat (projection) matrix. This produces an unbiased estimate of the covariance matrix if the model is homoskedastic. hc2 tends to produce slightly more conservative confidence intervals than robust.

hc3 uses $u_j^2/(1-h_{jj})^2$ as suggested by Davidson and MacKinnon (1993 and 2004), who report that this often produces better results when the model is heteroskedastic. hc3 produces confidence intervals that tend to be even more conservative.

See, in particular, Davidson and MacKinnon (2004, 239), who advocate the use of hc2 or hc3 instead of the plain robust estimator for nonlinear least squares.

Specifying either hc2 or hc3 implies robust.

Reporting

level(*#*); see [R] **estimation options**.

leave leaves behind after estimation a set of new variables with the same names as the estimated parameters containing the derivatives of $\mathrm{E}(y)$ with respect to the parameters. If the dataset contains an existing variable with the same name as a parameter, then using leave causes nl to issue an error message with return code 110.

title(*string*) specifies an optional title that will be displayed just above the table of parameter estimates.

title2(*string*) specifies an optional subtitle that will be displayed between the title specified in title() and the table of parameter estimates. If title2() is specified but title() is not, title2() has the same effect as title().

Opt options

optimization_options: iterate(*#*), [no]log, trace. iterate() specifies the maximum number of iterations; log/nolog specifies whether or not to show the iteration log; and trace specifies that the iteration log should include the current parameter vector. These options are seldom used.

eps(*#*) specifies the convergence criterion for successive parameter estimates and for the residual sum of squares. The default is eps(1e-5).

delta(*#*) specifies the relative change in a parameter to be used in computing the numeric derivatives. The derivative for parameter β_i is computed as $\{f(X, \beta_1, \beta_2, \ldots, \beta_i + d, \beta_{i+1}, \ldots) - f(X, \beta_1, \beta_2, \ldots, \beta_i, \beta_{i+1}, \ldots)\}/d$, where d is $\delta(\beta_i + \delta)$. The default is delta(4e-7).

Remarks

Remarks are presented under the headings

> *Substitutable expressions*
> *Substitutable expression programs*
> *Built-in functions*
> *Log-normal errors*
> *Other uses*
> *Weights*
> *Potential errors*
> *General comments on fitting nonlinear models*
> *Function evaluator programs*

`nl` fits an arbitrary nonlinear function by least squares. The interactive version allows you to enter the function directly on the command line or dialog box using *substitutable expressions*. You can write a *substitutable expression program* for functions that you fit frequently to save yourself time. Finally, *function evaluator programs* give you the most flexibility in defining your nonlinear function, though they are more complicated to use.

The next section explains the substitutable expressions that are used to define the regression function, and the section thereafter explains how to write substitutable expression program files so that you do not need to type in commonly used functions over and over. Later sections highlight other features of `nl`.

The final section discusses function evaluator programs. If you find substitutable expressions adequate to define your nonlinear function, then you can skip that section entirely. Function evaluator programs are generally needed only for complicated problems, such as multistep estimators. The program receives a vector of parameters at which it is to compute the function and a variable into which the results are to be placed.

Substitutable expressions

You define the nonlinear function to be fitted by `nl` using a substitutable expression. Substitutable expressions are just like any other mathematical expressions involving scalars and variables, such as those you would use with Stata's `generate` command, except that the parameters to be estimated are bound in braces. See [U] **13.2 Operators** and [U] **13.3 Functions** for more information on expressions.

For example, suppose that you wish to fit the function

$$y_i = \beta_0(1 - e^{-\beta_1 x_i}) + \epsilon_i$$

where β_0 and β_1 are the parameters to be estimated and ϵ_i is an error term. You would simply type

    . nl (y = {b0}*(1 - exp(-1*{b1}*x)))

Note that you must enclose the entire equation in parentheses. Because `b0` and `b1` are enclosed in braces, `nl` knows that they are parameters in the model. `nl` will initialize `b0` and `b1` to zero by default. To request that `nl` initialize `b0` to one and `b1` to 0.25, you would type

    . nl (y = {b0=1}*(1 - exp(-1*{b1=0.25}*x)))

That is, inside the braces denoting a parameter, you put the parameter name followed by an equals sign and the initial value. If a parameter appears in your function multiple times, you need only specify an initial value once (or never, if you wish to set the initial value to zero). If you do specify more than one initial value for the same parameter, `nl` will use the *last* value given. Parameter names must follow the same conventions as variable names. See [U] **11.3 Naming conventions**.

Frequently, even nonlinear functions contain linear combinations of variables. As an example, suppose that you wish to fit the function

$$y_i = \beta_0 \left\{ 1 - e^{-(\beta_1 x_{1i} + \beta_2 x_{2i} + \beta_3 x_{3i})} \right\} + \epsilon_i$$

nl allows you to declare a linear combination of variables using the shorthand notation

```
. nl (y = {b0=1}*(1 - exp(-1*{xb: x1 x2 x3})))
```

In the syntax {xb: x1 x2 x3}, you are telling nl that you are declaring a linear combination named xb that is a function of three variables, x1, x2, and x3. nl will create three parameters, named xb_x1, xb_x2, and xb_x3, and initialize them to zero. Instead of typing the previous command, you could have typed

```
. nl (y = {b0=1}*(1 - exp(-1*({xb_x1}*x1 + {xb_x2}*x2 + {xb_x3}*x3))))
```

and yielded the same result. You can refer to the parameters created by nl in the linear combination later in the function, though you must declare the linear combination first if you intend to do that. When creating linear combinations, nl ensures that the parameter names it chooses are unique and have not yet been used in the function.

In general, there are three rules to follow when defining substitutable expressions:

1. Parameters of the model are bound in braces: {b0}, {param}, etc.
2. Initial values for parameters are given by including an equal sign and the initial value inside the braces: {b0=1}, {param=3.571}, etc.
3. Linear combinations of variables can be included using the notation {*eqname*:*varlist*}: {xb: mpg price weight}, {score: w x z}, etc. Parameters of linear combinations are initialized to zero.

If you specify initial values using the initial() option, they override whatever initial values are given within the substitutable expression. Substitutable expressions are so named because, once values are assigned to the parameters, the resulting expression can be handled by generate and replace.

▷ Example 1

We wish to fit the CES production function

$$\ln Q_i = \beta_0 - \frac{1}{\rho} \ln \left\{ \delta K_i^{-\rho} + (1 - \delta) L_i^{-\rho} \right\} + \epsilon_i \tag{1}$$

where $\ln Q_i$ is the log of output for firm i; K_i and L_i are firm i's capital and labor usage, respectively; and ϵ_i is a regression error term. Because ρ appears in the denominator of a fraction, zero is not a feasible initial value; for a CES production function, $\rho = 1$ is a reasonable choice. Setting $\delta = 0.5$ implies labor and capital have equal impacts on output, which is also a reasonable choice for an initial value. We type

```
. use http://www.stata-press.com/data/r9/production, clear

. nl (lnq = {b0} - 1/{rho=1}*ln({delta=0.5}*capital^(-1*{rho}) +
> (1 - {delta})*labor^(-1*{rho})))
(obs = 100)
Iteration 0:  residual SS =   1407.299
Iteration 1:  residual SS =   29.38631
Iteration 2:  residual SS =   29.36637
Iteration 3:  residual SS =   29.36583
Iteration 4:  residual SS =   29.36581
Iteration 5:  residual SS =   29.36581
Iteration 6:  residual SS =   29.36581
Iteration 7:  residual SS =   29.36581
```

Source	SS	df	MS
Model	91.1449924	2	45.5724962
Residual	29.3658055	97	.302740263
Total	120.510798	99	1.21728079

```
Number of obs =        100
F(  2,     97) =     150.53
Prob > F       =     0.0000
R-squared      =     0.7563
Adj R-squared  =     0.7513
Root MSE       =   .5502184
Res. dev.      =   161.2538
```

lnoutput	Coef.	Std. Err.	t	P>|t|	[95% Conf. Interval]	
b0	3.792158	.099682	38.04	0.000	3.594316	3.989999
rho	1.386993	.472584	2.93	0.004	.4490443	2.324941
delta	.4823616	.0519791	9.28	0.000	.3791975	.5855258

```
* Parameter b0 taken as constant term in model & ANOVA table
  (SEs, P values, CIs, and correlations are asymptotic approximations)
```

nl will attempt to find a constant term in the model, and if one is found, mention it at the bottom of the output. nl found b0 to be a constant because the partial derivative $\partial \ln Q_i / \partial b0$ has a coefficient of variation less than eps() in the estimation sample.

The elasticity of substitution for the CES production function is $\sigma = 1/(1 + \rho)$; and, having fitted the model, we can use nlcom to estimate it:

```
. nlcom (1/(1 + _b[rho]))
      _nl_1:  1/(1 + _b[rho])
```

lnq	Coef.	Std. Err.	t	P>|t|	[95% Conf. Interval]	
_nl_1	.4189372	.0829424	5.05	0.000	.2543194	.583555

See [R] **nlcom** and [U] **13.5 Accessing coefficients and standard errors** for more information.

◁

nl's output closely mimics that of regress; see [R] **regress** for more information. The model F test, R^2, sums of squares, and similar statistics are calculated in the same way that regress calculates them. If no "constant" term is specified, the usual caveats apply to the interpretation of the F and R^2 statistics; see the comments and references in Goldstein (1992).

As with regress, if robust, cluster(), hc2, or hc3 is specified, the F statistic is a Wald test based on the robustly estimated covariance matrix. Moreover, the ANOVA table is suppressed because it is no longer appropriate in a statistical sense. The R^2 statistic remains valid as a goodness-of-fit statistic, though the relationship between the F and R^2 statistics that holds in the classical regression model does not apply when a robust covariance matrix is used.

Substitutable expression programs

If you fit the same model frequently or if you want to write an estimator that will operate on whatever variables you specify, then you will want to write a substitutable expression program. That program will return a macro containing a substitutable expression that `nl` can then evaluate, and it may optionally calculate initial values as well. The name of the program must begin with the letters `nl`.

To illustrate, suppose that you use the CES production function often in your work. Instead of typing in the formula each time, you can write a program like this:

```
program nlces, rclass
        version 9
        syntax varlist(min=3 max=3) if
        local logout : word 1 of `varlist'
        local capital : word 2 of `varlist'
        local labor : word 3 of `varlist'
        // Initial value for b0 given delta=0.5 and rho=1
        tempvar y
        generate double `y' = `logout' + ln(0.5*`capital'^-1 + 0.5*`labor'^-1)
        summarize `y' `if', meanonly
        local b0val = r(mean)
        // Terms for substitutable expression
        local capterm "{delta=0.5}*`capital'^(-1*{rho})"
        local labterm "(1-{delta})*`labor'^(-1*{rho})"
        local term2    "1/{rho=1}*ln(`capterm' + `labterm')"
        // Return substitutable expression and title
        return local eq "`logout' = {b0=`b0val'} - `term2'"
        return local title "CES ftn., ln Q=`logout', K=`capital', L=`labor'"
end
```

The program accepts three variables for log output, capital, and labor, and it accepts an `if` *exp* qualifier to restrict the estimation sample. All programs that you write to use with `nl` must accept an `if` *exp* qualifier because, when `nl` calls the program, it passes a binary variable that marks the estimation sample (the variable equals one if the observation is in the sample and zero otherwise). When calculating initial values, you will want to restrict your computations to the estimation sample, and you can do so by using `if` with any commands that accept `if` *exp* qualifiers. Even if your program does not calculate initial values or otherwise use the `if` qualifier, the `syntax` statement must still allow it. See [P] **syntax** for more information on the `syntax` command and the use of `if`.

As in the previous example, reasonable initial values for δ and ρ are 0.5 and 1, respectively. Conditional on those values, (1) can be rewritten as

$$\beta_0 = \ln Q_i + \ln(0.5K_i^{-1} + 0.5L_i^{-1}) - \epsilon_i \tag{2}$$

so a good initial value for β_0 is the mean of the right-hand side of (2) ignoring ϵ_i. Lines seven through ten of the function evaluator program calculate that mean and store it in a local macro. Notice the use of `if` in the `summarize` statement so that the mean is calculated only for the estimation sample.

The final part of the program returns two macros. The macro `title` is optional and defines a short description of the model that will be displayed in the output immediately above the table of parameter estimates. The macro `eq` is required and defines the substitutable expression that `nl` will use. If the expression is short, you can define it all at once. However, because the expression used here is somewhat lengthy, defining local macros and then building up the final expression from them is easier.

To verify that there are no errors in your program, you can call it directly and then use `return list`:

```
. use http://www.stata-press.com/data/r9/production, clear
. nlces lnoutput capital labor
. return list
macros:
            r(title) : "CES ftn., ln Q=lnoutput, K=capital, L=labor"
               r(eq) : "lnoutput = {b0=3.711606264663641} - 1/{rho=1}*ln({delt
> a=0.5}*capital^(-1*{rho}) + (1-{delta})*labor^(-1*{rho}))"
```

Notice that the macro r(eq) contains the same substitutable expression that we specified at the command line in the preceding example, except for the initial value for b0. In short, an nl substitutable expression program should return in r(eq) the same substitutable expression you would type at the command line. The only difference is that when writing a substitutable expression program, you do not bind the entire expression inside parentheses.

Having written the program, you can use it by typing

```
. nl ces: lnoutput capital labor
```

(There is a space between nl and ces.) The output is identical to that shown in example 1, save for the title defined in the function evaluator program that appears immediately above the table of parameter estimates.

❏ Technical Note

You will want to store nlces as an ado-file called nlces.ado. The alternative is to type the code into Stata interactively or to place the code in a do-file. While those alternatives are adequate for occasional use, if you save the program as an ado-file, you can use the function anytime you use Stata without having to redefine the program. When nl attempts to execute nlces, if the program is not in Stata's memory, Stata will search the disk(s) for an ado-file of the same name, and, if found, automatically load it. All you have to do is name the file with the .ado suffix and then place it in a directory where Stata will find it. You should put the file in the directory Stata reserves for user-written ado-files, which, depending on your operating system, is c:\ado\personal (Windows), ~/ado/personal (Unix), or ~:ado:personal (Macintosh). See [U] **17 Ado-files**.   ❏

In some cases you may want to pass additional options to the substitutable expression program. You can modify the syntax statement of your program to accept whatever options you wish. Then when you call nl with the syntax

```
. nl func_prog: varlist, options
```

any *options* that are not recognized by nl (see the table of options at the beginning of this entry) are passed on to your function evaluator program. The only other restriction is that your program cannot accept an option named at because nl uses that option with function evaluator programs.

Built-in functions

Some functions are used so frequently that nl has them built-in so that you do not need to write them yourself. nl automatically chooses initial values for the parameters, though you can use the initial(...) option to override them.

Three alternatives are provided for exponential regression with one asymptote:

exp3	$y_i = \beta_0 + \beta_1 \beta_2^{x_i} + \epsilon_i$
exp2	$y_i = \beta_1 \beta_2^{x_i} + \epsilon_i$
exp2a	$y_i = \beta_1 \left(1 - \beta_2^{x_i}\right) + \epsilon_i$

For instance, typing `nl exp3: ras dvl` fits the three-parameter exponential model (parameters β_0, β_1, and β_2) using $y_i = $ `ras` and $x_i = $ `dvl`.

Two alternatives are provided for the logistic function (symmetric sigmoid shape; not to be confused with logistic regression):

log4	$y_i = \beta_0 + \beta_1 \Big/ \Big[1 + \exp\big\{-\beta_2(x_i - \beta_3)\big\}\Big] + \epsilon_i$
log3	$y_i = \beta_1 \Big/ \Big[1 + \exp\big\{-\beta_2(x_i - \beta_3)\big\}\Big] + \epsilon_i$

Finally, two alternatives are provided for the Gompertz function (asymmetric sigmoid shape):

gom4	$y_i = \beta_0 + \beta_1 \exp\Big[-\exp\big\{-\beta_2(x_i - \beta_3)\big\}\Big] + \epsilon_i$
gom3	$y_i = \beta_1 \exp\Big[-\exp\big\{-\beta_2(x_i - \beta_3)\big\}\Big] + \epsilon_i$

Log-normal errors

A nonlinear model with errors that are independently and identically distributed normal may be written

$$y_i = f(\mathbf{x}_i, \boldsymbol{\beta}) + u_i, \qquad u_i \sim \mathrm{N}(0, \sigma^2) \tag{3}$$

for $i = 1, \dots, n$. If the y_i are thought to have a k-shifted lognormal instead of a normal distribution— that is, $\ln(y_i - k) \sim \mathrm{N}(\zeta_i, \tau^2)$, and the systematic part $f(\mathbf{x}_i, \boldsymbol{\beta})$ of the original model is still thought appropriate for y_i—the model becomes

$$\ln(y_i - k) = \zeta_i + v_i = \ln\big\{f(\mathbf{x}_i, \boldsymbol{\beta}) - k\big\} + v_i, \quad v_i \sim \mathrm{N}(0, \tau^2) \tag{4}$$

This model is fitted if `lnlsq(k)` is specified.

If model (4) is correct, the variance of $(y_i - k)$ is proportional to $\big\{f(\mathbf{x}_i, \boldsymbol{\beta}) - k\big\}^2$. Probably the most common case is $k = 0$, sometimes called "proportional errors" since the standard error of y_i is proportional to its expectation, $f(\mathbf{x}_i, \boldsymbol{\beta})$. Assuming that the value of k is known, (4) is just another nonlinear model in $\boldsymbol{\beta}$, and it may be fitted as usual. However, we may wish to compare the fit of (3) with that of (4) using the residual sum of squares or the deviance D, $D = -2 \times$ log-likelihood, from each model. To do so, we must allow for the change in scale introduced by the log transformation.

Assuming, then, the y_i to be normally distributed, Atkinson (1985, 85–87, 184), by considering the Jacobian $\prod |\partial \ln(y_i - k)/\partial y_i|$, showed that multiplying both sides of (4) by the geometric mean of $y_i - k$, $\dot{y}$, gives residuals on the same scale as those of y_i. The geometric mean is given by

$$\dot{y} = e^{n^{-1} \sum \ln(y_i - k)}$$

which is a constant for a given dataset. The residual deviance for (3) and for (4) may be expressed as

$$D(\widehat{\boldsymbol{\beta}}) = \Big\{1 + \ln(2\pi\widehat{\sigma}^2)\Big\}n \tag{5}$$

where $\widehat{\boldsymbol{\beta}}$ is the maximum likelihood estimate (MLE) of $\boldsymbol{\beta}$ for each model and $n\widehat{\sigma}^2$ is the RSS from (3), or that from (4) multiplied by $\dot{y}^2$.

Since (3) and (4) are models with different error structures but the same functional form, the arithmetic difference in their RSS or deviances is not easily tested for statistical significance. However, if the deviance difference is "large" (> 4, say), we would naturally prefer the model with the smaller deviance. Of course, the residuals for each model should be examined for departures from assumptions (nonconstant variance, non-normality, serial correlations, etc.) in the usual way.

Alternatively, consider modeling

$$E(y_i) = 1/(C + Ae^{Bx_i}) \tag{6}$$

$$E(1/y_i) = E(y_i') = C + Ae^{Bx_i} \tag{7}$$

where C, A, and B are parameters to be estimated. Using the data $(y, x) = (.04, 5)$, $(.06, 12)$, $(.08, 25)$, $(.1, 35)$, $(.15, 42)$, $(.2, 48)$, $(.25, 60)$, $(.3, 75)$, and $(.5, 120)$ (Danuso 1991), fitting the models yields

Model	C	A	B	RSS	Deviance
(6)	1.781	25.74	−.03926	−.001640	−51.95
(6) with `lnlsq(0)`	1.799	25.45	−.04051	−.001431	−53.18
(7)	1.781	25.74	−.03926	8.197	24.70
(7) with `lnlsq(0)`	1.799	27.45	−.04051	3.651	17.42

There is little to choose between the two versions of the logistic model (6), whereas for the exponential model (7), the fit using `lnlsq(0)` is much better (a deviance difference of 7.28). The reciprocal transformation has introduced heteroskedasticity into y_i', which is countered by the proportional errors property of the lognormal distribution implicit in `lnlsq(0)`. The deviances are not comparable between the logistic and exponential models because the change of scale has not been allowed for, although in principle it could be.

Other uses

Even if you are fitting linear regression models, you may find that `nl` can save you some typing. Since you specify the parameters of your model explicitly, you can impose constraints on them directly.

▷ Example 2

In [R] **cnsreg**, we showed how to fit the model

$$\text{mpg} = \beta_0 + \beta_1 \text{price} + \beta_2 \text{weight} + \beta_3 \text{displ} + \beta_4 \text{gear\_ratio} + \beta_5 \text{foreign} + \beta_6 \text{length} + u$$

subject to the constraints

$$\beta_1 = \beta_2 = \beta_3 = \beta_6$$
$$\beta_4 = -\beta_5 = \beta_0/20$$

An alternative way is to use `nl`:

```
. sysuse auto, clear
. nl (mpg = {b0} + {b1}*price + {b1}*weight + {b1}*displ +
> {b0}/20*gear_ratio - {b0}/20*foreign + {b1}*length)
(obs = 74)

Iteration 0:  residual SS =      36008
Iteration 1:  residual SS =   1578.522
```

Source	SS	df	MS
Model	34429.4777	2	17214.7389
Residual	1578.52226	72	21.9239203
Total	36008	74	486.594595

```
                                    Number of obs =        74
                                    F(  2,    72) =    785.20
                                    Prob > F      =    0.0000
                                    R-squared     =    0.9562
                                    Adj R-squared =    0.9549
                                    Root MSE      =  4.682299
                                    Res. dev.     =  436.4562
```

mpg	Coef.	Std. Err.	t	P>|t|	[95% Conf.	Interval]
b0	26.52229	1.375178	19.29	0.000	23.78092	29.26365
b1	-.000923	.0001534	-6.02	0.000	-.0012288	-.0006172

(SEs, P values, CIs, and correlations are asymptotic approximations)

The point estimates and standard errors for β_0 and β_1 are identical to those reported in [R] **cnsreg**. To get the estimate for β_4, we can use **nlcom**:

```
. nlcom _b[b0]/20

       _nl_1:  _b[b0]/20
```

mpg	Coef.	Std. Err.	t	P>|t|	[95% Conf.	Interval]
_nl_1	1.326114	.0687589	19.29	0.000	1.189046	1.463183

The advantage to using **nl** is that we do not need to use the **constraint** command six times.

◁

nl is also a useful tool when doing exploratory data analysis. For example, you may want to run a regression of y on a function of x, though you have not decided whether to use sqrt(x) or ln(x). You can use **nl** to run both regressions without having first to generate two new variables:

```
. nl (y = {b0} + {b1}*ln(x))
. nl (y = {b0} + {b1}*sqrt(x))
```

Weights

Weights are specified the usual way—analytic and frequency weights as well as **iweights** are supported; see [U] **20.16 Weighted estimation**. Use of analytic weights implies that the y_i have different variances. Therefore, model (3) may be rewritten as

$$y_i = f(\mathbf{x}_i, \boldsymbol{\beta}) + u_i, \qquad u_i \sim \mathrm{N}(0, \sigma^2/w_i) \tag{3a}$$

where w_i are (positive) weights, assumed to be known and normalized such that their sum equals the number of observations. The residual deviance for (3a) is

$$D(\widehat{\boldsymbol{\beta}}) = \left\{1 + \ln(2\pi\widehat{\sigma}^2)\right\}n - \sum \ln(w_i) \tag{5a}$$

(compare with (5)), where

$$n\widehat{\sigma}^2 = \text{RSS} = \sum w_i \big\{ y_i - f(\mathbf{x}_i, \widehat{\boldsymbol{\beta}}) \big\}^2$$

Defining and fitting a model equivalent to (4) when weights have been specified as in (3a) is not straightforward and has not been attempted. Thus deviances using and not using the `lnlsq()` option may not be strictly comparable when analytic weights (other than 0 and 1) are used.

You do not need to modify your substitutable expression in any way to use weights. If, however, you write a substitutable expression program, then you should account for weights when obtaining initial values. When `nl` calls your program, it passes whatever weight expression (if any) was specified by the user. Here is an outline of a substitutable expression program that accepts weights:

```
program nl name, rclass
        version 9
        syntax varlist [aw fw iw] if
        ...
        // Obtain initial values allowing weights
        // Use the syntax ['weight''exp'].  For example,
        summarize varname ['weight''exp'] 'if'
        regress depvar varlist ['weight''exp'] 'if'
        ...
        // Return substitutable expression
        return local eq "substitutable expression"
        return local title "description of estimator"
end
```

For details on how the `syntax` command processes weight expressions, see [P] **syntax**.

Potential errors

`nl` is reasonably robust to the inability of your nonlinear function to be evaluated at some parameter values. `nl` does assume that your function can be evaluated at the initial values of the parameters. If your function cannot be evaluated at the initial values, an error message is issued with return code 480. Recall that if you do not specify an initial value for a parameter, then `nl` initializes it to zero. Many nonlinear functions cannot be evaluated when some parameters are zero, so in those cases specifying alternative initial values is crucial.

Thereafter, as `nl` changes the parameter values, it monitors your function for unexpected missing values. If these are detected, `nl` backs up. That is, `nl` finds a point between the previous, known-to-be-good parameter vector and the new, known-to-be-bad vector at which the function can be evaluated and continues its iterations from that point.

`nl` requires that once a parameter vector is found where the predictions can be calculated, small changes to the parameter vector be made in order to calculate numeric derivatives. If a boundary is encountered at this point, an error message is issued with return code 481.

When specifying `lnlsq()`, an attempt to take logarithms of $y_i - k$ when $y_i \leq k$ results in an error message with return code 482.

If `iterate()` iterations are performed and estimates still have not converged, results are presented with a warning, and the return code is set to 430.

If you use the programmed substitutable expression version of `nl` with a function evaluator program, or *vice versa*, Stata issues an error message. Verify that you are using the syntax appropriate for the program you have.

General comments on fitting nonlinear models

In many cases, achieving convergence is problematic. For example, a unique minimum of the sum-of-squares function may not exist. A large amount of literature exists on different algorithms that have been used, on strategies for obtaining good initial parameter values, and on tricks for parameterizing the model to make its behavior as "linear-like" as possible. Selected references are Kennedy and Gentle (1980, ch. 10) for computational matters and Ross (1990) and Ratkowsky (1983) for all three aspects. Ratkowsky's book is particularly clear and approachable, with useful discussion on the meaning and practical implications of "intrinsic" and "parameter-effects" nonlinearity. An excellent text on nonlinear estimation is Gallant (1987). Also see Davidson and MacKinnon (1993 and 2004).

To enhance the success of `nl`, pay attention to the form of the model fitted, along the lines of Ratkowsky and Ross. For example, Ratkowsky (1983, 49–59) analyzes three possible three-parameter "yield-density" models for plant growth:

$$E(y_i) = \begin{cases} (\alpha + \beta x_i)^{-1/\theta} \\ (\alpha + \beta x_i + \gamma x_i^2)^{-1} \\ (\alpha + \beta x_i^\phi)^{-1} \end{cases}$$

All three models give similar fits. However, he shows that the second formulation is dramatically more "linear-like" than the other two, and therefore has better convergence properties. In addition, the parameter estimates are virtually unbiased and normally distributed, and the asymptotic approximation to the standard errors, correlations, and confidence intervals is much more accurate than for the other models. Even within a given model, the way the parameters are expressed (e.g., ϕ^{x_i} or $e^{\theta x_i}$) affects the degree of linearity and convergence behavior.

Function evaluator programs

Occasionally a nonlinear function may be so complex that writing a substitutable expression for it is impractical. For example, there could be a large number of parameters in the model. Alternatively, if you are implementing a two-step estimator, writing a substitutable expression may be altogether impossible. Function evaluator programs can be used in these situations.

`nl` will pass to your function evaluator program a list of variables, a weight expression, a variable marking the estimation sample, and a vector of parameters. Your program is to replace the dependent variable, which is the first variable in the variables list, with the values of the nonlinear function evaluated at those parameters. As with substitutable expression programs, the first two letters of the name must be `nl`.

To focus on the mechanics of the function evaluator program, again let's compare the CES production function to the previous examples. The function evaluator program is

```
program nlces2
        version 9
        syntax varlist(min=3 max=3) if , at(name)
        local logout : word 1 of 'varlist'
        local capital : word 2 of 'varlist'
        local labor : word 3 of 'varlist'
        // Retrieve parameters out of at matrix
        tempname b0 rho delta
        scalar 'b0' = 'at'[1, 1]
        scalar 'rho' = 'at'[1, 2]
        scalar 'delta' = 'at'[1, 3]
        tempvar kterm lterm
        generate double 'kterm' = 'delta'*'capital'^(-1*'rho') 'if'
        generate double 'lterm' = (1-'delta')*'labor'^(-1*'rho') 'if'
        // Fill in dependent variable
        replace 'logout' = 'b0' - 1/'rho'*ln('kterm' + 'lterm') 'if'
end
```

Unlike the previous `nlces` program, this one is not declared to be r-class. The `syntax` statement again accepts three variables: one for log output, one for capital, and one for labor. An `if` *exp* is again required because `nl` will pass a binary variable marking the estimation sample. All function evaluator programs must accept an option named `at()` that takes a `name` as an argument—that is how `nl` passes the parameter vector to your program.

The next part of the program retrieves the output, labor, and capital variables from the variables list. It then breaks up the temporary matrix `at` and retrieves the parameters `b0`, `rho`, and `delta`. Pay careful attention to the order in which the parameters refer to the columns of the `at` matrix because that will affect the syntax you use with `nl`. The temporary names you use inside this program are immaterial, however.

The rest of the program computes the nonlinear function, using a couple of temporary variables to hold intermediate results. The final line of the program then replaces the dependent variable with the values of the function. Notice the use of `'if'` to restrict attention to the estimation sample. `nl` makes a copy of your dependent variable so that when the command is finished your data is left unchanged.

To use the program and fit your model, you type

```
. use http://www.stata-press.com/data/r9/production, clear
. nl ces2 @ lnoutput capital labor, parameters(b0 rho delta)
> initial(b0 0 rho 1 delta 0.5)
```

The output is again identical to that shown in example 1. The order in which the parameters were specified in the `parameters()` option is the same in which they are retrieved from the `at` matrix in the program. To initialize them, you simply list the parameter name, a space, the initial value, and so on.

If you use the `nparameters()` option instead of the `parameters()` option, the parameters are named b1, b2, ..., bk, where k is the number of parameters. Thus, you could have typed

```
. nl ces2 @ lnoutput capital labor, nparameters(3) initial(b1 0 b2 1 b3 0.5)
```

With that syntax, the parameters called `b0`, `rho`, and `delta` in the program will be labeled b1, b2, and b3, respectively. In programming situations or if there are many parameters, instead of listing the parameter names and initial values in the `initial()` option, you may find it more convenient to pass a column vector. In those cases, you could type

```
. matrix myvals = (0, 1, 0.5)
. nl ces2 @ lnoutput capital labor, nparameters(3) initial(myvals)
```

In summary, a function evaluator program receives a list of variables, the first of which is the dependent variable that you are to replace with the values of your nonlinear function. Additionally, it must accept an if *exp*, as well as an option named at that will contain the vector of parameters at which nl wants the function evaluated. You are then free to do whatever is necessary to evaluate your function and replace the dependent variable.

If you wish to use weights, your function evaluator program's syntax statement must accept them. If your program consists only of, for example, generate statements, you need not do anything with the weights passed to your program. However, if in calculating the nonlinear function you use commands such as summarize or regress, then you will want to use the weights with those commands.

As with substitutable expression programs, nl will pass to it any options specified that nl does not accept, providing you with a way to pass additional information to your function.

❏ Technical Note

Before version 9 of Stata, the nl command used a different syntax, which required you to write a *nlfcn* program, and it did not have a syntax for interactive use other than the seven functions that were built-in. The old syntax of nl still works, and you can still use those *nlfcn* programs. If nl does not see a colon, at sign, or a set of parentheses surrounding the equation in your command, it assumes that the old syntax is being used.

The current version of nl uses scalars and matrices to store intermediate calculations instead of local and global macros as the old version did, so the current version produces more accurate results. In practice, however, any discrepancies are likely to be very small.

❏

Saved Results

nl saves in e():

Scalars

e(N)	number of observations	e(r2_a)	adjusted R-squared
e(k)	number of parameters	e(F)	F statistic
e(mss)	model sum of squares	e(rmse)	root mean squared error
e(tss)	total sum of squares	e(converge)	0 if no convergence; 1 otherwise
e(df_m)	model degrees of freedom	e(df_t)	total degrees of freedom
e(rss)	residual sum of squares	e(dev)	residual deviance
e(df_r)	residual degrees of freedom	e(N_clust)	number of clusters
e(mms)	model mean square	e(ic)	number of iterations
e(msr)	residual mean square	e(lnlsq)	value of lnlsq if specified
e(r2)	R-squared	e(log_t)	1 if lnlsq specified; 0 otherwise
e(cj)	position of constant in e(b) or 0 if no constant	e(gm_2)	square of geometric mean of $(y-k)$ if lnlsq; 1 otherwise

Macros

e(cmd)	nl		e(sexp)	substitutable expression
e(depvar)	name of dependent variable		e(params)	names of parameters
e(title)	title in estimation output		e(clustvar)	name of cluster variable
e(title_2)	secondary title in estimation output		e(funcprog)	function evaluator program
e(wtype)	weight type		e(vcetype)	covariance estimation method
e(wexp)	weight expression		e(predict)	program used to implement predict
e(type)	1 = interactively entered expression		e(rhs)	right-hand side variables (function
	2 = substitutable expression program			evaluator program version only)
	3 = function evaluator program			

Matrices

e(b)	coefficient vector		e(V)	variance–covariance matrix of the
e(init)	initial values vector			estimators

Functions

e(sample)	marks estimation sample

Methods and Formulas

nl is implemented as an ado-file.

The derivation here is based on Davidson and MacKinnon (2004, chapter 6). Let β denote the $k \times 1$ vector of parameters, and write the regression function using matrix notation as $\mathbf{y} = \mathbf{f}(\mathbf{x}, \beta) + \mathbf{u}$ so that the objective function can be written as

$$\text{SSR}(\beta) = \{\mathbf{y} - \mathbf{f}(\mathbf{x}, \beta)\}' \, \mathbf{D} \, \{\mathbf{y} - \mathbf{f}(\mathbf{x}, \beta)\}$$

The $\mathbf{D}$ matrix contains the weights and is defined in [R] **regress**; if no weights are specified, then $\mathbf{D}$ is the $N \times N$ identity matrix. Taking a second-order Taylor series expansion centered at β_0 yields

$$\text{SSR}(\beta) \approx \text{SSR}(\beta_0) + \mathbf{g}'(\beta_0)(\beta - \beta_0) + \frac{1}{2}(\beta - \beta_0)' \mathbf{H}(\beta_0)(\beta - \beta_0) \tag{8}$$

where $\mathbf{g}(\beta_0)$ denotes the $k \times 1$ gradient of $\text{SSR}(\beta)$ evaluated at β_0 and $\mathbf{H}(\beta_0)$ denotes the $k \times k$ Hessian of $\text{SSR}(\beta)$ evaluated at β_0. Letting $\mathbf{X}$ denote the $N \times k$ matrix of derivatives of $\mathbf{f}(\mathbf{x}, \beta)$ with respect to β, the gradient $\mathbf{g}(\beta)$ is

$$\mathbf{g}(\beta) = -2\mathbf{X}'\mathbf{D}\mathbf{u} \tag{9}$$

$\mathbf{X}$ and $\mathbf{u}$ are obviously functions of β, though for notational simplicity that dependence is not shown explicitly. The (m, n) element of the Hessian can be written

$$H_{mn}(\beta) = -2 \sum_{i=1}^{i=N} d_{ii} \left[\frac{\partial^2 f_i}{\partial \beta_m \partial \beta_n} u_i - X_{im} X_{in} \right] \tag{10}$$

where d_{ii} is the ith diagonal element of $\mathbf{D}$. As discussed in Davidson and MacKinnon (2004, chapter 6), the first term inside the brackets of (10) has expectation zero, so the Hessian can be approximated as

$$\mathbf{H}(\beta) = 2\mathbf{X}'\mathbf{D}\mathbf{X} \tag{11}$$

Differentiating the Taylor series expansion of $\mathrm{SSR}(\beta)$ shown in (8) yields the first-order condition for a minimum

$$\mathbf{g}(\beta_0) + \mathbf{H}(\beta_0)(\beta - \beta_0) = \mathbf{0}$$

which suggests the iterative procedure

$$\beta_{j+1} = \beta_j - \alpha\mathbf{H}^{-1}(\beta_j)\mathbf{g}(\beta_j) \tag{12}$$

where α is a "step size" parameter chosen at each iteration to improve convergence. Using (9) and (11), (12) can be written

$$\beta_{j+1} = \beta_j + \alpha(\mathbf{X'DX})^{-1}\mathbf{X'Du} \tag{13}$$

where $\mathbf{X}$ and $\mathbf{u}$ are evaluated at β_j. Apart from the scalar α, the second term on the right-hand side of (13) can be computed via a (weighted) regression of the columns of $\mathbf{X}$ on the errors. `nl` computes the derivatives numerically and then calls `regress`. At each iteration, α is set to one, and a candidate value β_{j+1}^* is computed by (13). If $\mathrm{SSR}(\beta_{j+1}^*) < \mathrm{SSR}(\beta_j)$, then $\beta_{j+1} = \beta_{j+1}^*$ and the iteration is complete. Otherwise, α is halved, a new β_{j+1}^* is calculated, and the process is repeated. Convergence is declared when $\alpha|\beta_{j+1,m}| \leq \epsilon(|\beta_{jm}| + \tau)$ for all $m = 1 \ldots k$. `nl` uses $\tau = 10^{-3}$ and, by default, $\epsilon = 10^{-5}$, though you can specify an alternative value of ϵ with the `eps()` option.

As derived, for example, in Davidson and MacKinnon (2004, chapter 6), an expedient way to obtain the covariance matrix is to compute $\mathbf{u}$ and the columns of $\mathbf{X}$ at the final estimate $\widehat{\beta}$, and then regress that $\mathbf{u}$ on $\mathbf{X}$. The covariance matrix of the estimated parameters of that regression serves as an estimate of $\mathrm{Var}(\widehat{\beta})$. A robust covariance matrix can be estimated by passing the `robust`, `cluster()`, `hc2`, or `hc3` options to `regress` when performing that regression.

All other statistics are calculated analogously to those in linear regression, except that the nonlinear function $f(\mathbf{x}_i, \beta)$ plays the role of the linear function $\mathbf{x}_i'\beta$. See [R] **regress**.

Acknowledgments

The original version of `nl` was written by Patrick Royston of the MRC Clinical Trials Unit, London, and published in Royston (1992). Francesco Danuso's menu-driven nonlinear regression program (1991) provided the inspiration.

References

Atkinson, A. C. 1985. *Plots, Transformations and Regression.* Oxford: Oxford University Press.

Danuso, F. 1991. sg1: Nonlinear regression command. *Stata Technical Bulletin* 1: 17–19. Reprinted in *Stata Technical Bulletin Reprints*, vol. 1, pp. 96–98.

Davidson, R. and J. G. MacKinnon. 1993. *Estimation and Inference in Econometrics.* New York: Oxford University Press.

——. 2004. *Econometric Theory and Methods.* New York: Oxford University Press.

Gallant, A. R. 1987. *Nonlinear Statistical Models.* New York: Wiley.

Goldstein, R. 1992. srd7: Adjusted summary statistics for logarithmic regressions. *Stata Technical Bulletin* 5: 17–21. Reprinted in *Stata Technical Bulletin Reprints*, vol. 1, pp. 178–183.

Kennedy, W. J., Jr., and J. E. Gentle. 1980. *Statistical Computing.* New York: Dekker.

Ratkowsky, D. A. 1983. *Nonlinear Regression Modeling.* New York: Dekker.

Ross, G. J. S. 1987. *MLP User Manual, release 3.08.* Oxford: Numerical Algorithms Group.

——. 1990. *Nonlinear Estimation.* New York: Springer.

Royston, P. 1992. sg1.2: Nonlinear regression command. *Stata Technical Bulletin* 7: 11–18. Reprinted in *Stata Technical Bulletin Reprints*, vol. 2, pp. 112–120.

——. 1993. sg1.4: Standard nonlinear curve fits. *Stata Technical Bulletin* 11: 17. Reprinted in *Stata Technical Bulletin Reprints*, vol. 2, p. 121.

Also See

Complementary: [R] **nl postestimation**; [R] **ml**, [R] **nlcom**

Background: [U] **11.1.10 Prefix commands**,

[U] **20 Estimation and postestimation commands**,

[R] **estimation options**

Title

nl postestimation — Postestimation tools for nl

Description

The following postestimation commands are available for nl:

command	description
estat	VCE and estimation sample summary
estimates	cataloging estimation results
nlcom	point estimates, standard errors, testing, and inference for nonlinear combinations of coefficients
predict	predictions and residuals
predictnl	point estimates, standard errors, testing, and inference for generalized predictions
test	Wald tests for simple and composite linear hypotheses
testnl	Wald tests of nonlinear hypotheses

See the corresponding entries in the *Stata Base Reference Manual* for details.

Syntax for predict

predict [*type*] *newvar* [*if*] [*in*] [, [yhat|residuals]]

These statistics are available both in and out of sample; type predict ... if e(sample) ... if wanted only for the estimation sample.

Options for predict

yhat, the default, calculates the predicted value of *depvar*.

residuals calculates the residuals.

Methods and Formulas

All postestimation commands listed above are implemented as ado-files.

Also See

Complementary: [R] **nl**; [R] **estimates**, [R] **nlcom**, [R] **predictnl**, [R] **test**, [R] **testnl**

Background: [U] **13.5 Accessing coefficients and standard errors**,

[U] **20 Estimation and postestimation commands**,

[R] **estat**, [R] **predict**

Title

nlcom — Nonlinear combinations of estimators

Syntax

Nonlinear combination of estimators—one expression

nlcom [*name:*]*exp* [, *options*]

Nonlinear combinations of estimators—more than one expression

nlcom ([*name:*]*exp*) [([*name:*]*exp*) ...] [, *options*]

The second syntax means that if more than one expression is specified, each must be surrounded by parentheses. *exp* is any function of the parameter estimates that is valid syntax for testnl; see [R] **testnl**. Note, however, that *exp* may not contain an equal sign or a comma. The optional *name* is any valid Stata name and labels the transformations.

options	description
post	post estimation results
<u>level</u>(#)	set confidence level; default is level(95)
<u>iterate</u>(#)	maximum number of iterations

Description

nlcom computes point estimates, standard errors, test statistics, significance levels, and confidence intervals for (possibly) nonlinear combinations of parameter estimates after any Stata estimation command. Results are displayed in the usual table format used for displaying estimation results. Calculations are based on the "delta method", an approximation appropriate in large samples.

nlcom supports svy estimation commands (svy: regress, svy: logit, etc.).

Options

post causes nlcom to behave like a Stata estimation (eclass) command. When post is specified, nlcom will post the vector of transformed estimators and its estimated variance–covariance matrix to e(). This option, in essence, makes the transformation permanent. Thus you could, after posting, treat the transformed estimation results in the same way as you would treat results from other Stata estimation commands. For example, after posting, you could redisplay the results by typing nlcom without any arguments, or use test to perform simultaneous tests of hypotheses on linear combinations of the transformed estimators.

Note that specifying post clears out the previous estimation results, which can be recovered only by refitting the original model or by storing the estimation results before running nlcom and then restoring them; see [R] **estimates**.

level(#) specifies the confidence level, as a percentage, for confidence intervals. The default is level(95) or as set by set level; see [U] **20.6 Specifying the width of confidence intervals**.

`iterate(#)` specifies the maximum number of iterations used to find the optimal step size in calculating numerical derivatives of the transformation(s) with respect to the original parameters. By default, the maximum number of iterations is 100, but convergence is usually achieved after only a few iterations. You should rarely have to use this option.

Remarks

Remarks are presented under the headings

> *Introduction*
> *Basics*
> *Using the post option*
> *Reparameterizing ML estimators for univariate data*
> *nlcom versus eform*

Introduction

`nlcom` and `predictnl` are Stata's "delta method" commands—they take nonlinear transformations of the estimated parameter vector from some fitted model and apply the delta method to calculate the variance, standard error, Wald test statistic, etc., of the transformations. `nlcom` is designed for functions of the parameters, and `predictnl` is designed for functions of the parameters and of the data, that is, for predictions.

`nlcom` generalizes `lincom` (see [R] **lincom**) in two ways. First, `nlcom` allows the transformations to be nonlinear. Second, `nlcom` can be used to simultaneously estimate many transformations (whether linear or nonlinear) and to obtain the estimated variance–covariance matrix of these transformations.

Basics

In [R] **lincom**, the following regression was performed:

```
. use http://www.stata-press.com/data/r9/regress
. regress y x1 x2 x3
```

Source	SS	df	MS		Number of obs =	148
					F(3, 144) =	96.12
Model	3259.3561	3	1086.45203		Prob > F =	0.0000
Residual	1627.56282	144	11.3025196		R-squared =	0.6670
					Adj R-squared =	0.6600
Total	4886.91892	147	33.2443464		Root MSE =	3.3619

y	Coef.	Std. Err.	t	P>|t|	[95% Conf. Interval]	
x1	1.457113	1.07461	1.36	0.177	-.666934	3.581161
x2	2.221682	.8610358	2.58	0.011	.5197797	3.923583
x3	-.006139	.0005543	-11.08	0.000	-.0072345	-.0050435
_cons	36.10135	4.382693	8.24	0.000	27.43863	44.76407

Then `lincom` was used to estimate the difference between the coefficients of `x1` and `x2`:

```
. lincom _b[x2] - _b[x1]
 ( 1)  - x1 + x2 = 0
```

y	Coef.	Std. Err.	t	P>|t|	[95% Conf. Interval]	
(1)	.7645682	.9950282	0.77	0.444	-1.20218	2.731316

It was noted, however, that nonlinear expressions are not allowed with lincom:

```
. lincom _b[x2]/_b[x1]
not possible with test
r(131);
```

Nonlinear transformations are instead estimated using nlcom:

```
. nlcom _b[x2]/_b[x1]
      _nl_1:  _b[x2]/_b[x1]
```

y	Coef.	Std. Err.	t	P>|t|	[95% Conf. Interval]	
_nl_1	1.524714	.9812848	1.55	0.122	-.4148688	3.464297

❑ Technical Note

The notation _b[*name*] is the standard way in Stata to refer to regression coefficients; see
[U] **13.5 Accessing coefficients and standard errors**. Some commands, such as lincom and test,
allow you to drop the _b[] and just refer to the coefficients by *name*. nlcom, however, requires the
full specification _b[*name*].   ❑

Returning to our linear regression example, nlcom also allows simultaneous estimation of more
than one combination:

```
. nlcom (_b[x2]/_b[x1]) (_b[x3]/_b[x1]) (_b[x3]/_b[x2])
      _nl_1:  _b[x2]/_b[x1]
      _nl_2:  _b[x3]/_b[x1]
      _nl_3:  _b[x3]/_b[x2]
```

y	Coef.	Std. Err.	t	P>|t|	[95% Conf. Interval]	
_nl_1	1.524714	.9812848	1.55	0.122	-.4148688	3.464297
_nl_2	-.0042131	.0033483	-1.26	0.210	-.0108313	.002405
_nl_3	-.0027632	.0010695	-2.58	0.011	-.0048772	-.0006493

We can also label the transformations to produce more informative names in the estimation table:

```
. nlcom (ratio21:_b[x2]/_b[x1]) (ratio31:_b[x3]/_b[x1]) (ratio32:_b[x3]/_b[x2])
      ratio21:  _b[x2]/_b[x1]
      ratio31:  _b[x3]/_b[x1]
      ratio32:  _b[x3]/_b[x2]
```

y	Coef.	Std. Err.	t	P>|t|	[95% Conf. Interval]	
ratio21	1.524714	.9812848	1.55	0.122	-.4148688	3.464297
ratio31	-.0042131	.0033483	-1.26	0.210	-.0108313	.002405
ratio32	-.0027632	.0010695	-2.58	0.011	-.0048772	-.0006493

nlcom saves the vector of estimated combinations and its estimated variance–covariance matrix
in r().

```
. matrix list r(b)

r(b)[1,3]
        ratio21      ratio31      ratio32
c1    1.5247143   -.00421315   -.00276324

. matrix list r(V)

symmetric r(V)[3,3]
              ratio21      ratio31      ratio32
ratio21    .96291982
ratio31   -.00287781    .00001121
ratio32   -.00014234    2.137e-06    1.144e-06
```

Using the post option

When used with the `post` option, `nlcom` saves the estimation vector and variance–covariance matrix in `e()`, making the transformation permanent:

```
. quietly nlcom (ratio21:_b[x2]/_b[x1]) (ratio31:_b[x3]/_b[x1])
> (ratio32:_b[x3]/_b[x2]), post

. matrix list e(b)

e(b)[1,3]
        ratio21      ratio31      ratio32
y1    1.5247143   -.00421315   -.00276324

. matrix list e(V)

symmetric e(V)[3,3]
              ratio21      ratio31      ratio32
ratio21    .96291982
ratio31   -.00287781    .00001121
ratio32   -.00014234    2.137e-06    1.144e-06
```

After posting, we can proceed as if we had just run a Stata estimation (`eclass`) command. For instance, we can replay the results,

```
. nlcom
```

y	Coef.	Std. Err.	t	P>|t|	[95% Conf. Interval]	
ratio21	1.524714	.9812848	1.55	0.122	-.4148688	3.464297
ratio31	-.0042131	.0033483	-1.26	0.210	-.0108313	.002405
ratio32	-.0027632	.0010695	-2.58	0.011	-.0048772	-.0006493

or perform other postestimation tasks in the transformed metric, this time making reference to the new "coefficients":

```
. display _b[ratio31]
-.00421315

. correlate, _coef
              ratio21   ratio31   ratio32
  ratio21 |    1.0000
  ratio31 |   -0.8759    1.0000
  ratio32 |   -0.1356    0.5969    1.0000

. test _b[ratio21] = 1

 ( 1)   ratio21 = 1

       F(  1,   144) =    0.29
            Prob > F =    0.5937
```

We see that testing _b[ratio21]=1 in the transformed metric is equivalent to testing using testnl _b[x2]/_b[x1]=1 in the original metric:

```
. quietly reg y x1 x2 x3
. testnl _b[x2]/_b[x1] = 1
  (1)  _b[x2]/_b[x1] = 1
             F(1, 144) =        0.29
              Prob > F =      0.5937
```

Note that we needed to refit the regression model to recover the original parameter estimates.

❑ Technical Note

In a previous technical note, we mentioned that commands such as lincom and test permit reference to *name* instead of _b[*name*]. This is not the case when lincom and test are used after nlcom, post. In the above, we used

```
. test _b[ratio21] = 1
```

rather than

```
. test ratio21 = 1
```

which would have returned an error. Consider this a limitation of Stata. For the shorthand notation to work, you need a variable named *name* in the data. In nlcom, however, *name* is just a coefficient label that does not necessarily corresponding to any variable in the data.

❑

Reparameterizing ML estimators for univariate data

When run using only a response and no covariates, Stata's maximum-likelihood (ML) estimation commands will produce ML estimates of the parameters of some assumed univariate distribution for the response. The parameterization, however, is usually not one we are used to dealing with in a nonregression setting. In such cases, nlcom can be used to transform the estimation results from a regression model to those from a maximum likelihood estimation of the parameters of a univariate probability distribution in a more familiar metric.

▷ Example 1

Consider the following univariate data on $Y = $ # of traffic accidents at a certain intersection in a given year:

```
. use http://www.stata-press.com/data/r9/trafint, clear
. summarize accidents
```

Variable	Obs	Mean	Std. Dev.	Min	Max
accidents	12	13.83333	14.47778	0	41

A quick glance of the output from summarize leads us to quickly reject the assumption that Y is distributed as Poisson since the estimated variance of Y is much greater than the estimated mean of Y.

Instead, we choose to model the data as univariate negative binomial, of which a common parameterization is

$$\Pr(Y = y) = \frac{\Gamma(r + y)}{\Gamma(r)\Gamma(y + 1)} p^r (1 - p)^y \qquad 0 \le p \le 1, \quad r > 0, \quad y = 0, 1, \dots$$

with

$$E(Y) = \frac{r(1 - p)}{p} \qquad \text{Var}(Y) = \frac{r(1 - p)}{p^2}$$

There exist no closed-form solutions for the maximum likelihood estimates of p and r, yet they may be estimated by the iterative method of Newton–Raphson. One way to get these estimates would be to write our own Newton–Raphson program for the negative binomial. Another way would be to write our own ML evaluator; see [R] **ml**.

The easiest solution, however, would be to use Stata's existing negative binomial ML regression command, nbreg. The only problem with this solution is that nbreg estimates a different parameterization of the negative binomial, but we can worry about that later.

```
. nbreg accidents

Fitting Poisson model:

Iteration 0:   log likelihood = -105.05361
Iteration 1:   log likelihood = -105.05361

Fitting constant-only model:

Iteration 0:   log likelihood = -43.948619
Iteration 1:   log likelihood = -43.891483
Iteration 2:   log likelihood =  -43.89144
Iteration 3:   log likelihood =  -43.89144

Fitting full model:

Iteration 0:   log likelihood =  -43.89144
Iteration 1:   log likelihood =  -43.89144
```

```
Negative binomial regression                Number of obs   =         12
                                            LR chi2(0)      =       0.00
Dispersion     = mean                       Prob > chi2     =          .
Log likelihood = -43.89144                  Pseudo R2       =     0.0000
```

accidents	Coef.	Std. Err.	z	P>|z|	[95% Conf. Interval]	
_cons	2.627081	.3192233	8.23	0.000	2.001415	3.252747
/lnalpha	.1402425	.4187147			-.6804233	.9609083
alpha	1.150553	.4817534			.5064026	2.61407

```
Likelihood-ratio test of alpha=0:   chibar2(01) =  122.32 Prob>=chibar2 = 0.000

. matrix list e(b)

e(b)[1,2]
        accidents:     lnalpha:
            _cons        _cons
y1      2.6270811    .14024253
```

From this output, we see that, when used with univariate data, nbreg estimates a regression intercept, β_0, and the logarithm of some parameter α. This parameterization is useful in regression models: β_0 is the intercept meant to be augmented with other terms of the linear predictor, and α is an overdispersion parameter used for comparison with the Poisson regression model.

However, we need to transform $(\beta_0, \ln \alpha)$ to (p, r). Examining *Methods and Formulas* of [R] **nbreg** reveals the transformation as

$$p = \{1 + \alpha \exp(\beta_0)\}^{-1} \qquad r = \alpha^{-1}$$

which we apply using `nlcom`:

```
. nlcom (p:1/(1 + exp([lnalpha]_b[_cons] + _b[_cons]))) (r:exp(-[lnalpha]_b[_cons]))
         p:  1/(1 + exp([lnalpha]_b[_cons] + _b[_cons]))
         r:  exp(-[lnalpha]_b[_cons])
```

accidents	Coef.	Std. Err.	z	P>|z|	[95% Conf. Interval]	
p	.0591157	.0292857	2.02	0.044	.0017168	.1165146
r	.8691474	.3639248	2.39	0.017	.1558679	1.582427

Given the invariance of maximum likelihood estimators and the properties of the delta method, the above parameter estimates, standard errors, etc. are precisely those we would have obtained had we instead performed the Newton–Raphson optimization in the (p, r) metric.

◁

❑ Technical Note

Note how we referred to the estimate of $\ln \alpha$ in above as `[lnalpha]_b[_cons]`. This is not entirely evident from the output of `nbreg`, which is why we listed the elements of `e(b)` so that we could examine the column labels; see [U] **13.5 Accessing coefficients and standard errors**.

❑

nlcom versus eform

Many Stata estimation commands allow you to display exponentiated regression coefficients, some by default, some optionally. Known as "`eform`" in Stata terminology, this reparameterization serves many uses: it gives odds ratios for logistic models, hazard ratios in survival models, incidence-rate ratios in Poisson models, and relative risk ratios in multinomial logit models, to name a few.

For example, consider the following estimation taken directly from [R] **poisson**:

```
. use http://www.stata-press.com/data/r9/airline
. gen lnN = ln(n)
. poisson injuries XYZowned lnN
Iteration 0:   log likelihood = -22.333875
Iteration 1:   log likelihood = -22.332276
Iteration 2:   log likelihood = -22.332276
```

Poisson regression				Number of obs	=	9
				LR chi2(2)	=	19.15
				Prob > chi2	=	0.0001
Log likelihood = -22.332276				Pseudo R2	=	0.3001

injuries	Coef.	Std. Err.	z	P>|z|	[95% Conf. Interval]	
XYZowned	.6840667	.3895877	1.76	0.079	-.0795111	1.447645
lnN	1.424169	.3725155	3.82	0.000	.6940517	2.154285
_cons	4.863891	.7090501	6.86	0.000	3.474178	6.253603

When we replay results and specify the `irr` (incidence-rate ratios) option,

```
. poisson, irr
Poisson regression                                  Number of obs  =          9
                                                    LR chi2(2)     =      19.15
                                                    Prob > chi2    =     0.0001
Log likelihood = -22.332276                         Pseudo R2      =     0.3001
```

injuries	IRR	Std. Err.	z	P>|z|	[95% Conf. Interval]	
XYZowned	1.981921	.7721322	1.76	0.079	.9235678	4.253085
lnN	4.154402	1.547579	3.82	0.000	2.00181	8.621728

we obtain the exponentiated regression coefficients and their estimated standard errors.

Contrast this with what we obtain if we exponentiate the coefficients manually using `nlcom`:

```
. nlcom (E_XYZowned:exp(_b[XYZowned])) (E_lnN:exp(_b[lnN]))
  E_XYZowned:  exp(_b[XYZowned])
      E_lnN:  exp(_b[lnN])
```

injuries	Coef.	Std. Err.	z	P>|z|	[95% Conf. Interval]	
E_XYZowned	1.981921	.7721322	2.57	0.010	.4685701	3.495273
E_lnN	4.154402	1.547579	2.68	0.007	1.121203	7.187602

There are three things to note when comparing `poisson, irr` (and `eform` in general) with `nlcom`:

1. The exponentiated coefficients and standard errors are identical. This is certainly good news.

2. The Wald test statistic (`z`) and level of significance are different. When using `poisson, irr` and other related `eform` options, the Wald test does not change from what you would have obtained without the `eform` option, and you can see this by comparing both versions of the `poisson` output given previously.

   When you use `eform`, Stata knows that what is usually desired is a test of

$$H_0 : \exp(\beta) = 1$$

and not the uninformative-by-comparison

$$H_0 : \exp(\beta) = 0$$

The test of $H_0 : \exp(\beta) = 1$ is asymptotically equivalent to a test of $H_0 : \beta = 0$, the Wald test in the original metric, but the latter has better small-sample properties. Thus if you specify `eform`, you get a test of $H_0 : \beta = 0$.

`nlcom`, however, is general. It does not attempt to infer the test of greatest interest for a given transformation, and so a test of

$$H_0 : \text{transformed coefficient} = 0$$

is always given, regardless of the transformation.

3. You may be surprised to see that, even though the coefficients and standard errors are identical, the confidence intervals (both 95%) are different.

eform confidence intervals are standard confidence intervals with the endpoints transformed. For example, the confidence interval for the coefficient on lnN is $(0.694, 2.154)$, whereas the confidence interval for the incidence-rate ratio due to lnN is $(\exp(0.694), \exp(2.154)) = (2.002, 8.619)$, which, except for some round-off error, is what we see from the output of poisson, irr. In the case of exponentiated coefficients, confidence intervals based on transform-the-endpoints methodology generally have better small-sample properties than their asymptotically equivalent counterparts.

The transform-the-endpoints method, however, only gives valid coverage when the transformation is monotonic. nlcom uses a more general and asymptotically equivalent method for calculating confidence intervals, as described in *Methods and Formulas*.

Saved Results

nlcom saves in r():

Scalars

r(N)	number of observations
r(df_r)	residual degrees of freedom

Matrices

r(b)	vector of transformed coefficients
r(V)	estimated variance–covariance matrix of the transformed coefficients

If post is specified, nlcom also saves in e():

Scalars

e(N)	number of observations
e(df_r)	residual degrees of freedom
e(N_strata)	number of strata L, if used after svy
e(N_psu)	number of sampled PSUs n, if used after svy

Macros

e(cmd)	nlcom
e(predict)	program used to implement predict
e(properties)	b V

Matrices

e(b)	vector of transformed coefficients
e(V)	estimated variance–covariance matrix of the transformed coefficients
e(V_srs)	simple-random-sampling-without-replacement (co)variance $\widehat{V}_{\text{srswor}}$, if svy
e(V_srswr)	simple-random-sampling-with-replacement (co)variance $\widehat{V}_{\text{srswr}}$, if svy and fpc()
e(V_msp)	misspecification (co)variance $\widehat{V}_{\text{msp}}$, if svy and available

Functions

e(sample)	marks estimation sample

Methods and Formulas

nlcom is implemented as an ado-file.

Given a $1 \times k$ vector of parameter estimates, $\widehat{\boldsymbol{\theta}} = (\widehat{\theta}_1, \ldots, \widehat{\theta}_k)$, consider the estimated p-dimensional transformation

$$g(\widehat{\boldsymbol{\theta}}) = [g_1(\widehat{\boldsymbol{\theta}}), g_2(\widehat{\boldsymbol{\theta}}), \ldots, g_p(\widehat{\boldsymbol{\theta}})]$$

The estimated variance–covariance of $g(\widehat{\boldsymbol{\theta}})$ is given by

$$\widehat{\text{Var}}\left\{g(\widehat{\boldsymbol{\theta}})\right\} = \mathbf{G}\mathbf{V}\mathbf{G}'$$

where $\mathbf{G}$ is the $p \times k$ matrix of derivatives for which

$$\mathbf{G}_{ij} = \left.\frac{\partial g_i(\boldsymbol{\theta})}{\partial \theta_j}\right|_{\boldsymbol{\theta}=\widehat{\boldsymbol{\theta}}} \qquad i = 1,\ldots,p \qquad j = 1,\ldots,k$$

and $\mathbf{V}$ is the estimated variance–covariance matrix of $\widehat{\boldsymbol{\theta}}$. Standard errors are obtained as the square roots of the variances.

The Wald test statistic for testing

$$H_0 : g_i(\boldsymbol{\theta}) = 0$$

versus the two-sided alternative is given by

$$Z_i = \frac{g_i(\widehat{\boldsymbol{\theta}})}{\left[\widehat{\text{Var}}_{ii}\left\{g(\widehat{\boldsymbol{\theta}})\right\}\right]^{1/2}}$$

In cases where the variance–covariance matrix of $\widehat{\boldsymbol{\theta}}$ is an asymptotic covariance matrix, Z_i is approximately distributed as Gaussian. In the case of linear regression, Z_i is taken to be approximately distributed as $t_{1,r}$ where r is the residual degrees of freedom from the original fitted model.

A $(1 - \alpha) \times 100\%$ confidence interval for $g_i(\boldsymbol{\theta})$ is given by

$$g_i(\widehat{\boldsymbol{\theta}}) \pm z_{\alpha/2}\left[\widehat{\text{Var}}_{ii}\left\{g(\widehat{\boldsymbol{\theta}})\right\}\right]^{1/2}$$

for those cases where Z_i is Gaussian and

$$g_i(\widehat{\boldsymbol{\theta}}) \pm t_{\alpha/2,r}\left[\widehat{\text{Var}}_{ii}\left\{g(\widehat{\boldsymbol{\theta}})\right\}\right]^{1/2}$$

for those cases where Z_i is t-distributed. z_p is the $1 - p$ quantile of the standard normal distribution, and $t_{p,r}$ is the $1 - p$ quantile of the t distribution with r degrees of freedom.

References

Feiveson, A. H. 1999. FAQ: What is the delta method, and how is it used to estimate the standard error of a transformed parameter? *http://www.stata.com/support/faqs/stat.*

Gould, W. W. 1996. crc43: Wald test of nonlinear hypotheses after model estimation. *Stata Technical Bulletin* 29: 2–4. Reprinted in *Stata Technical Bulletin Reprints*, vol. 5, pp. 15–18.

Oehlert, G. W. 1992. A note on the delta method. *The American Statistician* 46: 27–29.

Phillips, P. C. and J. Y. Park. 1988. On the formulation of Wald tests of nonlinear restrictions. *Econometrica* 56: 1065–1083.

Also See

Related:	[R] **lincom**, [R] **predictnl**, [R] **test**, [R] **testnl**
Background:	[U] **20 Estimation and postestimation commands**

Title

nlogit — Nested logit regression

Syntax

Nested logit regression

> nlogit *depvar* (*altsetvarB* = *indepvarsB*) [... (*altsetvar2* = *indepvars2*)
>
> (*altsetvar1* = *indepvars1*)] [*if*] [*in*] [*weight*] , group(*varname*) [*options*]

Create variable based on specification of branches

> nlogitgen *newvar* = *varname* (*branchlist*) [, nolog]

Display tree structure

> nlogittree *varlist* [, nolabel]

options	description
Model	
*group(*varname*)	group ID variable
ivconstraints(*string*)	apply specified linear constraints of the inclusive-value parameters
constraints(*constraints*)	apply specified linear constraints
SE/Robust	
vce(*vcetype*)	*vcetype* may be oim, robust, or opg
robust	synonym for vce(robust)
Reporting	
level(#)	set confidence level; default is level(95)
notree	suppress display of tree-structure output
nolabel	suppress value labels in tree-structure output
clogit	report clogit estimates
Max options	
maximize_options	control the maximization process; seldom used

*group(*varname*) is required.

bootstrap, by, jackknife, rolling, statsby, and xi are allowed; see [U] **11.1.10 Prefix commands**.

fweights and iweights are allowed, but are interpreted to apply to groups as a whole and not to individual observations; see [U] **11.1.6 weight**.

See [U] **20 Estimation and postestimation commands** for additional capabilities of estimation commands.

where

depvar	is a dichotomous variable coded as 0 for unselected alternatives and 1 for the selected alternative.
altsetvarB	is a categorical variable that identifies the bottom, or final, set of all alternatives.
indepvarsB	are the attributes of the bottom-level alternatives (absolute or perceived) and possibly interactions of individual attributes with the bottom-level alternatives.
altsetvar2	is a categorical variable that identifies the second-level set of alternatives—these must be mutually exclusive groups of the third-level alternatives.
indepvars2	are the attributes of the second-level alternatives (absolute or perceived) and possibly interactions of individual attributes with the second-level alternatives.
altsetvar1	is a categorical variable that identifies the top- or first-level set of alternatives—these alternatives must be mutually exclusive groups of the second-level alternatives.
indepvars1	are the attributes of the first-level alternatives—either of an alternative alone (absolute) or as the alternative is perceived by the chooser (perceived)—and possibly interactions of individual attributes with the first-level alternatives.

where *branchlist* is   *branch, branch* [*, branch, ...*]

branch is   [*label:*] *outcome* [| *outcome* [| *outcome* ...]]

and *outcome* is either a numerical value of the variable specified as *varname* or the value label associated with that numerical value.

Description

nlogit fits a nested logit model using full maximum likelihood. The model may contain one or more levels. For a single-level model, nlogit fits the same model as clogit; see [R] **clogit**.

nlogitgen generates a new categorical variable based on the specification of the branches. For instance,

```
. nlogitgen type = restaurant(fast: 1 | 2, family: 3 | 4 | 5, fancy: 6 | 7)
```

is equivalent to

```
. gen type = 1 if restaurant == 1 | restaurant == 2
. replace type = 2 if restaurant == 3 | restaurant == 4 | restaurant== 5
. replace type = 3 if restaurant == 6 | restaurant == 7
. label define lb_type 1 fast 2 family 3 fancy
. label value type lb_type
```

nlogittree displays the tree structure based on the *varlist*. Note that the bottom level should be specified first. For instance,

```
. nlogittree restaurant type
```

(Continued on next page)

Options

   Model

group(*varname*) is required; it specifies the identifier variable for the groups.

ivconstraints(*string*) specifies the linear constraints of the inclusive-value parameters. You can constrain inclusive-value parameters to be equal to each other, equal to fixed values, etc. Inclusive-value parameters are referred to by the corresponding level labels; for instance, ivconstraints(fast = family) or ivconstraints(fast=1).

constraints(*constraints*); see [R] **estimation options**.

   SE/Robust

vce(*vcetype*); see [R] *vce_option*.

robust; see [R] **estimation options**. This option is not allowed when the model contains more than three levels.

   Reporting

level(*#*); see [R] **estimation options**.

notree specifies that the tree structure of the nested logit model not be displayed.

nolabel causes the numeric codes rather than the label values to be displayed in the tree structure of the nested logit model.

clogit specifies that the initial values obtained from clogit be displayed.

   Max options

maximize_options: <u>difficult</u>, <u>tech</u>nique(*algorithm_spec*), <u>iter</u>ate(*#*), [<u>no</u>]<u>log</u>, <u>trace</u>, <u>grad</u>ient, showstep, <u>hess</u>ian, <u>shownr</u>tolerance, <u>tol</u>erance(*#*), <u>ltol</u>erance(*#*), <u>gtol</u>erance(*#*), <u>nrtol</u>erance(*#*), <u>nonrtol</u>erance; see [R] **maximize**. These options are seldom used.

Remarks

Remarks are presented under the headings

        Introduction
        Data setup and the tree structure
        Test of the independence of irrelevant alternatives (IIA)
        Estimation
        Inclusive-value parameters

Introduction

nlogit performs full maximum-likelihood estimation for nested logit models. These are models of a decision process that is made in stages (levels) in which the decisions in later stages (levels) are limited by those made in earlier stages (levels). In particular, the decision at each level partitions the choice set into more and more specific alternative sets or groupings of choices.

Let's look at an example and clarify some terminology. The tree structure of a family's decision about where to eat might look something like this:

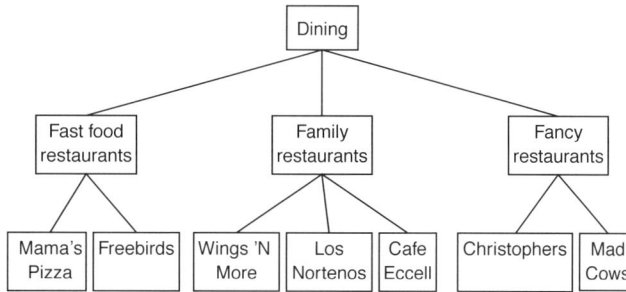

First the family decides whether to eat fast food, eat at a family restaurant, or eat at a fancy restaurant. This first-level decision limits their second-level decision to the alternatives available within the selected restaurant type. If they have chosen fast food, their second-level decision is between Mama's Pizza and Freebirds; if they have chosen a family restaurant, the second-level decision is between Cafe Eccell, Los Nortenos, and Wings 'N More. If they decide on a fancy restaurant, then the second-level decision is between Mad Cows and Christophers.

To be clear, we will use the following terms to describe these models.

level, or decision level, is the level or stage at which a decision is made. First-level decisions are made first, followed by second-level decisions, and so on. In the example above, there are only two levels. In the first level, a type of restaurant is chosen—fast food, family, or fancy—and in the second level, a specific restaurant is chosen.

bottom level is the level where the final decision is made. In our example, this is when we choose a specific restaurant.

alternative set is the set of all possible alternatives at any given decision level.

bottom alternative set is the set of all possible alternatives at the bottom level. This is often referred to as the choice set in the economics-choice literature. In our example, the bottom alternative set is all seven of the specific restaurants.

alternative is a specific alternative within an alternative set. In the first level of our example, "fast food" is an alternative. In the second or bottom level, "MadCows" is an alternative. Not all alternatives within an alternative set are available to someone making a choice at a specific stage, only those that are nested within all prior decisions.

chosen alternative is the alternative from an alternative set that we observe someone having chosen.

A one-level nested logit model is the same as a conditional logit model. The multinomial logit and conditional logit models assume the independence of irrelevant alternatives (IIA), which, basically, means that the relative probabilities of various alternatives remain constant, regardless of which alternatives are included in the model. When this assumption is violated, nested logit may be used. In a nested logit model, alternatives are grouped into subgroups such that the (IIA) assumption is valid merely within each group.

McFadden (1977, 1981) showed how this model can be derived from a rational choice framework. Amemiya (1985, chapter 9) contains a very nice discussion of how this model can be derived under the assumption of utility maximization.

For a two-level nested logit model, we index the first-level alternative as i and the bottom-level alternative as j. Let $\mathbf{x}_{ij}$ and $\mathbf{y}_i$ refer to the vectors of explanatory variables specific to categories (i, j) and (i), respectively. We write

$$\Pr_{ij} = \Pr_{j|i} \Pr_i$$

The conditional probability $\Pr_{j|i}$ will involve only the parameters $\boldsymbol{\beta}$:

$$\Pr_{j|i} = \frac{\exp(\mathbf{x}_{ij}\boldsymbol{\beta})}{\sum_k \exp(\mathbf{x}_{ik}\boldsymbol{\beta})}$$

We define the inclusive values for category (i) as

$$I_i = \ln\left\{\sum_k \exp(\mathbf{x}_{ik}\boldsymbol{\beta})\right\}$$

then

$$\Pr_i = \frac{\exp(\mathbf{y}_i\boldsymbol{\alpha} + \tau_i I_i)}{\sum_m \exp(\mathbf{y}_i\boldsymbol{\alpha} + \tau_m I_m)}$$

Data setup and the tree structure

`nlogitgen` and `nlogittree` are designed to help you specify and display the tree structure of the nested logit model.

▷ Example 1

Using fictional data, we have data on 300 families and their choice of seven local restaurants. The restaurants are Freebirds, MamasPizza, CafeEccell, LosNortenos, WingsNmore, Christophers, and MadCows. We want to explore the relationship of the decision about where to eat to the household income (`income` in 1000s of dollars), the number of kids in the household (`kids`), the rating of the restaurant according to the local restaurant guide (`rating` 0 to 5), the average meal cost per person (`cost`), and the distance between the household and the restaurant (`distance` in miles). `income` and `kids` are attributes of the family, `rating` is an attribute of the alternative (the restaurant), and `cost` and `distance` are attributes of the alternative as perceived by the families—that is, each family has its own cost and distance for each restaurant.

```
. use http://www.stata-press.com/data/r9/restaurant
. describe
Contains data from http://www.stata-press.com/data/r9/restaurant.dta
  obs:         2,100
  vars:            8                          10 Mar 2005 01:17
  size:       75,600 (99.0% of memory free)
```

	storage	display	value	
variable name	type	format	label	variable label
family_id	float	%9.0g		family id
restaurant	float	%12.0g	names	choices of restaurants
income	float	%9.0g		household income
cost	float	%9.0g		average meal cost per person
kids	float	%9.0g		number of kids in the household
rating	float	%9.0g		ratings in local restaurant guide
distance	float	%9.0g		distance between home and restaurant
chosen	float	%9.0g		0 no 1 yes

```
Sorted by:  family_id
```

```
. list family_id restaurant chosen kids rating distance in 1/21, sepby(fam) abbrev(10)
```

	family_id	restaurant	chosen	kids	rating	distance
1.	1	Freebirds	1	1	0	1.245553
2.	1	MamasPizza	0	1	1	2.82493
3.	1	CafeEccell	0	1	2	4.21293
4.	1	LosNortenos	0	1	3	4.167634
5.	1	WingsNmore	0	1	2	6.330531
6.	1	Christophers	0	1	4	10.19829
7.	1	MadCows	0	1	5	5.601388
8.	2	Freebirds	0	3	0	4.162657
9.	2	MamasPizza	0	3	1	2.865081
10.	2	CafeEccell	0	3	2	5.337799
11.	2	LosNortenos	1	3	3	4.282864
12.	2	WingsNmore	0	3	2	8.133914
13.	2	Christophers	0	3	4	8.664631
14.	2	MadCows	0	3	5	9.119597
15.	3	Freebirds	1	3	0	2.112586
16.	3	MamasPizza	0	3	1	2.215329
17.	3	CafeEccell	0	3	2	6.978715
18.	3	LosNortenos	0	3	3	5.117877
19.	3	WingsNmore	0	3	2	5.312941
20.	3	Christophers	0	3	4	9.551273
21.	3	MadCows	0	3	5	5.539806

Suppose that for each family, the decision about where to eat is a decision consisting of two steps. First, the family decides whether to eat fast food, eat at a family restaurant, or eat at a fancy restaurant. This first-level decision limits their second-level decision to the alternatives available within the selected restaurant type. If they have chosen fast food, their second-level decision is between MamasPizza and Freebirds; if they have chosen a family restaurant, the second-level decision is between CafeEccell, LosNortenos, and WingsNmore; if they have chosen a fancy restaurant, the second-level decision is between Christophers and MadCows.

To run nlogit, we need to generate a categorical variable that identifies the first-level set of alternatives: fast food, family restaurants, or fancy restaurants. This can be accomplished easily by using nlogitgen.

```
. nlogitgen type = restaurant(fast: Freebirds | MamasPizza, family: CafeEccell
> | LosNortenos | WingsNmore, fancy: Christophers | MadCows)
new variable type is generated with 3 groups
label list lb_type
lb_type:
          1 fast
          2 family
          3 fancy
```

(Continued on next page)

```
. nlogittree restaurant type
tree structure specified for the nested logit model
        top --> bottom
        type    restaurant
    _____
        fast    Freebirds
                MamasPizza
      family    CafeEccell
                LosNorte~s
                WingsNmore
       fancy    Christop~s
                MadCows
```

The new categorical variable is `type`, which takes value 1 (fast) if `restaurant` is Freebirds or MamasPizza, value 2 (family) if `restaurant` is CafeEccell, LosNortenos, or WingsNmore, and value 3 (fancy) otherwise. `nlogittree` displays the tree structure.

◁

❑ Technical Note

We could also use values instead of value labels of `restaurant` in `nlogitgen`. Value labels are optional, and the default value labels for `type` are `type1`, `type2`, and `type3`. The vertical bar is also optional.

```
. use http://www.stata-press.com/data/r9/restaurant, clear
. nlogitgen type = restaurant(1 2, 3 4 5, 6 7)
new variable type is generated with 3 groups
label list lb_type
lb_type:
           1 type1
           2 type2
           3 type3
. nlogittree restaurant type
tree structure specified for the nested logit model
        top --> bottom
        type    restaurant
    _____
       type1    Freebirds
                MamasPizza
       type2    CafeEccell
                LosNorte~s
                WingsNmore
       type3    Christop~s
                MadCows
```

❑

Test of the independence of irrelevant alternatives (IIA)

The property of the multinomial logit model and conditional logit model that odds ratios are independent of the other alternatives is referred to as the independence of irrelevant alternatives (IIA).

Hausman and McFadden (1984) suggest that if a subset of the choice set truly is irrelevant with respect to the other alternatives, omitting it from the model will not lead to inconsistent estimates. Therefore, Hausman's (1978) specification test can be used to test for IIA.

▷ Example 2

Suppose that we want to run `clogit` on our choice-of-restaurants dataset. We also want to test IIA for the subset of family restaurants against the alternatives of fast food places and fancy restaurants. To do so, we need to use Stata's `hausman` command; see [R] **hausman**.

We first run the estimation on the full bottom alternative set, save the results using `estimates store`, and then run the estimation on the bottom alternative set, excluding the alternatives of family restaurants. We then run the `hausman` test.

```
. gen incFast = (type == 1) * income
. gen incFancy = (type == 3) * income
. gen kidFast = (type == 1) * kids
. gen kidFancy = (type == 3) * kids
. clogit chosen cost rating dist incFast incFancy kidFast kidFancy, group(family_id)
Iteration 0:   log likelihood =  -490.4956
Iteration 1:   log likelihood = -488.91277
Iteration 2:   log likelihood = -488.90834
Iteration 3:   log likelihood = -488.90834
```

```
Conditional (fixed-effects) logistic regression       Number of obs   =       2100
                                                      LR chi2(7)      =     189.73
                                                      Prob > chi2     =     0.0000
Log likelihood = -488.90834                           Pseudo R2       =     0.1625
```

chosen	Coef.	Std. Err.	z	P>|z|	[95% Conf. Interval]	
cost	-.1367799	.0358479	-3.82	0.000	-.2070404	-.0665193
rating	.3066622	.1418291	2.16	0.031	.0286823	.584642
distance	-.1977505	.0471653	-4.19	0.000	-.2901927	-.1053082
incFast	-.0390183	.0094018	-4.15	0.000	-.0574455	-.0205911
incFancy	.0407053	.0080405	5.06	0.000	.0249462	.0564644
kidFast	-.2398757	.1063674	-2.26	0.024	-.448352	-.0313994
kidFancy	-.3893862	.1143797	-3.40	0.001	-.6135662	-.1652061

```
. estimates store fullset
. clogit chosen cost rating dist incFast kidFast if type !=2, group(family_id)
note: 222 groups (888 obs) dropped due to all positive or
      all negative outcomes.
Iteration 0:   log likelihood = -88.608092
Iteration 1:   log likelihood = -85.974978
Iteration 2:   log likelihood = -85.955347
Iteration 3:   log likelihood = -85.955324
Iteration 4:   log likelihood = -85.955324
```

```
Conditional (fixed-effects) logistic regression       Number of obs   =        312
                                                      LR chi2(5)      =      44.35
                                                      Prob > chi2     =     0.0000
Log likelihood = -85.955324                           Pseudo R2       =     0.2051
```

chosen	Coef.	Std. Err.	z	P>|z|	[95% Conf. Interval]	
cost	-.0616621	.067852	-0.91	0.363	-.1946496	.0713254
rating	.1659001	.2832041	0.59	0.558	-.3891698	.72097
distance	-.244396	.0995056	-2.46	0.014	-.4394234	-.0493687
incFast	-.0737506	.0177444	-4.16	0.000	-.108529	-.0389721
kidFast	.4105386	.2137051	1.92	0.055	-.0083157	.8293928

```
. hausman . fullset
```

	(b)	(B)	(b-B)	sqrt(diag(V_b-V_B))
	.	fullset	Difference	S.E.
cost	-.0616621	-.1367799	.0751178	.0576092
rating	.1659001	.3066622	-.1407621	.2451308
distance	-.244396	-.1977505	-.0466456	.0876173
incFast	-.0737506	-.0390183	-.0347323	.015049
kidFast	.4105386	-.2398757	.6504143	.1853533

```
                        b = consistent under Ho and Ha; obtained from clogit
            B = inconsistent under Ha, efficient under Ho; obtained from clogit
    Test:  Ho:  difference in coefficients not systematic
            chi2(5) = (b-B)'[(V_b-V_B)^(-1)](b-B)
                    =       10.70
            Prob>chi2 =     0.0577
            (V_b-V_B is not positive definite)
```

The small p-value indicates that the IIA assumption between the family restaurants' subset and the alternatives of other restaurants is weak, hinting that the more complex nested logit model should be utilized.

◁

Estimation

▷ Example 3

Let's examine how alternative-specific attributes apply to the bottom alternative set (all seven of the specific restaurants), and how family-specific attributes apply to the alternative set at the first decision level (all three types of restaurants).

```
. use http://www.stata-press.com/data/r9/restaurant, clear
. nlogitgen type = restaurant(fast: Freebirds | MamasPizza, family: CafeEccell
> | LosNortenos | WingsNmore, fancy: Christophers | MadCows)
new variable type is generated with 3 groups
label list lb_type
lb_type:
          1 fast
          2 family
          3 fancy
. gen incFast = (type == 1)*income
. gen incFancy = (type == 3)*income
. gen kidFast = (type == 1)*kids
. gen kidFancy = (type == 3)*kids
```

```
. nlogit chosen (restaurant = cost rating distance) (type = incFast incFancy
> kidFast kidFancy), group(family_id) nolog
tree structure specified for the nested logit model
        top --> bottom
        type    restaurant

        fast    Freebirds
                MamasPizza
      family    CafeEccell
                LosNorte~s
                WingsNmore
       fancy    Christop~s
                MadCows
```

```
Nested logit regression
Levels            =         2              Number of obs    =      2100
Dependent variable =     chosen           LR chi2(10)      =  199.6293
Log likelihood     = -483.9584            Prob > chi2      =    0.0000
```

	Coef.	Std. Err.	z	P>|z|	[95% Conf. Interval]
restaurant					
cost	-.0944352	.03402	-2.78	0.006	-.1611131 -.0277572
rating	.1793759	.126895	1.41	0.157	-.0693338 .4280855
distance	-.1745797	.0433352	-4.03	0.000	-.2595152 -.0896443
type					
incFast	-.0287502	.0116242	-2.47	0.013	-.0515332 -.0059672
incFancy	.0458373	.0089109	5.14	0.000	.0283722 .0633024
kidFast	-.0704164	.1394359	-0.51	0.614	-.3437058 .2028729
kidFancy	-.3626381	.1171277	-3.10	0.002	-.5922041 -.1330721
(incl. value parameters)					
type					
/fast	5.715758	2.332871	2.45	0.014	1.143415 10.2881
/family	1.721222	1.152002	1.49	0.135	-.5366608 3.979105
/fancy	1.466588	.4169075	3.52	0.000	.6494642 2.283711

```
LR test of homoskedasticity (iv = 1): chi2(3)=      9.90    Prob > chi2 = 0.0194
```

In this model,

$$\Pr(\text{restaurant} \mid \text{type}) = \Pr(\beta_{\text{cost}} \text{ cost} + \beta_{\text{rating}} \text{ rating} + \beta_{\text{dist}} \text{ distance})$$

$$\Pr(\text{type}) = \Pr(\alpha_{\text{iFast}} \text{ incFast} + \alpha_{\text{iFancy}} \text{ incFancy} + \alpha_{\text{kFast}} \text{ kidFast} + \alpha_{\text{kFast}} \text{ kidFast}$$

$$+ \tau_{\text{fast}} \text{ IV}_{\text{fast}} + \tau_{\text{family}} \text{ IV}_{\text{family}} + \tau_{\text{fancy}} \text{ IV}_{\text{fancy}})$$

The LR test against the constant-only model indicates that the model is significant (p-value $= 0.000$). The inclusive-value parameters for `fast`, `family`, and `fancy` are 5.715758, 1.721222, and 1.466588, respectively. The LR test reported at the bottom of the table is a test for the nesting (heteroskedasticity) against the null hypothesis of homoskedasticity. Computationally, it is the comparison of the likelihood of a non-nested `clogit` model against the nested logit model likelihood. The χ^2 value of 9.90 clearly supports the use of the nested logit model with these data.

◁

Inclusive-value parameters

nlogit allows you to apply linear constraints of the inclusive-value parameters. You can constrain inclusive-value parameters to, say, be equal to each other, or specify fixed values rather than allowing these parameters to be freely estimated.

▷ Example 4

Continuing with the example above, we fix the three inclusive-value parameters to be 1 to recover the model fitted by clogit.

```
. nlogit chosen (restaurant = cost rating distance) (type = incFast incFancy
> kidFast kidFancy), group(family_id) ivc(fast=1, family=1, fancy=1) notree nolog
User-defined constraints:
    IV constraints:
        [fast]_cons = 1
        [family]_cons = 1
        [fancy]_cons = 1
Nested logit regression
Levels           =           2            Number of obs   =       2100
Dependent variable =       chosen         LR chi2(7)      =   189.7294
Log likelihood   = -488.90834             Prob > chi2     =     0.0000
```

	Coef.	Std. Err.	z	P>|z|	[95% Conf. Interval]	
restaurant						
cost	-.1367799	.0358479	-3.82	0.000	-.2070404	-.0665193
rating	.3066626	.1418291	2.16	0.031	.0286827	.5846424
distance	-.1977508	.0471653	-4.19	0.000	-.2901931	-.1053085
type						
incFast	-.0390182	.0094018	-4.15	0.000	-.0574454	-.020591
incFancy	.0407053	.0080405	5.06	0.000	.0249462	.0564644
kidFast	-.2398756	.1063674	-2.26	0.024	-.4483517	-.0313994
kidFancy	-.3893868	.1143797	-3.40	0.001	-.6135669	-.1652067
(incl. value parameters) type						
/fast	1	.	.	.	.	.
/family	1	.	.	.	.	.
/fancy	1	.	.	.	.	.

```
LR test of homoskedasticity (iv = 1): chi2(0)=   -0.00    Prob > chi2 =       .
```

```
. clogit chosen cost rating distance incFast incFancy kidFast kidFancy,
> group(family_id) nolog
```

```
Conditional (fixed-effects) logistic regression        Number of obs   =        2100
                                                       LR chi2(7)      =      189.73
                                                       Prob > chi2     =      0.0000
Log likelihood = -488.90834                            Pseudo R2       =      0.1625
```

chosen	Coef.	Std. Err.	z	P>|z|	[95% Conf. Interval]	
cost	-.1367799	.0358479	-3.82	0.000	-.2070404	-.0665193
rating	.3066622	.1418291	2.16	0.031	.0286823	.584642
distance	-.1977505	.0471653	-4.19	0.000	-.2901927	-.1053082
incFast	-.0390183	.0094018	-4.15	0.000	-.0574455	-.0205911
incFancy	.0407053	.0080405	5.06	0.000	.0249462	.0564644
kidFast	-.2398757	.1063674	-2.26	0.024	-.448352	-.0313994
kidFancy	-.3893862	.1143797	-3.40	0.001	-.6135662	-.1652061

◁

Saved Results

nlogit saves in e():

Scalars

e(N)	number of observations	e(levels)	depth of the model
e(k_eq)	number of equations	e(chi2)	χ^2
e(N_g)	number of groups	e(chi2_c)	χ^2 for comparison test
e(df_m)	model degrees of freedom	e(p)	p-value for χ^2 test
e(df_mc)	model degrees of freedom for clogit	e(p_c)	p-value for comparison test
		e(rank)	rank of e(V)
e(ll)	log likelihood	e(ic)	number of iterations
e(ll_0)	log likelihood, constant-only model	e(rc)	return code
e(ll_c)	log likelihood, clogit model	e(converged)	1 if converged, 0 otherwise

Macros

e(cmd)	nlogit	e(vce)	*vcetype* specified in vce()
e(depvar)	name of dependent variable	e(vcetype)	title used to label Std. Err.
e(group)	name of group() variable	e(opt)	type of optimization
e(level#)	*altsetvar#*	e(ml_method)	type of ml method
e(wtype)	weight type	e(user)	likelihood-evaluator program
e(wexp)	weight expression	e(technique)	maximization technique
e(title)	title in estimation output	e(crittype)	optimization criterion
e(chi2type)	LR; type of model χ^2 test	e(properties)	b V
e(iv_names)	inclusive-value parameters	e(predict)	program used to implement predict

Matrices

e(b)	coefficient vector	e(V)	variance–covariance matrix of the estimators
e(ilog)	iteration log (up to 20 iterations)		
e(n_alters)	number of inclusive parameters for the first and subsequent choice levels, excluding the bottom level	e(gradient)	gradient vector

Functions

e(sample)	marks estimation sample

Methods and Formulas

`nlogit` is implemented as an ado-file. Greene (2003, 725–727) and Maddala (1983, 67–70) provide introductions to the nested logit model.

We will present the methods and formulas for a three-level nested logit model. The extension of this model to cases involving more levels is apparent but is more complicated.

Following Greene (2003), we index the first-level alternative as i, the second-level alternative as j, and the bottom-level alternative as k. Let $\mathbf{x}_{ijk}$, $\mathbf{y}_{ij}$ and $\mathbf{z}_i$ refer to row vectors of explanatory variables specific to categories (i,j,k), (i,j), and (i), respectively. We write

$$\Pr_{ijk} = \Pr_{k|ij} \Pr_{j|i} \Pr_i$$

The conditional probability $\Pr_{k|ij}$ will involve only the parameters $\boldsymbol{\beta}$:

$$\Pr_{k|ij} = \frac{\exp(\mathbf{x}_{ijk}\boldsymbol{\beta})}{\sum_n \exp(\mathbf{x}_{ijn}\boldsymbol{\beta})}$$

We define the inclusive values for category (i,j) as

$$I_{ij} = \ln\left\{\sum_n \exp(\mathbf{x}_{ijn}\boldsymbol{\beta})\right\}$$

and

$$\Pr_{j|i} = \frac{\exp(\mathbf{y}_{ij}\boldsymbol{\alpha} + \tau_{ij}I_{ij})}{\sum_m \exp(\mathbf{y}_{im}\boldsymbol{\alpha} + \tau_{im}I_{im})}$$

We define inclusive values for category (i) as

$$J_i = \ln\left\{\sum_m \exp(\mathbf{y}_{im}\boldsymbol{\alpha} + \tau_{im}I_{im})\right\}$$

then

$$\Pr_i = \frac{\exp(\mathbf{z}_i\boldsymbol{\gamma} + \delta_i J_i)}{\sum_l \exp(\mathbf{z}_l\boldsymbol{\gamma} + \delta_l J_l)}$$

If we restrict all the τ_{ij} and δ_i to be 1, we recover the conditional logit model of the following form:

$$\Pr_{ijk} = \frac{\exp(V_{ijk})}{\sum_l \sum_m \sum_n \exp(V_{ijk})}$$

where

$$V_{ijk} = \mathbf{x}_{ijk}\boldsymbol{\beta} + \mathbf{y}_{ij}\boldsymbol{\alpha} + \mathbf{z}_i\boldsymbol{\gamma}$$

There are two ways to fit the nested logit model: sequential estimation and full-information maximum-likelihood estimation. `nlogit` fits the model using the full-information maximum-likelihood method. If $g = 1, 2, \ldots, G$ denotes the groups, and $\Pr_{ijk}^g$ is the probability of category (i, j, k) being a positive outcome in group g, the log likelihood of the nested logit model is

$$
\ln L = \sum_g \ln\left(\Pr_{ijk}^g\right)
$$

$$
= \sum_g \left(\ln \Pr_{k|ij}^g + \ln \Pr_{j|i}^g + \ln \Pr_i^g \right)
$$

References

Amemiya, T. 1985. *Advanced Econometrics*. Cambridge, MA: Harvard University Press.

Greene, W. H. 2003. *Econometric Analysis*. 5th ed. Upper Saddle River, NJ: Prentice Hall.

Hausman, J. 1978. Specification tests in econometrics. *Econometrica* 46: 1251–1271.

Hausman, J. and D. McFadden. 1984. Specification tests in econometrics. *Econometrica* 52: 1219–1240.

Heiss, F. 2002. Structural choice analysis with nested logit models. *Stata Journal* 2: 227–252.

Maddala, G. S. 1983. *Limited-Dependent and Qualitative Variables in Econometrics*. Cambridge: Cambridge University Press.

McFadden, D. 1977. Quantitative methods for analyzing behavior of individuals: Some recent developments. Cowles Foundation Discussion Paper no. 474.

——. 1981. Econometric models of probabilistic choice. In *Structural Analysis of Discrete Data with Econometric Applications*, pp. 198–272. Cambridge, MA: MIT Press.

Also See

Complementary:	[R] **nlogit postestimation**; [R] **constraint**
Related:	[R] **clogit**, [R] **mlogit**, [R] **ologit**
Background:	[U] **11.1.10 Prefix commands**,
	[U] **20 Estimation and postestimation commands**,
	[R] **estimation options**, [R] **maximize**, [R] *vce_option*

Title

nlogit postestimation — Postestimation tools for nlogit

Description

The following postestimation commands are available for `nlogit`:

command	description
estat	AIC, BIC, VCE, and estimation sample summary
estimates	cataloging estimation results
hausman	Hausman's specification test
lincom	point estimates, standard errors, testing, and inference for linear combinations of coefficients
lrtest	likelihood-ratio test
mfx	marginal effects or elasticities
nlcom	point estimates, standard errors, testing, and inference for nonlinear combinations of coefficients
predict	predictions, residuals, influence statistics, and other diagnostic measures
predictnl	point estimates, standard errors, testing, and inference for generalized predictions
test	Wald tests for simple and composite linear hypotheses
testnl	Wald tests of nonlinear hypotheses

See the corresponding entries in the *Stata Base Reference Manual* for details.

(*Continued on next page*)

Syntax for predict

> predict [*type*] *newvar* [*if*] [*in*] [, *statistic*]

statistic	description
pb	predicted probability of choosing bottom-level, or choice-set, alternatives; each alternative identified by *altsetvarB*; the default
p1	predicted probability of choosing first-level alternatives; each alternative identified by *altsetvar1*
p2	predicted probability of choosing second-level alternatives; each choice identified by *altsetvar2*
...	
p#	predicted probability of choosing #-level alternatives; each alternative identified by *altsetvar#*
xbb	linear prediction for the bottom-level alternatives
xb1	linear prediction for the first-level alternatives
xb2	linear prediction for the second-level alternatives
...	
xb#	linear prediction for the #-level alternatives
condpb	Pr(each bottom alternative | alternative is available after all earlier choices)
condp1	Pr(each level 1 alternative) = p1
condp2	Pr(each level 2 alternative | alternative is available after level 1 decision)
...	
condp#	Pr(each level # alternative | alternative is available after all earlier choices)
ivb	inclusive value for the bottom-level alternatives
iv2	inclusive value for the second-level alternatives
...	
iv#	inclusive value for the #-level alternatives

The inclusive value for the first-level alternatives is not used in estimation; therefore, it is not calculated.

These statistics are available both in and out of sample; type predict ... if e(sample) ... if wanted only for the estimation sample.

Options for predict

Consider a nested logit model with 3 levels: $\Pr(ijk) = \Pr(k \mid ij)\Pr(j \mid i)\Pr(i)$, then

pb, the default, calculates the probability of choosing bottom-level alternatives; $pb = \Pr(ijk)$.

p1 calculates the probability of choosing first-level alternatives; $p1 = \Pr(i)$.

p2 calculates the probability of choosing second-level alternatives; $p2 = \Pr(ij) = \Pr(j \mid i)\Pr(i)$.

xbb calculates the linear prediction for the bottom-level alternatives.

xb1 calculates the linear prediction for the first-level alternatives.

xb2 calculates the linear prediction for the second-level alternatives.

condpb = $\Pr(k \mid ij)$.

condp1 = $\Pr(i)$.

condp2 = $\Pr(j \mid i)$.

ivb calculates the inclusive value for the bottom-level alternatives: ivb = $\ln\{\sum_{k \mid ij} \exp(xb_b)\}$, where xb_b is the linear prediction for the bottom-level alternatives.

iv2 calculates the inclusive value for the second-level alternatives:
iv2 = $\ln\{\sum_{j \mid i} \exp(xb_2 + \tau_j iv_b)\}$, where xb_2 is the linear prediction for the second-level alternatives, iv_b is the inclusive value for the bottom-level alternatives, and τ_j are the parameters for the inclusive value.

Remarks

predict may be used after nlogit to obtain the predicted values of the probabilities, the conditional probabilities, the linear predictions, and the inclusive values for each level of the nested logit model. Predicted probabilities for nlogit must be interpreted carefully. Probabilities are estimated for each group as a whole and not for individual observations.

▷ Example 1

Continuing with our model in example 3 of [R] **nlogit**, we can predict pb = $\Pr(\text{restaurant})$; p1 = $\Pr(\text{type})$; condpb = $\Pr(\text{restaurant} \mid \text{type})$; xbb, the linear prediction for the bottom-level alternatives; xb1, the linear prediction for the first-level alternatives; and ivb, the inclusive values for the bottom-level alternatives.

```
. quietly nlogit chosen (restaurant = cost rating distance ) (type = incFast
> incFancy kidFast kidFancy), group(family_id) nolog
. predict pb
(option pb assumed; Pr(restaurant))
. predict p1, p1
. predict condpb, condpb
. predict xbb, xbb
. predict xb1, xb1
. predict ivb, ivb
. list family_id chosen pb p1 condpb in 1/14, sep(7) divider abbrev(9)
```

	family_id	chosen	pb	p1	condpb
1.	1	1	.0831245	.1534534	.5416919
2.	1	0	.070329	.1534534	.4583081
3.	1	0	.2763391	.7266538	.3802899
4.	1	0	.284375	.7266538	.3913486
5.	1	0	.1659397	.7266538	.2283615
6.	1	0	.0399215	.1198928	.3329766
7.	1	0	.0799713	.1198928	.6670234
8.	2	0	.01176	.0286579	.4103599
9.	2	0	.0168978	.0286579	.5896401
10.	2	0	.2942401	.7521651	.3911909
11.	2	1	.2975767	.7521651	.3956268
12.	2	0	.1603483	.7521651	.2131824
13.	2	0	.1277234	.219177	.582741
14.	2	0	.0914536	.219177	.417259

. list family_id chosen xbb xb1 ivb in 1/14, sep(7) divider abbrev(9)

	family_id	chosen	xbb	xb1	ivb
1.	1	1	-.731619	-1.191674	-.1185611
2.	1	0	-.8987747	-1.191674	-.1185611
3.	1	0	-1.149417	0	-.1825957
4.	1	0	-1.120752	0	-.1825957
5.	1	0	-1.659421	0	-.1825957
6.	1	0	-3.514237	1.425016	-2.414554
7.	1	0	-2.819484	1.425016	-2.414554
8.	2	0	-1.22427	-1.878761	-.3335493
9.	2	0	-.8617923	-1.878761	-.3335493
10.	2	0	-1.239346	0	-.3007865
11.	2	1	-1.22807	0	-.3007865
12.	2	0	-1.846394	0	-.3007865
13.	2	0	-2.804756	1.570648	-2.264743
14.	2	0	-3.138791	1.570648	-2.264743

◁

Methods and Formulas

All postestimation commands listed above are implemented as ado-files.

Also See

Complementary: [R] **nlogit**; [R] **estimates**, [R] **hausman**, [R] **lincom**, [R] **lrtest**,
[R] **mfx**, [R] **nlcom**, [R] **predictnl**, [R] **test**, [R] **testnl**

Background: [U] **13.5 Accessing coefficients and standard errors**,
[U] **20 Estimation and postestimation commands**,
[R] **estat**, [R] **predict**

Title

nptrend — Test for trend across ordered groups

Syntax

nptrend *varname* $\left[\,if\,\right]$ $\left[\,in\,\right]$, by(*groupvar*) $\left[\,\underline{no}detail\ \underline{s}core(scorevar)\,\right]$

Description

nptrend performs a nonparametric test for trend across ordered groups.

Options

⌐ Main ⌐

by(*groupvar*) is required; it specifies the group on which the data is to be ordered.

nodetail suppresses the listing of group rank sums.

score(*scorevar*) defines scores for groups. When it is not specified, the values of *groupvar* are used for the scores.

Remarks

nptrend performs the nonparametric test for trend across ordered groups developed by Cuzick (1985), which is an extension of the Wilcoxon rank-sum test (ranksum; see [R] **signrank**). A correction for ties is incorporated into the test. nptrend is a useful adjunct to the Kruskal–Wallis test; see [R] **kwallis**.

In addition to nptrend, the signtest and spearman commands can be useful for nongrouped data; see [R] **signrank** and [R] **spearman**. The Cox and Stuart test, for instance, applies the sign test to differences between equally spaced observations of *varname*. The Daniels test calculates Spearman's rank correlation of *varname* with a time index. Under appropriate conditions, the Daniels test is more powerful than the Cox and Stuart test. See Conover (1999) for a discussion of these tests and their asymptotic relative efficiency.

▷ Example 1

The following data (Altman 1991, 217) show ocular exposure to ultraviolet radiation for 32 pairs of sunglasses classified into three groups according to the amount of visible light transmitted.

Group	Transmission of visible light	Ocular exposure to ultraviolet radiation
1	< 25%	1.4 1.4 1.4 1.6 2.3 2.3
2	25 to 35%	0.9 1.0 1.1 1.1 1.2 1.2 1.5 1.9 2.2 2.6 2.6 2.6 2.8 2.8 3.2 3.5 4.3 5.1
3	> 35%	0.8 1.7 1.7 1.7 3.4 7.1 8.9 13.5

Entering these data into Stata, we have

```
. use http://www.stata-press.com/data/r9/sg
. list, sep(6)
```

	group	exposure
1.	1	1.4
2.	1	1.4
3.	1	1.4
4.	1	1.6
5.	1	2.3
6.	1	2.3
7.	2	.9
	(output omitted)	
31.	3	8.9
32.	3	13.5

We use `nptrend` to test for a trend of (increasing) exposure across the three groups by typing

```
. nptrend exposure, by(group)
     group      score       obs      sum of ranks
         1          1         6                76
         2          2        18               290
         3          3         8               162

         z =   1.52
  Prob > |z| = 0.129
```

When the groups are given any equally spaced scores (such as -1, 0, 1), we will obtain the same answer as above. To illustrate the effect of changing scores, an analysis of these data with scores 1, 2, and 5 (admittedly not very sensible in this case) produces

```
. gen mysc = cond(group==3,5,group)
. nptrend exposure, by(group) score(mysc)
     group      score       obs      sum of ranks
         1          1         6                76
         2          2        18               290
         3          5         8               162

         z =   1.46
  Prob > |z| = 0.143
```

This example suggests that the analysis is not all that sensitive to the scores chosen.

◁

❏ Technical Note

The grouping variable may be either a string variable or a numeric variable. If it is a string variable and no score variable is specified, the natural numbers 1, 2, 3, ... are assigned to the groups in the sort order of the string variable. This may not always be what you expect. For example, the sort order of the strings "one", "two", "three" is "one", "three", "two".

❏

Saved Results

nptrend saves in r():

Scalars

r(N)	number of observations	r(z)	z statistic
r(p)	two-sided p-value	r(T)	test statistic

Methods and Formulas

nptrend is implemented as an ado-file.

nptrend is based on a method in Cuzick (1985). The following description of the statistic is from Altman (1991, 215–217). We have k groups of sample sizes n_i ($i = 1, \ldots, k$). The groups are given scores, l_i, which reflect their ordering, such as 1, 2, and 3. The scores do not have to be equally spaced, but they usually are. $N = \sum n_i$ observations are ranked from 1 to N, and the sums of the ranks in each group, R_i, are obtained. L, the weighted sum of all the group scores, is

$$L = \sum_{i=1}^{k} l_i n_i$$

The statistic T is calculated as

$$T = \sum_{i=1}^{k} l_i R_i$$

Under the null hypothesis, the expected value of T is $E(T) = .5(N+1)L$, and its standard error is

$$\mathrm{se}(T) = \sqrt{\frac{n+1}{12}\left(N\sum_{i=1}^{k} l_i^2 n_i - L^2\right)}$$

so that the test statistic, z, is given by $z = \{T - E(T)\}/\mathrm{se}(T)$, which has an approximately standard Normal distribution when the null hypothesis of no trend is true.

The correction for ties affects the standard error of T. Let $\widetilde{N}$ be the number of unique values of the variable being tested ($\widetilde{N} \leq N$), and let t_j be the number of times the jth unique value of the variable appears in the data. Define

$$a = \frac{\sum_{j=1}^{\widetilde{N}} t_j(t_j^2 - 1)}{N(N^2 - 1)}$$

The corrected standard error of T is $\widetilde{\mathrm{se}}(T) = \sqrt{1-a}\ \mathrm{se}(T)$.

Acknowledgments

nptrend was written by K. A. Stepniewska and D. G. Altman (1992) of the Imperial Cancer Research Fund, London.

References

Altman, D. G. 1991. *Practical Statistics for Medical Research*. London: Chapman & Hall.

Conover, W. J. 1999. *Practical Nonparametric Statistics*. 3rd ed. New York: Wiley.

Cuzick, J. 1985. A Wilcoxon-type test for trend. *Statistics in Medicine* 4: 87–90.

Sasieni, P. 1996. snp12: Stratified test for trend across ordered groups. *Stata Technical Bulletin* 33: 24–27. Reprinted in *Stata Technical Bulletin Reprints*, vol. 6, pp. 196–200.

Sasieni, P., K. A. Stepniewska, and D. G. Altman. 1996. snp11: Test for trend across ordered groups revisited. *Stata Technical Bulletin* 32: 27–29. Reprinted in *Stata Technical Bulletin Reprints*, vol. 6, pp. 193–196.

Stepniewska, K. A. and D. G. Altman. 1992. snp4: Nonparametric test for trend across ordered groups. *Stata Technical Bulletin* 9: 21–22. Reprinted in *Stata Technical Bulletin Reprints*, vol. 2, p. 169.

Also See

Related: [R] **kwallis**, [R] **signrank**, [R] **spearman**, [R] **symmetry**,
                    [ST] **epitab**

Title

ologit — Ordered logistic regression

Syntax

$$\underline{\text{olo}}\text{git } \textit{depvar } \left[\textit{indepvars}\right] \left[\textit{if}\right] \left[\textit{in}\right] \left[\textit{weight}\right] \left[, \textit{options}\right]$$

options	description
Model	
<u>off</u>set(*varname*)	include *varname* in model with coefficient constrained to 1
SE/Robust	
vce(*vcetype*)	*vcetype* may be <u>robust</u>, <u>boot</u>strap, or <u>jack</u>knife
<u>r</u>obust	synonym for vce(robust)
<u>cl</u>uster(*varname*)	adjust standard errors for intragroup correlation
Reporting	
<u>l</u>evel(#)	set confidence level; default is level(95)
or	report odds ratios
Max options	
maximize_options	control the maximization process; seldom used

bootstrap, by, jackknife, rolling, statsby, stepwise, svy, and xi are allowed; see
[U] **11.1.10 Prefix commands**.
fweights, iweights, and pweights are allowed; see [U] **11.1.6 weight**.
See [U] **20 Estimation and postestimation commands** for additional capabilities of estimation commands.

Description

ologit fits ordered logit models of ordinal variable *depvar* on the independent variables *indepvars*. The actual values taken on by the dependent variable are irrelevant, except that larger values are assumed to correspond to "higher" outcomes. Up to 50 outcomes are allowed in Stata/SE and Intercooled Stata, and up to 20 outcomes are allowed in Small Stata.

See [R] **logistic** for a list of related estimation commands.

Options

⌐ Model ⌐

offset(*varname*); see [R] **estimation options**.

⌐ SE/Robust ⌐

vce(*vcetype*); see [R] ***vce_option***.

robust, cluster(*varname*); see [R] **estimation options**. cluster() can be used with pweights to produce estimates for unstratified cluster-sampled data, but see [SVY] **svy: ologit** for a command especially designed for survey data.

Reporting

level(*#*); see [R] **estimation options**.

or reports the estimated coefficients transformed to odds ratios, i.e., e^b rather than b. Standard errors and confidence intervals are similarly transformed. This option affects how results are displayed, not how they are estimated. or may be specified at estimation or when replaying previously estimated results.

Max options

maximize_options: <u>iter</u>ate(*#*), [<u>no</u>]log, <u>trace</u>, <u>tol</u>erance(*#*), <u>ltol</u>erance(*#*); see [R] **maximize**. These options are seldom used.

Remarks

Ordered logit models are used to estimate relationships between an ordinal dependent variable and a set of independent variables. An *ordinal* variable is a variable that is categorical and ordered, for instance, "poor", "good", and "excellent", which might indicate a person's current health status or the repair record of a car. If there are only two outcomes, see [R] **logistic**, [R] **logit**, and [R] **probit**. This entry is concerned only with more than two outcomes. If the outcomes cannot be ordered (e.g., residency in the north, east, south, or west), see [R] **mlogit**. This entry is concerned only with models in which the outcomes can be ordered.

In ordered logit, an underlying score is estimated as a linear function of the independent variables and a set of cutpoints. The probability of observing outcome i corresponds to the probability that the estimated linear function, plus random error, is within the range of the cutpoints estimated for the outcome:

$$\Pr(\text{outcome}_j = i) = \Pr(\kappa_{i-1} < \beta_1 x_{1j} + \beta_2 x_{2j} + \cdots + \beta_k x_{kj} + u_j \leq \kappa_i)$$

u_j is assumed to be logistically distributed in ordered logit. In either case, we estimate the coefficients $\beta_1, \beta_2, \ldots, \beta_k$ together with the cutpoints $\kappa_1, \kappa_2, \ldots, \kappa_{k-1}$, where k is the number of possible outcomes. κ_0 is taken as $-\infty$, and κ_k is taken as $+\infty$. All of this is a direct generalization of the ordinary two-outcome logit model.

▷ Example 1

We wish to analyze the 1977 repair records of 66 foreign and domestic cars. The data are a variation of the automobile dataset described in [U] **1.2.1 Sample datasets**. The 1977 repair records, like those in 1978, take on values "Poor", "Fair", "Average", "Good", and "Excellent". Here is a cross-tabulation of the data:

```
. use http://www.stata-press.com/data/r9/fullauto
(Automobile Models)
```

```
. tabulate rep77 foreign, chi2
  Repair |
  Record |            Foreign
    1977 |  Domestic     Foreign  |     Total
---------+----------------------+----------
    Poor |         2           1  |         3
    Fair |        10           1  |        11
 Average |        20           7  |        27
    Good |        13           7  |        20
Excellent|         0           5  |         5
---------+----------------------+----------
   Total |        45          21  |        66

          Pearson chi2(4) =  13.8619   Pr = 0.008
```

Although it appears that `foreign` takes on the values "`Domestic`" and "`Foreign`", it is actually a numeric variable taking on the values 0 and 1. Similarly, `rep77` takes on the values 1, 2, 3, 4, and 5, corresponding to "`Poor`", "`Fair`", and so on. The more meaningful words appear because we have attached value labels to the data; see [U] **12.6.3 Value labels**.

Since the chi-squared value is significant, we could claim that there is a relationship between `foreign` and `rep77`. Literally, however, we can only claim that the distributions are different; the chi-squared test is not directional. One way to model these data is to model the categorization that took place when the data were created. Cars have a true frequency of repair, which we will assume is given by $S_j = \beta \, \mathtt{foreign}_j + u_j$, and a car is categorized as "poor" if $S_j \leq \kappa_0$, as "fair" if $\kappa_0 < S_j \leq \kappa_1$, and so on:

```
. ologit rep77 foreign

Iteration 0:   log likelihood = -89.895098
Iteration 1:   log likelihood = -85.951765
Iteration 2:   log likelihood = -85.908227
Iteration 3:   log likelihood = -85.908161

Ordered logistic regression                    Number of obs   =         66
                                               LR chi2(1)      =       7.97
                                               Prob > chi2     =     0.0047
Log likelihood = -85.908161                    Pseudo R2       =     0.0444
```

rep77	Coef.	Std. Err.	z	P>|z|	[95% Conf. Interval]	
foreign	1.455878	.5308946	2.74	0.006	.4153436	2.496412
/cut1	-2.765562	.5988207			-3.939229	-1.591895
/cut2	-.9963603	.3217704			-1.627019	-.3657019
/cut3	.9426153	.3136396			.3278929	1.557338
/cut4	3.123351	.5423237			2.060416	4.186286

Our model is $S_j = 1.46 \, \mathtt{foreign}_j + u_j$; the expected value for foreign cars is 1.46 and, for domestic cars, 0; foreign cars have better repair records.

The estimated cutpoints tell us how to interpret the score. For a foreign car, the probability of a poor record is the probability that $1.46 + u_j \leq -2.77$, or equivalently, $u_j \leq -4.23$. Making this calculation requires familiarity with the logistic distribution: the probability is $1/(1 + e^{4.23}) = .014$. On the other hand, for domestic cars, the probability of a poor record is the probability $u_j \leq -2.77$, which is .059.

This, it seems to us, is a far more reasonable prediction than we would have made based on the table alone. The table showed that 2 out of 45 domestic cars had poor records, while 1 out of 21 foreign cars had poor records—corresponding to probabilities $2/45 = .044$ and $1/21 = .048$. The predictions from our model imposed a smoothness assumption—foreign cars should not, overall, have better repair records without the difference revealing itself in each category. In our data, the fractions of foreign and domestic cars in the poor category are virtually identical only because of the randomness associated with small samples.

Thus if we were asked to predict the true fractions of foreign and domestic cars that would be classified in the various categories, we would choose the numbers implied by the ordered logit model:

	tabulate		logit	
	Domestic	Foreign	Domestic	Foreign
Poor	.044	.048	.059	.014
Fair	.222	.048	.210	.065
Average	.444	.333	.450	.295
Good	.289	.333	.238	.467
Excellent	.000	.238	.043	.159

See [R] **ologit postestimation** for a more complete explanation of how to generate predictions from an ordered logit model.

◁

❑ Technical Note

In this case, ordered logit provides an alternative to ordinary two-outcome logistic models with an arbitrary dichotomization, which might otherwise have been tempting. We could, for instance, have summarized these data by converting the five-outcome `rep77` variable to a two-outcome variable, combining cars in the average, fair, and poor categories to make one outcome and combining cars in the good and excellent categories to make the second.

Another even less appealing alternative would have been to use ordinary regression, arbitrarily labeling "excellent" as 5, "good" as 4, and so on. The problem is that with different but equally valid labelings (say 10 for "excellent"), we would obtain different estimates. We would have no way of choosing one metric over another. That is not, however, true of `ologit`. The actual values used to label the categories make no difference other than through the order they imply.

In fact, our labeling was 5 for "excellent", 4 for "good", and so on. The words "excellent" and "good" appear in our output because we attached a value label to the variables; see [U] **12.6.3 Value labels**. If we were to now go back and type `replace rep77=10 if rep77==5`, changing all the 5s to 10s, we would still obtain exactly the same results when we refitted our model.

❑

▷ Example 2

In the example above, we used ordered logit as a way to model a table. We are not, however, limited to including only a single explanatory variable or to including only categorical variables. We can explore the relationship of `rep77` with any of the variables in our data. We might, for instance, model `rep77` not only in terms of the origin of manufacture, but also including `length` (a proxy for size) and `mpg`:

(Continued on next page)

```
. ologit rep77 foreign length mpg

Iteration 0:   log likelihood = -89.895098
Iteration 1:   log likelihood = -78.775147
Iteration 2:   log likelihood = -78.256299
Iteration 3:   log likelihood = -78.250722
Iteration 4:   log likelihood = -78.250719

Ordered logistic regression                    Number of obs   =         66
                                               LR chi2(3)      =      23.29
                                               Prob > chi2     =     0.0000
Log likelihood = -78.250719                    Pseudo R2       =     0.1295
```

rep77	Coef.	Std. Err.	z	P>|z|	[95% Conf. Interval]	
foreign	2.896807	.7906411	3.66	0.000	1.347179	4.446435
length	.0828275	.02272	3.65	0.000	.0382972	.1273579
mpg	.2307677	.0704548	3.28	0.001	.0926788	.3688566
/cut1	17.92748	5.551191			7.047344	28.80761
/cut2	19.86506	5.59648			8.896161	30.83396
/cut3	22.10331	5.708935			10.914	33.29262
/cut4	24.69213	5.890754			13.14647	36.2378

foreign still plays a role—and an even larger role than previously. We find that larger cars tend to have better repair records, as do cars with better mileage ratings.

◁

Saved Results

ologit saves in e():

Scalars

e(N)	number of observations	e(ll)	log likelihood
e(k_cat)	number of categories	e(ll_0)	log likelihood, constant-only model
e(df_m)	model degrees of freedom	e(N_clust)	number of clusters
e(r2_p)	pseudo-R-squared	e(chi2)	χ^2
		e(converged)	1 if converged, 0 otherwise

Macros

e(cmd)	ologit	e(chi2type)	Wald or LR; type of model χ^2 test
e(depvar)	name of dependent variable	e(vce)	*vcetype* specified in vce()
e(wtype)	weight type	e(vcetype)	title used to label Std. Err.
e(wexp)	weight expression	e(crittype)	optimization criterion
e(title)	title in estimation output	e(predict)	program used to implement predict
e(clustvar)	name of cluster variable	e(properties)	b V
e(offset)	offset		

Matrices

e(b)	coefficient vector	e(V)	variance–covariance matrix of the
e(cat)	category values		estimators

Functions

e(sample)	marks estimation sample

Methods and Formulas

See Long and Freese (2003, chapter 5) for a discussion of models for ordinal outcomes and examples that use Stata. A straightforward textbook description of the model fit by `ologit`, as well as the models fit by `oprobit`, `clogit`, and `mlogit`, can be found in Greene (2003, chapter 21). When you have a qualitative dependent variable, several estimation procedures are available. A popular choice is multinomial logistic regression (see [R] **mlogit**), but if you use this procedure when the response variable is ordinal, you are discarding information because multinomial logit ignores the ordered aspect of the outcome. Ordered logit and probit models provide a means to exploit the ordering information.

There is more than one "ordered logit" model. The model fitted by `ologit`, which we will call the ordered logit model, is also known as the proportional odds model. Another popular choice, not fitted by `ologit`, is known as the stereotype model; see [R] **slogit**. All ordered logit models have been derived by starting with a binary logit/probit model and generalizing it to allow for more than two outcomes.

The proportional-odds ordered logit model is so called because, if we consider the odds $\text{odds}(k) = P(Y \leq k)/P(Y > k)$, then $\text{odds}(k_1)$ and $\text{odds}(k_2)$ have the same ratio for all independent variable combinations. The model is based on the principle that the only effect of combining adjoining categories in ordered categorical regression problems should be a loss of efficiency in estimating the regression parameters (McCullagh 1980). This model was also described by Zavoina and McKelvey (1975) and, previously, by Aitchison and Silvey (1957) in a different algebraic form. Brant (1990) offers a set of diagnostics for the model.

Peterson and Harrell (1990) suggest a model that allows nonproportional odds for a subset of the explanatory variables. `ologit` does not allow this, but a model similar to this was implemented by Fu (1998).

The stereotype model rejects the principle on which the ordered logit model is based. Anderson (1984) argues that there are two distinct types of ordered categorical variables: "grouped continuous", such as income, where the "type a" model applies; and "assessed", such as extent of pain relief, where the stereotype model applies. Greenland (1985) independently developed the same model. The stereotype model starts with a multinomial logistic regression model and imposes constraints on this model.

Goodness of fit for `ologit` can be evaluated by comparing the likelihood value with that obtained by fitting the model with `mlogit`. Let $\ln L_1$ be the log-likelihood value reported by `ologit`, and let $\ln L_0$ be the log-likelihood value reported by `mlogit`. If there are p independent variables (excluding the constant) and k categories, `mlogit` will estimate $p(k-1)$ additional parameters. We can then perform a "likelihood-ratio test", i.e., calculate $-2(\ln L_1 - \ln L_0)$, and compare it to $\chi^2\{p(k-2)\}$. This test is only suggestive because the ordered logit model is not nested within the multinomial logit model. A large value of $-2(\ln L_1 - \ln L_0)$ should, however, be taken as evidence of poorness of fit. Marginally large values, on the other hand, should not be taken too seriously.

The coefficients and cutpoints are estimated using maximum likelihood as described in [R] **maximize**. In our parameterization, no constant appears, as the effect is absorbed into the cutpoints.

`ologit` and `oprobit` begin by tabulating the dependent variable. Category $i = 1$ is defined as the minimum value of the variable, $i = 2$ as the next ordered value, and so on, for the empirically determined k categories.

The probability of a given observation in the case of ordered logit is

$$p_{ij} = \Pr(y_j = i) = \Pr\left(\kappa_{i-1} < \mathbf{x}_j\boldsymbol{\beta} + u \le \kappa_i\right)$$

$$= \frac{1}{1 + \exp(-\kappa_i + \mathbf{x}_j\boldsymbol{\beta})} - \frac{1}{1 + \exp(-\kappa_{i-1} + \mathbf{x}_j\boldsymbol{\beta})}$$

Note that κ_0 is defined as $-\infty$ and κ_k as $+\infty$.

In the case of ordered probit, the probability of a given observation is

$$p_{ij} = \Pr(y_j = i) = \Pr\left(\kappa_{i-1} < \mathbf{x}_j\boldsymbol{\beta} + u \le \kappa_i\right)$$

$$= \Phi\left(\kappa_i - \mathbf{x}_j\boldsymbol{\beta}\right) - \Phi\left(\kappa_{i-1} - \mathbf{x}_j\boldsymbol{\beta}\right)$$

where $\Phi()$ is the standard normal cumulative distribution function.

The log likelihood is

$$\ln L = \sum_{j=1}^{N} w_j \sum_{i=1}^{k} I_i(y_j) \ln p_{ij}$$

where w_j is an optional weight and

$$I_i(y_j) = \begin{cases} 1, & \text{if } y_j = i \\ 0, & \text{otherwise} \end{cases}$$

References

Aitchison, J. and S. D. Silvey. 1957. The generalization of probit analysis to the case of multiple responses. *Biometrika* 44: 131–140.

Anderson, J. A. 1984. Regression and ordered categorical variables (with discussion). *Journal of the Royal Statistical Society, Series B* 46: 1–30.

Brant, R. 1990. Assessing proportionality in the proportional odds model for ordinal logistic regression. *Biometrics* 46: 1171–1178.

Fu, V. K. 1998. sg88: Estimating generalized ordered logit models. *Stata Technical Bulletin* 44: 27–30. Reprinted in *Stata Technical Bulletin Reprints*, vol. 8, pp. 160–164.

Goldstein, R. 1997. sg59: Index of ordinal variation and Neyman–Barton GOF. *Stata Technical Bulletin* 33: 10–12. Reprinted in *Stata Technical Bulletin Reprints*, vol. 6, pp. 145–147.

Greene, W. H. 2003. *Econometric Analysis*. 5th ed. Upper Saddle River, NJ: Prentice Hall.

Greenland, S. 1985. An application of logistic models to the analysis of ordinal response. *Biometrical Journal* 27: 189–197.

Kleinbaum, D. G. and M. Klein. 2002. *Logistic Regression: A Self-Learning Text*. 2nd ed. New York: Springer.

Long, J. S. 1997. *Regression Models for Categorical and Limited Dependent Variables*. Thousand Oaks, CA: Sage.

Long, J. S. and J. Freese. 2003. *Regression Models for Categorical Dependent Variables Using Stata*. rev. ed. College Station, TX: Stata Press.

Lunt, M. 2001. sg163: Stereotype ordinal regression. *Stata Technical Bulletin* 61: 12–18. Reprinted in *Stata Technical Bulletin Reprints*, vol. 10, pp. 298–307.

McCullagh, P. 1977. A logistic model for paired comparisons with ordered categorical data. *Biometrika* 64: 449–453.

——. 1980. Regression models for ordinal data (with discussion). *Journal of the Royal Statistical Society, Series B* 42: 109–142.

McCullagh, P. and J. A. Nelder. 1989. *Generalized Linear Models*. 2nd ed. London: Chapman & Hall.

Peterson, B. and F. E. Harrell, Jr. 1990. Partial proportional odds models for ordinal response variables. *Applied Statistics* 39: 205–217.

Wolfe, R. 1998. sg86: Continuation-ratio models for ordinal response data. *Stata Technical Bulletin* 44: 18–21. Reprinted in *Stata Technical Bulletin Reprints*, vol. 8, pp. 149–153.

Wolfe, R. and W. W. Gould. 1998. sg76: An approximate likelihood-ratio test for ordinal response models. *Stata Technical Bulletin* 42: 24–27. Reprinted in *Stata Technical Bulletin Reprints*, vol. 7, pp. 199–204.

Zavoina, W. and R. D. McKelvey. 1975. A statistical model for the analysis of ordinal level dependent variables. *Journal of Mathematical Sociology* 4: 103–120.

Also See

Complementary:	[R] **ologit postestimation**
Related:	[R] **logistic**, [R] **logit**, [R] **mlogit**, [R] **oprobit**, [R] **slogit**, [SVY] **svy: ologit**
Background:	[U] **11.1.10 Prefix commands**, [U] **20 Estimation and postestimation commands**, [R] **estimation options**, [R] **maximize**, [R] *vce_option*

Title

> **ologit postestimation** — Postestimation tools for ologit

Description

The following postestimation commands are available for `ologit`:

command	description
adjust	adjusted predictions of $\mathbf{x}\beta$ or $\exp(\mathbf{x}\beta)$
estat	AIC, BIC, VCE, and estimation sample summary
estimates	cataloging estimation results
lincom	point estimates, standard errors, testing, and inference for linear combinations of coefficients
linktest	link test for model specification
lrtest	likelihood-ratio test
mfx	marginal effects or elasticities
nlcom	point estimates, standard errors, testing, and inference for nonlinear combinations of coefficients
predict	predictions, residuals, influence statistics, and other diagnostic measures
predictnl	point estimates, standard errors, testing, and inference for generalized predictions
suest	seemingly unrelated estimation
test	Wald tests for simple and composite linear hypotheses
testnl	Wald tests of nonlinear hypotheses

See the corresponding entries in the *Stata Base Reference Manual* for details.

Syntax for predict

> predict $\begin{bmatrix} type \end{bmatrix}$ *newvars* $\begin{bmatrix} if \end{bmatrix}$ $\begin{bmatrix} in \end{bmatrix}$ $\begin{bmatrix} , statistic \underline{\text{o}}\text{utcome}(outcome) \underline{\text{nooff}}\text{set} \end{bmatrix}$

> predict $\begin{bmatrix} type \end{bmatrix}$ $\{$ *stub** $|$ *newvar*$_{\text{reg}}$ *newvar*$_{\kappa_1}$ ... *newvar*$_{\kappa_{k-1}}$ $\}$ $\begin{bmatrix} if \end{bmatrix}$ $\begin{bmatrix} in \end{bmatrix}$, $\underline{\text{sc}}\text{ores}$

where k is the number of outcomes in the model.

statistic	description
pr	predicted probabilities; the default
xb	linear prediction
stdp	standard error of the linear prediction

Note that with the `pr` option, you specify one or k new variables depending upon whether the `outcome()` option is also specified (where k is the number of categories of *depvar*). With `xb` and `stdp`, one new variable is specified.

These statistics are available both in and out of sample; type `predict ... if e(sample) ...` if wanted only for the estimation sample.

Options for predict

pr, the default, calculates the predicted probabilities. If you do not also specify the outcome() option, you must specify k new variables, where k is the number of categories of the dependent variable. Say that you fitted a model by typing `ologit result x1 x2`, and `result` takes on three values. Then you could type `predict p1 p2 p3` to obtain all three predicted probabilities. If you specify the outcome() option, you must specify one new variable. Say that `result` takes on the values 1, 2, and 3. Typing `predict p1, outcome(1)` would produce the same p1.

xb calculates the linear prediction. You specify one new variable, for example, `predict linear, xb`. The linear prediction is defined, ignoring the contribution of the estimated cutpoints.

stdp calculates the standard error of the linear prediction. You specify one new variable, for example, `predict se, stdp`.

outcome(outcome**)** specifies for which outcome the predicted probabilities are to be calculated. outcome() should contain either a single value of the dependent variable or one of #1, #2, ..., with #1 meaning the first category of the dependent variable, #2 the second category, etc.

nooffset is relevant only if you specified offset(varname) for `ologit`. It modifies the calculations made by predict so that they ignore the offset variable; the linear prediction is treated as $\mathbf{x}_j\mathbf{b}$ rather than as $\mathbf{x}_j\mathbf{b} + \text{offset}_j$.

scores calculates equation-level score variables. The number of score variables created will equal the number of outcomes in the model. If the number of outcomes in the model were k, then

the first new variable will contain $\partial \ln L / \partial(\mathbf{x}_j\mathbf{b})$;

the second new variable will contain $\partial \ln L / \partial \kappa_1$;

the third new variable will contain $\partial \ln L / \partial \kappa_2$;

. . .

the kth new variable will contain $\partial \ln L / \partial \kappa_{k-1}$, where κ_i refers to the ith cutpoint.

Remarks

See [U] **20 Estimation and postestimation commands** for instructions on obtaining the variance–covariance matrix of the estimators, predicted values, and hypothesis tests. Also see [R] **lrtest** for performing likelihood-ratio tests.

▷ Example 1

In example 2 of [R] **ologit**, we fitted the model `ologit rep77 foreign length mpg`. The predict command can be used to obtain the predicted probabilities.

We type `predict` followed by the names of the new variables to hold the predicted probabilities, ordering the names from low to high. In our data, the lowest outcome is "poor", and the highest is "excellent". We have five categories, so we must type five names following `predict`; the choice of names is up to us:

```
. predict poor fair avg good exc
(option p assumed; predicted probabilities)
```

```
. list exc good make model rep78 if rep77>=., sep(4) divider
```

	exc	good	make	model	rep78
3.	.0033341	.0393056	AMC	Spirit	.
10.	.0098392	.1070041	Buick	Opel	.
32.	.0023406	.0279497	Ford	Fiesta	Good
44.	.015697	.1594413	Merc.	Monarch	Average
53.	.065272	.4165188	Peugeot	604	.
56.	.005187	.059727	Plym.	Horizon	Average
57.	.0261461	.2371826	Plym.	Sapporo	.
63.	.0294961	.2585825	Pont.	Phoenix	.

The eight cars listed were introduced after 1977, so they do not have 1977 repair records in our data. We predicted what their 1977 repair records might have been using the fitted model. We see that, based on its characteristics, the Peugeot 604 had about a $41.65 + 6.53 \approx 48.2$ percent chance of a good or excellent repair record. The Ford Fiesta, which had only a 3 percent chance of a good or excellent repair record, in fact, had a good record when it was introduced in the following year.

◁

❑ Technical Note

For ordered logit, `predict, xb` produces $S_j = x_{1j}\beta_1 + x_{2j}\beta_2 + \cdots + x_{kj}\beta_k$. The ordered-logit predictions are then the probability that $S_j + u_j$ lies between a pair of cutpoints, κ_{i-1} and κ_i. Some handy formulas are

$$\Pr(S_j + u_j < \kappa) = 1/(1 + e^{S_j - \kappa})$$
$$\Pr(S_j + u_j > \kappa) = 1 - 1/(1 + e^{S_j - \kappa})$$
$$\Pr(\kappa_1 < S_j + u_j < \kappa_2) = 1/(1 + e^{S_j - \kappa_2}) - 1/(1 + e^{S_j - \kappa_1})$$

Rather than using `predict` directly, we could calculate the predicted probabilities by hand. If we wished to obtain the predicted probability that the repair record is excellent and the probability that it is good, we look back at `ologit`'s output to obtain the cutpoints. We find that "good" corresponds to the interval $/\texttt{cut3} < S_j + u < /\texttt{cut4}$ and "excellent" to the interval $S_j + u > /\texttt{cut4}$:

```
. predict score, xb
. generate probgood = 1/(1+exp(score-_b[/cut4])) - 1/(1+exp(score-_b[/cut3]))
. generate probexc = 1 - 1/(1+exp(score-_b[/cut4]))
```

The results of our calculation will be exactly the same as those produced in the previous example. Note that we refer to the estimated cutpoints just as we would any coefficient, so `_b[/cut3]` refers to the value of the `/cut3` coefficient; see [U] **13.5 Accessing coefficients and standard errors**.

❑

Methods and Formulas

All postestimation commands listed above are implemented as ado-files.

Also See

Complementary:	[R] **ologit**; [R] **adjust**, [R] **estimates**, [R] **lincom**, [R] **linktest**, [R] **lrtest**, [R] **mfx**, [R] **nlcom**, [R] **predictnl**, [R] **suest**, [R] **test**, [R] **testnl**
Background:	[U] **13.5 Accessing coefficients and standard errors**, [U] **20 Estimation and postestimation commands**, [R] **estat**, [R] **predict**

Title

> **oneway** — One-way analysis of variance

Syntax

> <u>on</u>eway *response_var factor_var* [*if*] [*in*] [*weight*] [, *options*]

options	description
Main	
<u>b</u>onferroni	Bonferroni multiple-comparison test
<u>sc</u>heffe	Schéffe multiple-comparison test
<u>si</u>dak	Šidák multiple-comparison test
<u>t</u>abulate	produce summary table
[<u>no</u>]<u>m</u>eans	include or suppress means; default is means
[<u>no</u>]<u>s</u>tandard	include or suppress standard deviations; default is standard
[<u>no</u>]<u>f</u>req	include or suppress frequencies; default is freq
[<u>no</u>]<u>o</u>bs	include or suppress number of obs; default is obs if data are weighted
<u>noa</u>nova	suppress the ANOVA table
<u>nol</u>abel	show numeric codes, not labels
<u>wr</u>ap	do not break wide tables
<u>m</u>issing	treat missing values as categories

by may be used with oneway; see [D] **by**.

aweights and fweights are allowed; see [U] **11.1.6 weight**.

Description

The oneway command reports one-way analysis-of-variance (ANOVA) models and performs multiple-comparison tests.

If you wish to fit more complicated ANOVA layouts or wish to fit analysis-of-covariance (ANCOVA) models, see [R] **anova**.

See [D] **encode** for examples of fitting ANOVA models on string variables.

See [R] **loneway** for an alternative oneway command with slightly different features.

Options

⌐ Main ⌐

bonferroni reports the results of a Bonferroni multiple-comparison test.

scheffe reports the results of a Scheffé multiple-comparison test.

sidak reports the results of a Šidák multiple-comparison test.

tabulate produces a table of summary statistics of the *response_var* by levels of the *factor_var*. The table includes the mean, standard deviation, frequency, and, if the data are weighted, the number of observations. Individual elements of the table may be included or suppressed by using the [no]means, [no]standard, [no]freq, and [no]obs options. For example, typing

```
oneway response factor, tabulate means standard
```

produces a summary table that contains only the means and standard deviations. You could achieve the same result by typing

```
oneway response factor, tabulate nofreq
```

[no]means includes or suppresses only the means from the table produced by the tabulate option. See tabulate above.

[no]standard includes or suppresses only the standard deviations from the table produced by the tabulate option. See tabulate above.

[no]freq includes or suppresses only the frequencies from the table produced by the tabulate option. See tabulate above.

[no]obs includes or suppresses only the reported number of observations from the table produced by the tabulate option. If the data are not weighted, only the frequency is reported. If the data are weighted, the frequency refers to the sum of the weights. See tabulate above.

noanova suppresses the display of the ANOVA table.

nolabel causes the numeric codes to be displayed rather than the value labels in the ANOVA and multiple-comparison test tables.

wrap requests that Stata not break up wide tables to make them more readable.

missing requests that missing values of *factor_var* be treated as a category rather than as observations to be omitted from the analysis.

Remarks

Remarks are presented under the headings

Introduction
Obtaining observed means
Multiple-comparison tests
Weighted data

Introduction

The oneway command reports one-way analysis-of-variance (ANOVA) models. To perform a one-way layout of a variable called endog on exog, type oneway endog exog.

▷ Example 1

We run an experiment varying the amount of fertilizer used in growing apple trees. We test four concentrations, using each concentration in three groves of twelve trees each. Later in the year, we measure the average weight of the fruit.

If all had gone well, we would have had three observations on the average weight for each of the four concentrations. Instead, two of the groves were mistakenly leveled by a confused man on a large bulldozer. We are left with the following dataset:

```
. use http://www.stata-press.com/data/r9/apple
(Apple trees)
. describe
Contains data from http://www.stata-press.com/data/r9/apple.dta
  obs:            10                          Apple trees
  vars:            2                          16 Jan 2005 11:23
  size:          140 (99.9% of memory free)
```

variable name	storage type	display format	value label	variable label
treatment	int	%8.0g		Fertilizer
weight	double	%10.0g		Average weight in grams

```
Sorted by:

. list, abbreviate(10)
```

	treatment	weight
1.	1	117.5
2.	1	113.8
3.	1	104.4
4.	2	48.9
5.	2	50.4
6.	2	58.9
7.	3	70.4
8.	3	86.9
9.	4	87.7
10.	4	67.3

To obtain the one-way analysis-of-variance results, we type

```
. oneway weight treatment
                      Analysis of Variance
    Source              SS         df      MS            F     Prob > F
```

Source	SS	df	MS	F	Prob > F
Between groups	5295.54433	3	1765.18144	21.46	0.0013
Within groups	493.591667	6	82.2652778		
Total	5789.136	9	643.237333		

```
Bartlett's test for equal variances:  chi2(3) =   1.3900  Prob>chi2 = 0.708
```

We find significant (at better than the 1% level) differences among the four concentrations.

◁

❑ Technical Note

Rather than using the `oneway` command, we could have performed this analysis using `anova`. The first example in [R] **anova** repeats this same analysis. You may wish to compare the output.

You will find the `oneway` command quicker than the `anova` command, and, as you will learn, `oneway` allows you to perform multiple-comparison tests. On the other hand, `anova` will let you generate predictions, examine the covariance matrix of the estimators, and perform more general hypothesis tests.

❑

❏ Technical Note

Although the output is a usual analysis-of-variance table, let's run through it anyway. The between-group sum of squares for the model is 5295.5 with 3 degrees of freedom, resulting in a mean square of $5295.5/3 \approx 1765.2$. The corresponding F statistic is 21.46 and has a significance level of 0.0013. Thus the model appears to be significant at the 0.13% level.

The second line summarizes the within-group (residual) variation. The within-group sum of squares is 493.59 with 6 degrees of freedom, resulting in a mean squared error of 82.27.

The between- and residual-group variations sum to the total sum of squares, which is reported as 5789.1 in the last line of the table. This is the total sum of squares of `weight` after removal of the mean. Similarly, the between plus residual degrees of freedom sum to the total degrees of freedom, 9. Remember that there are 10 observations. Subtracting 1 for the mean, we are left with 9 total degrees of freedom.

At the bottom of the table, Bartlett's test for equal variances is reported. The value of the statistic is 1.39. The corresponding significance level (χ^2 with 3 degrees of freedom) is 0.708, so we cannot reject the assumption that the variances are homogeneous.

❏

Obtaining observed means

▷ Example 2

We typed `oneway weight treatment` to obtain an ANOVA table of weight of fruit by fertilizer concentration. Although we obtained the table, we did not obtain any information on which fertilizer seems to work the best. If we add the `tabulate` option, we obtain that additional information:

```
. oneway weight treatment, tabulate
```

Fertilizer	Summary of Average weight in grams		
	Mean	Std. Dev.	Freq.
1	111.9	6.7535176	3
2	52.733333	5.3928966	3
3	78.65	11.667262	2
4	77.5	14.424978	2
Total	80.62	25.362124	10

	Analysis of Variance				
Source	SS	df	MS	F	Prob > F
Between groups	5295.54433	3	1765.18144	21.46	0.0013
Within groups	493.591667	6	82.2652778		
Total	5789.136	9	643.237333		

Bartlett's test for equal variances: chi2(3) = 1.3900 Prob>chi2 = 0.708

We find that the average weight was largest when we used fertilizer concentration 1.

◁

Multiple-comparison tests

▷ Example 3

oneway can also perform multiple-comparison tests using either Bonferroni, Scheffé, or Šidák normalizations. For instance, to obtain the Bonferroni multiple-comparison test, we specify the bonferroni option:

```
. oneway weight treatment, bonferroni
                         Analysis of Variance
      Source              SS         df      MS            F      Prob > F

Between groups       5295.54433      3   1765.18144     21.46     0.0013
Within groups        493.591667      6   82.2652778

      Total          5789.136        9   643.237333
Bartlett's test for equal variances:  chi2(3) =   1.3900  Prob>chi2 = 0.708
```

```
              Comparison of Average weight in grams by Fertilizer
                                (Bonferroni)
Row Mean-|
Col Mean |        1          2          3

       2 |  -59.1667
         |    0.001

       3 |   -33.25    25.9167
         |    0.042     0.122

       4 |    -34.4    24.7667     -1.15
         |    0.036     0.146     1.000
```

The results of the Bonferroni test are presented as a matrix. The first entry, -59.17, represents the difference between fertilizer concentrations 2 and 1 (labeled "Row Mean – Col Mean" in the upper stub of the table). Remember that in the previous example we requested the tabulate option. Looking back, we find that the means of concentrations 1 and 2 are 111.90 and 52.73, respectively. Thus $52.73 - 111.90 = -59.17$.

Underneath that number is reported "0.001". This is the Bonferroni-adjusted significance of the difference. The difference is significant at the 0.1% level. Looking down the column, we see that concentration 3 is also worse than concentration 1 (4.2% level), as is concentration 4 (3.6% level).

Based on this evidence, we would use concentration 1 if we grew apple trees.

◁

▷ Example 4

We can just as easily obtain the Scheffé-adjusted significance levels. Rather than specifying the bonferroni option, we specify the scheffe option.

We will also add the `noanova` option to prevent Stata from redisplaying the ANOVA table:

```
. oneway weight treatment, noanova scheffe
            Comparison of Average weight in grams by Fertilizer
                               (Scheffe)
Row Mean-|
Col Mean |          1          2          3
---------+---------------------------------
       2 |   -59.1667
         |     0.001
         |
       3 |    -33.25    25.9167
         |     0.039     0.101
         |
       4 |     -34.4    24.7667      -1.15
         |     0.034     0.118      0.999
```

The differences are the same as those we obtained in the Bonferroni output, but the significance levels are not. According to the Bonferroni-adjusted numbers, the significance of the difference between fertilizer concentrations 1 and 3 is 4.2%. The Scheffé-adjusted significance level is 3.9%.

We will leave it to you to decide which results are more accurate.

◁

▷ Example 5

Let's conclude this example by obtaining the Šidák-adjusted multiple-comparison tests. We do this to illustrate Stata's capabilities to calculate these results, as searching across adjustment methods until you find the results you want is not a valid technique for obtaining significance levels.

```
. oneway weight treatment, noanova sidak
            Comparison of Average weight in grams by Fertilizer
                               (Sidak)
Row Mean-|
Col Mean |          1          2          3
---------+---------------------------------
       2 |   -59.1667
         |     0.001
         |
       3 |    -33.25    25.9167
         |     0.041     0.116
         |
       4 |     -34.4    24.7667      -1.15
         |     0.035     0.137      1.000
```

We find results that are similar to the Bonferroni-adjusted numbers.

◁

Henry Scheffé (1907–1977) was born in New York. He studied mathematics at the University of Wisconsin, gaining a doctorate with a dissertation on differential equations. He taught mathematics at Wisconsin, Oregon State University, and Reed College, but his interests changed to statistics and he joined Wilks at Princeton. After periods at Syracuse, UCLA, and Columbia, Scheffé settled in Berkeley from 1953. His research increasingly focused on linear models and particularly ANOVA, on which he produced a celebrated monograph. His death was the result of a bicycle accident.

Weighted data

▷ Example 6

oneway can work with both weighted and unweighted data. Let's assume that we wish to perform a one-way layout of the death rate on the four census regions of the United States using state data. Our data contain three variables, drate (the death rate), region (the region), and pop (the population of the state).

To fit the model, we type oneway drate region [weight=pop], although we typically abbreviate weight as w. We will also add the tabulate option to demonstrate how the table of summary statistics differs for weighted data:

```
. use http://www.stata-press.com/data/r9/census8
(1980 Census data by state)

. oneway drate region [w=pop], tabulate
(analytic weights assumed)
```

Census region	Summary of Death Rate Mean	Std. Dev.	Freq.	Obs.
NE	97.15	5.82	49135283	9
N Cntrl	88.10	5.58	58865670	12
South	87.05	10.40	74734029	16
West	75.65	8.23	43172490	13
Total	87.34	10.43	2.259e+08	50

Analysis of Variance Source	SS	df	MS	F	Prob > F
Between groups	2360.92281	3	786.974272	12.17	0.0000
Within groups	2974.09635	46	64.6542685		
Total	5335.01916	49	108.877942		

Bartlett's test for equal variances: chi2(3) =   5.4971 Prob>chi2 = 0.139

When the data are weighted, the summary table has four columns rather than three. The column labeled "Freq." reports the sum of the weights. The overall frequency is $2.259 \cdot 10^8$, meaning that there are approximately 226 million people in the U.S.

The ANOVA table is appropriately weighted. Also see [U] **11.1.6 weight**.

◁

Saved Results

oneway saves in r():

Scalars

r(N)	number of observations	r(df_m)	between-group degrees of freedom
r(F)	F statistic	r(rss)	within-group sum of squares
r(df_r)	within-group degrees of freedom	r(chi2bart)	Bartlett's χ^2
r(mss)	between-group sum of squares	r(df_bart)	Bartlett's degrees of freedom

Methods and Formulas

The model of one-way analysis of variance is

$$y_{ij} = \mu + \alpha_i + \epsilon_{ij}$$

for levels $i = 1, \ldots, k$ and observations $j = 1, \ldots, n_i$. Define $\overline{y}_i$ as the (weighted) mean of y_{ij} over j and $\overline{y}$ as the overall (weighted) mean of y_{ij}. Define w_{ij} as the weight associated with y_{ij}, which is 1 if the data are unweighted. w_{ij} is normalized to sum to $n = \sum_i n_i$ if aweights are used and is otherwise not normalized. w_i refers to $\sum_j w_{ij}$ and w refers to $\sum_i w_i$.

The between-group sum of squares is then

$$S_1 = \sum_i w_i (\overline{y}_i - \overline{y})^2$$

The total sum of squares is

$$S = \sum_i \sum_j w_{ij} (y_{ij} - \overline{y})^2$$

The within-group sum of squares is given by $S_e = S - S_1$.

The between-group mean square is $s_1^2 = S_1/(k-1)$, and the within-group mean square is $s_e^2 = S_e/(w-k)$. The test statistic is $F = s_1^2/s_e^2$. See, for instance, Snedecor and Cochran (1989).

Bartlett's test

Bartlett's test assumes that you have m independent, normal, random samples and tests the hypothesis $\sigma_1^2 = \sigma_2^2 = \cdots = \sigma_m^2$. The test statistic, M, is defined as

$$M = \frac{(T-m)\ln\widehat{\sigma}^2 - \sum(T_i-1)\ln\widehat{\sigma}_i^2}{1 + \frac{1}{3(m-1)}\sum \frac{1}{T_i-1} - \frac{1}{T-m}}$$

where there are T overall observations, T_i observations in the ith group, and

$$(T_i - 1)\widehat{\sigma}_i^2 = \sum_{j=1}^{T_i} (y_{ij} - \overline{y}_i)^2$$

$$(T - m)\widehat{\sigma}^2 = \sum_{i=1}^{m} (T_i - 1)\widehat{\sigma}_i^2$$

An approximate test of the homogeneity of variance is based on the statistic M with critical values obtained from the χ^2 distribution of $m - 1$ degrees of freedom. See Bartlett (1937) or Judge et al. (1985, 447–449).

Multiple-comparison tests

Let's begin by reviewing the logic behind these adjustments. The "standard" t statistic for the comparison of two means is

$$t = \frac{\overline{y}_i - \overline{y}_j}{s\sqrt{\frac{1}{n_i} + \frac{1}{n_j}}}$$

where s is the overall standard deviation, $\overline{y}_i$ is the measured average of y in group i, and n_i is the number of observations in the group. We perform hypothesis tests by calculating this t statistic. We simultaneously choose a critical level α and look up the t statistic corresponding to that level in a table. We reject the hypothesis if our calculated t exceeds the value we looked up. Alternatively, since we have a computer at our disposal, we calculate the significance-level e corresponding to our calculated t statistic, and if $e < \alpha$, we reject the hypothesis.

This logic works well when we are performing a *single* test. Now consider what happens when we perform a number of separate tests, say, n of them. Let's assume, just for discussion, that we set α equal to 0.05 and that we will perform six tests. For each test, we have a 0.05 probability of falsely rejecting the equality-of-means hypothesis. Overall, then, our chances of falsely rejecting *at least one* of the hypotheses is $1 - (1 - .05)^6 \approx .26$ if the tests are independent.

The idea behind multiple-comparison tests is to control for the fact that we will perform multiple tests and to reduce our overall chances of falsely rejecting each hypothesis to α rather than letting it increase with each additional test. (See Miller 1981 and Hochberg and Tamhane 1987 for rather advanced texts on multiple-comparison procedures.)

The Bonferroni adjustment (see Miller 1981; also see Winer, Brown, and Michels 1991, 158–166) does this by (falsely but approximately) asserting that the critical level we should use, a, is the true critical level α divided by the number of tests n; that is, $a = \alpha/n$. For instance, if we are going to perform six tests, each at the .05 significance level, we want to adopt a critical level of $.05/6 \approx .00833$.

We can just as easily apply this logic to e, the significance level associated with our t statistic, as to our critical level α. If a comparison has a calculated significance of e, then its "real" significance, adjusted for the fact of n comparisons, is $n \cdot e$. If a comparison has a significance level of, say, .012, and we perform 6 tests, then its "real" significance is .072. If we adopt a critical level of .05, we cannot reject the hypothesis. If we adopt a critical level of .10, we can reject it.

Of course, this calculation can go above 1, but that just means that there is no $\alpha < 1$ for which we could reject the hypothesis. (This situation arises due to the crude nature of the Bonferroni adjustment.) Stata handles this case by simply calling the significance level 1. Thus the formula for the Bonferroni significance level is

$$e_b = \min(1, en)$$

where $n = k(k-1)/2$ is the number of comparisons.

The Šidák adjustment (Šidák 1967; also see Winer, Brown, and Michels 1991, 165–166) is slightly different and provides a tighter bound. It starts with the assertion that

$$a = 1 - (1 - \alpha)^{1/n}$$

Turning this formula around and substituting calculated significance levels, we obtain

$$e_s = \min\left\{1, 1 - (1 - e)^n\right\}$$

For example, if the calculated significance is 0.012 and we perform 6 tests, the "real" significance is approximately 0.07.

The Scheffé test (Scheffé 1953, 1959; also see Winer, Brown, and Michels 1991, 191–195) differs in derivation, but it attacks the same problem. Let there be k means for which we want to make all the pairwise tests. Two means are declared significantly different if

$$t \geq \sqrt{(k-1)F(\alpha; k-1, \nu)}$$

where $F(\alpha; k-1, \nu)$ is the α-critical value of the F distribution with $k-1$ numerator and ν denominator degrees of freedom. Scheffé's test has the nicety that it never declares a contrast significant if the overall F test is not significant.

Turning the test around, Stata calculates a significance level

$$\hat{e} = F\left(\frac{t^2}{k-1}, k-1, \nu\right)$$

For instance, you have a calculated t statistic of 4.0 with 50 degrees of freedom. The simple t test says that the significance level is .00021. The F test equivalent, 16 with 1 and 50 degrees of freedom, says the same. If you are comparing three means, however, you calculate an F test of 8.0 with 2 and 50 degrees of freedom, which says the significance level is .0010.

References

Altman, D. G. 1991. *Practical Statistics for Medical Research.* London: Chapman & Hall.

Bartlett, M. S. 1937. Properties of sufficiency and statistical tests. *Proceedings of the Royal Society, Series A* 160: 268–282.

Daniel, C. and E. L. Lehmann. 1979. Henry Scheffé 1907–1977. *Annals of Statistics* 7: 1149–1161.

Hochberg, Y. and A. C. Tamhane. 1987. *Multiple Comparison Procedures.* New York: Wiley.

Judge, G. G., W. E. Griffiths, R. C. Hill, H. Lütkepohl, and T.-C. Lee. 1985. *The Theory and Practice of Econometrics.* 2nd ed. New York: Wiley.

Miller, R. G., Jr. 1981. *Simultaneous Statistical Inference.* 2nd ed. New York: Springer.

Scheffé, H. 1953. A method for judging all contrasts in the analysis of variance. *Biometrika* 40: 87–104.

——. 1959. *The Analysis of Variance.* New York: Wiley.

Šidák, Z. 1967. Rectangular confidence regions for the means of multivariate normal distributions. *Journal of the American Statistical Association* 62: 626–633.

Snedecor, G. W. and W. G. Cochran. 1989. *Statistical Methods.* 8th ed. Ames, IA: Iowa State University Press.

Winer, B. J., D. R. Brown, and K. M. Michels. 1991. *Statistical Principles in Experimental Design.* 3rd ed. New York: McGraw–Hill.

Also See

Complementary:	[D] **encode**
Related:	[R] **anova**, [R] **loneway**, [R] **table**
Background:	[U] **18.8 Accessing results calculated by other programs**

Title

> **oprobit** — Ordered probit regression

Syntax

> $\underline{\text{opro}}\text{bit}$ *depvar* $\left[\textit{indepvars}\right]$ $\left[\textit{if}\right]$ $\left[\textit{in}\right]$ $\left[\textit{weight}\right]$ $\left[\text{, }\textit{options}\right]$

options	description
Model	
$\underline{\text{off}}\text{set}(\textit{varname})$	include *varname* in model with coefficient constrained to 1
SE/Robust	
vce(*vcetype*)	*vcetype* may be $\underline{\text{robust}}$, $\underline{\text{boot}}\text{strap}$, or $\underline{\text{jackknife}}$
$\underline{\text{robust}}$	synonym for vce(robust)
$\underline{\text{cl}}\text{uster}(\textit{varname})$	adjust standard errors for intragroup correlation
Reporting	
$\underline{\text{level}}(\#)$	set confidence level; default is level(95)
Max options	
maximize_options	control the maximization process; seldom used

bootstrap, by, jackknife, rolling, statsby, stepwise, svy, and xi are allowed; see
  [U] **11.1.10 Prefix commands**.
fweights, iweights, and pweights are allowed; see [U] **11.1.6 weight**.
See [U] **20 Estimation and postestimation commands** for additional capabilities of estimation commands.

Description

oprobit fits ordered probit models of ordinal variable *depvar* on the independent variables *indepvars*. The actual values taken on by the dependent variable are irrelevant, except that larger values are assumed to correspond to "higher" outcomes. Up to 50 outcomes are allowed in Stata/SE and Intercooled Stata, and up to 20 are allowed in Small Stata.

See [R] **logistic** for a list of related estimation commands.

Options

> Model

offset(*varname*); see [R] **estimation options**.

> SE/Robust

vce(*vcetype*); see [R] *vce_option*.

robust, cluster(*varname*); see [R] **estimation options**. cluster() can be used with pweights to produce estimates for unstratified cluster-sampled data, but see [SVY] **svy: oprobit** for a command especially designed for survey data.

Reporting

level(#); see [R] **estimation options**.

Max options

maximize_options: <u>iterate</u>(#), [<u>no</u>]<u>log</u>, <u>trace</u>, <u>tol</u>erance(#), <u>ltol</u>erance(#); see [R] **maximize**. These options are seldom used.

Remarks

An ordered probit model is used to estimate relationships between an ordinal dependent variable and a set of independent variables. An *ordinal* variable is a variable that is categorical and ordered, for instance, "poor", "good", and "excellent", which might indicate a person's current health status or the repair record of a car. If there are only two outcomes, see [R] **logistic**, [R] **logit**, and [R] **probit**. This entry is concerned only with more than two outcomes. If the outcomes cannot be ordered (e.g., residency in the north, east, south, or west), see [R] **mlogit**. This entry is concerned only with models in which the outcomes can be ordered.

In ordered probit, an underlying score is estimated as a linear function of the independent variables and a set of cutpoints. The probability of observing outcome i corresponds to the probability that the estimated linear function, plus random error, is within the range of the cutpoints estimated for the outcome:

$$\Pr(\text{outcome}_j = i) = \Pr(\kappa_{i-1} < \beta_1 x_{1j} + \beta_2 x_{2j} + \cdots + \beta_k x_{kj} + u_j \leq \kappa_i)$$

u_j is assumed to be normally distributed. In either case, we estimate the coefficients β_1, β_2, ..., β_k together with the cutpoints κ_1, κ_2, ..., κ_{I-1}, where I is the number of possible outcomes. κ_0 is taken as $-\infty$, and κ_I is taken as $+\infty$. All of this is a direct generalization of the ordinary two-outcome probit model.

▷ Example 1

In [R] **ologit**, we use a variation of the automobile dataset (see [U] **1.2.1 Sample datasets**) to analyze the 1977 repair records of 66 foreign and domestic cars. We use ordered logit to explore the relationship of rep77 in terms of foreign (origin of manufacture), length (a proxy for size), and mpg. Here we fit the same model using ordered probit rather than ordered logit:

(Continued on next page)

```
. use http://www.stata-press.com/data/r9/fullauto
(Automobile Models)
. oprobit rep77 foreign length mpg
Iteration 0:    log likelihood = -89.895098
Iteration 1:    log likelihood = -78.141221
Iteration 2:    log likelihood = -78.020314
Iteration 3:    log likelihood = -78.020025
```

Ordered probit regression

Log likelihood = -78.020025

				Number of obs	=	66
				LR chi2(3)	=	23.75
				Prob > chi2	=	0.0000
				Pseudo R2	=	0.1321

rep77	Coef.	Std. Err.	z	P>|z|	[95% Conf. Interval]	
foreign	1.704861	.4246786	4.01	0.000	.8725057	2.537215
length	.0468675	.012648	3.71	0.000	.022078	.0716571
mpg	.1304559	.0378627	3.45	0.001	.0562464	.2046654
/cut1	10.1589	3.076749			4.128586	16.18922
/cut2	11.21003	3.107522			5.119399	17.30066
/cut3	12.54561	3.155228			6.361476	18.72974
/cut4	13.98059	3.218786			7.671888	20.2893

We find that foreign cars have better repair records, as do larger cars and cars with better mileage ratings.

◁

Saved Results

oprobit saves in e():

Scalars

e(N)	number of observations	e(ll)	log likelihood
e(k_cat)	number of categories	e(ll_0)	log likelihood, constant-only model
e(df_m)	model degrees of freedom	e(N_clust)	number of clusters
e(r2_p)	pseudo-R-squared	e(chi2)	χ^2
		e(converged)	1 if converged, 0 otherwise

Macros

e(cmd)	oprobit	e(chi2type)	Wald or LR; type of model χ^2 test
e(depvar)	name of dependent variable	e(crittype)	optimization criterion
e(wtype)	weight type	e(vce)	*vcetype* specified in vce()
e(wexp)	weight expression	e(vcetype)	title used to label Std. Err.
e(title)	title in estimation output	e(predict)	program used to implement predict
e(clustvar)	name of cluster variable	e(properties)	b V
e(offset)	offset		

Matrices

e(b)	coefficient vector	e(V)	variance–covariance matrix of the
e(cat)	category values		estimators

Functions

e(sample)	marks estimation sample

Methods and Formulas

Please see the *Methods and Formulas* section of [R] **ologit**.

References

Aitchison, J. and S. D. Silvey. 1957. The generalization of probit analysis to the case of multiple responses. *Biometrika* 44: 131–140.

Goldstein, R. 1997. sg59: Index of ordinal variation and Neyman–Barton GOF. *Stata Technical Bulletin* 33: 10–12. Reprinted in *Stata Technical Bulletin Reprints*, vol. 6, pp. 145–147.

Greene, W. H. 2003. *Econometric Analysis*. 5th ed. Upper Saddle River, NJ: Prentice Hall.

Long, J. S. 1997. *Regression Models for Categorical and Limited Dependent Variables*. Thousand Oaks, CA: Sage.

Long, J. S. and J. Freese. 2003. *Regression Models for Categorical Dependent Variables Using Stata*. rev. ed. College Station, TX: Stata Press.

Lunt, M. 2001. sg163: Stereotype ordinal regression. *Stata Technical Bulletin* 61: 12–18. Reprinted in *Stata Technical Bulletin Reprints*, vol. 10, pp. 298–307.

Stewart, M. B. 2004. Semi-nonparametric estimation of extended ordered probit models. *Stata Journal* 4: 27–39.

Wolfe, R. 1998. sg86: Continuation-ratio models for ordinal response data. *Stata Technical Bulletin* 44: 18–21. Reprinted in *Stata Technical Bulletin Reprints*, vol. 8, pp. 149–153.

Wolfe, R. and W. W. Gould. 1998. sg76: An approximate likelihood-ratio test for ordinal response models. *Stata Technical Bulletin* 42: 24–27. Reprinted in *Stata Technical Bulletin Reprints*, vol. 7, pp. 199–204.

Also See

Complementary:	[R] **oprobit postestimation**
Related:	[R] **logistic**, [R] **mlogit**, [R] **ologit**, [R] **probit**, [R] **slogit**, [SVY] **svy: oprobit**
Background:	[U] **11.1.10 Prefix commands**, [U] **20 Estimation and postestimation commands**, [R] **estimation options**, [R] **maximize**, [R] *vce_option*

Title

oprobit postestimation — Postestimation tools for oprobit

Description

The following postestimation commands are available for `oprobit`:

command	description
adjust	adjusted predictions of $\mathbf{x}\beta$
estat	AIC, BIC, VCE, and estimation sample summary
estimates	cataloging estimation results
lincom	point estimates, standard errors, testing, and inference for linear combinations of coefficients
linktest	link test for model specification
lrtest	likelihood-ratio test
mfx	marginal effects or elasticities
nlcom	point estimates, standard errors, testing, and inference for nonlinear combinations of coefficients
predict	predictions, residuals, influence statistics, and other diagnostic measures
predictnl	point estimates, standard errors, testing, and inference for generalized predictions
suest	seemingly unrelated estimation
test	Wald tests for simple and composite linear hypotheses
testnl	Wald tests of nonlinear hypotheses

See the corresponding entries in the *Stata Base Reference Manual* for details.

Syntax for predict

predict $\left[\textit{type}\right]$ *newvars* $\left[\textit{if}\right]$ $\left[\textit{in}\right]$ $\left[\text{, }\textit{statistic} \underline{\text{o}}\text{utcome}(\textit{outcome}) \underline{\text{nooff}}\text{set}\right]$

predict $\left[\textit{type}\right]$ $\left\{ \textit{stub*} \mid \textit{newvar}_{\text{reg}} \textit{newvar}_{\kappa_1} \ldots \textit{newvar}_{\kappa_{k-1}} \right\}$ $\left[\textit{if}\right]$ $\left[\textit{in}\right]$, $\underline{\text{sc}}\text{ores}$

where k is the number of outcomes in the model.

statistic	description
<u>p</u>r	predicted probabilities; the default
xb	linear prediction
stdp	standard error of the linear prediction

Note that with the pr option, you specify either one or k new variables depending upon whether the outcome() option is also specified (where k is the number of categories of *depvar*). With xb and stdp, one new variable is specified.

These statistics are available both in and out of sample; type predict ... if e(sample) ... if wanted only for the estimation sample.

Options for predict

pr, the default, calculates the predicted probabilities. If you do not also specify the outcome() option, you must specify k new variables, where k is the number of categories of the dependent variable. Say that you fitted a model by typing oprobit result x1 x2, and result takes on three values. Then you could type predict p1 p2 p3 to obtain all three predicted probabilities. If you specify the outcome() option, you must specify one new variable. Say that result takes on values 1, 2, and 3. Typing predict p1, outcome(1) would produce the same p1.

xb calculates the linear prediction. You specify one new variable, for example, predict linear, xb. The linear prediction is defined ignoring the contribution of the estimated cutpoints.

stdp calculates the standard error of the linear prediction. You specify one new variable, for example, predict se, stdp.

outcome(*outcome*) specifies for which outcome the predicted probabilities are to be calculated. outcome() should contain either a single value of the dependent variable or one of #1, #2, ..., with #1 meaning the first category of the dependent variable, #2 the second category, etc.

nooffset is relevant only if you specified offset(*varname*) for oprobit. It modifies the calculations made by predict so that they ignore the offset variable; the linear prediction is treated as $\mathbf{x}_j\mathbf{b}$ rather than as $\mathbf{x}_j\mathbf{b} + \text{offset}_j$.

scores calculates equation-level score variables. The number of score variables created will equal the number of outcomes in the model. If the number of outcomes in the model were k, then

the first new variable will contain $\partial \ln L/\partial(\mathbf{x}_j\mathbf{b})$;

the second new variable will contain $\partial \ln L/\partial \kappa_1$;

the third new variable will contain $\partial \ln L/\partial \kappa_2$;

...

the kth new variable will contain $\partial \ln L/\partial \kappa_{k-1}$, where κ_i refers to the ith cutpoint.

Remarks

See [U] **20 Estimation and postestimation commands** for instructions on obtaining the variance–covariance matrix of the estimators, predicted values, and hypothesis tests. Also see [R] **lrtest** for performing likelihood-ratio tests.

▷ Example 1

In example 1 of [R] **oprobit**, we fitted the model oprobit rep77 foreign length mpg. The predict command can be used to obtain the predicted probabilities. We type predict followed by the names of the new variables to hold the predicted probabilities, ordering the names from low to high. In our data, the lowest outcome is "poor", and the highest is "excellent". We have five categories, so we must type five names following predict; the choice of names is up to us:

(Continued on next page)

```
. predict poor fair avg good exc
(option p assumed; predicted probabilities)
. list make model exc good if rep77>=., sep(4) divider
```

	make	model	exc	good
3.	AMC	Spirit	.0006044	.0351813
10.	Buick	Opel	.0043803	.1133763
32.	Ford	Fiesta	.0002927	.0222789
44.	Merc.	Monarch	.0093209	.1700846
53.	Peugeot	604	.0734199	.4202766
56.	Plym.	Horizon	.001413	.0590294
57.	Plym.	Sapporo	.0197543	.2466034
63.	Pont.	Phoenix	.0234156	.266771

◁

❑ Technical Note

For ordered probit, `predict, xb` produces $S_j = x_{1j}\beta_1 + x_{2j}\beta_2 + \cdots + x_{kj}\beta_k$. Ordered probit is identical to ordered logit, except that we use different distribution functions for calculating probabilities. The ordered-probit predictions are then the probability that $S_j + u_j$ lies between a pair of cutpoints κ_{i-1} and κ_i. The formulas in the case of ordered probit are

$$\Pr(S_j + u < \kappa) = \Phi(\kappa - S_j)$$

$$\Pr(S_j + u > \kappa) = 1 - \Phi(\kappa - S_j) = \Phi(S_j - \kappa)$$

$$\Pr(\kappa_1 < S_j + u < \kappa_2) = \Phi(\kappa_2 - S_j) - \Phi(\kappa_1 - S_j)$$

Rather than using `predict` directly, we could calculate the predicted probabilities by hand.

```
. predict pscore, xb
. generate probexc = normal(pscore-_b[/cut4])
. generate probgood = normal(_b[/cut4]-pscore) - normal(_b[/cut3]-pscore)
```

❑

Methods and Formulas

All postestimation tools listed above are implemented as ado-files.

Also See

Complementary:	[R] **oprobit**; [R] **adjust**, [R] **estimates**, [R] **lincom**, [R] **linktest**, [R] **lrtest**, [R] **mfx**, [R] **nlcom**, [R] **predictnl**, [R] **suest**, [R] **test**, [R] **testnl**
Background:	[U] **13.5 Accessing coefficients and standard errors**, [U] **20 Estimation and postestimation commands**, [R] **estat**, [R] **predict**

Title

---
orthog — Orthogonalize variables and compute orthogonal polynomials
---

Syntax

Orthogonalize variables

orthog $\left[\,varlist\,\right]$ $\left[\,if\,\right]$ $\left[\,in\,\right]$ $\left[\,weight\,\right]$, generate(*newvarlist*) $\left[\,\underline{\text{matr}}\text{ix}(matname)\,\right]$

Compute orthogonal polynomial

orthpoly *varname* $\left[\,if\,\right]$ $\left[\,in\,\right]$ $\left[\,weight\,\right]$,

$\left\{\,\underline{\text{generate}}(newvarlist)\,|\,\underline{\text{poly}}(matname)\,\right\}$ $\left[\,\underline{\text{degree}}(\#)\,\right]$

orthpoly requires that generate(*newvarlist*) or poly(*matname*), or both, be specified.

varlist may contain time-series operators; see [U] **11.4.3 Time-series varlists**.

iweights, fweights, pweights, and aweights are allowed, see [U] **11.1.6 weight**.

Description

orthog orthogonalizes a set of variables, creating a new set of orthogonal variables (all of type double), using a modified Gram–Schmidt procedure (Golub and Van Loan 1996). Note that the order of the variables determines the orthogonalization; hence, the "most important" variables should be listed first.

Note that execution time is proportional to the square of the number of variables. With a large number (> 10) of variables, orthog will be fairly slow.

orthpoly computes orthogonal polynomials for a single variable.

Options for orthog

    ┌─ Main ───

generate(*newvarlist*) is required. generate() creates new orthogonal variables of type double. For orthog, *newvarlist* will contain the orthogonalized *varlist*. If *varlist* contains d variables, then so will *newvarlist*. *newvarlist* can be specified by giving a list of exactly d new variable names, or it can be abbreviated using the styles *newvar1*- *newvard* or *newvar**. For these two styles of abbreviation, new variables *newvar1*, *newvar2*, ..., *newvard* are generated.

matrix(*matname*) creates a $(d+1) \times (d+1)$ matrix containing the matrix R defined by $X = QR$, where X is the $N \times (d+1)$ matrix representation of *varlist* plus a column of ones and Q is the $N \times (d+1)$ matrix representation of *newvarlist* plus a column of ones (d = number of variables in *varlist*, and N = number of observations).

Options for orthpoly

generate(*newvarlist*) or poly(), or both, must be specified. generate() creates new orthogonal variables of type double. *newvarlist* will contain orthogonal polynomials of degree 1, 2, ..., d evaluated at *varname*, where d is as specified by degree(d). *newvarlist* can be specified by giving a list of exactly d new variable names, or it can be abbreviated using the styles *newvar1–newvard* or *newvar**. For these two styles of abbreviation, new variables *newvar1*, *newvar2*, ..., *newvard* are generated.

poly(*matname*) creates a $(d + 1) \times (d + 1)$ matrix called *matname* containing the coefficients of the orthogonal polynomials. The orthogonal polynomial of degree $i \leq d$ is

$$matname[i, d+1] + matname[i, 1]*varname + matname[i, 2]*varname^2$$
$$+ \cdots + matname[i, i]*varname^i$$

Note that the coefficients corresponding to the constant term are placed in the last column of the matrix. The last row of the matrix is all zero, except for the last column, which corresponds to the constant term.

degree(*#*) specifies the highest-degree polynomial to include. Orthogonal polynomials of degree 1, 2, ..., $d = \#$ are computed. The default is $d = 1$.

Remarks

Orthogonal variables are useful for two reasons. The first is numerical accuracy for highly collinear variables. Stata's regress and other estimation commands can face a large amount of collinearity and still produce accurate results. But, at some point, these commands will drop variables due to collinearity. If you know with certainty that the variables are not perfectly collinear, you may want to retain all of their effects in the model. If you use orthog or orthpoly to produce a set of orthogonal variables, all variables will be present in the estimation results.

Users are more likely to find orthogonal variables useful for the second reason: ease of interpreting results. orthog and orthpoly create a set of variables such that the "effects" of all the preceding variables have been removed from each variable. For example, if we issue the command

```
. orthog x1 x2 x3, generate(q1 q2 q3)
```

the effect of the constant is removed from x1 to produce q1; the constant and x1 are removed from x2 to produce q2; and finally the constant, x1, and x2 are removed from x3 to produce q3. Hence,

$$q1 = r_{01} + r_{11}\,x1$$
$$q2 = r_{02} + r_{12}\,x1 + r_{22}\,x2$$
$$q3 = r_{03} + r_{13}\,x1 + r_{23}\,x2 + r_{33}\,x3$$

This can be generalized and written in matrix notation as

$$X = QR$$

where X is the $N \times (d + 1)$ matrix representation of *varlist* plus a column of ones, and Q is the $N \times (d + 1)$ matrix representation of *newvarlist* plus a column of ones (d = number of variables in *varlist* and N = number of observations). The $(d + 1) \times (d + 1)$ matrix R is a permuted upper-triangular matrix, i.e., R would be upper triangular if the constant were first, but the constant is last, so the first row/column has been permuted with the last row/column. Since Stata's estimation commands list the constant term last, this allows R, obtained via the matrix() option, to be used to transform estimation results.

▷ Example 1

Consider Stata's `auto.dta` dataset. Suppose that we postulate a model in which `price` depends on the car's `length`, `weight`, headroom (`headroom`), and trunk size (`trunk`). These predictors are collinear, but not extremely so—the correlations are not that close to 1:

```
. use http://www.stata-press.com/data/r9/auto
(1978 Automobile Data)

. correlate length weight headroom trunk
(obs=74)
```

	length	weight	headroom	trunk
length	1.0000			
weight	0.9460	1.0000		
headroom	0.5163	0.4835	1.0000	
trunk	0.7266	0.6722	0.6620	1.0000

`regress` certainly has no trouble fitting this model:

```
. regress price length weight headroom trunk
```

Source	SS	df	MS
Model	236016580	4	59004145
Residual	399048816	69	5783316.17
Total	635065396	73	8699525.97

```
Number of obs =      74
F(  4,    69) =   10.20
Prob > F      =  0.0000
R-squared     =  0.3716
Adj R-squared =  0.3352
Root MSE      =  2404.9
```

price	Coef.	Std. Err.	t	P>|t|	[95% Conf. Interval]
length	-101.7092	42.12534	-2.41	0.018	-185.747 -17.67147
weight	4.753066	1.120054	4.24	0.000	2.518619 6.987512
headroom	-711.5679	445.0204	-1.60	0.114	-1599.359 176.2236
trunk	114.0859	109.9488	1.04	0.303	-105.2559 333.4277
_cons	11488.47	4543.902	2.53	0.014	2423.638 20553.31

However, we may believe *a priori* that `length` is the most important predictor, followed by `weight`, `headroom`, and `trunk`. We would like to remove the "effect" of `length` from all the other predictors, remove `weight` from `headroom` and `trunk`, and remove `headroom` from `trunk`. We can do this by running `orthog`, and then we fit the model again using the orthogonal variables:

```
. orthog length weight headroom trunk, gen(olength oweight oheadroom otrunk) matrix(R)

. regress price olength oweight oheadroom otrunk
```

Source	SS	df	MS
Model	236016580	4	59004145
Residual	399048816	69	5783316.17
Total	635065396	73	8699525.97

```
Number of obs =      74
F(  4,    69) =   10.20
Prob > F      =  0.0000
R-squared     =  0.3716
Adj R-squared =  0.3352
Root MSE      =  2404.9
```

price	Coef.	Std. Err.	t	P>|t|	[95% Conf. Interval]
olength	1265.049	279.5584	4.53	0.000	707.3454 1822.753
oweight	1175.765	279.5584	4.21	0.000	618.0617 1733.469
oheadroom	-349.9916	279.5584	-1.25	0.215	-907.6955 207.7122
otrunk	290.0776	279.5584	1.04	0.303	-267.6262 847.7815
_cons	6165.257	279.5584	22.05	0.000	5607.553 6722.961

Using the matrix R, we can transform the results obtained using the orthogonal predictors back to the metric of original predictors:

```
. matrix b = e(b)*inv(R)'
. matrix list b

b[1,5]
        length      weight     headroom        trunk        _cons
y1   -101.70924   4.7530659   -711.56789    114.08591    11488.475
```

◁

❏ Technical Note

The matrix R obtained using the matrix() option with orthog can also be used to recover X (the original *varlist*) from Q (the orthogonalized *newvarlist*), one variable at a time. Continuing with the previous example, we illustrate how to recover the trunk variable:

```
. matrix C = R[1...,"trunk"]'
. matrix score double rtrunk = C
. compare rtrunk trunk
```

	count	minimum	difference average	maximum
rtrunk>trunk	74	8.88e-15	1.92e-14	3.55e-14
jointly defined	74	8.88e-15	1.92e-14	3.55e-14
total	74			

In this example, the recovered variable rtrunk is almost exactly the same as the original trunk variable. When you are orthogonalizing many variables, this procedure can be performed to check the numerical soundness of the orthogonalization. Because of the ordering of the orthogonalization procedure, the last variable and the variables near the end of the *varlist* are the most important ones to check.

❏

The orthpoly command effectively does for polynomial terms what the orthog command does for an arbitrary set of variables.

▷ Example 2

Again consider the auto.dta dataset. Suppose that we wish to fit the model

$$\texttt{mpg} = \beta_0 + \beta_1 \texttt{weight} + \beta_2 \texttt{weight}^2 + \beta_3 \texttt{weight}^3 + \beta_4 \texttt{weight}^4 + \epsilon$$

We will first compute the regression with natural polynomials:

```
. gen double w1 = weight
. gen double w2 = w1*w1
. gen double w3 = w2*w1
. gen double w4 = w3*w1
```

```
. correlate w1-w4
(obs=74)
```

	w1	w2	w3	w4
w1	1.0000			
w2	0.9915	1.0000		
w3	0.9665	0.9916	1.0000	
w4	0.9279	0.9679	0.9922	1.0000

```
. regress mpg w1-w4
```

Source	SS	df	MS
Model	1652.73666	4	413.184164
Residual	790.722803	69	11.4597508
Total	2443.45946	73	33.4720474

Number of obs = 74
$F(4, 69) = 36.06$
Prob > F = 0.0000
R-squared = 0.6764
Adj R-squared = 0.6576
Root MSE = 3.3852

mpg	Coef.	Std. Err.	t	P>|t|	[95% Conf.	Interval]
w1	.0289302	.1161939	0.25	0.804	-.2028704	.2607307
w2	-.0000229	.0000566	-0.40	0.687	-.0001359	.0000901
w3	5.74e-09	1.19e-08	0.48	0.631	-1.80e-08	2.95e-08
w4	-4.86e-13	9.14e-13	-0.53	0.596	-2.31e-12	1.34e-12
_cons	23.94421	86.60667	0.28	0.783	-148.8314	196.7198

Some of the correlations among the powers of `weight` are very large, but this does not create any problems for `regress`. However, we may wish to look at the quadratic trend with the constant removed, the cubic trend with the quadratic and constant removed, etc. `orthpoly` will generate polynomial terms with this property:

```
. orthpoly weight, generate(pw*) deg(4) poly(P)
. regress mpg pw1-pw4
```

Source	SS	df	MS
Model	1652.73666	4	413.184164
Residual	790.722803	69	11.4597508
Total	2443.45946	73	33.4720474

Number of obs = 74
$F(4, 69) = 36.06$
Prob > F = 0.0000
R-squared = 0.6764
Adj R-squared = 0.6576
Root MSE = 3.3852

mpg	Coef.	Std. Err.	t	P>|t|	[95% Conf.	Interval]
pw1	-4.638252	.3935245	-11.79	0.000	-5.423312	-3.853192
pw2	.8263545	.3935245	2.10	0.039	.0412947	1.611414
pw3	-.3068616	.3935245	-0.78	0.438	-1.091921	.4781982
pw4	-.209457	.3935245	-0.53	0.596	-.9945168	.5756028
_cons	21.2973	.3935245	54.12	0.000	20.51224	22.08236

Compare the p-values of the terms in the natural-polynomial regression with those in the orthogonal-polynomial regression. With orthogonal polynomials, it is easy to see that the pure cubic and quartic trends are not significant and that the constant, linear, and quadratic terms each have $p < 0.05$.

The matrix P obtained with the `poly()` option can be used to transform coefficients for orthogonal polynomials to coefficients for natural polynomials:

```
. orthpoly weight, poly(P) deg(4)
. matrix b = e(b)*P
```

```
. matrix list b
b[1,5]
          deg1        deg2        deg3        deg4       _cons
y1    .02893016   -.00002291   5.745e-09   -4.862e-13   23.944212
```

◁

Methods and Formulas

orthog and orthpoly are implemented as ado-files.

orthog's orthogonalization can be written in matrix notation as

$$X = QR$$

where X is the $N \times (d + 1)$ matrix representation of *varlist* plus a column of ones and Q is the $N \times (d + 1)$ matrix representation of *newvarlist* plus a column of ones (d = number of variables in *varlist*, and N = number of observations). The $(d + 1) \times (d + 1)$ matrix R is a permuted upper-triangular matrix, i.e., R would be upper triangular if the constant were first, but the constant is last, so the first row/column has been permuted with the last row/column.

Q and R are obtained using a modified Gram–Schmidt procedure; see Golub and Van Loan (1996, 218–219) for details. Note that the traditional Gram–Schmidt procedure is notoriously unsound, but the modified procedure is quite good. orthog performs two passes of this procedure.

orthpoly uses the Christoffel–Darboux recurrence formula (Abramowitz and Stegun 1968).

Both orthog and orthpoly normalize the orthogonal variables such that

$$Q'WQ = MI$$

where $W = \text{diag}(w_1, w_2, \ldots, w_N)$ with weights $w_1, w_2, \ldots, w_N$ (all 1 if weights are not specified), and M is the sum of the weights (the number of observations if weights are not specified).

References

Abramowitz, M. and I. A. Stegun, ed. 1968. *Handbook of Mathematical Functions*, 7th printing. Washington, DC: National Bureau of Standards.

Golub, G. H. and C. F. Van Loan. 1996. *Matrix Computations*. 3rd ed. Baltimore: Johns Hopkins University Press.

Sribney, W. M. 1995. sg37: Orthogonal polynomials. *Stata Technical Bulletin* 25: 17–18. Reprinted in *Stata Technical Bulletin Reprints*, vol. 5, pp. 96–98.

Also See

Complementary:	[P] **matrix**
Related:	[R] **regress**
Background:	[U] **20 Estimation and postestimation commands**

Title

> **pcorr** — Partial correlation coefficients

Syntax

pcorr *varname₁* *varlist* [*if*] [*in*] [*weight*]

*varname*₁ and *varlist* may contain time-series operators; see [U] **11.4.3 Time-series varlists**.

by may be used with pcorr; see [D] **by**.

aweights and fweights are allowed; see [U] **11.1.6 weight**.

Description

pcorr displays the partial correlation coefficient of *varname₁* with each variable in *varlist*, holding the other variables in *varlist* constant.

Remarks

Assume that y is determined by x_1, x_2, ..., x_k. The partial correlation between y and x_1 is an attempt to estimate the correlation that would be observed by y and x_1 if the other xs did not vary.

▷ Example 1

Using our automobile dataset (described in [U] **1.2.1 Sample datasets**), we can obtain the simple correlations between price, mpg, weight, and foreign from correlate (see [R] **correlate**):

```
. use http://www.stata-press.com/data/r9/auto
(1978 Automobile Data)

. correlate price mpg weight foreign
(obs=74)
```

	price	mpg	weight	foreign
price	1.0000			
mpg	-0.4686	1.0000		
weight	0.5386	-0.8072	1.0000	
foreign	0.0487	0.3934	-0.5928	1.0000

Although correlate gave us the full correlation matrix, our interest is in just the first column. We find, for instance, that the higher the mpg, the lower the price. We obtain the partial correlation coefficients using pcorr:

```
. pcorr price mpg weight foreign
(obs=74)
```

Partial correlation of price with

Variable	Corr.	Sig.
mpg	0.0352	0.769
weight	0.5488	0.000
foreign	0.5402	0.000

We now find that, holding `weight` and `foreign` constant, the partial correlation of `price` with `mpg` is virtually zero. Similarly, in the simple correlations, we found that `price` and `foreign` were virtually uncorrelated. In the partial correlations—holding `mpg` and `weight` constant—we find that `price` and `foreign` are positively correlated.

◁

❏ Technical Note

Some caution is in order when interpreting the above results. As we said at the outset, the partial correlation coefficient is an *attempt* to estimate the correlation that would be observed if the other variables were held constant. `pcorr` makes it too easy to ignore the fact that we are fitting a model. In the example above, the model is

$$\text{price} = \beta_0 + \beta_1 \text{mpg} + \beta_2 \text{weight} + \beta_3 \text{foreign} + \epsilon$$

which is, in all honesty, a rather silly model. Even if we accept the implied economic assumptions of the model—that consumers value `mpg`, `weight`, and `foreign`—do we really believe that consumers place equal value on every extra 1,000 pounds of weight? That is, have we correctly parameterized the model? If we have not, then the estimated partial correlation coefficients may not represent what they claim to represent. Partial correlation coefficients are a reasonable way to summarize data if we are convinced that the underlying model is reasonable. We should not, however, pretend that there is no underlying model and that the partial correlation coefficients are unaffected by the assumptions and parameterization.

❏

Methods and Formulas

`pcorr` is implemented as an ado-file.

Results are obtained by fitting a linear regression of *varname*$_1$ on *varlist*; see [R] **regress**. The partial correlation coefficient between *varname*$_1$ and each variable in *varlist* is then defined as

$$\frac{t}{\sqrt{t^2 + n - k}}$$

(Theil 1971, 174), where t is the t statistic, n is the number of observations, and k is the number of independent variables, including the constant but excluding any dropped variables. The significance is given by $2 * \text{ttail}(n - k, \text{abs}(t))$.

References

Theil, H. 1971. *Principles of Econometrics*. New York: Wiley.

Also See

Related: [R] **correlate**, [R] **spearman**

Title

> **permute** — Monte Carlo permutation tests

Syntax

Compute permutation test

> permute *permvar* *exp_list* [, *options*] : *command*

Report saved results

> permute [*varlist*] [using *filename*] [, *display_options*]

options	description
Main	
<u>r</u>eps(*#*)	perform *#* random permutations; default is reps(100)
<u>l</u>eft	<u>r</u>ight	compute one-sided *p*-values; default is two-sided
Options	
<u>str</u>ata(*varlist*)	permute within strata
<u>sa</u>ving(*filename*, …)	save results to *filename*; save statistics in double precision; save results to *filename* every *#* replications
Reporting	
<u>l</u>evel(*#*)	set confidence level; default is level(95)
<u>noh</u>eader	suppress the table header
<u>nol</u>egend	suppress the table legend
<u>v</u>erbose	display the full table legend
nodots	suppress the replication dots
<u>noi</u>sily	display any output from *command*
<u>trace</u>	trace the *command*
<u>title</u>(*text*)	use *text* as title for permutation results
Advanced	
eps(*#*)	numerical tolerance; seldom used
nodrop	do not drop observations
nowarn	do not warn when e(sample) is not set
force	do not check for *weights* or svy commands; seldom used
reject(*exp*)	identify invalid results
seed(*#*)	set random-number seed to *#*

weights are not allowed in *command*.

(Continued on next page)

display_options	description
<u>l</u>eft | <u>r</u>ight	compute one-sided p-values; default is two-sided
<u>l</u>evel(*#*)	set confidence level; default is level(95)
<u>no</u>header	suppress the table header
<u>no</u>legend	suppress the table legend
<u>v</u>erbose	display the full table legend
<u>t</u>itle(*text*)	use *text* as title for results
eps(*#*)	numerical tolerance; seldom used

exp_list contains	(*name*: *elist*)
	elist
	eexp
elist contains	*newvar* = (*exp*)
	(*exp*)
eexp is	*specname*
	[*eqno*]*specname*
specname is	_b
	_b[]
	_se
	_se[]
eqno is	# #
	name

exp is a standard Stata expression; see [U] **13 Functions and expressions**.

Distinguish between [ ], which are to be typed, and [ ], which indicate optional arguments.

Description

permute estimates p-values for permutation tests based on Monte Carlo simulations. Typing

    . permute *permvar* *exp_list*, reps(*#*): *command*

randomly permutes the values in *permvar* # times, each time executing *command* and collecting the associated values from the expression in *exp_list*.

These p-value estimates can be one-sided: $\Pr(T^* \leq T)$ or $\Pr(T^* \geq T)$. The default is two-sided: $\Pr(|T^*| \geq |T|)$. Here T^* denotes the value of the statistic from a randomly permuted dataset, and T denotes the statistic as computed on the original data.

permvar identifies the variable whose observed values will be randomly permuted.

command defines the statistical command to be executed. Most Stata commands and user-written programs can be used with permute, as long as they follow standard Stata syntax; see [U] **11 Language syntax**. The by prefix may not be part of *command*.

exp_list specifies the statistics to be retrieved after the execution of *command*.

permute may be used for replaying results, but this feature is only appropriate when a dataset generated by permute is currently in memory or is identified by the using option. The variables specified in *varlist* in this context must be present in the respective dataset.

Options

reps(#) specifies the number of random permutations to perform. The default is 100.

left or right request that one-sided *p*-values be computed. If left is specified, an estimate of $\Pr(T^* \leq T)$ is produced, where T^* is the test statistic and T is its observed value. If right is specified, an estimate of $\Pr(T^* \geq T)$ is produced. By default, two-sided *p*-values are computed; that is, $\Pr(|T^*| \geq |T|)$ is estimated.

strata(*varlist*) specifies that the permutations be performed within each stratum defined by the values of *varlist*.

saving(*filename*[, *suboptions*]) creates a Stata data file (.dta file) consisting of, for each statistic in *exp_list*, a variable containing the permutation replicates.

  double specifies that the results for each replication be stored as doubles, meaning 8-byte reals. By default, they are stored as floats, meaning 4-byte reals.

  every(#) specifies that results are to be written to disk every #th replication. every() should only be specified in conjunction with saving() when *command* takes a long time for each replication. This will allow recovery of partial results should some other software crash your computer. See [P] **postfile**.

  replace indicates that *filename* may exist, and, if it does, it should be overwritten. This option is not shown in the dialog box.

level(#) specifies the confidence level, as a percentage, for confidence intervals. The default is level(95) or as set by set level; see [R] **level**.

noheader suppresses display of the table header. This option implies nolegend.

nolegend suppresses display of the table legend. The table legend identifies the rows of the table with the expressions they represent.

verbose requests that the full table legend be displayed. By default, coefficients and standard errors are not displayed.

nodots suppresses display of the replication dots. By default, a single dot character is displayed for each successful replication. A single red 'x' is displayed if *command* returns an error or if one of the values in *exp_list* is missing.

noisily requests that any output from *command* be displayed. This option implies nodots.

trace causes a trace of the execution of *command* to be displayed. This option implies noisily.

title(*text*) specifies a title to be displayed above the table of permutation results; the default title is Monte Carlo permutation results.

```
            Advanced
```

eps(#) specifies the numerical tolerance for testing $|T^*| \geq |T|$, $T^* \leq T$, or $T^* \geq T$. These are considered true if, respectively, $|T^*| \geq |T|-\#$, $T^* \leq T+\#$, or $T^* \geq T-\#$. The default is 1e-7. You will not have to specify eps() under normal circumstances.

nodrop is a rarely used option to keep from dropping observations outside the if and in qualifiers. This option has no effect when neither if nor in is supplied.

nowarn suppresses the printing of a warning message when *command* does not set e(sample).

force suppresses the restriction that *command* may not specify weights or be a svy command. permute is not suited for weighted estimation, thus permute should not be used with weights or svy. permute reports an error when it encounters weights or svy in *command* if the force option is not specified. This is a seldom used option, so use it only if you know what you are doing!

reject(*exp*) identifies an expression that indicates when results should be rejected. When *exp* is true, the resulting values are reset to missing values.

seed(#) sets the random-number seed. Specifying this option is equivalent to typing the following command prior to calling permute:

. set seed #

Remarks

Permutation tests determine the significance of the observed value of a test statistic in light of rearranging the order (permuting) of the observed values of a variable.

▷ Example 1

Suppose that we conducted an experiment to determine the effect of a treatment on the development of cells. Further suppose that we are restricted to six experimental units due to the extreme cost of the experiment. Thus three units are to be given a placebo, and three units are given the treatment. The measurement is the number of newly developed healthy cells. The following listing gives the hypothetical data, along with some summary statistics.

```
. input y treatment
          y  treatment
1. 7 0
2. 9 0
3. 11 0
4. 10 1
5. 12 1
6. 14 1
7. end

. sort treatment

. summarize y
```

Variable	Obs	Mean	Std. Dev.	Min	Max
y	6	10.5	2.428992	7	14

```
. by treatment: summarize y
```

```
-> treatment = 0
    Variable |        Obs        Mean    Std. Dev.        Min        Max
-------------+-----------------------------------------------------------
           y |          3           9           2           7         11
```

```
-> treatment = 1
    Variable |        Obs        Mean    Std. Dev.        Min        Max
-------------+-----------------------------------------------------------
           y |          3          12           2          10         14
```

Clearly, there are more cells in the treatment group than in the placebo group, but a statistical test is needed to conclude that the treatment does affect the development of cells. If the sum of the treatment measures is our test statistic, we can use `permute` to determine the probability of observing 36 or more cells, given the observed data and assuming that there is no effect due to the treatment.

```
. set seed 1234

. permute y sum=r(sum), saving(permdish) right nodrop nowarn: sum y if treatment
(running summarize on estimation sample)

Permutation replications (100)
----+--- 1 ---+--- 2 ---+--- 3 ---+--- 4 ---+--- 5
..................................................    50
..................................................   100

Monte Carlo permutation results                Number of obs    =          6
        command:  summarize y if treatment
            sum:  r(sum)
    permute var:  y
```

T	T(obs)	c	n	p=c/n	SE(p)	[95% Conf. Interval]
sum	36	9	100	0.0900	0.0286	.0419836 .1639823

```
Note:  Confidence interval is with respect to p=c/n.
Note:  c = #{T >= T(obs)}
```

We see that 9 of the 100 randomly permuted datasets yielded sums from the treatment group larger than or equal to the observed sum of 36. Thus the evidence is not strong enough, at the 5% level, to reject the null hypothesis that there is no effect of the treatment.

Note that due to the very small size of this experiment, we could have calculated the exact permutation p-value from all possible permutations. There are 6 units, but we want the sum of the treatment units. Thus there are $\binom{6}{3} = 20$ permutation sums from the possible unique permutations.

$$
\begin{array}{llll}
7+9+10 = 26 & 7+10+12 = 29 & 9+10+11 = 30 & 9+12+14 = 35 \\
7+9+11 = 27 & 7+10+14 = 31 & 9+10+12 = 31 & 10+11+12 = 33 \\
7+9+12 = 28 & 7+11+12 = 30 & 9+10+14 = 33 & 10+11+14 = 35 \\
7+9+14 = 30 & 7+11+14 = 32 & 9+11+12 = 32 & 10+12+14 = 36 \\
7+10+11 = 28 & 7+12+14 = 33 & 9+11+14 = 34 & 11+12+14 = 37
\end{array}
$$

Two of the 20 permutation sums are greater than or equal to 36. Thus the exact p-value for this permutation test is 0.1. Note that tied values will decrease the number of unique permutations.

When the `saving()` option is supplied, `permute` saves the values of the permutation statistic to the indicated file, in our case, `permdish.dta`. This file can be used to replay the result of `permute`. Note that the `level()` option controls the confidence level of the confidence interval for the permutation p-value. This confidence interval is calculated using `cii` with the reported n (number of nonmissing replications) and c (the counter for events of significance).

```
. permute using permdish, level(80)
Monte Carlo permutation results              Number of obs    =        6
       command:  summarize y if treatment
           sum:  r(sum)
    permute var:  y
```

T		T(obs)	c	n	p=c/n	SE(p)	[80% Conf. Interval]
	sum	36	9	100	0.0900	0.0286	.0550468	.1383518

```
Note:  Confidence interval is with respect to p=c/n.
Note:  c = #{|T| >= |T(obs)|}
```

◁

▷ Example 2

Consider some fictional data from a randomized complete-block design in which we wish to determine the significance of five treatments.

```
. use http://www.stata-press.com/data/r9/permute1
. list y treatment in 1/10, abbrev(10)
```

	y	treatment
1.	4.407557	1
2.	5.693386	1
3.	7.099699	1
4.	3.12132	1
5.	5.242648	1
6.	4.280349	2
7.	4.508785	2
8.	4.079967	2
9.	5.904368	2
10.	3.010556	2

These data may be analyzed using `anova`.

```
. anova y treatment subject
```

| | | Number of obs = | 50 | R-squared | = | 0.3544 |
| | | Root MSE | = .914159 | Adj R-squared = | | 0.1213 |

Source	Partial SS	df	MS	F	Prob > F
Model	16.5182188	13	1.27063221	1.52	0.1574
treatment	13.0226706	9	1.44696341	1.73	0.1174
subject	3.49554813	4	.873887032	1.05	0.3973
Residual	30.0847503	36	.835687509		
Total	46.6029691	49	.951081002		

Suppose that we want to compute the significance of the F statistic for `treatment` by using `permute`. All we need to do is write a short program that will save the result of this statistic for `permute` to use. For example,

```
program panova, rclass
        version 9
        args response fac_intrst fac_other
        anova 'response' 'fac_intrst' 'fac_other'
        return scalar Fmodel = e(F)
        test 'fac_intrst'
        return scalar F = r(F)
end
```

Now in `panova`, `test` saves the F statistic for the factor of interest in `r(F)`. This is different from `e(F)`, which is the overall model F statistic for the model fitted by `anova` that `panova` saves in `r(Fmodel)`. In the following example, we use the `strata()` option so that the treatments are randomly rearranged within each `subject`. It should not be too surprising that the estimated p-values are equal for this example, since the two F statistics are equivalent when controlling for differences between subjects. However, we would not expected to always get the same p-values every time we reran `permute`.

```
. set seed 1234
. permute treatment treatmentF=r(F) modelF=r(Fmodel), reps(1000) ///
> strata(subject) saving(permanova) nodots: panova y treatment subject
Monte Carlo permutation results
Number of strata =          5                    Number of obs   =          50
      command:  panova y treatment subject
   treatmentF:  r(F)
       modelF:  r(Fmodel)
   permute var:  treatment
```

T	T(obs)	c	n	p=c/n	SE(p)	[95% Conf. Interval]	
treatmentF	1.731465	117	1000	0.1170	0.0102	.097729	.1385575
modelF	1.520463	117	1000	0.1170	0.0102	.097729	.1385575

Note: Confidence intervals are with respect to p=c/n.
Note: c = #{|T| >= |T(obs)|}

◁

▷ Example 3

As a final example, let's consider estimating the p-value of the Z statistic returned by `ranksum`. Suppose that we collected data from some experiment: `y` is some measure we took on 17 individuals, and `group` identifies the group that an individual belongs to.

(Continued on next page)

```
. use http://www.stata-press.com/data/r9/permute2, clear
. list
```

	group	y
1.	1	6
2.	1	11
3.	1	20
4.	1	2
5.	1	9
6.	1	5
7.	0	2
8.	0	1
9.	0	6
10.	0	0
11.	0	2
12.	0	3
13.	0	3
14.	0	12
15.	0	4
16.	0	1
17.	0	5

Next we analyze the data using `ranksum` and notice that the observed value of the test statistic (saved as `r(z)`) is -2.02 with an approximate p-value of 0.0434.

```
. ranksum y, by(group)
Two-sample Wilcoxon rank-sum (Mann-Whitney) test
```

group	obs	rank sum	expected
0	11	79	99
1	6	74	54
combined	17	153	153

```
unadjusted variance       99.00
adjustment for ties       -0.97
                         ------
adjusted variance         98.03
Ho: y(group==0) = y(group==1)
          z =   -2.020
   Prob > |z| =    0.0434
```

The observed value of the rank-sum statistic is 79, with an expected value (under the null hypothesis of no group effect) of 99. There are 17 observations, so the permutation distribution contains $\binom{17}{6} = 12{,}376$ possible values of the rank-sum statistic if we ignore ties. With ties, we have fewer possible values, but still too many to want to enumerate them. Thus we use `permute` with 10,000 replications and see that the Monte Carlo permutation test agrees with the result of the test based on the normal approximation.

```
. set seed 18385766
```

```
. permute y z=r(z), reps(10000) nowarn nodots: ranksum y, by(group)
Monte Carlo permutation results                    Number of obs    =         17
       command:  ranksum y, by(group)
             z:  r(z)
    permute var:  y
```

T		T(obs)	c	n	p=c/n	SE(p)	[95% Conf. Interval]
	z	-2.020002	444	10000	0.0444	0.0021	.040446	.048622

```
Note:  Confidence interval is with respect to p=c/n.
Note:  c = #{|T| >= |T(obs)|}
```

◁

❏ Technical Note

permute reports confidence intervals for p to emphasize that it is based on the binomial estimator for proportions. In cases where the variability implied by the confidence interval makes conclusions difficult, you may increase the number of replications to determine more precisely the significance of the test statistic of interest. In other words, the value of p from permute will converge to the true permutation p-value as the number of replications gets arbitrarily large.   ❏

Saved Results

permute saves in r():

Scalars
r(N)	sample size	r(level)	confidence level
r(N_reps)	number of requested replications		

Macros
r(cmd)	permute	r(left)	left or empty
r(command)	*command* following colon	r(right)	right or empty
r(permvar)	permutation variable	r(seed)	initial random-number seed
r(title)	title in output	r(event)	T <= T(obs), T >= T(obs),
			or |T| <= |T(obs)|

Matrices
r(b)	observed statistics	r(p)	observed proportions
r(c)	count when r(event) is true	r(se)	standard errors of observed proportions
r(reps)	number of nonmissing results	r(ci)	confidence intervals of observed proportions

Methods and Formulas

permute is implemented as an ado-file.

References

Good, P. I. 2001. *Resampling Methods: A Practical Guide to Data Analysis.* 2nd ed. Boston: Birkhäuser.

Also See

Complementary: [P] **postfile**

Related: [R] **bootstrap**, [R] **ci**, [R] **jackknife**, [R] **simulate**

Title

> **pk** — Pharmacokinetic (biopharmaceutical) data

Description

The term pk refers to pharmacokinetic data and the Stata commands, all of which begin with the letters pk, designed to do some of the analyses commonly performed in the pharmaceutical industry. The system is intended for the analysis of pharmacokinetic data, although some of the commands are of general use.

The pk commands are

pkexamine	[R] **pkexamine**	Calculate pharmacokinetic measures
pksumm	[R] **pksumm**	Summarize pharmacokinetic data
pkshape	[R] **pkshape**	Reshape (pharmacokinetic) Latin-square data
pkcross	[R] **pkcross**	Analyze crossover experiments
pkequiv	[R] **pkequiv**	Perform bioequivalence tests
pkcollapse	[R] **pkcollapse**	Generate pharmacokinetic measurement dataset

Remarks

Several types of clinical trials are commonly performed in the pharmaceutical industry. Examples include combination trials, multicenter trials, equivalence trials, and active control trials. For each type of trial, there is an optimal study design for estimating the effects of interest. Currently, the pk system can be used to analyze equivalence trials, which are usually conducted using a crossover design; however, it is possible to use a parallel design and still draw conclusions about equivalence.

Equivalence trials assess bioequivalence between two drugs. While it is impossible to prove that two drugs behave exactly the same way, the United States Food and Drug Administration believes that if the absorption properties of two drugs are similar, the two drugs will produce similar effects and have similar safety profiles. Generally, the goal of an equivalence trial is to assess the equivalence of a generic drug with an existing drug. This is commonly accomplished by comparing a confidence interval about the difference between a pharmacokinetic measurement of two drugs with a confidence limit constructed from U.S. federal regulations. If the confidence interval is entirely within the confidence limit, the drugs are declared bioequivalent. An alternative approach to assessing bioequivalence is to use the method of interval hypotheses testing. pkequiv is used to conduct these tests of bioequivalence.

Several pharmacokinetic measures can be used to ascertain how available a drug is for cellular absorption. The most common measure is the area under the time-versus-concentration curve (AUC). Another common measure of drug availability is the maximum concentration (C_{max}) achieved by the drug during the follow-up period. Stata reports these and other less-common measures of drug availability, including the time at which the maximum drug concentration was observed and the duration of the period during which the subject was being measured. Stata also reports the elimination rate, that is, the rate at which the drug is metabolized; and the drug's half-life, that is, the time it takes for the drug concentration to fall to one-half of its maximum concentration.

pkexamine computes and reports all the pharmacokinetic measures that Stata produces, including four calculations of the area under the time-versus-concentration curve. The standard area under the curve from 0 to the maximum observed time ($AUC_{0,t_{max}}$) is computed using cubic splines or the trapezoidal rule. Additionally, pkexamine also computes the area under the curve from 0 to infinity by extending the standard time-versus-concentration curve from the maximum observed time using three different methods. The first method simply extends the standard curve using a least-squares linear fit through the last few data points. The second method extends the standard curve by fitting a decreasing exponential curve through the last few data points. Lastly, the third method extends the curve by fitting a least-squares linear regression line on the log concentration. The mathematical details of these extensions are described in *Methods and Formulas* of [R] **pkexamine**.

Data from an equivalence trial may also be analyzed using methods appropriate to the particular study design. When you have a crossover design, pkcross can be used to fit an appropriate ANOVA model. As an aside, a crossover design is simply a restricted Latin square; therefore, pkcross can also be used to analyze any Latin-square design.

There are some practical concerns when dealing with data from equivalence trials. Primarily, the data must be organized in a manner that Stata can use. The pk commands include pkcollapse and pkshape, which are designed to help transform data from a common format to one that is suitable for analysis with Stata.

In the following example, we illustrate several different data formats that are frequently encountered in pharmaceutical research and describe how these formats can be transformed to formats that can be analyzed with Stata.

▷ Example 1

Assume that we have one subject and are interested in determining the drug profile for that subject. A reasonable experiment would be to give the subject the drug and then measure the concentration of the drug in the subject's blood over a time period. For example, here is a portion of a dataset from Chow and Liu (2000, 11):

```
. use http://www.stata-press.com/data/r9/auc
. list, abbrev(14)
```

	id	time	concentration
1.	1	0	0
2.	1	.5	0
3.	1	1	2.8
4.	1	1.5	4.4
5.	1	2	4.4
6.	1	3	4.7
7.	1	4	4.1
8.	1	6	4
9.	1	8	3.6
10.	1	12	3
11.	1	16	2.5
12.	1	24	2
13.	1	32	1.6

Examining these data, we notice that the concentration quickly increases, plateaus for a short period, and then slowly decreases over time. pkexamine is used to calculate the pharmacokinetic measures of interest. pkexamine is explained in detail in [R] **pkexamine**. The output is

```
. pkexamine time conc
                              Maximum concentration =        4.7
                       Time of maximum concentration =          3
                    Time of last observation (Tmax) =         32
                                   Elimination rate =     0.0279
                                          Half life =    24.8503
```

Area under the curve

| | AUC [0, inf.) | AUC [0, inf.) | AUC [0, inf.) |
AUC [0, Tmax]	Linear of log conc.	Linear fit	Exponential fit
85.24	142.603	107.759	142.603

Fit based on last 3 points.

Clinical trials, however, require that data be collected on more than one subject. There are several ways to enter raw measured data collected on several subjects. It would be reasonable to enter for each subject the drug concentration value at specific points in time. Such data could be

```
id   conc1   conc2   conc3   conc4   conc5   conc6   conc7
 1     0       1       4       7       5       3       1
 2     0       2       6       5       4       3       2
 3     0       1       2       3       5       4       1
```

where conc1 is the concentration at the first measured time, conc2 is the concentration at the second measured time, etc. This format requires that each drug concentration measurement be made at the same time on each subject. Another more flexible way to enter the data is to have an observation with three variables for each time measurement on a subject. Each observation would have a subject ID, the time at which the measurement was made, and the corresponding drug concentration at that time. The data would be

(Continued on next page)

```
. use http://www.stata-press.com/data/r9/pkdata

. list id concA time, sepby(id)
```

	id	concA	time
1.	1	0	0
2.	1	3.073403	.5
3.	1	5.188444	1
4.	1	5.898577	1.5
5.	1	5.096378	2
6.	1	6.094085	3
7.	1	5.158772	4
8.	1	5.7065	6
9.	1	5.272467	8
10.	1	4.4576	12
11.	1	5.146423	16
12.	1	4.947427	24
13.	1	1.920421	32
14.	2	0	0
15.	2	2.48462	.5
16.	2	4.883569	1
17.	2	7.253442	1.5
18.	2	5.849345	2
19.	2	6.761085	3
20.	2	4.33839	4
21.	2	5.04199	6
22.	2	4.25128	8
23.	2	6.205004	12
24.	2	5.566165	16
25.	2	3.689007	24
26.	2	3.644063	32
27.	3	0	0
		(output omitted)	
207.	20	4.673281	24
208.	20	3.487347	32

Stata expects the data to be organized in the second form. If your data are organized as described in the first dataset, you will need to use reshape to change the data to the second form; see [D] **reshape**. Because the data in the second (or long) format contain information for one drug on several subjects, pksumm can be used to produce summary statistics of the pharmacokinetic measurements. The output is

```
. pksumm id time concA
................

Summary statistics for the pharmacokinetic measures
```

				Number of observations =	16	
Measure	Mean	Median	Variance	Skewness	Kurtosis	p-value
auc	151.63	152.18	127.58	-0.34	2.07	0.55
aucline	397.09	219.83	178276.59	2.69	9.61	0.00
aucexp	668.60	302.96	720356.98	2.67	9.54	0.00
auclog	665.95	298.03	752573.34	2.71	9.70	0.00
half	90.68	29.12	17750.70	2.36	7.92	0.00
ke	0.02	0.02	0.00	0.88	3.87	0.08
cmax	7.37	7.42	0.40	-0.64	2.75	0.36
tomc	3.38	3.00	7.25	2.27	7.70	0.00
tmax	32.00	32.00	0.00	.	.	.

Until now, we have been concerned with the profile of only one drug. We have characterized the profile of that drug by individual subjects using `pkexamine` and by a group of subjects using `pksumm`. The goal of an equivalence trial, however, is to compare two drugs, which we will do in the remainder of this example.

In the case of equivalence trials, the study design most often used is the crossover design. For a complete discussion of crossover designs, see Ratkowsky, Evans, and Alldredge (1993).

Briefly, crossover designs require that each subject be given both treatments at two different times. The order in which the treatments are applied changes between groups. For example, if we had 20 subjects numbered 1 through 20, the first 10 would receive treatment A during the first period of the study, and then they would be given treatment B. The second 10 subjects would be given treatment B during the first period of the study, and then they would be given treatment A. Each subject in the study will have four variables that describe the observation: a subject identifier, a sequence identifier that indicates the order of treatment, and two outcome variables, one for each treatment. The outcome variables for each subject are the pharmacokinetic measures. The data must be transformed from a series of measurements on individual subjects to data containing the pharmacokinetic measures for each subject. In Stata parlance, this is referred to as a collapse, which can be done with `pkcollapse`; see [R] **pkcollapse**.

Here is a portion of our data:

```
. list, sepby(id)
```

	id	seq	time	concA	concB
1.	1	1	0	0	0
2.	1	1	.5	3.073403	3.712592
3.	1	1	1	5.188444	6.230602
4.	1	1	1.5	5.898577	7.885944
5.	1	1	2	5.096378	9.241735
6.	1	1	3	6.094085	13.10507
7.	1	1	4	5.158772	.169429
8.	1	1	6	5.7065	8.759894
9.	1	1	8	5.272467	7.985409
10.	1	1	12	4.4576	7.740126
11.	1	1	16	5.146423	7.607208
12.	1	1	24	4.947427	7.588428
13.	1	1	32	1.920421	2.791115
14.	2	1	0	0	0
15.	2	1	.5	2.48462	.9209593
16.	2	1	1	4.883569	5.925818
17.	2	1	1.5	7.253442	8.710549
18.	2	1	2	5.849345	10.90552
19.	2	1	3	6.761085	8.429898
20.	2	1	4	4.33839	5.573152
21.	2	1	6	5.04199	6.32341
22.	2	1	8	4.25128	.5251224
23.	2	1	12	6.205004	7.415988
24.	2	1	16	5.566165	6.323938
25.	2	1	24	3.689007	1.133553
26.	2	1	32	3.644063	5.759489
27.	3	1	0	0	0
			(output omitted)		
207.	20	2	24	4.673281	6.059818
208.	20	2	32	3.487347	5.213639

This format is similar to the second format described above, except that now we have measurements for two drugs at each time for each subject. We transform these data using `pkcollapse`:

```
. pkcollapse time concA concB, id(id) keep(seq) stat(auc)
..............................
. list, sep(8) abbrev(10)
```

	id	seq	auc_concA	auc_concB
1.	1	1	150.9643	218.5551
2.	2	1	146.7606	133.3201
3.	3	1	160.6548	126.0635
4.	4	1	157.8622	96.17461
5.	5	1	133.6957	188.9038
6.	7	1	160.639	223.6922
7.	8	1	131.2604	104.0139
8.	9	1	168.5186	237.8962
9.	10	2	137.0627	139.7382
10.	12	2	153.4038	202.3942
11.	13	2	163.4593	136.7848
12.	14	2	146.0462	104.5191
13.	15	2	158.1457	165.8654
14.	18	2	147.1977	139.235
15.	19	2	164.9988	166.2391
16.	20	2	145.3823	158.5146

For this example, we chose to use the AUC for two drugs as our pharmacokinetic measure. Note that we could have used any of the measures computed by `pkexamine`. In addition to the AUCs, the dataset also contains a sequence variable for each subject indicating when each treatment was administered.

The data produced by `pkcollapse` are in what Stata calls wide format. That is, there is one observation per subject containing two or more outcomes. To use `pkcross` and `pkequiv`, we need to transform these data to long format. This can be accomplished using `pkshape`; see [R] **pkshape**.

Consider the first subject in the dataset. This subject is in sequence one, which means that treatment A was applied during the first period of the study and treatment B was applied in the second period of the study. We need to split the first observation into two observations so that the outcome measure is only in a single variable. In addition, we need two new variables, one indicating the treatment the subject received and another recording the period of the study when the subject received that treatment. We might expect the expansion of the first subject to be

```
id   sequence        auc    treat    period
 1          1   150.9643       A         1
 1          1   218.5551       B         2
```

We see that subject number 1 was in sequence 1, had an AUC of 150.9643 when treatment A was applied in the first period of the study and had an AUC of 218.5551 when treatment B was applied.

Similarly, the expansion of subject 10 (the first subject in sequence 2) would be

```
id   sequence        auc    treat    period
10          2   137.0627       B         1
10          2   139.7382       A         2
```

In this case, treatment B was applied to the subject during the first period of the study, and treatment A was applied to the subject during the second period of the study.

An additional complication is common in crossover study designs. The treatment applied in the first period of the study might still have some effect on the outcome in the second period. In this example, each subject was given one treatment followed by another treatment. In order to get accurate estimates of treatment effects, it is necessary to account for the effect that the first treatment has in the second period of the study. This is called the carryover effect. We must, therefore, have a variable that indicates which treatment was applied in the first treatment period. pkshape creates a variable that indicates the carryover effect. Note that for treatments applied during the first treatment period, there will never be a carryover effect. Thus the expanded data created by pkshape for subject 1 will be

id	sequence	outcome	treat	period	carry
1	1	150.9643	A	1	0
1	1	218.5551	B	2	A

and the data for subject 10 will be

id	sequence	outcome	treat	period	carry
10	2	137.0627	B	1	0
10	2	139.7382	A	2	B

We pkshape the data:

```
. pkshape id seq auc*, order(ab ba)

. sort id sequence period

. list, sep(16)
```

	id	sequence	outcome	treat	carry	period
1.	1	1	150.9643	1	0	1
2.	1	1	218.5551	2	1	2
3.	2	1	146.7606	1	0	1
4.	2	1	133.3201	2	1	2
5.	3	1	160.6548	1	0	1
6.	3	1	126.0635	2	1	2
7.	4	1	157.8622	1	0	1
8.	4	1	96.17461	2	1	2
9.	5	1	133.6957	1	0	1
10.	5	1	188.9038	2	1	2
11.	7	1	160.639	1	0	1
12.	7	1	223.6922	2	1	2
13.	8	1	131.2604	1	0	1
14.	8	1	104.0139	2	1	2
15.	9	1	168.5186	1	0	1
16.	9	1	237.8962	2	1	2
17.	10	2	137.0627	2	0	1
18.	10	2	139.7382	1	2	2
19.	12	2	153.4038	2	0	1
20.	12	2	202.3942	1	2	2
21.	13	2	163.4593	2	0	1
22.	13	2	136.7848	1	2	2
23.	14	2	146.0462	2	0	1
24.	14	2	104.5191	1	2	2
25.	15	2	158.1457	2	0	1
26.	15	2	165.8654	1	2	2
27.	18	2	147.1977	2	0	1
28.	18	2	139.235	1	2	2
29.	19	2	164.9988	2	0	1
30.	19	2	166.2391	1	2	2
31.	20	2	145.3823	2	0	1
32.	20	2	158.5146	1	2	2

As an aside, crossover designs do not require that each subject receive each treatment, but if they do, the crossover design is referred to as a complete crossover design.

The last dataset is organized in a manner that can be analyzed with Stata. To fit an ANOVA model to these data, we can use `anova` or `pkcross`. To conduct equivalence tests, we can use `pkequiv`. This example is further analyzed in [R] **pkcross** and [R] **pkequiv**.

◁

References

Chow, S. C. and J. P. Liu. 2000. *Design and Analysis of Bioavailability and Bioequivalence Studies.* 2nd ed. New York: Dekker.

Ratkowsky, D. A., M. A. Evans, and J. R. Alldredge. 1993. *Cross-over Experiments: Design, Analysis and Application.* New York: Dekker.

Also See

Complementary:	[R] **anova**,
	[D] **reshape**, [D] **statsby**
Related:	[R] **pkcollapse**, [R] **pkequiv**, [R] **pkexamine**, [R] **pkshape**, [R] **pksumm**

Title

> **pkcollapse** — Generate pharmacokinetic measurement dataset

Syntax

> pkcollapse *time concentration* $\big[\,if\,\big]$, id(*id_var*) $\big[\,options\,\big]$

options	description
Main	
* id(*id_var*)	subject ID variable
stat(*measures*)	create specified *measures*; default is all
trapezoid	use trapezoidal rule; default is cubic splines
fit(#)	use # points to estimate $\mathrm{AUC}_{0,\infty}$; default is fit(3)
keep(*varlist*)	keep variables in *varlist*
force	force collapse
nodots	suppress dots during calculation

*id(*id_var*) is required.

measures	description
auc	area under the concentration-time curve ($\mathrm{AUC}_{0,\infty}$)
aucline	area under the concentration-time curve from 0 to ∞ using a linear extension
aucexp	area under the concentration-time curve from 0 to ∞ using an exponential extension
auclog	area under the log-concentration-time curve extended with a linear fit
half	half-life of the drug
ke	elimination rate
cmax	maximum concentration
tmax	time at last concentration
tomc	time of maximum concentration

Description

pkcollapse generates new variables with the pharmacokinetic summary measures of interest.

pkcollapse is one of the pk commands. Please read [R] **pk** before reading this entry.

Options

Main

id(*id_var*) is required and specifies the variable that contains the subject ID over which pkcollapse is to operate.

stat(*measures*) specifies the measures to be generated. The default is to generate all the measures.

trapezoid tells Stata to use the trapezoidal rule when calculating the AUC. The default is to use cubic splines, which give better results for most functions. When the curve is very irregular, trapezoid may give better results.

fit(*#*) specifies the number of points to use in estimating the $AUC_{0,\infty}$. The default is fit(3), the last 3 points. This should be viewed as a minimum; the appropriate number of points will depend on your data.

keep(*varlist*) specifies the variables to be kept during the collapse. Variables not specified with the keep() option will be dropped. When keep() is specified, the keep variables are checked to ensure that all values of the variables are the same within *id_var*.

force forces the collapse, even in cases where the values of the keep() variables are different within the *id_var*.

nodots suppresses the display of dots during calculation.

Remarks

pkcollapse generates all the summary pharmacokinetic measures.

▷ Example 1

We demonstrate the use of pkcollapse with the data described in [R] **pk**. We have drug concentration data on 15 subjects. Each subject is measured at 13 time points over a 32-hour period. Some of the records are

```
. use http://www.stata-press.com/data/r9/pkdata
. list, sep(0)
```

	id	seq	time	concA	concB
1.	1	1	0	0	0
2.	1	1	.5	3.073403	3.712592
3.	1	1	1	5.188444	6.230602
4.	1	1	1.5	5.898577	7.885944
5.	1	1	2	5.096378	9.241735
6.	1	1	3	6.094085	13.10507
			(output omitted)		
14.	2	1	0	0	0
15.	2	1	.5	2.48462	.9209593
16.	2	1	1	4.883569	5.925818
17.	2	1	1.5	7.253442	8.710549
18.	2	1	2	5.849345	10.90552
19.	2	1	3	6.761085	8.429898
			(output omitted)		
207.	20	2	24	4.673281	6.059818
208.	20	2	32	3.487347	5.213639

Although pksumm allows us to view all the pharmacokinetic measures, we can create a dataset with the measures using pkcollapse.

```
. pkcollapse time concA concB, id(id) stat(auc) keep(seq)
. . . . . . . . . . . . . . . . . . . . . . . . . . .
```

```
. list, sep(8) abbrev(10)
```

	id	seq	auc_concA	auc_concB
1.	1	1	150.9643	218.5551
2.	2	1	146.7606	133.3201
3.	3	1	160.6548	126.0635
4.	4	1	157.8622	96.17461
5.	5	1	133.6957	188.9038
6.	7	1	160.639	223.6922
7.	8	1	131.2604	104.0139
8.	9	1	168.5186	237.8962
9.	10	2	137.0627	139.7382
10.	12	2	153.4038	202.3942
11.	13	2	163.4593	136.7848
12.	14	2	146.0462	104.5191
13.	15	2	158.1457	165.8654
14.	18	2	147.1977	139.235
15.	19	2	164.9988	166.2391
16.	20	2	145.3823	158.5146

The resulting dataset, which we will call pkdata2, contains one observation per subject. This dataset is in wide format. If we want to use pkcross or pkequiv, we must transform these data to long format, which we do in the last example of [R] **pkshape**.

◁

Methods and Formulas

pkcollapse is implemented as an ado-file.

The statistics generated by pkcollapse are described in [R] **pkexamine**.

Also See

Related: [R] **pkcross**, [R] **pkequiv**, [R] **pkexamine**, [R] **pkshape**, [R] **pksumm**

Background: [R] **pk**

Title

pkcross — Analyze crossover experiments

Syntax

pkcross *outcome* $\left[\,if\,\right]$ $\left[\,in\,\right]$ $\left[\,,\ options\,\right]$

options	description
Model	
sequence(*varname*)	sequence variable; default is sequence(sequence)
treatment(*varname*)	treatment variable; default is treatment(treat)
period(*varname*)	period variable; default is period(period)
id(*varname*)	ID variable
carryover(*varname*)	name of carryover variable; default is carryover(carry)
carryover(none)	omit carryover effects from model; default is carryover(carry)
model(*string*)	specify the model to fit
sequential	estimate sequential instead of partial sums of squares
Parameterization	
param(3)	estimate mean and the period, treatment, and sequence effects; assume no carryover effects exist; the default
param(1)	estimate mean and the period, treatment, and carryover effects; assume no sequence effects exist
param(2)	estimate mean, period and treatment effects, and period-by-treatment interaction; assume no sequence or carryover effects exist
param(4)	estimate mean, period and treatment effects, and period-by-treatment interaction; assume no period or crossover effects exist

Description

pkcross analyzes data from a crossover design experiment. When analyzing pharmaceutical trial data, if the treatment, carryover, and sequence variables are known, the omnibus test for separability of the treatment and carryover effects is calculated.

pkcross is one of the pk commands. Please read [R] **pk** before reading this entry.

Options

⌐ Model ⌐

sequence(*varname*) specifies the variable that contains the sequence in which the treatment was administered. If this option is not specified, sequence(sequence) is assumed.

treatment(*varname*) specifies the variable that contains the treatment information. If this option is not specified, treatment(treat) is assumed.

period(*varname*) specifies the variable that contains the period information. If this option is not specified, period(period) is assumed.

id(*varname*) specifies the variable that contains the subject identifiers. If this option is not specified, id(id) is assumed.

carryover(*varname* | none) specifies the variable that contains the carryover information. If carry(none) is specified, the carryover effects are omitted from the model. If this option is not specified, carryover(carry) is assumed.

model(*string*) specifies the model to be fitted. For higher-order crossover designs, this can be useful if you want to fit a model other than the default. However, anova (see [R] **anova**) can also be used to fit a crossover model. The default model for higher-order crossover designs is outcome predicted by sequence, period, treatment, and carryover effects. By default, the model statement is model(sequence period treat carry).

sequential specifies that sequential sums of squares be estimated.

_____⌐ Parameterization ⌐_____

param(#) specifies which of the four parameterizations to use for the analysis of a 2×2 crossover experiment. This option is ignored with higher-order crossover designs. The default is param(3). See the technical note for 2×2 crossover designs for more details.

param(1) estimates the overall mean, the period effects, the treatment effects, and the carryover effects, assuming that no sequence effects exist.

param(2) estimates the overall mean, the period effects, the treatment effects, and the period-by-treatment interaction, assuming that no sequence or carryover effects exist.

param(3) estimates the overall mean, the period effects, the treatment effects, and the sequence effects, assuming that no carryover effects exist. This is the default parameterization.

param(4) estimates the overall mean, the sequence effects, the treatment effects, and the sequence-by-treatment interaction, assuming that no period or carryover effects exist. When the sequence by treatment is equivalent to the period effect, this reduces to the third parameterization.

Remarks

pkcross is designed to analyze crossover experiments. Use pkshape first to reshape your data; see [R] **pkshape**. pkcross assumes that the data were reshaped by pkshape or are organized in the same manner as produced with pkshape. Washout periods are indicated by the number 0. See the technical note in this entry for more information on analyzing 2×2 crossover experiments.

❏ Technical Note

The 2×2 crossover design cannot be used to estimate more than four parameters because there are only four pieces of information (the four cell means) collected. pkcross uses ANOVA models to analyze the data, so one of the four parameters must be the overall mean of the model, leaving just three degrees of freedom to estimate the remaining effects (period, sequence, treatment, and carryover). Thus the model is overparameterized. Estimation of treatment and carryover effects requires the assumption of either no period effects or no sequence effects. Some researchers maintain that it is a bad idea to estimate carryover effects at the expense of other effects. This is a limitation of this design. pkcross implements four parameterizations for this model. They are numbered sequentially from one to four and are described in the *Options* section of this entry.

❏

▷ Example 1

Consider the example data published in Chow and Liu (2000, 73) and described in [R] **pkshape**. We have entered and reshaped the data with pkshape and have variables that identify the subjects, periods, treatments, sequence, and carryover treatment. To compute the ANOVA table, use pkcross:

```
. use http://www.stata-press.com/data/r9/chowliu
. pkshape id seq period1 period2, order(ab ba)
. pkcross outcome
```

```
                                        sequence variable = sequence
                                          period variable = period
                                       treatment variable = treat
                                       carryover variable = carry
                                               id variable = id
```

Analysis of variance (ANOVA) for a 2x2 crossover study

Source of Variation	SS	df	MS	F	Prob > F
Intersubjects					
Sequence effect	276.00	1	276.00	0.37	0.5468
Residuals	16211.49	22	736.89	4.41	0.0005
Intrasubjects					
Treatment effect	62.79	1	62.79	0.38	0.5463
Period effect	35.97	1	35.97	0.22	0.6474
Residuals	3679.43	22	167.25		
Total	20265.68	47			

Omnibus measure of separability of treatment and carryover =   29.2893%

There is evidence of intersubject variability, but there are no other significant effects. The omnibus test for separability is a measure reflecting the degree to which the study design allows the treatment effects to be estimated independently of the carryover effects. The measure of separability of the treatment and carryover effects indicates approximately 29% separability. This can be interpreted as the degree to which the treatment and carryover effects are orthogonal; that is, the treatment and carryover effects are about 29% orthogonal. This is a characteristic of the design of the study. For a complete discussion, see Ratkowsky, Evans, and Alldredge (1993). Compared to the output in Chow and Liu (2000), the sequence effect is mislabeled as a carryover effect. See Ratkowsky, Evans, and Alldredge (1993, section 3.2) for a complete discussion of the mislabeling.

By specifying param(1), we obtain parameterization 1 for this model.

```
. pkcross outcome, param(1)
```

```
                                        sequence variable = sequence
                                          period variable = period
                                       treatment variable = treat
                                       carryover variable = carry
                                               id variable = id
```

Analysis of variance (ANOVA) for a 2x2 crossover study

Source of Variation	Partial SS	df	MS	F	Prob > F
Treatment effect	301.04	1	301.04	0.67	0.4189
Period effect	255.62	1	255.62	0.57	0.4561
Carryover effect	276.00	1	276.00	0.61	0.4388
Residuals	19890.92	44	452.07		
Total	20265.68	47			

Omnibus measure of separability of treatment and carryover =   29.2893%

◁

▷ Example 2

Consider the case of a two-treatment, four-sequence, two-period crossover design. This design is commonly referred to as Balaam's design. Ratkowsky et al. (1993) published the following data from an amantadine trial:

```
. use http://www.stata-press.com/data/r9/balaam, clear
. list, sep(0)
```

	id	seq	period1	period2	period3
1.	1	-ab	9	8.75	8.75
2.	2	-ab	12	10.5	9.75
3.	3	-ab	17	15	18.5
4.	4	-ab	21	21	21.5
5.	1	-ba	23	22	18
6.	2	-ba	15	15	13
7.	3	-ba	13	14	13.75
8.	4	-ba	24	22.75	21.5
9.	5	-ba	18	17.75	16.75
10.	1	-aa	14	12.5	14
11.	2	-aa	27	24.25	22.5
12.	3	-aa	19	17.25	16.25
13.	4	-aa	30	28.25	29.75
14.	1	-bb	21	20	19.51
15.	2	-bb	11	10.5	10
16.	3	-bb	20	19.5	20.75
17.	4	-bb	25	22.5	23.5

The sequence identifier must be a string with zeros to indicate washout or baseline periods, or a number. If the sequence identifier is numeric, the order option must be specified with pkshape. If the sequence identifier is a string, pkshape will create sequence, period, and treatment identifiers without the order option. In this example, the dash is used to indicate a baseline period, which is an invalid code for this purpose. As a result, the data must be encoded; see [D] **encode**.

```
. encode seq, gen(num_seq)
. pkshape id num_seq period1 period2 period3, order(0aa 0ab 0ba 0bb)
. pkcross outcome, se
```
```
                              sequence variable = sequence
                                period variable = period
                             treatment variable = treat
                             carryover variable = carry
                                    id variable = id
```

	Analysis of variance (ANOVA) for a crossover study				
Source of Variation	SS	df	MS	F	Prob > F
Intersubjects					
Sequence effect	285.82	3	95.27	1.01	0.4180
Residuals	1221.49	13	93.96	59.96	0.0000
Intrasubjects					
Period effect	15.13	2	7.56	6.34	0.0048
Treatment effect	8.48	1	8.48	8.86	0.0056
Carryover effect	0.11	1	0.11	0.12	0.7366
Residuals	29.56	30	0.99		
Total	1560.59	50			

Omnibus measure of separability of treatment and carryover =   64.6447%

In this example, the sequence specifier used dashes instead of zeros to indicate a baseline period during which no treatment was given. For `pkcross` to work, we need to encode the string sequence variable and then use the `order` option with `pkshape`. A word of caution: `encode` does not necessarily choose the first sequence to be sequence 1, as in this example. Always double-check the sequence numbering when using `encode`.

◁

▷ Example 3

Continuing with the example from [R] **pkshape**, we fit an ANOVA model.

```
. use http://www.stata-press.com/data/r9/pkdata3
. list, sep(8)
```

	id	sequence	outcome	treat	carry	period
1.	1	1	150.9643	A	0	1
2.	2	1	146.7606	A	0	1
3.	3	1	160.6548	A	0	1
4.	4	1	157.8622	A	0	1
5.	5	1	133.6957	A	0	1
6.	7	1	160.639	A	0	1
7.	8	1	131.2604	A	0	1
8.	9	1	168.5186	A	0	1
9.	10	2	137.0627	B	0	1
10.	12	2	153.4038	B	0	1
11.	13	2	163.4593	B	0	1
12.	14	2	146.0462	B	0	1
13.	15	2	158.1457	B	0	1
14.	18	2	147.1977	B	0	1
15.	19	2	164.9988	B	0	1
16.	20	2	145.3823	B	0	1
17.	1	1	218.5551	B	A	2
18.	2	1	133.3201	B	A	2
19.	3	1	126.0635	B	A	2
20.	4	1	96.17461	B	A	2
21.	5	1	188.9038	B	A	2
22.	7	1	223.6922	B	A	2
23.	8	1	104.0139	B	A	2
24.	9	1	237.8962	B	A	2
25.	10	2	139.7382	A	B	2
26.	12	2	202.3942	A	B	2
27.	13	2	136.7848	A	B	2
28.	14	2	104.5191	A	B	2
29.	15	2	165.8654	A	B	2
30.	18	2	139.235	A	B	2
31.	19	2	166.2391	A	B	2
32.	20	2	158.5146	A	B	2

The ANOVA model is fitted using pkcross:

```
. pkcross outcome
```

```
                                     sequence variable = sequence
                                       period variable = period
                                    treatment variable = treat
                                    carryover variable = carry
                                           id variable = id
```

```
              Analysis of variance (ANOVA) for a 2x2 crossover study
      Source of Variation  |      SS        df        MS         F      Prob > F
```

Source of Variation	SS	df	MS	F	Prob > F
Intersubjects					
Sequence effect	378.04	1	378.04	0.29	0.5961
Residuals	17991.26	14	1285.09	1.40	0.2691
Intrasubjects					
Treatment effect	455.04	1	455.04	0.50	0.4931
Period effect	419.47	1	419.47	0.46	0.5102
Residuals	12860.78	14	918.63		
Total	32104.59	31			

Omnibus measure of separability of treatment and carryover =   29.2893%

◁

▷ Example 4

Consider the case of a six-treatment crossover trial in which the squares are not variance balanced. The following dataset is from a partially balanced crossover trial published by Ratkowsky et al. (1993):

```
. use http://www.stata-press.com/data/r9/nobalance
. list, sep(4)
```

	cow	seq	period1	period2	period3	period4	block
1.	1	adbe	38.7	37.4	34.3	31.3	1
2.	2	baed	48.9	46.9	42	39.6	1
3.	3	ebda	34.6	32.3	28.5	27.1	1
4.	4	deab	35.2	33.5	28.4	25.1	1
5.	1	dafc	32.9	33.1	27.5	25.1	2
6.	2	fdca	30.4	29.4	26.7	23.1	2
7.	3	cfda	30.8	29.3	26.4	23.2	2
8.	4	acdf	25.7	26.1	23.4	18.7	2
9.	1	efbc	25.4	26	23.9	19.9	3
10.	2	becf	21.8	23.9	21.7	17.6	3
11.	3	fceb	21.4	22	19.4	16.6	3
12.	4	cbfe	22.8	21	18.6	16.1	3

(*Continued on next page*)

In cases in which there is no variance balance in the design, a square or blocking variable is needed to indicate in which treatment cell a sequence was observed, but the mechanical steps are the same.

```
. set matsize 42
. pkshape cow seq period1 period2 period3 period4
. pkcross outcome, model(block cow|block period|block treat carry) se
```

| | Number of obs = | 48 | R-squared | = | 0.9966 |
| | Root MSE | = .730751 | Adj R-squared = | | 0.9906 |

Source	Seq. SS	df	MS	F	Prob > F
Model	2650.0419	30	88.3347302	165.42	0.0000
block	1607.17045	2	803.585226	1504.85	0.0000
cow|block	628.621899	9	69.8468777	130.80	0.0000
period|block	407.531876	9	45.2813195	84.80	0.0000
treat	2.48979215	5	.497958429	0.93	0.4846
carry	4.22788534	5	.845577068	1.58	0.2179
Residual	9.07794631	17	.533996842		
Total	2659.11985	47	56.5770181		

When the model statement is used and the omnibus measure of separability is desired, specify the variables in the `treatment()`, `carryover()`, and `sequence()` options to `pkcross`.   ◁

Methods and Formulas

`pkcross` is implemented as an ado-file.

`pkcross` uses ANOVA to fit models for crossover experiments; see [R] **anova**.

The omnibus measure of separability is

$$S = 100(1 - V)\%$$

where V is Cramér's V and is defined as

$$V = \left\{ \frac{\frac{\chi^2}{N}}{\min(r - 1, c - 1)} \right\}^{\frac{1}{2}}$$

The χ^2 is calculated as

$$\chi^2 = \sum_i \sum_j \left\{ \frac{(O_{ij} - E_{ij})^2}{E_{ij}} \right\}$$

where O and E are the observed and expected counts in a table of the number of times each treatment is followed by the other treatments.

References

Chow, S. C. and J. P. Liu. 2000. *Design and Analysis of Bioavailability and Bioequivalence Studies.* 2nd ed. New York: Dekker.

Neter, J., M. H. Kutner, C. J. Nachtsheim, and W. Wasserman. 1996. *Applied Linear Statistical Models.* 4th ed. Chicago: Irwin.

Ratkowsky, D. A., M. A. Evans, and J. R. Alldredge. 1993. *Cross-over Experiments: Design, Analysis and Application.* New York: Dekker.

Also See

Complementary:	[D] **statsby**
Related:	[R] **pkcollapse**, [R] **pkequiv**, [R] **pkexamine**, [R] **pkshape**, [R] **pksumm**
Background:	[R] **pk**

Title

pkequiv — Perform bioequivalence tests

Syntax

pkequiv *outcome treatment period sequence id* [*if*] [*in*] [, *options*]

options	description
Options	
compare(*string*)	compare the two specified values of the treatment variable
limit(*#*)	equivalence limit (between .10 and .99); default is .2
level(*#*)	set confidence level; default is level(90)
fieller	calculate confidence interval by Fieller's theorem
symmetric	calculate symmetric equivalence interval
anderson	Anderson and Hauck hypothesis test for bioequivalence
tost	two one-sided hypothesis tests for bioequivalence
noboot	do not estimate probability that CI lies within confidence limits

Description

pkequiv performs bioequivalence testing for two treatments. By default, pkequiv calculates a standard confidence interval symmetric about the difference between the two treatment means. pkequiv also calculates confidence intervals symmetric about zero and intervals based on Fieller's theorem. Additionally, pkequiv can perform interval hypothesis tests for bioequivalence.

pkequiv is one of the pk commands. Please read [R] **pk** before reading this entry.

Options

     Options

compare(*string*) specifies the two treatments to be tested for equivalence. In some cases, there may be more than two treatments, but the equivalence can only be determined between any two treatments.

limit(*#*) specifies the equivalence limit. The default is .2. The equivalence limit can only be changed symmetrically; that is, it is not possible to have a .15 lower limit and a .2 upper limit in the same test.

level(*#*) specifies the confidence level, as a percentage, for confidence intervals. The default is level(90). Note that this is not controlled by the set level command.

fieller specifies that an equivalence interval based on Fieller's theorem be calculated.

symmetric specifies that a symmetric equivalence interval be calculated.

anderson specifies that the Anderson and Hauck hypothesis test for bioequivalence be computed. This option is ignored when calculating equivalence intervals based on Fieller's theorem or when calculating a confidence interval that is symmetric about zero.

tost specifies that the two one-sided hypothesis tests for bioequivalence be computed. This option is ignored when calculating equivalence intervals based on Fieller's theorem or when calculating a confidence interval that is symmetric about zero.

noboot prevents the estimation of the probability that the confidence interval lies within the confidence limits. If this option is not specified, this probability is estimated by resampling the data.

Remarks

pkequiv is designed to conduct tests for bioequivalence based on data from a crossover experiment. pkequiv requires that the user specify the *outcome*, *treatment*, *period*, *sequence*, and *id* variables. The data must be in the same format as that produced by pkshape; see [R] **pkshape**.

(*Continued on next page*)

▷ Example 1

Continuing with the example from [R] **pkshape**, we will conduct equivalence testing.

```
. use http://www.stata-press.com/data/r9/pkdata3
. list, sep(4)
```

	id	sequence	outcome	treat	carry	period
1.	1	1	150.9643	A	0	1
2.	2	1	146.7606	A	0	1
3.	3	1	160.6548	A	0	1
4.	4	1	157.8622	A	0	1
5.	5	1	133.6957	A	0	1
6.	7	1	160.639	A	0	1
7.	8	1	131.2604	A	0	1
8.	9	1	168.5186	A	0	1
9.	10	2	137.0627	B	0	1
10.	12	2	153.4038	B	0	1
11.	13	2	163.4593	B	0	1
12.	14	2	146.0462	B	0	1
13.	15	2	158.1457	B	0	1
14.	18	2	147.1977	B	0	1
15.	19	2	164.9988	B	0	1
16.	20	2	145.3823	B	0	1
17.	1	1	218.5551	B	A	2
18.	2	1	133.3201	B	A	2
19.	3	1	126.0635	B	A	2
20.	4	1	96.17461	B	A	2
21.	5	1	188.9038	B	A	2
22.	7	1	223.6922	B	A	2
23.	8	1	104.0139	B	A	2
24.	9	1	237.8962	B	A	2
25.	10	2	139.7382	A	B	2
26.	12	2	202.3942	A	B	2
27.	13	2	136.7848	A	B	2
28.	14	2	104.5191	A	B	2
29.	15	2	165.8654	A	B	2
30.	18	2	139.235	A	B	2
31.	19	2	166.2391	A	B	2
32.	20	2	158.5146	A	B	2

Now we can conduct a bioequivalence test between treat = A and treat = B.

```
. set seed 1
. pkequiv outcome treat period seq id
      Classic confidence interval for bioequivalence
```

	[equivalence limits]		[test limits]	
difference:	-30.296	30.296	-11.332	26.416
ratio:	80%	120%	92.519%	117.439%

```
probability test limits are within equivalence limits =    0.6410
note: reference treatment = 1
```

The default output for pkequiv shows a confidence interval for the difference of the means (test limits), the ratio of the means, and the federal equivalence limits. The classic confidence interval can be constructed around the difference between the average measure of effect for the two drugs or around the ratio of the average measure of effect for the two drugs. pkequiv reports both the difference measure and the ratio measure. For these data, U.S. federal government regulations state that the confidence interval for the difference must be entirely contained within the range $[-30.296, 30.296]$, and between 80% and 120% for the ratio. In this case, the test limits are within the equivalence limits. Although the test limits are inside the equivalence limits, there is only a 64% assurance that the observed confidence interval will be within the equivalence limits in the long run. This is an interesting case because, although this sample shows bioequivalence, the evaluation of the long-run performance indicates possible problems. These fictitious data were generated with high intersubject variability, which causes poor long-run performance.

If we conduct a bioequivalence test with the data published in Chow and Liu (2000, 73), which we introduced in [R] **pk** and fully described in [R] **pkshape**, we observe that the probability that the test limits are within the equivalence limits is very high.

```
. use http://www.stata-press.com/data/r9/chowliu2
. set seed 1
. pkequiv outcome treat period seq id
      Classic confidence interval for bioequivalence
```

	[equivalence limits]		[test limits]	
difference:	-16.512	16.512	-8.698	4.123
ratio:	80%	120%	89.464%	104.994%

```
probability test limits are within equivalence limits =    0.9980
note: reference treatment = 1
```

For these data, the test limits are well within the equivalence limits, and the probability that the test limits are within the equivalence limits is 99.8%.

◁

(Continued on next page)

▷ Example 2

We compute a confidence interval that is symmetric about zero:

```
. pkequiv outcome treat period seq id, symmetric
```

Westlake's symmetric confidence interval for bioequivalence

	[Equivalence limits]		[Test mean]
Test formulation:	75.145	89.974	80.272

note: reference treatment = 1

The reported equivalence limit is constructed symmetrically about the reference mean, which is equivalent to constructing a confidence interval symmetric about zero for the difference in the two drugs. In the output above, we see that the test formulation mean of 80.272 is within the equivalence limits, indicating that the test drug is bioequivalent to the reference drug.

pkequiv displays interval hypothesis tests of bioequivalence if you specify the tost and/or the anderson options. For example,

```
. set seed 1
. pkequiv outcome treat period seq id, tost anderson
```

Classic confidence interval for bioequivalence

	[equivalence limits]		[test limits]	
difference:	−16.512	16.512	−8.698	4.123
ratio:	80%	120%	89.464%	104.994%

probability test limits are within equivalence limits =   0.9980
Schuirmann's two one-sided tests

upper test statistic =	−5.036	p-value =	0.000
lower test statistic =	3.810	p-value =	0.001

Anderson and Hauck's test

noncentrality parameter =	4.423		
test statistic =	−0.613	empirical p-value =	0.0005

note: reference treatment = 1

Both of Schuirmann's one-sided tests are highly significant, suggesting that the two drugs are bioequivalent. A similar conclusion is drawn from the Anderson and Hauck test of bioequivalence.

◁

Saved Results

pkequiv saves in r():

Scalars

r(stddev)	pooled-sample standard deviation of period differences from both sequences
r(uci)	upper confidence interval for a classic interval
r(lci)	lower confidence interval for a classic interval
r(delta)	delta value used in calculating a symmetric confidence interval
r(u3)	upper confidence interval for Fieller's confidence interval
r(l3)	lower confidence interval for Fieller's confidence interval

Methods and Formulas

pkequiv is implemented as an ado-file.

The lower confidence interval for the difference in the two treatments for the classic shortest confidence interval is

$$L_1 = \left(\overline{Y}_T - \overline{Y}_R\right) - t_{(\alpha, n_1 + n_2 - 2)} \widehat{\sigma}_d \sqrt{\frac{1}{n_1} + \frac{1}{n_2}}$$

The upper limit is

$$U_1 = \left(\overline{Y}_T - \overline{Y}_R\right) + t_{(\alpha, n_1 + n_2 - 2)} \widehat{\sigma}_d \sqrt{\frac{1}{n_1} + \frac{1}{n_2}}$$

The limits for the ratio measure are

$$L_2 = \left(\frac{L_1}{\overline{Y}_R} + 1\right) 100\%$$

and

$$U_2 = \left(\frac{U_1}{\overline{Y}_R} + 1\right) 100\%$$

where $\overline{Y}_T$ is the mean of the test formulation of the drug, $\overline{Y}_R$ is the mean of the reference formulation of the drug, and $t_{(\alpha, n_1 + n_2 - 2)}$ is the t distribution with $n_1 + n_2 - 2$ degrees of freedom. $\widehat{\sigma}_d$ is the pooled sample variance of the period differences from both sequences, defined as

$$\widehat{\sigma}_d = \frac{1}{n_1 + n_2 - 2} \sum_{k=1}^{2} \sum_{i=1}^{n_k} \left(d_{ik} - \overline{d}_{.k}\right)^2$$

The upper and lower limits for the symmetric confidence interval are $\overline{Y}_R + \Delta$ and $\overline{Y}_R - \Delta$, where

$$\Delta = k_1 \widehat{\sigma}_d \sqrt{\frac{1}{n_1} + \frac{1}{n_2}} - \left(\overline{Y}_T - \overline{Y}_R\right)$$

and (simultaneously)

$$\Delta = -k_2 \widehat{\sigma}_d \sqrt{\frac{1}{n_1} + \frac{1}{n_2}} + 2\left(\overline{Y}_T - \overline{Y}_R\right)$$

and k_1 and k_2 are computed iteratively to satisfy the above equalities and the condition

$$\int_{k_1}^{k_2} f(t)dt = 1 - 2\alpha$$

where $f(t)$ is the probability density function of the t distribution with $n_1 + n_2 - 2$ degrees of freedom.

See Chow and Liu (2000, 88) for details about calculating the confidence interval based on Fieller's theorem.

The two test statistics for the two one-sided tests of equivalence are

$$T_L = \frac{(\overline{Y}_T - \overline{Y}_R) - \theta_L}{\widehat{\sigma}_d \sqrt{\frac{1}{n_1} + \frac{1}{n_2}}}$$

and

$$T_U = \frac{(\overline{Y}_T - \overline{Y}_R) - \theta_U}{\widehat{\sigma}_d \sqrt{\frac{1}{n_1} + \frac{1}{n_2}}}$$

where $-\theta_L = \theta_U$ and are the regulated confidence limits.

The logic of the Anderson and Hauck test is tricky; see Chow and Liu (2000) for a complete explanation. However, the test statistic is

$$T_{AH} = \frac{(\overline{Y}_T - \overline{Y}_R) - \left(\frac{\theta_L + \theta_U}{2}\right)}{\widehat{\sigma}_d \sqrt{\frac{1}{n_1} + \frac{1}{n_2}}}$$

and the noncentrality parameter is estimated by

$$\widehat{\delta} = \frac{\theta_U - \theta_L}{2\widehat{\sigma}_d \sqrt{\frac{1}{n_1} + \frac{1}{n_2}}}$$

The empirical p-value is calculated as

$$p = F_t\left(|T_{AH}| - \widehat{\delta}\right) - F_t\left(-|T_{AH}| - \widehat{\delta}\right)$$

where F_t is the cumulative distribution function of the t distribution with $n_1 + n_2 - 2$ degrees of freedom.

References

Chow, S. C. and J. P. Liu. 2000. *Design and Analysis of Bioavailability and Bioequivalence Studies*. 2nd ed. New York: Dekker.

Neter, J., M. H. Kutner, C. J. Nachtsheim, and W. Wasserman. 1996. *Applied Linear Statistical Models*. 4th ed. Chicago: Irwin.

Ratkowsky, D. A., M. A. Evans, and J. R. Alldredge. 1993. *Cross-over Experiments: Design, Analysis and Application*. New York: Dekker.

Also See

Complementary:	[D] **statsby**
Related:	[R] **pkcollapse**, [R] **pkcross**, [R] **pkexamine**, [R] **pkshape**, [R] **pksumm**
Background:	[R] **pk**

Title

> **pkexamine** — Calculate pharmacokinetic measures

Syntax

pkexamine *time concentration* $\left[\,if\,\right]$ $\left[\,in\,\right]$ $\left[\,,\ options\,\right]$

options	description
Main	
<u>trap</u>ezoid	use trapezoidal rule; default is cubic splines
fit(#)	use # points to estimate $AUC_{0,\infty}$; default is fit(3)
<u>g</u>raph	graph the AUC
line	graph the linear extension
log	graph the log extension
exp(#)	plot the exponential fit for the $AUC_{0,\infty}$
AUC plot	
cline_options	affect rendition of plotted points connected by lines
Add plot	
addplot(*plot*)	add other plots to the generated graph
Y-Axis, X-Axis, Title, Caption, Legend, Overall	
twoway_options	any options other than by() documented in [G] *twoway_options*

by may be used with pkexamine; see [D] **by**.

Description

pkexamine calculates pharmacokinetic measures from time-and-concentration subject-level data. pkexamine computes and displays the maximum measured concentration, the time at the maximum measured concentration, the time of the last measurement, the elimination time, the half-life, and the area under the concentration-time curve (AUC). Three estimates of the area under the concentration-time curve from 0 to infinity ($AUC_{0,\infty}$) are also calculated.

pkexamine is one of the pk commands. Please read [R] **pk** before reading this entry.

Options

> **Main**

trapezoid specifies that the trapezoidal rule be used to calculate the AUC. The default is cubic splines, which give better results for most functions. In cases where the curve is very irregular, trapezoid may give better results.

fit(#) specifies the number of points, counting back from the last measurement, to use in fitting the extension to estimate the $AUC_{0,\infty}$. The default is fit(3), or the last three points. This should be viewed as a minimum; the appropriate number of points will depend on your data.

413

graph tells pkexamine to graph the concentration-time curve.

line and log specify the estimates of the $AUC_{0,\infty}$ to display when graphing the $AUC_{0,\infty}$. These options are ignored, unless they are specified with the graph option.

exp(#) specifies that the exponential fit for the $AUC_{0,\infty}$ be plotted. You must specify the maximum time value to which you want to plot the curve, and this time value must be greater than the maximum time measurement in the data. If you specify 0, the curve will be plotted to the point at which the linear extension would cross the x-axis. This option is not valid with the line or log options and is ignored, unless the graph option is also specified.

⌐ AUC plot ⌐

cline_options affect the rendition of the plotted points connected by lines; see [G] *cline_options*.

⌐ Add plot ⌐

addplot(*plot*) provides a way to add other plots to the generated graph. See [G] *addplot_option*.

⌐ Y-Axis, X-Axis, Title, Caption, Legend, Overall ⌐

twoway_options are any of the options documented in [G] *twoway_options*, excluding by(). These include options for titling the graph (see [G] *title_options*) and options for saving the graph to disk (see [G] *saving_option*).

Remarks

pkexamine computes summary statistics for a given patient in a pharmacokinetic trial. If by *idvar*: is specified, statistics will be displayed for each subject in the data.

▷ Example 1

Chow and Liu (2000, 11) present data on a study examining primidone concentrations versus time for a subject over a 32-hour period after dosing.

```
. use http://www.stata-press.com/data/r9/auc
. list, abbrev(14)
```

	id	time	concentration
1.	1	0	0
2.	1	.5	0
3.	1	1	2.8
4.	1	1.5	4.4
5.	1	2	4.4
6.	1	3	4.7
7.	1	4	4.1
8.	1	6	4
9.	1	8	3.6
10.	1	12	3
11.	1	16	2.5
12.	1	24	2
13.	1	32	1.6

We use `pkexamine` to produce the summary statistics.

```
. pkexamine time conc, graph
                       Maximum concentration =        4.7
               Time of maximum concentration =          3
          Time of last observation (Tmax) =           32
                       Elimination rate =         0.0279
                              Half life =        24.8503
```

Area under the curve

AUC [0, Tmax]	AUC [0, inf.) Linear of log conc.	AUC [0, inf.) Linear fit	AUC [0, inf.) Exponential fit
85.24	142.603	107.759	142.603

Fit based on last 3 points.

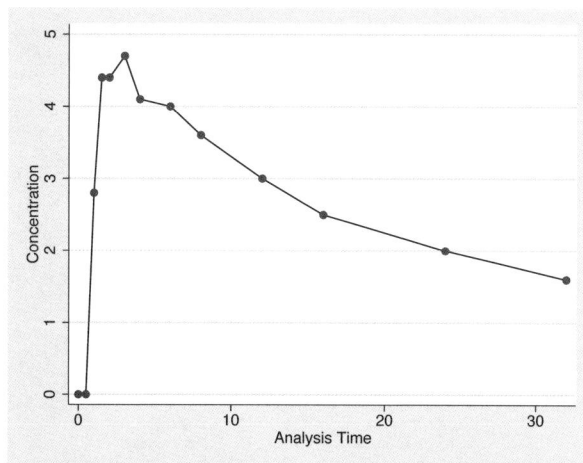

The maximum concentration of 4.7 occurs at time 3, and the time of the last observation (Tmax) is 32. In addition to the AUC, which is calculated from 0 to the maximum value of `time`, `pkexamine` also reports the area under the curve, computed by extending the curve using each of three methods: a linear fit to the log of the concentration, a linear regression line, and a decreasing exponential regression line. See *Methods and Formulas* for details on these three methods.

By default, all extensions to the AUC are based on the last three points. Looking at the graph for these data, it seems more appropriate to use the last seven points to estimate the $AUC_{0,\infty}$:

```
. pkexamine time conc, fit(7)
                       Maximum concentration =        4.7
               Time of maximum concentration =          3
          Time of last observation (Tmax) =           32
                       Elimination rate =         0.0349
                              Half life =        19.8354
```

Area under the curve

AUC [0, Tmax]	AUC [0, inf.) Linear of log conc.	AUC [0, inf.) Linear fit	AUC [0, inf.) Exponential fit
85.24	131.027	96.805	129.181

Fit based on last 7 points.

This decreased the estimate of the $AUC_{0,\infty}$ for all extensions. To see a graph of the $AUC_{0,\infty}$ using a linear extension, specify the graph and line options.

```
. pkexamine time conc, fit(7) graph line
                               Maximum concentration =       4.7
                        Time of maximum concentration =         3
                        Time of last observation (Tmax) =        32
                               Elimination rate =    0.0349
                                      Half life =   19.8354
```

Area under the curve

AUC [0, Tmax]	AUC [0, inf.) Linear of log conc.	AUC [0, inf.) Linear fit	AUC [0, inf.) Exponential fit
85.24	131.027	96.805	129.181

Fit based on last 7 points.

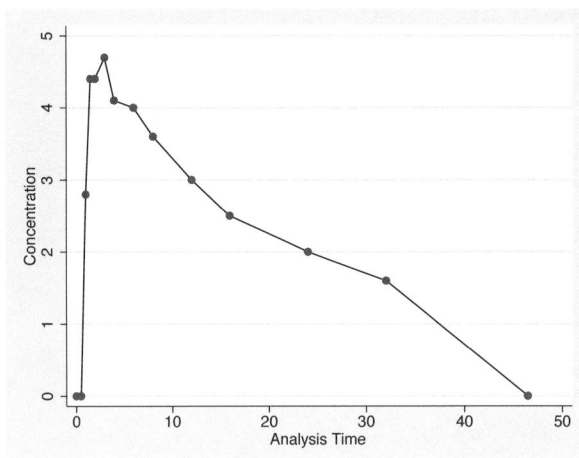

Saved Results

pkexamine saves in r():

Scalars

r(auc)	area under the concentration curve
r(half)	half-life of the drug
r(ke)	elimination rate
r(tmax)	time at last concentration measurement
r(cmax)	maximum concentration
r(tomc)	time of maximum concentration
r(auc_line)	$AUC_{0,\infty}$ estimated with a linear fit
r(auc_exp)	$AUC_{0,\infty}$ estimated with an exponential fit
r(auc_ln)	$AUC_{0,\infty}$ estimated with a linear fit of the natural log

Methods and Formulas

pkexamine is implemented as an ado-file.

The $\text{AUC}_{0,t_{\max}}$ is defined as

$$\text{AUC}_{0,t_{\max}} = \int_0^{t_{\max}} C_t \, dt$$

where C_t is the concentration at time t. By default, the integral is calculated numerically using cubic splines. However, if the trapezoidal rule is used, the $\text{AUC}_{0,t_{\max}}$ is given as

$$\text{AUC}_{0,t_{\max}} = \sum_{i=2}^{k} \frac{C_{i-1} + C_i}{2} (t_i - t_{i-1})$$

The $\text{AUC}_{0,\infty}$ is the $\text{AUC}_{0,t_{\max}} + \text{AUC}_{t_{\max},\infty}$, or

$$\text{AUC}_{0,\infty} = \int_0^{t_{\max}} C_t \, dt + \int_{t_{\max}}^{\infty} C_t \, dt$$

When using the linear extension to the $\text{AUC}_{0,t_{\max}}$, the integration is cut off when the line crosses the x-axis. The log extension is a linear extension on the log concentration scale. The area for the exponential extension is

$$\text{AUC}_{0,\infty} = \int_{t_{\max}}^{\infty} e^{-(\beta_0 + t\beta_1)} \, dt = -\frac{e^{-(\beta_0 + t_{\max}\beta_1)}}{\beta_1}$$

Finally, the elimination rate, K_{eq}, is the negative of the parameter estimate for a linear regression of log time on concentration and is given in the standard manner:

$$K_{\text{eq}} = -\frac{\sum_{i=1}^{k} \left(C_i - \overline{C}\right) \left(\ln t_i - \overline{\ln t}\right)}{\sum_{i=1}^{k} \left(C_i - \overline{C}\right)^2}$$

and

$$t_{1/2} = \frac{\ln 2}{K_{\text{eq}}}$$

References

Chow, S. C. and J. P. Liu. 2000. *Design and Analysis of Bioavailability and Bioequivalence Studies*. 2nd ed. New York: Dekker.

Also See

Complementary:	[D] **statsby**
Related:	[R] **pkcollapse**, [R] **pkcross**, [R] **pkequiv**, [R] **pkshape**, [R] **pksumm**
Background:	[R] **pk**

Title

pkshape — Reshape (pharmacokinetic) Latin-square data

Syntax

pkshape *id sequence period1 period2* [*period list*] [*, options*]

options	description
order(*string*)	apply treatments in specified order
outcome(*newvar*)	name for outcome variable; default is outcome(outcome)
treatment(*newvar*)	name for treatment variable; default is treatment(treat)
carryover(*newvar*)	name for carryover variable; default is carryover(carry)
sequence(*newvar*)	name for sequence variable; default is sequence(sequence)
period(*newvar*)	name for period variable; default is period(period)

Description

pkshape reshapes the data for use with anova, pkcross, and pkequiv. Latin-square and crossover data are often organized in a manner that cannot be analyzed easily with Stata. pkshape reorganizes the data in memory for use in Stata.

pkshape is one of the pk commands. Please read [R] **pk** before reading this entry.

Options

order(*string*) specifies the order in which treatments were applied. If the sequence() specifier is a string variable that specifies the order, this option is not necessary. Otherwise, order() specifies how to generate the treatment and carryover variables. Any string variable can be used to specify the order. In the case of crossover designs, any washout periods can be indicated with the number 0.

outcome(*newvar*) specifies the name for the outcome variable in the reorganized data. By default, outcome(outcome) is used.

treatment(*newvar*) specifies the name for the treatment variable in the reorganized data. By default, treatment(treat) is used.

carryover(*newvar*) specifies the name for the carryover variable in the reorganized data. By default, carryover(carry) is used.

sequence(*newvar*) specifies the name for the sequence variable in the reorganized data. By default, sequence(sequence) is used.

period(*newvar*) specifies the name for the period variable in the reorganized data. By default, period(period) is used.

Remarks

Often data from a Latin-square experiment are naturally organized in a manner that Stata cannot manage easily. pkshape reorganizes Latin-square data so that it can be used with anova (see [R] **anova**) or any pk command. This includes the classic 2×2 crossover design commonly used in pharmaceutical research, as well as many other Latin-square designs.

▷ Example 1

Consider the example data published in Chow and Liu (2000, 73). There are 24 patients, 12 in each sequence. Sequence 1 consists of the reference formulation followed by the test formulation; sequence 2 is the test formulation followed by the reference formulation. The measurements reported are the $AUC_{0-t_{max}}$ for each patient and for each period.

```
. use http://www.stata-press.com/data/r9/chowliu
. list, sep(4)
```

	id	seq	period1	period2
1.	1	1	74.675	73.675
2.	4	1	96.4	93.25
3.	5	1	101.95	102.125
4.	6	1	79.05	69.45
5.	11	1	79.05	69.025
6.	12	1	85.95	68.7
7.	15	1	69.725	59.425
8.	16	1	86.275	76.125
9.	19	1	112.675	114.875
10.	20	1	99.525	116.25
11.	23	1	89.425	64.175
12.	24	1	55.175	74.575
13.	2	2	74.825	37.35
14.	3	2	86.875	51.925
15.	7	2	81.675	72.175
16.	8	2	92.7	77.5
17.	9	2	50.45	71.875
18.	10	2	66.125	94.025
19.	13	2	122.45	124.975
20.	14	2	99.075	85.225
21.	17	2	86.35	95.925
22.	18	2	49.925	67.1
23.	21	2	42.7	59.425
24.	22	2	91.725	114.05

Since the outcome for a single person is in two different variables, the treatment that was applied to an individual is a function of the period and the sequence. To analyze this using anova, all the outcomes must be in one variable, and each covariate must be in its own variable. To reorganize these data, use pkshape:

```
. pkshape id seq period1 period2, order(ab ba)
. sort seq id treat
```

```
. list, sep(8)
```

	id	sequence	outcome	treat	carry	period
1.	1	1	74.675	1	0	1
2.	1	1	73.675	2	1	2
3.	4	1	96.4	1	0	1
4.	4	1	93.25	2	1	2
5.	5	1	101.95	1	0	1
6.	5	1	102.125	2	1	2
7.	6	1	79.05	1	0	1
8.	6	1	69.45	2	1	2
9.	11	1	79.05	1	0	1
10.	11	1	69.025	2	1	2
11.	12	1	85.95	1	0	1
12.	12	1	68.7	2	1	2
13.	15	1	69.725	1	0	1
14.	15	1	59.425	2	1	2
15.	16	1	86.275	1	0	1
16.	16	1	76.125	2	1	2
17.	19	1	112.675	1	0	1
18.	19	1	114.875	2	1	2
19.	20	1	99.525	1	0	1
20.	20	1	116.25	2	1	2
21.	23	1	89.425	1	0	1
22.	23	1	64.175	2	1	2
23.	24	1	55.175	1	0	1
24.	24	1	74.575	2	1	2
25.	2	2	37.35	1	2	2
26.	2	2	74.825	2	0	1
27.	3	2	51.925	1	2	2
28.	3	2	86.875	2	0	1
29.	7	2	72.175	1	2	2
30.	7	2	81.675	2	0	1
31.	8	2	77.5	1	2	2
32.	8	2	92.7	2	0	1
33.	9	2	71.875	1	2	2
34.	9	2	50.45	2	0	1
35.	10	2	94.025	1	2	2
36.	10	2	66.125	2	0	1
37.	13	2	124.975	1	2	2
38.	13	2	122.45	2	0	1
39.	14	2	85.225	1	2	2
40.	14	2	99.075	2	0	1
41.	17	2	95.925	1	2	2
42.	17	2	86.35	2	0	1
43.	18	2	67.1	1	2	2
44.	18	2	49.925	2	0	1
45.	21	2	59.425	1	2	2
46.	21	2	42.7	2	0	1
47.	22	2	114.05	1	2	2
48.	22	2	91.725	2	0	1

Now the data are organized into separate variables that indicate each factor level for each of the covariates, so the data may be used with anova or pkcross; see [R] **anova** and [R] **pkcross**.

◁

▷ Example 2

Consider the study of background music on bank teller productivity published in Neter et al. (1996). The data are

Week	Monday	Tuesday	Wednesday	Thursday	Friday
1	18(D)	17(C)	14(A)	21(B)	17(E)
2	13(C)	34(B)	21(E)	16(A)	15(D)
3	7(A)	29(D)	32(B)	27(E)	13(C)
4	17(E)	13(A)	24(C)	31(D)	25(B)
5	21(B)	26(E)	26(D)	31(C)	7(A)

The numbers are the productivity scores, and the letters represent the treatment. We entered the data into Stata:

```
. use http://www.stata-press.com/data/r9/music, clear
. list
```

	id	seq	day1	day2	day3	day4	day5
1.	1	dcabe	18	17	14	21	17
2.	2	cbead	13	34	21	16	15
3.	3	adbec	7	29	32	27	13
4.	4	eacdb	17	13	24	31	25
5.	5	bedca	21	26	26	31	7

We reshape these data with pkshape:

```
. pkshape id seq day1 day2 day3 day4 day5
. list, sep(0)
```

	id	sequence	outcome	treat	carry	period
1.	3	1	7	1	0	1
2.	5	2	21	3	0	1
3.	2	3	13	5	0	1
4.	1	4	18	2	0	1
5.	4	5	17	4	0	1
6.	3	1	29	2	1	2
7.	5	2	26	4	3	2
8.	2	3	34	3	5	2
9.	1	4	17	5	2	2
10.	4	5	13	1	4	2
11.	3	1	32	3	2	3
12.	5	2	26	2	4	3
13.	2	3	21	4	3	3
14.	1	4	14	1	5	3
15.	4	5	24	5	1	3
16.	3	1	27	4	3	4
17.	5	2	31	5	2	4
18.	2	3	16	1	4	4
19.	1	4	21	3	1	4
20.	4	5	31	2	5	4
21.	3	1	13	5	4	5
22.	5	2	7	1	5	5
23.	2	3	15	2	1	5
24.	1	4	17	4	3	5
25.	4	5	25	3	2	5

In this case, the `sequence` variable is a string variable that specifies how the treatments were applied, so the `order` option is not used. In cases where the sequence variable is a string and the `order` is specified, the arguments from the `order` option are used. We could now produce an ANOVA table:

```
. anova outcome seq period treat
```

	Number of obs =	25	R-squared	=	0.8666
	Root MSE = 3.96232		Adj R-squared =		0.7331

Source	Partial SS	df	MS	F	Prob > F
Model	1223.6	12	101.966667	6.49	0.0014
sequence	82	4	20.5	1.31	0.3226
period	477.2	4	119.3	7.60	0.0027
treat	664.4	4	166.1	10.58	0.0007
Residual	188.4	12	15.7		
Total	1412	24	58.8333333		

◁

▷ Example 3

Consider the Latin-square crossover example published in Neter et al. (1996). The example is about apple sales given different methods for displaying apples.

Pattern	Store	Week 1	Week 2	Week 3
1	1	9(B)	12(C)	15(A)
	2	4(B)	12(C)	9(A)
2	1	12(A)	14(B)	3(C)
	2	13(A)	14(B)	3(C)
3	1	7(C)	18(A)	6(B)
	2	5(C)	20(A)	4(B)

We entered the data into Stata:

```
. use http://www.stata-press.com/data/r9/applesales, clear
. list, sep(2)
```

	id	seq	p1	p2	p3	square
1.	1	1	9	12	15	1
2.	2	1	4	12	9	2
3.	3	2	12	14	3	1
4.	4	2	13	14	3	2
5.	5	3	7	18	6	1
6.	6	3	5	20	4	2

Now the data can be reorganized using descriptive names for the outcome variables.

```
. pkshape id seq p1 p2 p3, order(bca abc cab) seq(pattern) period(order)
> treat(displays)
```

```
. anova outcome pattern order display id|pattern
                              Number of obs =       18    R-squared     =  0.9562
                              Root MSE      = 1.59426    Adj R-squared =  0.9069

                 Source |   Partial SS      df      MS            F     Prob > F

                  Model |  443.666667       9  49.2962963       19.40    0.0002

                pattern |  .333333333       2  .166666667        0.07    0.9370
                  order |  233.333333       2  116.666667       45.90    0.0000
               displays |         189       2        94.5       37.18    0.0001
             id|pattern |          21       3           7        2.75    0.1120

               Residual |  20.3333333       8  2.54166667

                  Total |         464      17  27.2941176
```

These are the same results reported by Neter et al. (1996).

◁

▷ Example 4

Continuing with the example from [R] **pkcollapse**, the data are

```
. use http://www.stata-press.com/data/r9/pkdata2, clear
. list, sep(4) abbrev(10)
```

	id	seq	auc_concA	auc_concB
1.	1	1	150.9643	218.5551
2.	2	1	146.7606	133.3201
3.	3	1	160.6548	126.0635
4.	4	1	157.8622	96.17461
5.	5	1	133.6957	188.9038
6.	7	1	160.639	223.6922
7.	8	1	131.2604	104.0139
8.	9	1	168.5186	237.8962
9.	10	2	137.0627	139.7382
10.	12	2	153.4038	202.3942
11.	13	2	163.4593	136.7848
12.	14	2	146.0462	104.5191
13.	15	2	158.1457	165.8654
14.	18	2	147.1977	139.235
15.	19	2	164.9988	166.2391
16.	20	2	145.3823	158.5146

```
. pkshape id seq auc_concA auc_concB, order(ab ba)
. sort period id
```

(Continued on next page)

```
. list, sep(4)
```

	id	sequence	outcome	treat	carry	period
1.	1	1	150.9643	1	0	1
2.	2	1	146.7606	1	0	1
3.	3	1	160.6548	1	0	1
4.	4	1	157.8622	1	0	1
5.	5	1	133.6957	1	0	1
6.	7	1	160.639	1	0	1
7.	8	1	131.2604	1	0	1
8.	9	1	168.5186	1	0	1
9.	10	2	137.0627	2	0	1
10.	12	2	153.4038	2	0	1
11.	13	2	163.4593	2	0	1
12.	14	2	146.0462	2	0	1
13.	15	2	158.1457	2	0	1
14.	18	2	147.1977	2	0	1
15.	19	2	164.9988	2	0	1
16.	20	2	145.3823	2	0	1
17.	1	1	218.5551	2	1	2
18.	2	1	133.3201	2	1	2
19.	3	1	126.0635	2	1	2
20.	4	1	96.17461	2	1	2
21.	5	1	188.9038	2	1	2
22.	7	1	223.6922	2	1	2
23.	8	1	104.0139	2	1	2
24.	9	1	237.8962	2	1	2
25.	10	2	139.7382	1	2	2
26.	12	2	202.3942	1	2	2
27.	13	2	136.7848	1	2	2
28.	14	2	104.5191	1	2	2
29.	15	2	165.8654	1	2	2
30.	18	2	139.235	1	2	2
31.	19	2	166.2391	1	2	2
32.	20	2	158.5146	1	2	2

◁

We call the resulting dataset pkdata3. We conduct equivalence testing on the data in [R] **pkequiv**, and we fit an ANOVA model to these data in the third example of [R] **pkcross**.

Methods and Formulas

pkshape is implemented as an ado-file.

References

Chow, S. C. and J. P. Liu. 2000. *Design and Analysis of Bioavailability and Bioequivalence Studies*. 2nd ed. New York: Dekker.

Neter, J., M. H. Kutner, C. J. Nachtsheim, and W. Wasserman. 1996. *Applied Linear Statistical Models*. 4th ed. Chicago: Irwin.

Also See

Related:	[R] **pkcollapse**, [R] **pkcross**, [R] **pkequiv**, [R] **pkexamine**, [R] **pksumm**; [R] **anova**
Background:	[R] **pk**

Title

> **pksumm** — Summarize pharmacokinetic data

Syntax

pksumm *id time concentration* $\left[\,if\,\right]$ $\left[\,in\,\right]$ $\left[\,,\ options\,\right]$

options	description
Main	
trapezoid	use trapezoidal rule to calculate AUC; default is cubic splines
fit(#)	use # points to estimate AUC; default is fit(3)
notimechk	do not check whether follow-up time for all subjects is the same
nodots	suppress the dots during calculation
graph	graph the distribution of *statistic*
stat(*statistic*)	graph the specified statistic; default is stat(auc)
Histogram, Density plots, Y-Axis, X-Axis, Title, Caption, Legend, Overall	
histogram_options	any of the options allowed with histogram; see [R] **histogram**

statistic	description
auc	area under the concentration-time curve ($AUC_{0,\infty}$); the default
aucline	area under the concentration-time curve from 0 to ∞ using a linear extension
aucexp	area under the concentration-time curve from 0 to ∞ using an exponential extension
auclog	area under the log-concentration-time curve extended with a linear fit
half	half-life of the drug
ke	elimination rate
cmax	maximum concentration
tmax	time at last concentration
tomc	time of maximum concentration

Description

pksumm obtains summary measures based on the first four moments from the empirical distribution of each pharmacokinetic measurement and tests the null hypothesis that the distribution of that measurement is normally distributed.

pksumm is one of the pk commands. Please read [R] **pk** before reading this entry.

Options

trapezoid specifies that the trapezoidal rule be used to calculate the AUC. The default is cubic splines, which give better results for most situations. In cases where the curve is very irregular, the trapezoidal rule may give better results.

fit(#) specifies the number of points, counting back from the last time measurement, to use in fitting the extension to estimate the $AUC_{0,\infty}$. The default is fit(3), the last 3 points. This should be viewed as a minimum; the appropriate number of points will depend on the data.

notimechk suppresses the check that the follow-up time for all subjects is the same. By default, pksumm expects the maximum follow-up time to be equal for all subjects.

nodots suppresses the progress dots during calculation. By default, a period is displayed for every call to calculate the pharmacokinetic measures.

graph requests a graph of the distribution of the statistic specified with stat().

stat(*statistic*) specifies the statistic that pksumm should graph. The default is stat(auc). If the graph option is not specified, this option is ignored.

histogram_options are any of the options allowed with histogram; see [R] **histogram**. Note that for pksumm, fraction is the default, not density.

Remarks

pksumm produces summary statistics for the distribution of nine common pharmacokinetic measurements. If there are more than eight subjects, pksumm also computes a test for normality on each measurement. The nine measurements summarized by pksumm are listed above and are described in *Methods and Formulas* of [R] **pkexamine** and [R] **pk**.

▷ Example 1

We demonstrate the use of pksumm on a variation of the data described in [R] **pk**. We have drug concentration data on 15 subjects, each measured at 13 time points over a 32-hour period. A few of the records are

(Continued on next page)

```
. use http://www.stata-press.com/data/r9/pksumm

. list, sep(0)
```

	id	time	conc
1.	1	0	0
2.	1	.5	3.073403
3.	1	1	5.188444
4.	1	1.5	5.898577
5.	1	2	5.096378
6.	1	3	6.094085
	(output omitted)		
183.	15	0	0
184.	15	.5	3.86493
185.	15	1	6.432444
186.	15	1.5	6.969195
187.	15	2	6.307024
188.	15	3	6.509584
189.	15	4	6.555091
190.	15	6	7.318319
191.	15	8	5.329813
192.	15	12	5.411624
193.	15	16	3.891397
194.	15	24	5.167516
195.	15	32	2.649686

We can use `pksumm` to view the summary statistics for all the pharmacokinetic parameters.

```
. pksumm id time conc
. . . . . . . . . . . . . .
```

Summary statistics for the pharmacokinetic measures

Number of observations = 15

Measure	Mean	Median	Variance	Skewness	Kurtosis	p-value
auc	150.74	150.96	123.07	−0.26	2.10	0.69
aucline	408.30	214.17	188856.87	2.57	8.93	0.00
aucexp	691.68	297.08	762679.94	2.56	8.87	0.00
auclog	688.98	297.67	797237.24	2.59	9.02	0.00
half	94.84	29.39	18722.13	2.26	7.37	0.00
ke	0.02	0.02	0.00	0.89	3.70	0.09
cmax	7.36	7.42	0.42	−0.60	2.56	0.44
tomc	3.47	3.00	7.62	2.17	7.18	0.00
tmax	32.00	32.00	0.00	.	.	.

For the 15 subjects, the mean $AUC_{0,t_{max}}$ is 150.74, and $\sigma^2 = 123.07$. The skewness of -0.26 indicates that the distribution is slightly skewed left. The p-value of 0.69 for the χ^2 test of normality indicates that we cannot reject the null hypothesis that the distribution is normal.

If we were to consider any of the three variants of the $AUC_{0,\infty}$, we would see that there is huge variability and that the distribution is heavily skewed. A skewness different from 0 and a kurtosis different from 3 are expected because the distribution of the $AUC_{0,\infty}$ is not normal.

We now graph the distribution of $AUC_{0,t_{max}}$ by specifying the `graph` option.

```
. pksumm id time conc, graph bin(20)
...............

Summary statistics for the pharmacokinetic measures
```

| | | | | Number of observations = | | 15 |
Measure	Mean	Median	Variance	Skewness	Kurtosis	p-value
auc	150.74	150.96	123.07	-0.26	2.10	0.69
aucline	408.30	214.17	188856.87	2.57	8.93	0.00
aucexp	691.68	297.08	762679.94	2.56	8.87	0.00
auclog	688.98	297.67	797237.24	2.59	9.02	0.00
half	94.84	29.39	18722.13	2.26	7.37	0.00
ke	0.02	0.02	0.00	0.89	3.70	0.09
cmax	7.36	7.42	0.42	-0.60	2.56	0.44
tomc	3.47	3.00	7.62	2.17	7.18	0.00
tmax	32.00	32.00	0.00	.	.	.

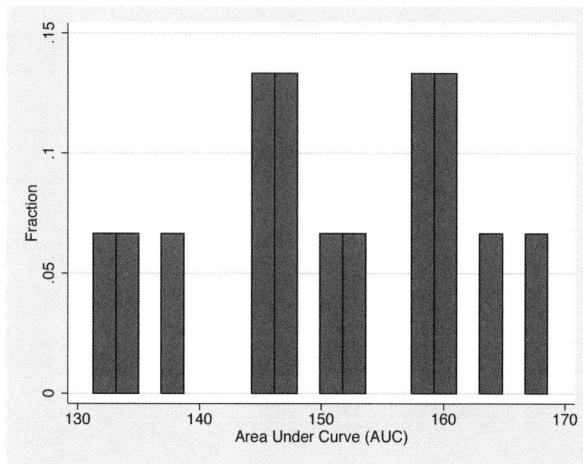

`graph`, by default, plots $AUC_{0,t_{max}}$. To plot a graph of one of the other pharmacokinetic measurements, we need to specify the `stat()` option. For example, we can ask Stata to produce a plot of the $AUC_{0,\infty}$ using the log extension:

```
. pksumm id time conc, stat(auclog) graph bin(20)
...............

Summary statistics for the pharmacokinetic measures
```

| | | | | Number of observations = | | 15 |
Measure	Mean	Median	Variance	Skewness	Kurtosis	p-value
auc	150.74	150.96	123.07	-0.26	2.10	0.69
aucline	408.30	214.17	188856.87	2.57	8.93	0.00
aucexp	691.68	297.08	762679.94	2.56	8.87	0.00
auclog	688.98	297.67	797237.24	2.59	9.02	0.00
half	94.84	29.39	18722.13	2.26	7.37	0.00
ke	0.02	0.02	0.00	0.89	3.70	0.09
cmax	7.36	7.42	0.42	-0.60	2.56	0.44
tomc	3.47	3.00	7.62	2.17	7.18	0.00
tmax	32.00	32.00	0.00	.	.	.

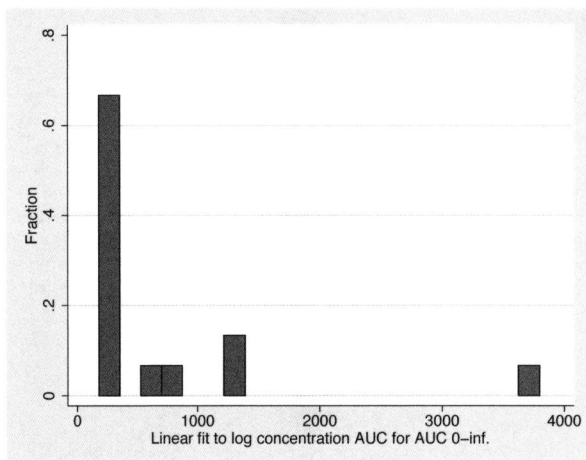

Linear fit to log concentration AUC for AUC 0–inf.

$\triangleleft$

Methods and Formulas

pksumm is implemented as an ado-file.

The χ^2 test for normality is conducted with sktest; see [R] **sktest** for more information on the test of normality.

The statistics reported by pksumm are identical to those reported by summarize and sktest; see [R] **summarize** and [R] **sktest**.

Also See

Related: [R] **pkcollapse**, [R] **pkcross**, [R] **pkequiv**, [R] **pkexamine**, [R] **pkshape**

Background: [R] **pk**

Title

poisson — Poisson regression

Syntax

poisson *depvar* $\left[\textit{indepvars}\right]$ $\left[\textit{if}\right]$ $\left[\textit{in}\right]$ $\left[\textit{weight}\right]$ $\left[\textit{, options}\right]$

options	description
Model	
<u>nocon</u>stant	suppress constant term
<u>exp</u>osure(*varname_e*)	include ln(*varname_e*) in model with coefficient constrained to 1
<u>off</u>set(*varname_o*)	include *varname_o* in model with coefficient constrained to 1
<u>constra</u>ints(*constraints*)	apply specified linear constraints
SE/Robust	
vce(*vcetype*)	*vcetype* may be oim, <u>r</u>obust, opg, <u>boot</u>strap, or <u>jack</u>knife
<u>r</u>obust	synonym for vce(robust)
<u>cl</u>uster(*varname*)	adjust standard errors for intragroup correlation
Reporting	
<u>level</u>(#)	set confidence level; default is level(95)
<u>irr</u>	report incidence-rate ratios
Max options	
maximize_options	control the maximization process; seldom used

depvar, *indepvars*, *varname_e*, and *varname_o* may contain time-series operators; see
[U] **11.4.3 Time-series varlists**.
bootstrap, by, jackknife, rolling, statsby, stepwise, svy, and xi are allowed; see
[U] **11.1.10 Prefix commands**.
fweights, iweights, and pweights are allowed; see [U] **11.1.6 weight**.
See [U] **20 Estimation and postestimation commands** for additional capabilities of estimation commands.

Description

poisson fits a Poisson regression of *depvar* on *indepvars*, where *depvar* is a non-negative count variable.

If you have panel data, see [XT] **xtpoisson**.

Options

> **Model**

noconstant, exposure(*varname_e*), offset(*varname_o*), constraints(*constraints*); see [R] **estimation options**.

431

SE/Robust

vce(*vcetype*); see [R] **vce_option**.

robust, cluster(*varname*); see [R] **estimation options**. cluster() can be used with pweights to produce estimates for unstratified cluster-sampled data, but see [SVY] **svy: poisson** for a command especially designed for survey data.

Reporting

level(#); see [R] **estimation options**.

irr reports estimated coefficients transformed to incidence-rate ratios, that is, e^{β_i} rather than β_i. Standard errors and confidence intervals are similarly transformed. This option affects how results are displayed, not how they are estimated or stored. irr may be specified at estimation or when replaying previously estimated results.

Max options

maximize_options: difficult, technique(*algorithm_spec*), iterate(#), [no]log, trace, gradient, showstep, hessian, shownrtolerance, tolerance(#), ltolerance(#), gtolerance(#), nrtolerance(#), nonrtolerance, from(*init_specs*); see [R] **maximize**. These options are seldom used.

Remarks

The basic idea of Poisson regression was outlined by Coleman (1964, 378–379). See Cameron and Trivedi (1998) and Feller (1968, 156–164) for information about the Poisson distribution. See Cameron and Trivedi (1998), Long (1997, chapter 8), Long and Freese (2003, chapter 7), McNeil (1996, chapter 6), and Selvin (2004, chapter 9) for an introduction to Poisson regression. Also see Selvin (2004, chapter 5) for a discussion of the analysis of spatial distributions, which includes a discussion of the Poisson distribution. An early example of Poisson regression was Cochran (1940).

Poisson regression fits models of the number of occurrences (counts) of an event. The Poisson distribution has been applied to diverse events, such as the number of soldiers kicked to death by horses in the Prussian army (Bortkewitsch 1898); the pattern of hits by buzz bombs launched against London during World War II (Clarke 1946); telephone connections to a wrong number (Thorndike 1926); and disease incidence, typically with respect to time, but occasionally with respect to space. The basic assumptions are as follows:

1. There is a quantity called the *incidence rate* that is the rate at which events occur. Examples are 5 per second, 20 per 1,000 person-years, 17 per square meter, and 38 per cubic centimeter.

2. The incidence rate can be multiplied by *exposure* to obtain the expected number of observed events. For example, a rate of 5 per second multiplied by 30 seconds means that 150 events are expected; a rate of 20 per 1,000 person-years multiplied by 2,000 person-years means that 40 events are expected; and so on.

3. Over very small exposures ϵ, the probability of finding more than one event is small compared with ϵ.

4. Nonoverlapping exposures are mutually independent.

With these assumptions, to find the probability of k events in an exposure of size E, you divide E into n subintervals $E_1, E_2, \ldots, E_n$, and approximate the answer as the binomial probability of observing k successes in n trials. If you let $n \to \infty$, you obtain the Poisson distribution.

In the Poisson regression model, the incidence rate for the jth observation is assumed to be given by

$$r_j = e^{\beta_0 + \beta_1 x_{1,j} + \cdots + \beta_k x_{k,j}}$$

If E_j is the exposure, the expected number of events, C_j, will be

$$C_j = E_j e^{\beta_0 + \beta_1 x_{1,j} + \cdots + \beta_k x_{k,j}}$$
$$= e^{\ln(E_j) + \beta_0 + \beta_1 x_{1,j} + \cdots + \beta_k x_{k,j}}$$

This model is fitted by `poisson`. Without the `exposure()` or `offset()` options, E_j is assumed to be 1 (equivalent to assuming that exposure is unknown), and controlling for exposure, if necessary, is your responsibility.

Comparing rates is most easily done by calculating *incidence-rate ratios* (IRR). For instance, what is the relative incidence rate of chromosome interchanges in cells as the intensity of radiation increases; the relative incidence rate of telephone connections to a wrong number as load increases; or the relative incidence rate of deaths due to cancer for females relative to males? That is, you want to hold all the xs in the model constant except one, say, the ith. The incidence-rate ratio for a one-unit change in x_i is

$$\frac{e^{\ln(E) + \beta_1 x_1 + \cdots + \beta_i (x_i + 1) + \cdots + \beta_k x_k}}{e^{\ln(E) + \beta_1 x_1 + \cdots + \beta_i x_i + \cdots + \beta_k x_k}} = e^{\beta_i}$$

More generally, the incidence-rate ratio for a Δx_i change in x_i is $e^{\beta_i \Delta x_i}$. The `lincom` command can be used after `poisson` to display incidence-rate ratios for any group relative to another; see [R] **lincom**.

▷ Example 1

Chatterjee, Hadi, and Price (2000, 164) give the number of injury incidents and the proportion of flights for each airline out of the total number of flights from New York for nine major U.S. airlines in a single year:

```
. use http://www.stata-press.com/data/r9/airline
. list
```

	airline	injuries	n	XYZowned
1.	1	11	0.0950	1
2.	2	7	0.1920	0
3.	3	7	0.0750	0
4.	4	19	0.2078	0
5.	5	9	0.1382	0
6.	6	4	0.0540	1
7.	7	3	0.1292	0
8.	8	1	0.0503	0
9.	9	3	0.0629	1

To their data we have added a fictional variable, XYZowned. We will imagine that an accusation is made that the airlines owned by XYZ Company have a higher injury rate.

```
. poisson injuries XYZowned, exposure(n) irr
Iteration 0:   log likelihood = -23.027197
Iteration 1:   log likelihood = -23.027177
Iteration 2:   log likelihood = -23.027177
```

```
Poisson regression                              Number of obs   =            9
                                                LR chi2(1)      =         1.77
                                                Prob > chi2     =       0.1836
Log likelihood = -23.027177                     Pseudo R2       =       0.0370
```

injuries	IRR	Std. Err.	z	P>|z|	[95% Conf. Interval]
XYZowned	1.463467	.406872	1.37	0.171	.8486578 2.523675
n	(exposure)				

We specified `irr` to see the incidence-rate ratios rather than the underlying coefficients. We estimate that XYZ Airlines' injury rate is 1.46 times larger than that for other airlines, but the 95% confidence interval is .85 to 2.52; we cannot even reject the hypothesis that XYZ Airlines has a lower injury rate.

◁

❏ Technical Note

In example 1, we assumed that each airline's exposure was proportional to its fraction of flights out of New York. What if "large" airlines, however, also used larger planes, and so had even more passengers than would be expected, given this measure of exposure? A better measure would be each airline's fraction of passengers on flights out of New York, a number that we do not have. Even so, we suppose that n represents this number to some extent, so a better estimate of the effect might be

```
. gen lnN=ln(n)

. poisson injuries XYZowned lnN

Iteration 0:    log likelihood = -22.333875
Iteration 1:    log likelihood = -22.332276
Iteration 2:    log likelihood = -22.332276

Poisson regression                              Number of obs   =            9
                                                LR chi2(2)      =        19.15
                                                Prob > chi2     =       0.0001
Log likelihood = -22.332276                     Pseudo R2       =       0.3001
```

injuries	Coef.	Std. Err.	z	P>|z|	[95% Conf. Interval]
XYZowned	.6840667	.3895877	1.76	0.079	-.0795111 1.447645
lnN	1.424169	.3725155	3.82	0.000	.6940517 2.154285
_cons	4.863891	.7090501	6.86	0.000	3.474178 6.253603

In this case, rather than specifying the `exposure()` option, we explicitly included the variable that would normalize for exposure in the model. We did not specify the `irr` option, so we see coefficients rather than incidence-rate ratios. We started with the model

$$\text{rate} = e^{\beta_0 + \beta_1 \text{XYZowned}}$$

The observed counts are therefore

$$\text{count} = n e^{\beta_0 + \beta_1 \text{XYZowned}} = e^{\ln(n) + \beta_0 + \beta_1 \text{XYZowned}}$$

which amounts to constraining the coefficient on $\ln(n)$ to 1. This is what was estimated when we specified the `exposure(n)` option. In the above model, we included the normalizing exposure ourselves and, rather than constraining the coefficient to be 1, estimated the coefficient.

The estimated coefficient is 1.42, a respectable distance away from 1, and is consistent with our speculation that larger airlines also use larger airplanes. With this small amount of data, however, we also have a wide confidence interval that includes 1.

Our estimated *coefficient* on XYZowned is now .684, and the implied incidence-rate ratio is $e^{.684} \approx 1.98$ (which we could also see by typing poisson, irr). The 95% confidence interval for the coefficient still includes 0 (the interval for the incidence-rate ratio includes 1), so while the point estimate is now larger, we still cannot be very certain of our results.

Our expert opinion would be that, while there is insufficient evidence to support the charge, there is enough evidence to justify collecting more data.

❏

▷ Example 2

In a famous age-specific study of coronary disease deaths among male British doctors, Doll and Hill (1966) reported the following data (reprinted in Rothman and Greenland 1998, 259):

Age	Smokers		Nonsmokers	
	Deaths	Person-years	Deaths	Person-years
35–44	32	52,407	2	18,790
45–54	104	43,248	12	10,673
55–64	206	28,612	28	5,710
65–74	186	12,663	28	2,585
75–84	102	5,317	31	1,462

The first step is to enter these data into Stata, which we have done:

```
. use http://www.stata-press.com/data/r9/dollhill3, clear
. list
```

	agecat	smokes	deaths	pyears
1.	1	1	32	52,407
2.	2	1	104	43,248
3.	3	1	206	28,612
4.	4	1	186	12,663
5.	5	1	102	5,317
6.	1	0	2	18,790
7.	2	0	12	10,673
8.	3	0	28	5,710
9.	4	0	28	2,585
10.	5	0	31	1,462

agecat 1 corresponds to 35–44, agecat 2 to 45–54, and so on. The most "natural" analysis of these data would begin by introducing indicator variables for each age category and a single indicator for smoking:

(Continued on next page)

```
. tab agecat, gen(a)
    agecat |      Freq.     Percent        Cum.
-----------+-----------------------------------
         1 |         2       20.00       20.00
         2 |         2       20.00       40.00
         3 |         2       20.00       60.00
         4 |         2       20.00       80.00
         5 |         2       20.00      100.00
-----------+-----------------------------------
     Total |        10      100.00
```

```
. poisson deaths smokes a2-a5, exposure(pyears) irr

Iteration 0:   log likelihood = -33.823284
Iteration 1:   log likelihood = -33.600471
Iteration 2:   log likelihood = -33.600153
Iteration 3:   log likelihood = -33.600153

Poisson regression                               Number of obs   =          10
                                                 LR chi2(5)      =      922.93
                                                 Prob > chi2     =      0.0000
Log likelihood = -33.600153                      Pseudo R2       =      0.9321
```

deaths	IRR	Std. Err.	z	P>|z|	[95% Conf. Interval]	
smokes	1.425519	.1530638	3.30	0.001	1.154984	1.759421
a2	4.410584	.8605197	7.61	0.000	3.009011	6.464997
a3	13.8392	2.542638	14.30	0.000	9.654328	19.83809
a4	28.51678	5.269878	18.13	0.000	19.85177	40.96395
a5	40.45121	7.775511	19.25	0.000	27.75326	58.95885
pyears	(exposure)					

In the above, we began by using `tabulate` to create the indicator variables. `tabulate` created a1 equal to 1 when `agecat` = 1 and 0, otherwise; a2 equal to 1 when `agecat` = 2 and 0, otherwise; and so on. See [U] **25 Dealing with categorical variables**.

We then fitted our model, specifying `irr` to obtain incidence-rate ratios. We estimate that smokers have 1.43 times the mortality rate of nonsmokers. See, however, example 1 in [R] **poisson postestimation**.

◁

Siméon-Denis Poisson (1781–1840) was a French mathematician and physicist who contributed to several fields: his name is perpetuated in Poisson brackets, Poisson's constant, Poisson's differential equation, Poisson's integral, and Poisson's ratio. Among many other results, he produced a version of the law of large numbers. His rather misleadingly titled *Recherches sur la probabilité des jugements* embraces a complete treatise on probability, as the subtitle indicates, including what is now known as the Poisson distribution. That, however, was discovered earlier by the Huguenot–British mathematician Abraham de Moivre (1667–1754).

Saved Results

poisson saves in e():

Scalars

e(N)	number of observations	e(N_clust)	number of clusters
e(k)	number of parameters	e(chi2)	χ^2
e(k_eq)	number of equations	e(p)	significance
e(k_dv)	number of dependent variables	e(rank)	rank of e(V)
e(df_m)	model degrees of freedom	e(ic)	number of iterations
e(r2_p)	pseudo-R-squared	e(rc)	return code
e(ll)	log likelihood	e(converged)	1 if converged, 0 otherwise
e(ll_0)	log likelihood, constant-only model		

Macros

e(cmd)	poisson	e(vcetype)	title used to label Std. Err.
e(depvar)	name of dependent variable	e(opt)	type of optimization
e(wtype)	weight type	e(ml_method)	type of ml method
e(wexp)	weight expression	e(user)	name of likelihood-evaluator program
e(title)	title in estimation output	e(technique)	maximization technique
e(clustvar)	name of cluster variable	e(crittype)	optimization criterion
e(offset)	offset	e(properties)	b V
e(chi2type)	Wald or LR; type of model χ^2 test	e(estat_cmd)	program used to implement estat
e(vce)	*vcetype* specified in vce()	e(predict)	program used to implement predict

Matrices

e(b)	coefficient vector	e(V)	variance–covariance matrix of
e(ilog)	iteration log (up to 20 iterations)		the estimators
e(gradient)	gradient vector		

Functions

e(sample)	marks estimation sample

Methods and Formulas

poisson is implemented as an ado-file.

The log likelihood (with weights w_j and offsets) is given by

$$\Pr(Y = y) = \frac{e^{-\lambda}\lambda^y}{y!}$$

$$\xi_j = \mathbf{x}_j\boldsymbol{\beta} + \text{offset}_j$$

$$f(y_j) = \frac{e^{-\exp(\xi_j)}e^{\xi_j y_j}}{y_j!}$$

$$\ln L = \sum_{j=1}^{n} w_j \left\{ -e^{\xi_j} + \xi_j y_j - \ln(y_j!) \right\}$$

References

Bortkewitsch, L. von. 1898. *Das Gesetz der Kleinen Zahlen*. Leipzig: Teubner.

Bru, B. 2001. Siméon-Denis Poisson. In *Statisticians of the Centuries*, ed. C. C. Heyde and E. Seneta, 123–126. New York: Springer.

Cameron, A. C. and P. K. Trivedi. 1998. *Regression Analysis of Count Data*. Cambridge: Cambridge University Press.

Chatterjee, S., A. S. Hadi, and B. Price. 2000. *Regression Analysis by Example*. 3rd ed. New York: Wiley.

Clarke, R. D. 1946. An application of the Poisson distribution. *Journal of the Institute of Actuaries* 22: 48.

Cochran, W. G. 1940. The analysis of variance when experimental errors follow the Poisson or binomial laws. *Annals of Mathematical Statistics* 11: 335–347. Reprinted as paper 22 in Cochran (1982).

——. 1982 *Contributions to Statistics*. New York: Wiley.

Coleman, J. S. 1964. *Introduction to Mathematical Sociology*. New York: Free Press.

Doll, R. and A. B. Hill. 1966. Mortality of British doctors in relation to smoking; observations on coronary thrombosis. In *Epidemiological Approaches to the Study of Cancer and Other Chronic Diseases*, ed. W. Haenszel. *National Cancer Institute Monograph* 19: 204–268.

Feller, W. 1968. *An Introduction to Probability Theory and Its Applications*, vol. 1. 3rd ed. New York: Wiley.

Hilbe, J. 1998. sg91: Robust variance estimators for MLE Poisson and negative binomial regression. *Stata Technical Bulletin* 45: 26–28. Reprinted in *Stata Technical Bulletin Reprints*, vol. 8, pp. 177–180.

——. 1999. sg102: Zero-truncated Poisson and negative binomial regression. *Stata Technical Bulletin* 47: 37–40. Reprinted in *Stata Technical Bulletin Reprints*, vol. 8, pp. 233–236.

Hilbe, J. and D. H. Judson. 1998. sg94: Right, left, and uncensored Poisson regression. *Stata Technical Bulletin* 46: 18–20. Reprinted in *Stata Technical Bulletin Reprints*, vol. 8, pp. 186–189.

Long, J. S. 1997. *Regression Models for Categorical and Limited Dependent Variables*. Thousand Oaks, CA: Sage.

Long, J. S. and J. Freese. 2001. Predicted probabilities for count models. *Stata Journal* 1: 51–57.

——. 2003. *Regression Models for Categorical Dependent Variables Using Stata*. rev. ed. College Station, TX: Stata Press.

McNeil, D. 1996. *Epidemiological Research Methods*. Chichester, UK: Wiley.

Newman, S. C. 2001. *Biostatistical Methods in Epidemiology*. New York: Wiley.

Poisson, S. D. 1837. *Recherches sur la probabilité des jugements en matière criminelle et en matière civile, précédés des règles générales du calcul des probabilités*. Paris: Bachelier.

Rodríguez, G. 1993. sbe10: An improvement to poisson. *Stata Technical Bulletin* 11: 11–14. Reprinted in *Stata Technical Bulletin Reprints*, vol. 2, pp. 94–98.

Rogers, W. H. 1991. sbe1: Poisson regression with rates. *Stata Technical Bulletin* 1: 11–12. Reprinted in *Stata Technical Bulletin Reprints*, vol. 1, pp. 62–64.

Rothman, K. J. and S. Greenland. 1998. *Modern Epidemiology*. 2nd ed. Philadelphia: Lippincott–Raven.

Rutherford, E., J. Chadwick, and C. D. Ellis. 1930. *Radiations from Radioactive Substances*. Cambridge: Cambridge University Press.

Selvin, S. 2004. *Statistical Analysis of Epidemiologic Data*. 3rd ed. New York: Oxford University Press.

Thorndike, F. 1926. Applications of Poisson's probability summation. *Bell System Technical Journal* 5: 604–624.

Tobias, A. and M. J. Campbell. 1998. sts13: Time-series regression for counts allowing for autocorrelation. *Stata Technical Bulletin* 46: 33–37. Reprinted in *Stata Technical Bulletin Reprints*, vol. 8, pp. 291–296.

Also See

Complementary:	[R] **poisson postestimation**; [R] **constraint**
Related:	[R] **glm**, [R] **nbreg**,
	[ST] **epitab**, [SVY] **svy: poisson**, [XT] **xtpoisson**
Background:	[U] **11.1.10 Prefix commands**,
	[U] **20 Estimation and postestimation commands**,
	[R] **estimation options**, [R] **maximize**, [R] *vce_option*

Title

poisson postestimation — Postestimation tools for poisson

Description

The following postestimation command is of special interest after `poisson`:

command	description
estat gof	goodness-of-fit test

For information about `estat gof`, see below.

In addition, the following standard postestimation commands are available:

command	description
adjust[1]	adjusted predictions of $\mathbf{x}\beta$ or $\exp(\mathbf{x}\beta)$
estat	AIC, BIC, VCE, and estimation sample summary
estimates	cataloging estimation results
lincom	point estimates, standard errors, testing, and inference for linear combinations of coefficients
linktest	link test for model specification
lrtest	likelihood-ratio test
mfx	marginal effects or elasticities
nlcom	point estimates, standard errors, testing, and inference for nonlinear combinations of coefficients
predict	predictions, residuals, influence statistics, and other diagnostic measures
predictnl	point estimates, standard errors, testing, and inference for generalized predictions
suest	seemingly unrelated estimation
test	Wald tests for simple and composite linear hypotheses
testnl	Wald tests of nonlinear hypotheses

[1] `adjust` does not work with time-series operators.

See the corresponding entries in the *Stata Base Reference Manual* for details.

Special-interest postestimation command

`estat gof` performs a goodness-of-fit test of the model. The default is the deviance statistic; specifying option `pearson` will give the Pearson statistic. If the test is significant, the Poisson regression model is inappropriate. In this case, you could try a negative binomial model; see [R] **nbreg**.

Syntax for predict

predict $[$ *type* $]$ *newvar* $[$ *if* $]$ $[$ *in* $]$ $[$, *statistic* <u>nooff</u>set $]$

statistic	description
n	predicted number of events; the default
ir	incidence rate
xb	linear prediction
stdp	standard error of the linear prediction
<u>sc</u>ore	first derivative of the log likelihood with respect to $\mathbf{x}_j\boldsymbol{\beta}$

These statistics are available both in and out of sample; type predict ... if e(sample) ... if wanted only for the estimation sample.

Options for predict

n, the default, calculates the predicted number of events, which is $\exp(\mathbf{x}_j\boldsymbol{\beta})$ if neither offset() nor exposure() was specified when the model was fitted; $\exp(\mathbf{x}_j\boldsymbol{\beta} + \text{offset}_j)$ if offset() was specified; or $\exp(\mathbf{x}_j\boldsymbol{\beta}) \times \text{exposure}_j$ if exposure() was specified.

ir calculates the incidence rate, $\exp(\mathbf{x}_j\boldsymbol{\beta})$, which is the predicted number of events when exposure is 1. Specifying ir is equivalent to specifying n when neither offset() nor exposure() was specified with the model was fitted.

xb calculates the linear prediction, which is $\mathbf{x}_j\boldsymbol{\beta}$ if neither offset() nor exposure() was specified; $\mathbf{x}_j\boldsymbol{\beta} + \text{offset}_j$ if offset() was specified; or $\mathbf{x}_j\boldsymbol{\beta} + \ln(\text{exposure}_j)$ if exposure() was specified; see nooffset below.

stdp calculates the standard error of the linear prediction.

score calculates the equation-level score, $\partial\ln L/\partial(\mathbf{x}_j\boldsymbol{\beta})$.

nooffset is relevant only if you specified offset() or exposure() when you fitted the model. It modifies the calculations made by predict so that they ignore the offset or exposure variable; the linear prediction is treated as $\mathbf{x}_j\boldsymbol{\beta}$ rather than as $\mathbf{x}_j\boldsymbol{\beta} + \text{offset}_j$ or $\mathbf{x}_j\boldsymbol{\beta} + \ln(\text{exposure}_j)$. Specifying predict ... , nooffset is equivalent to specifying predict ... , ir.

Syntax for estat gof

estat gof $[$, <u>pe</u>arson $]$

Option for estat gof

pearson requests that estat gof calculate the Pearson statistic rather than the deviance statistic.

(*Continued on next page*)

Remarks

▷ Example 1

Continuing with example 2 of [R] **poisson**, we use `estat gof` to determine whether the model fits the data well.

```
. estat gof
        Goodness-of-fit chi2  =   12.13244
        Prob > chi2(4)        =    0.0164
```

The goodness-of-fit χ^2 tells us that, given the model, we can reject the hypothesis that these data are Poisson distributed at the 1.64% significance level.

So let us now back up and be more careful. We can most easily obtain the incidence-rate ratios within age categories using `ir`; see [ST] **epitab**:

```
. ir deaths smokes pyears, by(agecat) nocrude nohet
```

agecat	IRR	[95% Conf. Interval]		M-H Weight	
1	5.736638	1.463519	49.39901	1.472169	(exact)
2	2.138812	1.173666	4.272307	9.624747	(exact)
3	1.46824	.9863626	2.264174	23.34176	(exact)
4	1.35606	.9082155	2.09649	23.25315	(exact)
5	.9047304	.6000946	1.399699	24.31435	(exact)
M-H combined	1.424682	1.154703	1.757784		

We find that the mortality incidence ratios are greatly different within age category, being highest for the youngest categories and actually dropping below 1 for the oldest. (In the last case, we might argue that those who smoke and who have not died by age 75 are self-selected to be particularly robust.)

Seeing this, we will now parameterize the smoking effects separately for each age category, although we will begin by combining age categories 3 and 4:

```
. gen sa1 = smokes*(agecat==1)
. gen sa2 = smokes*(agecat==2)
. gen sa34 = smokes*(agecat==3 | agecat==4)
. gen sa5 = smokes*(agecat==5)
```

(Continued on next page)

```
. poisson deaths sa1 sa2 sa34 sa5 a2-a5, exposure(pyears) irr
Iteration 0:   log likelihood = -31.635422
Iteration 1:   log likelihood = -27.788819
Iteration 2:   log likelihood = -27.573604
Iteration 3:   log likelihood = -27.572645
Iteration 4:   log likelihood = -27.572645
```

```
Poisson regression                              Number of obs    =          10
                                                LR chi2(8)       =      934.99
                                                Prob > chi2      =      0.0000
Log likelihood = -27.572645                     Pseudo R2        =      0.9443
```

deaths	IRR	Std. Err.	z	P>|z|	[95% Conf. Interval]	
sa1	5.736638	4.181257	2.40	0.017	1.374811	23.93711
sa2	2.138812	.6520701	2.49	0.013	1.176691	3.887609
sa34	1.412229	.2017485	2.42	0.016	1.067343	1.868557
sa5	.9047304	.1855513	-0.49	0.625	.6052658	1.35236
a2	10.5631	8.067702	3.09	0.002	2.364153	47.19624
a3	47.671	34.3741	5.36	0.000	11.60056	195.8978
a4	98.22766	70.85013	6.36	0.000	23.89324	403.8245
a5	199.21	145.3357	7.26	0.000	47.67694	832.365
pyears	(exposure)					

```
. estat gof
        Goodness-of-fit chi2  =   .0774185
        Prob > chi2(1)        =     0.7808
```

Note that the goodness-of-fit χ^2 is now small; we are no longer running roughshod over the data. Let us now consider simplifying the model. The point estimate of the incidence-rate ratio for smoking in age category 1 is much larger than that for smoking in age category 2, but the confidence interval for sa1 is similarly wide. Is the difference real?

```
. test sa1=sa2
 ( 1)  [deaths]sa1 - [deaths]sa2 = 0
           chi2(  1) =      1.56
         Prob > chi2 =      0.2117
```

The point estimates may be far apart, but there is insufficient data, and we may be observing random differences. With that success, might we also combine the smokers in age categories 3 and 4 with those in 1 and 2?

```
. test sa34=sa2, accum
 ( 1)  [deaths]sa1 - [deaths]sa2 = 0
 ( 2)  - [deaths]sa2 + [deaths]sa34 = 0
           chi2(  2) =      4.73
         Prob > chi2 =      0.0938
```

Combining age categories 1 through 4 may be overdoing it—the 9.38% significance level is enough to stop us, although others may disagree.

Thus we now fit our final model:

```
. generate sa12 = (sa1|sa2)
. poisson deaths sa12 sa34 sa5 a2-a5, exposure(pyears) irr
Iteration 0:   log likelihood = -31.967194
Iteration 1:   log likelihood = -28.524666
Iteration 2:   log likelihood = -28.514535
Iteration 3:   log likelihood = -28.514535
```

```
Poisson regression                              Number of obs   =          10
                                                LR chi2(7)      =      933.11
                                                Prob > chi2     =      0.0000
Log likelihood = -28.514535                     Pseudo R2       =      0.9424
```

deaths	IRR	Std. Err.	z	P>|z|	[95% Conf. Interval]	
sa12	2.636259	.7408403	3.45	0.001	1.519791	4.572907
sa34	1.412229	.2017485	2.42	0.016	1.067343	1.868557
sa5	.9047304	.1855513	-0.49	0.625	.6052658	1.35236
a2	4.294559	.8385329	7.46	0.000	2.928987	6.296797
a3	23.42263	7.787716	9.49	0.000	12.20738	44.94164
a4	48.26309	16.06939	11.64	0.000	25.13068	92.68856
a5	97.87965	34.30881	13.08	0.000	49.24123	194.561
pyears	(exposure)					

The above strikes us as a fair representation of the data.

◁

Methods and Formulas

All postestimation commands listed above are implemented as ado-files.

In the following, we use the same notation as in [R] **poisson**.

The equation-level scores are given by

$$\text{score}(\mathbf{x}\boldsymbol{\beta})_j = y_j - e^{\xi_j}$$

The deviance (D) and Pearson (P) goodness-of-fit statistics are given by

$$\ln L_{\max} = \sum_{j=1}^{n} w_j \left[-y_j \{\ln(y_j) - 1\} - \ln(y_j!) \right]$$

$$\chi_D^2 = -2\{\ln L - \ln L_{\max}\}$$

$$\chi_P^2 = \sum_{j=1}^{n} \frac{w_j (y_j - e^{\xi_j})^2}{e^{\xi_j}}$$

Also See

Complementary: [R] **poisson**; [R] **adjust**, [R] **estimates**, [R] **lincom**, [R] **linktest**, [R] **lrtest**,
[R] **mfx**, [R] **nlcom**, [R] **predictnl**, [R] **suest**, [R] **test**, [R] **testnl**

Background: [U] **13.5 Accessing coefficients and standard errors**,
[U] **20 Estimation and postestimation commands**,
[R] **estat**, [R] **predict**

Title

> **predict** — Obtain predictions, residuals, etc., after estimation

Syntax

After single-equation (SE) *models*

> predict [*type*] *newvar* [*if*] [*in*] [, *single_options*]

After multiple-equation (ME) *models*

> predict [*type*] *newvar* [*if*] [*in*] [, *multiple_options*]

> predict [*type*] { *stub** | *newvar*$_1$ ... *newvar*$_q$ } [*if*] [*in*] , <u>sc</u>ores

single_options	description
Main	
xb	calculate linear prediction
stdp	calculate standard error of the prediction
<u>sc</u>ore	calculate first derivative of the log likelihood with respect to $\mathbf{x}_j\beta$
Options	
<u>noo</u>ffset	ignore any offset() or exposure() variable
other_options	command-specific options

multiple_options	description
Main	
<u>e</u>quation(*eqno* [,*eqno*])	specify equations
xb	calculate linear prediction
stdp	calculate standard error of the prediction
stddp	calculate the difference in linear predictions
Options	
<u>noo</u>ffset	ignore any offset() or exposure() variable
other_options	command-specific options

Description

predict calculates predictions, residuals, influence statistics, and the like after estimation. Exactly what predict can do is determined by the previous estimation command; command-specific options are documented with each estimation command. Regardless of command-specific options, the actions of predict share certain similarities across estimation commands:

1) predict *newvar* creates *newvar* containing "predicted values"—numbers related to the $E(y_j|\mathbf{x}_j)$. For instance, after linear regression, predict *newvar* creates $\mathbf{x}_j\mathbf{b}$ and, after probit, creates the probability $\Phi(\mathbf{x}_j\mathbf{b})$.

2) predict *newvar*, xb creates *newvar* containing $\mathbf{x}_j\mathbf{b}$. This may be the same result as (1) (e.g., linear regression) or different (e.g., probit), but regardless, option xb is allowed.

3) predict *newvar*, stdp creates *newvar* containing the standard error of the linear prediction $\mathbf{x}_j\mathbf{b}$.

4) predict *newvar*, *other_options* may create *newvar* containing other useful quantities; see help or the *Reference* manual entry for the particular estimation command to find out about other available options.

5) nooffset added to any of the above commands requests that the calculation ignore any offset or exposure variable specified by including the offset(*varname*) or exposure(*varname*) options when you fitted the model.

predict can be used to make in-sample or out-of-sample predictions:

6) predict calculates the requested statistic for all possible observations, whether they were used in fitting the model or not. predict does this for standard options (1) through (3) and generally does this for estimator-specific options (4).

7) predict *newvar* if e(sample), ... restricts the prediction to the estimation subsample.

8) Some statistics make sense only with respect to the estimation subsample. In such cases, the calculation is automatically restricted to the estimation subsample, and the documentation for the specific option states this. Even so, you can still specify if e(sample) if you are uncertain.

9) predict can make out-of-sample predictions even using other datasets. In particular, you can

```
. use ds1
. (fit a model)
. use two                    /* another dataset       */
. predict yhat, ...          /* fill in the predictions */
```

Options

____| Main |____

xb calculates the linear prediction from the fitted model. All models can be thought of as estimating a set of parameters $b_1, b_2, \ldots, b_k$, and the linear prediction is $\widehat{y}_j = b_1 x_{1j} + b_2 x_{2j} + \cdots + b_k x_{kj}$, often written in matrix notation as $\widehat{\mathbf{y}}_j = \mathbf{x}_j\mathbf{b}$. In the case of linear regression, the values $\widehat{y}_j$ are called the predicted values or, for out-of-sample predictions, the forecast. In the case of logit and probit for example, $\widehat{y}_j$ is called the logit or probit index.

$x_{1j}, x_{2j}, \ldots, x_{kj}$ are obtained from the data currently in memory and do not necessarily correspond to the data on the independent variables used to fit the model (obtaining $b_1, b_2, \ldots, b_k$).

stdp calculates the standard error of the prediction after any estimation command. Here the prediction means the same thing as the "index", namely $\mathbf{x}_j\mathbf{b}$. The statistic produced by stdp can be thought of as the standard error of the predicted expected value, or mean index, for the observation's covariate pattern. This is also commonly referred to as the standard error of the fitted value. The calculation can be made in or out of sample.

stddp is allowed only after you have previously fitted a multiple-equation model. The standard error of the difference in linear predictions $(\mathbf{x}_{1j}\mathbf{b} - \mathbf{x}_{2j}\mathbf{b})$ between equations 1 and 2 is calculated. This option requires that equation(*eqno*$_1$,*eqno*$_2$) be specified.

score calculates the equation-level score, $\partial\ln L/\partial(\mathbf{x}_j\boldsymbol{\beta})$. Here $\ln L$ refers to the log-likelihood function.

scores is the ME model equivalent of the score option, resulting in multiple equation-level score variables. An equation-level score variable is created for each equation in the model; ancillary parameters—such as $\ln \sigma$ and atanhρ—make up separate equations.

equation($eqno$ [, $eqno$])—synonym outcome()—is relevant only when you have previously fitted a multiple-equation model. It specifies the equation to which you are referring.

equation() is typically filled in with one $eqno$—it would be filled in that way with options xb and stdp, for instance. equation(#1) would mean the calculation is to be made for the first equation, equation(#2) would mean the second, and so on. Alternatively, you could refer to the equations by their names. equation(income) would refer to the equation named income and equation(hours) to the equation named hours.

If you do not specify equation(), results are the same as if you specified equation(#1).

Other statistics, such as stddp, refer to between-equation concepts. In those cases, you might specify equation(#1,#2) or equation(income,hours). When two equations must be specified, equation() is required.

___

⌐ Options ∟

nooffset may be combined with most statistics and specifies that the calculation should be made, ignoring any offset or exposure variable specified when the model was fitted.

This option is available, even if it is not documented for predict after a specific command. If neither the offset($varname$) option nor the exposure($varname$) option was specified when the model was fitted, specifying nooffset does nothing.

other_options refers to command-specific options that are documented with each command.

Remarks

Remarks are presented under the headings

> *Estimation-sample predictions*
> *Out-of-sample predictions*
> *Residuals*
> *Single-equation (SE) models*
> *SE model scores*
> *Multiple-equation (ME) models*
> *ME model scores*

Most of the examples are presented using linear regression, but the general syntax is applicable to all estimators.

You can think of any estimation command as estimating a set of coefficients b_1, b_2, ..., b_k corresponding to the variables x_1, x_2, ..., x_k, along with a (possibly empty) set of ancillary statistics γ_1, γ_2, ..., γ_m. All estimation commands save the b_is and γ_is. predict accesses that saved information and combines it with the data currently in memory to make various calculations. For instance, predict can calculate the linear prediction, $\widehat{y}_j = b_1 x_{1j} + b_2 x_{2j} + \cdots + b_k x_{kj}$. The data on which predict makes the calculation can be the same data used to fit the model or a different dataset—it does not matter. predict uses the saved parameter estimates from the model, obtains the corresponding values of x for each observation in the data, and then combines them to produce the desired result.

Estimation-sample predictions

▷ Example 1

We have a 74-observation dataset on automobiles, including the mileage rating (mpg), the car's weight (weight), and whether the car is foreign (foreign). We fit the model

```
. use http://www.stata-press.com/data/r9/auto
(1978 Automobile Data)

. regress mpg weight if foreign
```

Source	SS	df	MS		Number of obs =	22
					F(1, 20) =	17.47
Model	427.990298	1	427.990298		Prob > F =	0.0005
Residual	489.873338	20	24.4936669		R-squared =	0.4663
					Adj R-squared =	0.4396
Total	917.863636	21	43.7077922		Root MSE =	4.9491

mpg	Coef.	Std. Err.	t	P>|t|	[95% Conf. Interval]	
weight	-.010426	.0024942	-4.18	0.000	-.0156287	-.0052232
_cons	48.9183	5.871851	8.33	0.000	36.66983	61.16676

If we were to type `predict pmpg` now, we would obtain the linear predictions for all 74 observations. To obtain the predictions just for the sample on which we fitted the model, we could type

```
. predict pmpg if e(sample)
(option xb assumed; fitted values)
(52 missing values generated)
```

In this example, `e(sample)` is true only for foreign cars because we typed `if foreign` when we fitted the model and because there are no missing values among the relevant variables. If there had been missing values, `e(sample)` would also account for those.

By the way, the `if e(sample)` restriction can be used with any Stata command, so we could obtain summary statistics on the estimation sample by typing

```
. summarize if e(sample)
(output omitted)
```

◁

Out-of-sample predictions

By out-of-sample predictions, we mean predictions extending beyond the estimation sample. In the example above, typing `predict pmpg` would generate linear predictions using all 74 observations.

`predict` will work on other datasets, too. You can use a new dataset and type `predict` to obtain results for that sample.

▷ Example 2

Using the same auto dataset, assume that we wish to fit the model

$$\text{mpg} = \beta_1 \text{weight} + \beta_2 \text{weight}^2 + \beta_3 \text{foreign} + \beta_4$$

We first create the `weight`^{2} variable, and then type the `regress` command:

```
. use http://www.stata-press.com/data/r9/auto, clear
(1978 Automobile Data)

. generate weight2=weight^2

. regress mpg weight weight2 foreign
```

Source	SS	df	MS		
Model	1689.15372	3	563.05124		
Residual	754.30574	70	10.7757963		
Total	2443.45946	73	33.4720474		

```
Number of obs =      74
F(  3,     70) =   52.25
Prob > F       =  0.0000
R-squared      =  0.6913
Adj R-squared  =  0.6781
Root MSE       =  3.2827
```

mpg	Coef.	Std. Err.	t	P>|t|	[95% Conf. Interval]	
weight	-.0165729	.0039692	-4.18	0.000	-.0244892	-.0086567
weight2	1.59e-06	6.25e-07	2.55	0.013	3.45e-07	2.84e-06
foreign	-2.2035	1.059246	-2.08	0.041	-4.3161	-.0909002
_cons	56.53884	6.197383	9.12	0.000	44.17855	68.89913

If we typed `predict pmpg` now, we would obtain predictions for all 74 cars in the current data. Instead, we are going to use a new dataset.

The dataset `newautos.dta` contains the make, weight, and place of manufacture of two cars, the Pontiac Sunbird and the Volvo 260. Let's use the dataset and create the predictions:

```
. use http://www.stata-press.com/data/r9/newautos, clear
(New Automobile Models)

. list
```

	make	weight	foreign
1.	Pont. Sunbird	2690	Domestic
2.	Volvo 260	3170	Foreign

```
. predict mpg
(option xb assumed; fitted values)
variable weight2 not found
r(111);
```

Things did not work. We typed `predict mpg`, and Stata responded with the message "variable weight2 not found". `predict` can calculate predicted values on a different dataset only if that dataset contains the variables that went into the model. In this case, our dataset does not contain a variable called `weight2`. `weight2` is just the square of weight, so we can create it and try again:

```
. generate weight2=weight^2

. predict mpg
(option xb assumed; fitted values)

. list
```

	make	weight	foreign	weight2	mpg
1.	Pont. Sunbird	2690	Domestic	7236100	23.47137
2.	Volvo 260	3170	Foreign	1.00e+07	17.78846

We obtained our predicted values. The Pontiac Sunbird has a predicted mileage rating of 23.5 mpg, whereas the Volvo 260 has a predicted rating of 17.8 mpg.

◁

Residuals

▷ Example 3

With many estimators, `predict` can calculate more than predicted values. With most regression-type estimators, we can, for instance, obtain residuals. Using our regression example, we return to our original data and obtain residuals by typing

```
. use http://www.stata-press.com/data/r9/auto, clear
(1978 Automobile Data)
. generate weight2=weight^2
. regress mpg weight weight2 foreign
  (output omitted )
. predict double resid, residuals
. summarize resid
```

Variable	Obs	Mean	Std. Dev.	Min	Max
resid	74	-1.78e-15	3.214491	-5.636126	13.85172

Notice that we could do this without refitting the model. Stata always remembers the last set of estimates, even as we use new datasets.

It was not necessary to type the `double` in `predict double resid, residuals`, but we wanted to remind you that you can specify the type of a variable in front of the variable's name; see [U] **11.4.2 Lists of new variables**. We made the new variable `resid` a `double` rather than the default `float`.

If you want your residuals to have a mean as close to zero as possible, remember to request the extra precision of `double`. If we had not specified `double`, the mean of `resid` would have been roughly 10^{-8} rather than 10^{-14}. Although 10^{-14} sounds more precise than 10^{-8}, the difference really does not matter.

◁

For linear regression, `predict` can also calculate standardized residuals and studentized residuals with the options `rstandard` and `rstudent`; for examples, see [R] **regress postestimation**.

Single-equation (SE) models

If you have not read the discussion above on using `predict` after linear regression, please do so. Also note that `predict`'s default calculation almost always produces a statistic in the same metric as the dependent variable of the fitted model—e.g., predicted counts for Poisson regression. In any case, `xb` can always be specified to obtain the linear prediction.

`predict` can calculate the standard error of the prediction, which is obtained by using the covariance matrix of the estimators.

▷ Example 4

After most binary outcome models (e.g., logistic, logit, probit, cloglog, scobit), predict calculates the probability of a positive outcome if we do not tell it otherwise. We can specify the xb option if we want the linear prediction (also known as the logit or probit index). The odd abbreviation xb is meant to suggest $\mathbf{x}\beta$. In logit and probit models, for example, the predicted probability is $p = F(\mathbf{x}\beta)$, where $F()$ is the logistic or normal cumulative distribution function, respectively.

```
. logistic foreign mpg weight
  (output omitted)
. predict phat
(option p assumed; Pr(foreign))
. predict idxhat, xb
. summarize foreign phat idxhat
```

Variable	Obs	Mean	Std. Dev.	Min	Max
foreign	74	.2972973	.4601885	0	1
phat	74	.2972973	.3052979	.000729	.8980594
idxhat	74	-1.678202	2.321509	-7.223107	2.175845

Since this is a logit model, we could obtain the predicted probabilities ourselves from the predicted index

```
. generate phat2 = exp(idxhat)/(1+exp(idxhat))
```

but using predict without options is easier.

◁

▷ Example 5

For all models, predict attempts to produce a predicted value in the same metric as the dependent variable of the model. We have seen that for dichotomous outcome models, the default statistic produced by predict is the probability of a success. Similarly, for Poisson regression, the default statistic produced by predict is the predicted count for the dependent variable. You can always specify the xb option to obtain the linear combination of the coefficients with an observation's x values (the inner product of the coefficients and x values). For poisson (without an explicit exposure), this is the natural log of the count.

```
. use http://www.stata-press.com/data/r9/airline, clear
. poisson injuries XYZowned
  (output omitted)
. predict injhat
(option n assumed; predicted number of events)
. predict idx, xb
. generate exp_idx = exp(idx)
. summarize injuries injhat exp_idx idx
```

Variable	Obs	Mean	Std. Dev.	Min	Max
injuries	9	7.111111	5.487359	1	19
injhat	9	7.111111	.8333333	6	7.666667
exp_idx	9	7.111111	.8333333	6	7.666667
idx	9	1.955174	.1225612	1.791759	2.036882

We note that our "hand-computed" prediction of the count (`exp_idx`) exactly matches what was produced by the default operation of `predict`.

If our model has an exposure-time variable, we can use `predict` to obtain the linear prediction with or without the exposure. Let's verify what we are getting by obtaining the linear prediction with and without exposure, transforming these predictions to count predictions and comparing them with the default count prediction from `predict`. We must remember to multiply by the exposure time when using `predict ... , nooffset`.

```
. use http://www.stata-press.com/data/r9/airline, clear
. poisson injuries XYZowned, exposure(n)
  (output omitted)
. predict double injhat
(option n assumed; predicted number of events)
. predict double idx, xb
. generate double exp_idx = exp(idx)
. predict double idxn, xb nooffset
. generate double exp_idxn = exp(idxn)*n
. summarize injuries injhat exp_idx exp_idxn idx idxn
```

Variable	Obs	Mean	Std. Dev.	Min	Max
injuries	9	7.111111	5.487359	1	19
injhat	9	7.111111	3.10936	2.919621	12.06158
exp_idx	9	7.111111	3.10936	2.919621	12.06158
exp_idxn	9	7.111111	3.10936	2.919621	12.06158
idx	9	1.869722	.4671044	1.071454	2.490025
idxn	9	4.18814	.1904042	4.061204	4.442013

Looking at the identical means and standard deviations for `injhat`, `exp_idx`, and `exp_idxn`, we see that we can reproduce the default computations of `predict` for `poisson` estimations. We have also demonstrated the relationship between the count predictions and the linear predictions with and without exposure.

◁

SE model scores

▷ Example 6

With most maximum likelihood estimators, `predict` can calculate equation-level scores. The first derivative of the log likelihood with respect to $x_j\beta$ is the equation-level score.

```
. use http://www.stata-press.com/data/r9/auto, clear
(1978 Automobile Data)
. logistic foreign mpg weight
  (output omitted)
. predict double sc, score
. summarize sc
```

Variable	Obs	Mean	Std. Dev.	Min	Max
sc	74	-1.05e-13	.3533133	-.8760856	.8821309

See [P] **_robust** and [SVY] **variance estimation** for details regarding the role equation-level scores play in linearization-based variance estimators.

◁

❑ Technical Note

`predict` after some estimation commands, such as `regress` and `ivreg`, allows the `score` option as a synonym for the `residuals` option.

❑

Multiple-equation (ME) models

If you have not read the above discussion on using `predict` after SE models, please do so. With the exception of the ability to select specific equations to predict from, the use of `predict` after ME models follows almost exactly the same form that it does for SE models.

▷ Example 7

The details of prediction statistics that are specific to particular ME models are documented with the estimation command. If you are using ME commands that do not have separate discussions on obtaining predictions, read the `predict` section in [R] **mlogit**, even if your interest is not in multinomial logistic regression. As a general introduction to the ME models, we will demonstrate `predict` after `sureg`:

```
. use http://www.stata-press.com/data/r9/auto, clear
(1978 Automobile Data)

. sureg (price foreign displ) (weight foreign length)

Seemingly unrelated regression
```

Equation	Obs	Parms	RMSE	"R-sq"	chi2	P
price	74	2	2202.447	0.4348	45.21	0.0000
weight	74	2	245.5238	0.8988	658.85	0.0000

	Coef.	Std. Err.	z	P>|z|	[95% Conf. Interval]	
price						
foreign	3137.894	697.3805	4.50	0.000	1771.054	4504.735
displacement	23.06938	3.443212	6.70	0.000	16.32081	29.81795
_cons	680.8438	859.8142	0.79	0.428	-1004.361	2366.049
weight						
foreign	-154.883	75.3204	-2.06	0.040	-302.5082	-7.257674
length	30.67594	1.531981	20.02	0.000	27.67331	33.67856
_cons	-2699.498	302.3912	-8.93	0.000	-3292.173	-2106.822

`sureg` estimated two equations, one called `price` and the other `weight`; see [R] **sureg**.

```
. predict pred_p, equation(price)
(option xb assumed; fitted values)

. predict pred_w, equation(weight)
(option xb assumed; fitted values)
```

```
. summarize price pred_p weight pred_w
```

Variable	Obs	Mean	Std. Dev.	Min	Max
price	74	6165.257	2949.496	3291	15906
pred_p	74	6165.257	1678.805	2664.81	10485.33
weight	74	3019.459	777.1936	1760	4840
pred_w	74	3019.459	726.0468	1501.602	4447.996

You may specify the equation by name, as we did above, or by number: equation(#1) means the same thing as equation(price) in this case.

◁

ME model scores

▷ Example 8

For ME models, predict allows you to specify a stub when generating equation-level score variables. predict generates new variables using this stub by appending an equation index. Depending upon the command, the index will start with 0 or 1. Here is an example where predict starts indexing the score variables with 0.

```
. ologit rep78 mpg weight
(output omitted )
. predict double sc*, scores
. summarize sc*
```

Variable	Obs	Mean	Std. Dev.	Min	Max
sc0	69	5.81e-18	.5337363	-.9854088	.921433
sc1	69	-1.69e-17	.186919	-.2738537	.9854088
sc2	69	3.30e-17	.4061637	-.5188487	1.130178
sc3	69	-1.41e-17	.5315368	-1.067351	.8194842
sc4	69	-4.83e-18	.360525	-.921433	.6140182

Although it involves much more typing, we could also specify the new variable names individually.

```
. predict double (sc_xb sc_1 sc_2 sc_3 sc_4), scores
. summarize sc_*
```

Variable	Obs	Mean	Std. Dev.	Min	Max
sc_xb	69	5.81e-18	.5337363	-.9854088	.921433
sc_1	69	-1.69e-17	.186919	-.2738537	.9854088
sc_2	69	3.30e-17	.4061637	-.5188487	1.130178
sc_3	69	-1.41e-17	.5315368	-1.067351	.8194842
sc_4	69	-4.83e-18	.360525	-.921433	.6140182

◁

Methods and Formulas

Denote the previously estimated coefficient vector as $\mathbf{b}$ and its estimated variance matrix as $\mathbf{V}$. predict works by recalling various aspects of the model, such as $\mathbf{b}$, and combining that information with the data currently in memory. Let us write $\mathbf{x}_j$ for the jth observation currently in memory.

The *predicted value* (**xb** option) is defined as $\widehat{y}_j = \mathbf{x}_j \mathbf{b} + \text{offset}_j$

The *standard error of the prediction* (**stdp**) is defined as $s_{p_j} = \sqrt{\mathbf{x}_j \mathbf{V} \mathbf{x}'_j}$

The *standard error of the difference in linear predictions* between equations 1 and 2 is defined as

$$s_{dp_j} = \left\{ (\mathbf{x}_{1j}, -\mathbf{x}_{2j}, \mathbf{0}, \ldots, \mathbf{0}) \; \mathbf{V} \; (\mathbf{x}_{1j}, -\mathbf{x}_{2j}, \mathbf{0}, \ldots, \mathbf{0})' \right\}^{\frac{1}{2}}$$

See the individual estimation commands for information about calculating command-specific `predict` statistics.

Also See

Related: [R] **predictnl**, [R] **regress**, [R] **regress postestimation**,
[P] **_predict**

Background: [U] **20 Estimation and postestimation commands**

Title

predictnl — Obtain nonlinear predictions, standard errors, etc., after estimation

Syntax

predictnl [*type*] *newvar* = *pnl_exp* [*if*] [*in*] [, *options*]

options	description
Main	
se(*newvar*)	create *newvar* containing standard errors
variance(*newvar*)	create *newvar* containing variances
wald(*newvar*)	create *newvar* containing the Wald test statistic
p(*newvar*)	create *newvar* containing the significance level (*p*-value) of the Wald test
ci(*newvars*)	create *newvars* containing lower and upper confidence intervals
level(#)	set confidence level; default is level(95)
g(*stub*)	create *stub*1, *stub*2, ..., *stub*k variables containing observation-specific derivatives
Advanced	
iterate(#)	maximum iterations for finding optimal step size; default is 100
force	calculate standard errors, etc., even when possibly inappropriate

Description

predictnl calculates (possibly) nonlinear predictions after any Stata estimation command and optionally calculates the variances, standard errors, Wald test statistics, significance levels, and confidence limits for these predictions. Unlike its companion "nonlinear" postestimation commands testnl and nlcom, predictnl generates functions of the data (i.e., predictions), not scalars. The quantities generated by predictnl are thus vectorized over the observations in the data.

Consider some general prediction, $g(\boldsymbol{\theta}, \mathbf{x}_i)$, for $i = 1, \ldots, n$, where $\boldsymbol{\theta}$ are the model parameters and $\mathbf{x}_i$ are some data for the ith observation; $\mathbf{x}_i$ is assumed fixed. Typically, $g(\boldsymbol{\theta}, \mathbf{x}_i)$ is estimated by $g(\widehat{\boldsymbol{\theta}}, \mathbf{x}_i)$, where $\widehat{\boldsymbol{\theta}}$ are the estimated model parameters, which are stored in e(b) following any Stata estimation command.

In its most common use, predictnl generates two variables: one containing the estimated prediction, $g(\widehat{\boldsymbol{\theta}}, \mathbf{x}_i)$, the other containing the estimated standard error of $g(\widehat{\boldsymbol{\theta}}, \mathbf{x}_i)$. The calculation of standard errors (and other obtainable quantities that are based on the standard errors, such as test statistics) is based on the "delta method", an approximation appropriate in large samples; see *Methods and Formulas*.

predictnl supports svy estimation commands (svy: regress, svy: logit, etc.).

The specification of $g(\widehat{\boldsymbol{\theta}}, \mathbf{x}_i)$ is handled by specifying *pnl_exp*, and the values of $g(\widehat{\boldsymbol{\theta}}, \mathbf{x}_i)$ are stored in the new variable *newvar* of storage type *type*. *pnl_exp* is any valid Stata expression and may also contain calls to two special functions unique to predictnl:

1. predict([*predict_options*]): When you are evaluating *pnl_exp*, predict() is a convenience function that replicates the calculation performed by the command

    predict ..., *predict_options*

    As such, the predict() function may be used either as a shorthand for the formula used to make this prediction or when the formula is not readily available. When used without arguments, predict() replicates the default prediction for that particular estimation command.

2. xb([*eqno*]): The xb() function replicates the calculation of the linear predictor $x_i b$ for equation *eqno*. If xb() is specified without *eqno*, the linear predictor for the first equation (or the only equation in single-equation estimation) is obtained.

    For example, xb(#1) (or equivalently, xb() with no arguments) translates to the linear predictor for the first equation, xb(#2) for the second, and so on. Alternatively, you could refer to the equations by their names, such as xb(income).

    When specifying *pnl_exp*, both of these functions may be used repeatedly, in combination, and in combination with other Stata functions and expressions. See *Remarks* for examples that utilize both of these functions.

Options

___Main___

se(*newvar*) adds *newvar* of storage type *type*, where for each i in the prediction sample, *newvar*[i] contains the estimated standard error of $g(\widehat{\theta}, x_i)$.

variance(*newvar*) adds *newvar* of storage type *type*, where for each i in the prediction sample, *newvar*[i] contains the estimated variance of $g(\widehat{\theta}, x_i)$.

wald(*newvar*) adds *newvar* of storage type *type*, where for each i in the prediction sample, *newvar*[i] contains the Wald test statistic for the test of the hypothesis $H_o: g(\theta, x_i) = 0$.

p(*newvar*) adds *newvar* of storage type *type*, where *newvar*[i] contains the significance level (p-value) of the Wald test of $H_o: g(\theta, x_i) = 0$ versus the two-sided alternative.

ci(*newvars*) requires the specification of two *newvars*, such that the ith observation of each will contain the left and right endpoints (respectively) of a confidence interval for $g(\theta, x_i)$. The level of the confidence intervals is determined by level(#).

level(#) specifies the confidence level, as a percentage, for confidence intervals. The default is level(95) or as set by set level; see [U] **20.6 Specifying the width of confidence intervals**.

g(*stub*) specifies that new variables, *stub1*, *stub2*, ..., *stubk* be created, where k is the dimension of θ. *stub1* will contain the observation-specific derivatives of $g(\theta, x_i)$ with respect to the first element, θ_1, of θ, *stub2* will contain the derivatives of $g(\theta, x_i)$ with respect to θ_2, etc. If the derivative of $g(\theta, x_i)$ with respect to a particular coefficient in θ equals zero for all observations in the prediction sample, the *stub* variable for that coefficient is not created. Note that the ordering of the parameters in θ is precisely that of the stored vector of parameter estimates e(b).

___Advanced___

iterate(#) specifies the maximum number of iterations used to find the optimal step size in the calculation of numerical derivatives of $g(\theta, x_i)$ with respect to θ. By default, the maximum number of iterations is 100, but convergence is usually achieved after only a few iterations. You should rarely have to use this option.

force forces the calculation of standard errors and other inference-related quantities in situations where predictnl would otherwise refuse to do so. The calculation of standard errors takes place by evaluating (at $\widehat{\theta}$) the numerical derivative of $g(\theta, \mathbf{x}_i)$ with respect to θ. If predictnl detects that $g()$ is possibly a function of random quantities other than $\widehat{\theta}$, it will refuse to calculate standard errors or any other quantity derived from them. The force option forces the calculation to take place, anyway. If you use the force option, there is no guarantee that any inference quantities (e.g., standard errors) will be correct or that the values obtained can be interpreted.

Remarks

Remarks are presented under the headings

Introduction
Nonlinear transformations and standard errors
Using xb() and predict()
Multiple-equation (ME) estimators
Test statistics and significance levels
Manipulability
Confidence intervals

Introduction

predictnl and nlcom are Stata's "delta method" commands—they take a nonlinear transformation of the estimated parameter vector from some fitted model and apply the delta method to calculate the variance, standard error, Wald test statistic, etc., of this transformation. nlcom is designed for scalar functions of the parameters, and predictnl is designed for functions of the parameters and of the data, that is, for predictions.

Nonlinear transformations and standard errors

We begin by fitting a probit model to the low birthweight data of Hosmer and Lemeshow (2000, 25). The data are described in detail in [R] **logistic**.

```
. use http://www.stata-press.com/data/r9/lbw
(Hosmer & Lemeshow data)

. probit low lwt smoke ptl ht

Iteration 0:   log likelihood =   -117.336
Iteration 1:   log likelihood = -106.76258
Iteration 2:   log likelihood = -106.67855
Iteration 3:   log likelihood = -106.67851
```

Probit regression				Number of obs	=	189
				LR chi2(4)	=	21.31
				Prob > chi2	=	0.0003
Log likelihood = -106.67851				Pseudo R2	=	0.0908

low	Coef.	Std. Err.	z	P>|z|	[95% Conf. Interval]	
lwt	-.0095164	.0036875	-2.58	0.010	-.0167438	-.0022891
smoke	.3487004	.2041771	1.71	0.088	-.0514794	.7488803
ptl	.365667	.1921201	1.90	0.057	-.0108814	.7422154
ht	1.082355	.410673	2.64	0.008	.2774504	1.887259
_cons	.4238985	.4823224	0.88	0.379	-.5214359	1.369233

After we fit such a model, we first would want to generate the predicted probabilities of a low birthweight, given the covariate values in the estimation sample. This is easily done using `predict` after `probit`, but it doesn't answer the question, "What are the standard errors of those predictions?"

For the time being, we will consider ourselves ignorant of any automated way to obtain the predicted probabilities after `probit`. The formula for the prediction is

$$P(y \neq 0 | \mathbf{x}_i) = \Phi(\mathbf{x}_i \boldsymbol{\beta})$$

where Φ is the standard cumulative normal. Thus for this example, $g(\boldsymbol{\theta}, \mathbf{x}_i) = \Phi(\mathbf{x}_i \boldsymbol{\beta})$. Armed with the formula, we can use `predictnl` to generate the predictions and their standard errors:

```
. predictnl phat = normal(_b[_cons] + _b[ht]*ht + _b[ptl]*ptl +
> _b[smoke]*smoke + _b[lwt]*lwt), se(phat_se)
. list phat phat_se lwt smoke ptl ht in -10/1
```

	phat	phat_se	lwt	smoke	ptl	ht
180.	.2363556	.042707	120	0	0	0
181.	.6577712	.1580714	154	0	1	1
182.	.2793261	.0519958	106	0	0	0
183.	.1502118	.0676338	190	1	0	0
184.	.5702871	.0819911	101	1	1	0
185.	.4477045	.079889	95	1	0	0
186.	.2988379	.0576306	100	0	0	0
187.	.4514706	.080815	94	1	0	0
188.	.5615571	.1551051	142	0	0	1
189.	.7316517	.1361469	130	1	0	1

Thus subject 180 in our data has an estimated probability of low birthweight of 23.6% with standard error 4.3%.

Used without options, `predictnl` is not much different from `generate`. By specifying the option `se(phat_se)`, we were able to obtain a variable containing the standard errors of the predictions; therein lies the utility of `predictnl`.

Using xb() and predict()

As was the case above, a prediction is often not a function of a few isolated parameters and their corresponding variables, but instead is some (possibly very elaborate) function of the entire linear predictor. For models with many predictors, the brute-force expression for the linear predictor can be quite cumbersome to type. An alternative is to use the inline function `xb()`. `xb()` is a shortcut for having to type `_b[_cons] + _b[ht]*ht + _b[ptl]*ptl + ...`,

```
. drop phat phat_se
. predictnl phat = normal(xb()), se(phat_se)
```

(Continued on next page)

```
. list phat phat_se lwt smoke ptl ht in -10/1
```

	phat	phat_se	lwt	smoke	ptl	ht
180.	.2363556	.042707	120	0	0	0
181.	.6577712	.1580714	154	0	1	1
182.	.2793261	.0519958	106	0	0	0
183.	.1502118	.0676338	190	1	0	0
184.	.5702871	.0819911	101	1	1	0
185.	.4477045	.079889	95	1	0	0
186.	.2988379	.0576306	100	0	0	0
187.	.4514706	.080815	94	1	0	0
188.	.5615571	.1551051	142	0	0	1
189.	.7316517	.1361469	130	1	0	1

which yields exactly the same results. This approach is easier, produces more readable code, and is less prone to errors, such as forgetting to include a term in the sum.

In this example, we used xb() without arguments since we only have one equation in our model. In multiple-equation (ME) settings, xb() (or equivalently xb(#1)) yields the linear predictor from the first equation, xb(#2) from the second, etc. Alternatively, you can refer to equations by their names, e.g., xb(income).

❏ Technical Note

Most estimation commands in Stata allow the postestimation calculation of linear predictors and their standard errors via predict. For example, to obtain these for the first (or only) equation in the model, you could type

```
predict xbvar, xb
predict stdpvar, stdp
```

Equivalently, you could type

```
predictnl xbvar = xb(), se(stdpvar)
```

but we recommend the first method, as it is faster. As we demonstrated above, however, predictnl is more general.

❏

Returning to our probit example, we can further simplify the calculation by using the inline function predict(). predict(*pred_options*) works by substituting, within our predictnl expression, the calculation performed by

```
predict ..., pred_options
```

In our example, we are interested in the predicted probabilities after a probit regression, normally obtained via

```
predict ..., p
```

We can obtain these predictions (and standard errors) using

```
. drop phat phat_se
. predictnl phat = predict(p), se(phat_se)
```

```
. list phat phat_se lwt smoke ptl ht in -10/1
```

	phat	phat_se	lwt	smoke	ptl	ht
180.	.2363556	.042707	120	0	0	0
181.	.6577712	.1580714	154	0	1	1
182.	.2793261	.0519958	106	0	0	0
183.	.1502118	.0676338	190	1	0	0
184.	.5702871	.0819911	101	1	1	0
185.	.4477045	.079889	95	1	0	0
186.	.2988379	.0576306	100	0	0	0
187.	.4514706	.080815	94	1	0	0
188.	.5615571	.1551051	142	0	0	1
189.	.7316517	.1361469	130	1	0	1

which again replicates what we have already done by other means. However, this version did not require knowledge of the formula for the predicted probabilities after a probit regression—predict(p) took care of that for us.

Note that since the predicted probability is the default prediction after probit, we could have just used predict() without arguments, namely,

```
. predictnl phat = predict(), se(phat_se)
```

Also note that the expression *pnl_exp* can be inordinately complicated, with multiple calls to predict() and xb(). For example,

```
. predictnl phat = normal(invnormal(predict()) + predict(xb)/xb() - 1), se(phat_se)
```

is perfectly valid and will give the same result as before, albeit a bit inefficiently.

❑ Technical Note

When using predict() and xb(), the *formula* for the calculation is substituted within *pnl_exp*, not the values that result from the application of that formula. To see this, note the subtle difference between

```
. predict xbeta, xb
. predictnl phat = normal(xbeta), se(phat_se)
```

and

```
. predictnl phat = normal(xb()), se(phat_se)
```

Both sequences will yield the same phat, yet for the first sequence, phat_se will equal zero for all observations. The reason is that, once evaluated, xbeta will contain the values of the linear predictor, yet these values are treated as fixed and nonstochastic as far as predictnl is concerned. By contrast, since xb() is shorthand for the formula used to calculate the linear predictor, it contains not values, but references to the estimated regression coefficients and corresponding variables. Thus the second method produces the desired result.

❑

Multiple-equation (ME) estimators

In [R] **mlogit**, data on insurance choice (Tarlov et al. 1989; Wells et al. 1989) were examined, and a multinomial logit was used to assess the effects of age, gender, race, and site of study (one of three sites) on the type of insurance:

```
. use http://www.stata-press.com/data/r9/sysdsn2
(Health insurance data)

. mlogit insure age male nonwhite site2 site3, nolog
Multinomial logistic regression                  Number of obs   =        615
                                                 LR chi2(10)     =      42.99
                                                 Prob > chi2     =     0.0000
Log likelihood = -534.36165                      Pseudo R2       =     0.0387
```

insure	Coef.	Std. Err.	z	P>|z|	[95% Conf. Interval]	
Prepaid						
age	-.011745	.0061946	-1.90	0.058	-.0238862	.0003962
male	.5616934	.2027465	2.77	0.006	.1643175	.9590693
nonwhite	.9747768	.2363213	4.12	0.000	.5115955	1.437958
site2	.1130359	.2101903	0.54	0.591	-.2989296	.5250013
site3	-.5879879	.2279351	-2.58	0.010	-1.034733	-.1412433
_cons	.2697127	.3284422	0.82	0.412	-.3740222	.9134476
Uninsure						
age	-.0077961	.0114418	-0.68	0.496	-.0302217	.0146294
male	.4518496	.3674867	1.23	0.219	-.268411	1.17211
nonwhite	.2170589	.4256361	0.51	0.610	-.6171725	1.05129
site2	-1.211563	.4705127	-2.57	0.010	-2.133751	-.2893747
site3	-.2078123	.3662926	-0.57	0.570	-.9257327	.510108
_cons	-1.286943	.5923219	-2.17	0.030	-2.447872	-.1260135

```
(insure==Indemnity is the base outcome)
```

Of particular interest is the estimation of the relative risk, which, for a given selection, is the ratio of the probability of making that selection to the probability of selecting the base category (insure==Indemnity in this case), given a set of covariate values. In a multinomial logit model, the relative risk (when comparing to the base category) simplifies to the exponentiated linear predictor for that selection.

Using this example, we can estimate the observation-specific relative risks of selecting a prepaid plan over the base category (with standard errors) by either referring to the Prepaid equation by name or number,

```
    . predictnl RRppaid = exp(xb(Prepaid)), se(SERRppaid)
```

or

```
    . predictnl RRppaid = exp(xb(#1)), se(SERRppaid)
```

since Prepaid is the first equation in the model.

Those of us for whom the simplified formula for the relative risk doesn't immediately come to mind may prefer to calculate the relative risk directly from its definition, that is, as a ratio of two predicted probabilities. After mlogit, the predicted probability for a category may be obtained using predict, but we must specify the category as the outcome:

```
. predictnl RRppaid = predict(outcome(Prepaid))/predict(outcome(Indemnity)),
> se(SERRppaid)
(1 missing value generated)
. list RRppaid SERRppaid age male nonwhite site2 site3 in 1/10
```

	RRppaid	SERRpp~d	age	male	nonwhite	site2	site3
1.	.6168578	.1503759	73.722107	0	0	1	0
2.	1.056658	.1790703	27.89595	0	0	1	0
3.	.8426442	.1511281	37.541397	0	0	0	0
4.	1.460581	.3671465	23.641327	0	1	0	1
5.	.9115747	.1324168	40.470901	0	0	1	0
6.	1.034701	.1696923	29.683777	0	0	1	0
7.	.9223664	.1344981	39.468857	0	0	1	0
8.	1.678312	.4216626	26.702255	1	0	0	0
9.	.9188519	.2256017	63.101974	0	1	0	1
10.	.5766296	.1334877	69.839828	0	0	0	0

The "(1 missing value generated)" message is not an error; further examination of the data would reveal that age is missing in one observation and that the offending observation (among others) is not in the estimation sample. Note that, just as with predict, predictnl can generate predictions in or out of the estimation sample.

Thus we estimate (among other things) that a white, female, 73-year-old from site 2 is less likely to choose a prepaid plan over an indemnity plan—her relative risk is about 62% with standard error 15%.

Test statistics and significance levels

Often a standard error calculation is just a means to an end, and what is really desired is a test of the hypothesis,

$$H_o : g(\boldsymbol{\theta}, \mathbf{x}_i) = 0$$

versus the two-sided alternative.

We can use predictnl to obtain the Wald test statistics and/or significance levels for the above tests, whether or not we want standard errors. To obtain the Wald test statistics, we use the wald() option; for significance levels, we use p().

Returning to our mlogit example, suppose that we wanted for each observation a test of whether the relative risk of choosing a prepaid plan over an indemnity plan is different from one. One way to do this would be to define $g()$ to be the relative risk minus one and then test whether $g()$ is different from zero.

(Continued on next page)

```
. predictnl RRm1 = exp(xb(Prepaid)) - 1, wald(W_RRm1) p(sig_RRm1)
(1 missing value generated)
Note: significance levels are with respect to the chi-squared(1) distribution.
. list RRm1 W_RRm1 sig_RRm1 age male nonwhite in 1/10
```

	RRm1	W_RRm1	sig_RRm1	age	male	nonwhite
1.	-.3831422	6.491778	.0108375	73.722107	0	0
2.	.0566578	.100109	.7516989	27.89595	0	0
3.	-.1573559	1.084116	.2977787	37.541397	0	0
4.	.4605812	1.573743	.2096643	23.641327	0	1
5.	-.0884253	.4459299	.5042742	40.470901	0	0
6.	.0347015	.0418188	.8379655	29.683777	0	0
7.	-.0776336	.3331707	.563798	39.468857	0	0
8.	.6783119	2.587788	.1076906	26.702255	1	0
9.	-.0811482	.1293816	.719074	63.101974	0	1
10.	-.4233705	10.05909	.001516	69.839828	0	0

The newly created variable W_RRm1 contains the Wald test statistic for each observation, and sig_RRm1 contains the level of significance. Thus our 73-year-old white female represented by the first observation would have a relative risk of choosing prepaid over indemnity that is significantly different from 1, at least at the 5% level. Note that, for this test, it was not necessary to generate a variable containing the standard error of the relative risk minus 1, but we could have done so had we wanted. We could have also omitted specifying wald(W_RRm1) if all we cared about were, say, the significance levels of the tests.

In this regard, predictnl acts as an observation-specific version of testnl, with the test results vectorized over the observations in the data. Note that the significance levels are pointwise—they are not adjusted to reflect any simultaneous testing over the observations in the data.

Manipulability

There are many ways to specify $g(\boldsymbol{\theta}, \mathbf{x}_i)$ to yield tests such that, for multiple specifications of $g()$, the theoretical conditions for which

$$H_o: g(\boldsymbol{\theta}, \mathbf{x}_i) = 0$$

is true will be equivalent. However, this does not mean that the tests themselves will be equivalent. This is known as the "manipulability" of the Wald test for nonlinear hypotheses; also see [R] **boxcox**.

As an example, consider the previous section where we defined $g()$ to be the relative risk between choosing a prepaid plan over an indemnity plan, minus 1. Alternatively, we could have defined $g()$ to be the risk difference—the probability of choosing a prepaid plan minus the probability of choosing an indemnity plan. Either specification of $g()$ yields a mathematically equivalent specification of $H_o: g() = 0$; that is, the risk difference will equal zero when the relative risk equals one. However, the tests themselves do not give the same results:

```
. predictnl RD = predict(outcome(Prepaid)) - predict(outcome(Indemnity)),
> wald(W_RD) p(sig_RD)
(1 missing value generated)
Note: significance levels are with respect to the chi-squared(1) distribution.

. list RD W_RD sig_RD RRm1 W_RRm1 sig_RRm1 in 1/10
```

	RD	W_RD	sig_RD	RRm1	W_RRm1	sig_RRm1
1.	-.2303744	4.230243	.0397097	-.3831422	6.491778	.0108375
2.	.0266902	.1058542	.7449144	.0566578	.100109	.7516989
3.	-.0768078	.9187646	.3377995	-.1573559	1.084116	.2977787
4.	.1710702	2.366535	.1239619	.4605812	1.573743	.2096643
5.	-.0448509	.4072922	.5233471	-.0884253	.4459299	.5042742
6.	.0165251	.0432816	.835196	.0347015	.0418188	.8379655
7.	-.0391535	.3077611	.5790573	-.0776336	.3331707	.563798
8.	.22382	4.539085	.0331293	.6783119	2.587788	.1076906
9.	-.0388409	.1190183	.7301016	-.0811482	.1293816	.719074
10.	-.2437626	6.151558	.0131296	-.4233705	10.05909	.001516

In certain cases (such as subject 8), the difference can be severe enough to potentially change the conclusion. The reason for this inconsistency is that the nonlinear Wald test is actually a standard Wald test of a first-order Taylor approximation of $g()$, and this approximation can differ according to how $g()$ is specified.

As such, keep in mind the manipulability of nonlinear Wald tests when drawing scientific conclusions.

Confidence intervals

We can also use `predictnl` to obtain confidence intervals for the observation-specific $g(\boldsymbol{\theta}, \mathbf{x}_i)$ by using the `ci()` option to specify two new variables to contain the left and right endpoints of the confidence interval, respectively. For example, we could generate confidence intervals for the risk differences calculated previously:

```
. drop RD

. predictnl RD = predict(outcome(Prepaid)) - predict(outcome(Indemnity)),
> ci(RD_lcl RD_rcl)
(1 missing value generated)
Note: Confidence intervals calculated using Z critical values.

. list RD RD_lcl RD_rcl age male nonwhite in 1/10
```

	RD	RD_lcl	RD_rcl	age	male	nonwhite
1.	-.2303744	-.4499073	-.0108415	73.722107	0	0
2.	.0266902	-.1340948	.1874752	27.89595	0	0
3.	-.0768078	-.2338625	.080247	37.541397	0	0
4.	.1710702	-.0468844	.3890248	23.641327	0	1
5.	-.0448509	-.1825929	.092891	40.470901	0	0
6.	.0165251	-.1391577	.1722078	29.683777	0	0
7.	-.0391535	-.177482	.099175	39.468857	0	0
8.	.22382	.0179169	.4297231	26.702255	1	0
9.	-.0388409	-.2595044	.1818226	63.101974	0	1
10.	-.2437626	-.4363919	-.0511332	69.839828	0	0

The confidence level, in this case, 95%, is either set using the `level()` option or obtained from the current default level, `c(level)`; see [U] **20.6 Specifying the width of confidence intervals**.

From the above output, we can see that, for subjects 1, 8, and 10, a 95% confidence interval for the risk difference does not contain zero, meaning that, for these subjects, there is some evidence of a significant difference in risks.

Note that the confidence intervals calculated by `predictnl` are pointwise; there is no adjustment (such as a Bonferroni correction) made so that these confidence intervals may be considered jointly at the specified level.

Methods and Formulas

`predictnl` is implemented as an ado-file.

For the ith observation, consider the transformation $g(\boldsymbol{\theta}, \mathbf{x}_i)$, estimated by $g(\widehat{\boldsymbol{\theta}}, \mathbf{x}_i)$, for the $1 \times k$ parameter vector $\boldsymbol{\theta}$ and data $\mathbf{x}_i$ ($\mathbf{x}_i$ is assumed fixed). The variance of $g(\widehat{\boldsymbol{\theta}}, \mathbf{x}_i)$ is estimated by

$$\widehat{\text{Var}}\left\{g(\widehat{\boldsymbol{\theta}}, \mathbf{x}_i)\right\} = \mathbf{G}\mathbf{V}\mathbf{G}'$$

where $\mathbf{G}$ is the vector of derivatives

$$\mathbf{G} = \left\{ \left. \frac{\partial g(\boldsymbol{\theta}, \mathbf{x}_i)}{\partial \boldsymbol{\theta}} \right|_{\boldsymbol{\theta} = \widehat{\boldsymbol{\theta}}} \right\}_{(1 \times k)}$$

and $\mathbf{V}$ is the estimated variance–covariance matrix of $\widehat{\boldsymbol{\theta}}$. Standard errors, $\widehat{\text{s.e.}}\{g(\widehat{\boldsymbol{\theta}}, \mathbf{x}_i)\}$, are obtained as the square roots of the variances.

The Wald test statistic for testing

$$H_o: g(\boldsymbol{\theta}, \mathbf{x}_i) = 0$$

versus the two-sided alternative is given by

$$W_i = \frac{\left\{g(\widehat{\boldsymbol{\theta}}, \mathbf{x}_i)\right\}^2}{\widehat{\text{Var}}\left\{g(\widehat{\boldsymbol{\theta}}, \mathbf{x}_i)\right\}}$$

In cases where the variance–covariance matrix of $\widehat{\boldsymbol{\theta}}$ is an asymptotic covariance matrix, W_i is approximately distributed as χ^2 with 1 degree of freedom. In the case of linear regression, W_i is taken to be approximately distributed as $F_{1,r}$, where r is the residual degrees of freedom from the original model fit. The levels of significance of the observation-by-observation tests of H_o versus the two-sided alternative are given by

$$p_i = \Pr(T > W_i)$$

where T is either a χ^2- or F-distributed random variable, as described above.

A $(1 - \alpha) \times 100\%$ confidence interval for $g(\boldsymbol{\theta}, \mathbf{x}_i)$ is given by

$$g(\widehat{\boldsymbol{\theta}}, \mathbf{x}_i) \pm z_{\alpha/2} \left[\widehat{\text{s.e.}}\left\{g(\widehat{\boldsymbol{\theta}}, \mathbf{x}_i)\right\}\right]$$

for those cases where W_i is χ^2-distributed, and

$$g(\widehat{\boldsymbol{\theta}}, \mathbf{x}_i) \pm t_{\alpha/2,r} \left[\widehat{\text{s.e.}}\left\{g(\widehat{\boldsymbol{\theta}}, \mathbf{x}_i)\right\}\right]$$

for those cases where W_i is F-distributed. z_p is the $1-p$ quantile of the standard normal distribution, and $t_{p,r}$ is the $1-p$ quantile of the t distribution with r degrees of freedom.

References

Gould, W. W. 1996. crc43: Wald test of nonlinear hypotheses after model estimation. *Stata Technical Bulletin* 29: 2–4. Reprinted in *Stata Technical Bulletin Reprints*, vol. 5, pp. 15–18.

Hosmer, D. W., Jr., and S. Lemeshow. 2000. *Applied Logistic Regression*. 2nd ed. New York: Wiley.

Phillips, P. C. and J. Y. Park. 1988. On the formulation of Wald tests of nonlinear restrictions. *Econometrica* 56: 1065–1083.

Tarlov, A. R., J. E. Ware, Jr., S. Greenfield, E. C. Nelson, E. Perrin, and M. Zubkoff. 1989. The medical outcomes study. *Journal of the American Medical Association* 262: 925–930.

Wells, K. E., R. D. Hays, M. A. Burnam, W. H. Rogers, S. Greenfield, and J. E. Ware, Jr. 1989. Detection of depressive disorder for patients receiving prepaid or fee-for-service care. *Journal of the American Medical Association* 262: 3298–3302.

Also See

Related: [R] **lincom**, [R] **lrtest**, [R] **nlcom**, [R] **predict**, [R] **test**, [R] **testnl**

Background: [U] **20 Estimation and postestimation commands**

Title

> **probit** — Probit regression

Syntax

Probit regression

> <u>pro</u>bit *depvar* [*indepvars*] [*if*] [*in*] [*weight*] [, *probit_options*]

Probit regression, reporting marginal effects

> dprobit [*depvar indepvars* [*if*] [*in*] [*weight*]] [, *dprobit_options*]

probit_options	description
Model	
<u>noc</u>onstant	suppress constant term
<u>off</u>set(*varname*)	include *varname* in model with coefficient constrained to 1
asis	retain perfect predictor variables
SE/Robust	
vce(*vcetype*)	*vcetype* may be <u>r</u>obust, <u>boot</u>strap or <u>jack</u>knife
<u>r</u>obust	synonym for vce(robust)
<u>cl</u>uster(*varname*)	adjust standard errors for intragroup correlation
Reporting	
<u>l</u>evel(#)	set confidence level; default is level(95)
Max options	
maximize_options	control the maximization process; seldom used
[†] <u>noco</u>ef	do not display the coefficient table; seldom used

(*Continued on next page*)

dprobit_options	description
Model	
offset(varname)	include varname in model with coefficient constrained to 1
at(matname)	point at which marginal effects are evaluated
asis	retain perfect predictor variables
classic	calculate mean effects for dummies like those for continuous variables
SE/Robust	
robust	compute standard errors using the robust/sandwich estimator
cluster(varname)	adjust standard errors for intragroup correlation
Reporting	
level(#)	set confidence level; default is level(95)
Max options	
maximize_options	control the maximization process; seldom used
† nocoef	do not display the coefficient table; seldom used

† nocoef does not appear in the dialog box.

depvar and *indepvars* for probit may contain time-series operators; see [U] **11.4.3 Time-series varlists**.

bootstrap, by, jackknife, rolling, statsby, stepwise, svy, and xi are allowed with probit, and
by, rolling, statsby, and xi are allowed with dprobit; see [U] **11.1.10 Prefix commands**.

fweights, iweights, and pweights are allowed; see [U] **11.1.6 weight**.

See [U] **20 Estimation and postestimation commands** for additional capabilities of estimation commands.

Description

probit fits a maximum-likelihood probit model.

dprobit fits maximum-likelihood probit models and is an alternative to probit. Rather than reporting the coefficients, dprobit reports the marginal effect, that is, the change in the probability for an infinitesimal change in each independent, continuous variable and, by default, reports the discrete change in the probability for dummy variables. probit may be typed without arguments after dprobit estimation to see the model in coefficient form.

If estimating on grouped data, see the bprobit command described in [R] **glogit**.

A number of auxiliary commands may be run after probit, logit, or logistic; see [R] **logistic postestimation** for a description of these commands.

See [R] **logistic** for a list of related estimation commands.

Options for probit

Model

noconstant, offset(varname); see [R] **estimation options**.

asis specifies that all specified variables and observations be retained in the maximization process. This option is typically not specified and may introduce numerical instability. Normally probit drops variables that perfectly predict success or failure in the dependent variable along with their associated observations. In those cases, the effective coefficient on the dropped variables is infinity (negative infinity) for variables that completely determine a success (failure). Dropping the variable and perfectly predicted observations has no effect on the likelihood or estimates of the remaining coefficients and increases the numerical stability of the optimization process. Specifying this option forces retention of perfect predictor variables and their associated observations.

___SE/Robust___

vce(*vcetype*); see [R] *vce_option*.

robust, cluster(*varname*); see [R] **estimation options**. cluster() can be used with pweights to produce estimates for unstratified cluster-sampled data, but see [SVY] **svy: probit** for a command especially designed for survey data.

___Reporting___

level(#); see [R] **estimation options**.

___Max options___

maximize_options: iterate(#), [no]log, trace, tolerance(#), ltolerance(#); see [R] **maximize**. These options are seldom used.

The following option is available with probit but is not shown in the dialog box:

nocoef specifies that the coefficient table not be displayed. This option is sometimes used by programmers but is of no use interactively.

Options for dprobit

___Model___

offset(*varname*); see [R] **estimation options**.

at(*matname*) specifies the point at which marginal effects are evaluated. The default is to evaluate at $\overline{x}$, the mean of the independent variables. If there are k independent variables, *matname* may be $1 \times k$ or $1 \times (k+1)$; that is, it may optionally include final element 1 reflecting the constant. at() may be specified when the model is fitted or when results are redisplayed.

asis; see *probit_options* above.

classic requests that the mean effects be calculated using the formula $f(\overline{x}\mathbf{b})b_i$ in all cases. If classic is not specified, $f(\overline{x}\mathbf{b})b_i$ is used for continuous variables, but the mean effects for dummy variables are calculated as $\Phi(\overline{x}_1\mathbf{b}) - \Phi(\overline{x}_0\mathbf{b})$. Here $\overline{x}_1 = \overline{x}$ but with element i set to 1, $\overline{x}_0 = \overline{x}$ but with element i set to 0, and $\overline{x}$ is the mean of the independent variables or the vector specified by at(). classic may be specified at estimation time or when the results are redisplayed. Results calculated without classic may be redisplayed with classic and vice versa.

robust, cluster(*varname*); see [R] **estimation options**.

level(*#*); see [R] **estimation options**.

maximize_options: <u>iterate</u>(*#*), [<u>no</u>]<u>log</u>, <u>trace</u>, <u>tol</u>erance(*#*), <u>ltol</u>erance(*#*); see [R] **maximize**. These options are seldom used.

The following option is available with dprobit but is not shown in the dialog box:

nocoef specifies that the coefficient table not be displayed. This option is sometimes used by programmers but is of no use interactively.

Remarks

Remarks are presented under the headings

> *Robust standard errors*
> *dprobit*
> *Model identification*

probit fits maximum likelihood models with dichotomous dependent (left-hand-side) variables coded as 0/1 (more precisely, coded as 0 and not 0).

▷ Example 1

Wu have data on the make, weight, and mileage rating of 22 foreign and 52 domestic automobiles. Wu wish to fit a probit model explaining whether a car is foreign based on its weight and mileage. Here is an overview of our data:

```
. use http://www.stata-press.com/data/r9/auto
(1978 Automobile Data)

. keep make mpg weight foreign

. describe

Contains data from http://www.stata-press.com/data/r9/auto.dta
  obs:            74                          1978 Automobile Data
  vars:            4                          13 Apr 2005 17:45
  size:         1,998 (99.7% of memory free)  (_dta has notes)
```

variable name	storage type	display format	value label	variable label
make	str18	%-18s		Make and Model
mpg	int	%8.0g		Mileage (mpg)
weight	int	%8.0gc		Weight (lbs.)
foreign	byte	%8.0g	origin	Car type

```
Sorted by:  foreign
    Note:  dataset has changed since last saved
```

```
. inspect foreign
foreign:  Car type                                   Number of Observations
                                                                        Non-
                                                     Total   Integers   Integers
      |   #                          Negative           -         -        -
      |   #                          Zero              52        52        -
      |   #                          Positive          22        22        -
      |   #
      |   #    #                     Total             74        74        -
      |   #    #                     Missing            -
      +--------------------------                      ----      ----     ----
      0                      1                         74
        (2 unique values)
              foreign is labeled and all values are documented in the label.
```

The variable `foreign` takes on two unique values, 0 and 1. The value 0 denotes a domestic car, and 1 denotes a foreign car.

The model that we wish to fit is

$$\Pr(\texttt{foreign} = 1) = \Phi(\beta_0 + \beta_1\texttt{weight} + \beta_2\texttt{mpg})$$

where Φ is the cumulative normal distribution.

To fit this model, we type

```
. probit foreign weight mpg
Iteration 0:  log likelihood = -45.03321
Iteration 1:  log likelihood = -29.244141
  (output omitted)
Iteration 5:  log likelihood = -26.844189
Probit regression                              Number of obs   =         74
                                               LR chi2(2)      =      36.38
                                               Prob > chi2     =     0.0000
Log likelihood = -26.844189                    Pseudo R2       =     0.4039
```

foreign	Coef.	Std. Err.	z	P>|z|	[95% Conf. Interval]	
weight	-.0023355	.0005661	-4.13	0.000	-.003445	-.0012261
mpg	-.1039503	.0515689	-2.02	0.044	-.2050235	-.0028772
_cons	8.275464	2.554142	3.24	0.001	3.269438	13.28149

We find that heavier cars are less likely to be foreign, and that cars yielding better gas mileage are also less likely to be foreign, at least holding the weight of the car constant.

See [R] **maximize** for an explanation of the output.

◁

❑ Technical Note

Stata interprets a value of 0 as a negative outcome (failure) and treats all other values (except missing) as positive outcomes (successes). Thus if your dependent variable takes on the values 0 and 1, 0 is interpreted as failure and 1 as success. If your dependent variable takes on the values 0, 1, and 2, 0 is still interpreted as failure, but both 1 and 2 are treated as successes.

If you prefer a more formal mathematical statement, when you type `probit` y x, Stata fits the model

$$\Pr(y_j \neq 0 \mid \mathbf{x}_j) = \Phi(\mathbf{x}_j\boldsymbol{\beta})$$

where Φ is the standard cumulative normal.

❏

Robust standard errors

If you specify the `robust` option, `probit` reports robust standard errors; see [U] **20.14 Obtaining robust variance estimates**.

▷ Example 2

In the case of the model from example 1, the robust calculation increases the standard error of the coefficient on `mpg` by almost 15 percent:

```
. probit foreign weight mpg, robust nolog
Probit regression                                Number of obs   =         74
                                                 Wald chi2(2)    =      30.26
                                                 Prob > chi2     =     0.0000
Log pseudolikelihood = -26.844189                Pseudo R2       =     0.4039
```

foreign	Coef.	Robust Std. Err.	z	P>|z|	[95% Conf. Interval]	
weight	-.0023355	.0004934	-4.73	0.000	-.0033025	-.0013686
mpg	-.1039503	.0593548	-1.75	0.080	-.2202836	.0123829
_cons	8.275464	2.539176	3.26	0.001	3.29877	13.25216

Without `robust`, the standard error for the coefficient on `mpg` was reported to be .052 with a resulting confidence interval of $[-.21, -.00]$.

◁

▷ Example 3

`robust` with the `cluster()` option can relax the independence assumption required by the probit estimator to independence between clusters. To demonstrate this, we will switch to a different dataset.

We are studying unionization of women in the United States and have a dataset with 26,200 observations on 4,434 women between 1970 and 1988. We will use the variables `age` (the women were 14–26 in 1968, and our data span the age range of 16–46), `grade` (years of schooling completed, ranging from 0 to 18), `not_smsa` (28% of the person-time was spent living outside an SMSA— standard metropolitan statistical area), `south` (41% of the person-time was in the South), and `southXt` (`south` interacted with year, treating 1970 as year 0). We also have variable `union`, indicating union membership. Overall, 22% of the person-time is marked as time under union membership, and 44% of these women have belonged to a union.

We fit the following model, ignoring that the women are observed an average of 5.9 times each in these data:

```
. use http://www.stata-press.com/data/r9/union
(NLS Women 14-24 in 1968)

. probit union age grade not_smsa south southXt

Iteration 0:  log likelihood =  -13864.23
Iteration 1:  log likelihood = -13548.436
Iteration 2:  log likelihood = -13547.308
Iteration 3:  log likelihood = -13547.308

Probit regression                          Number of obs   =      26200
                                           LR chi2(5)      =     633.84
                                           Prob > chi2     =     0.0000
Log likelihood = -13547.308                Pseudo R2       =     0.0229
```

union	Coef.	Std. Err.	z	P>|z|	[95% Conf. Interval]	
age	.0059461	.0015798	3.76	0.000	.0028496	.0090425
grade	.02639	.0036651	7.20	0.000	.0192066	.0335735
not_smsa	-.1303911	.0202523	-6.44	0.000	-.1700848	-.0906975
south	-.4027254	.033989	-11.85	0.000	-.4693426	-.3361081
southXt	.0033088	.0029253	1.13	0.258	-.0024247	.0090423
_cons	-1.113091	.0657808	-16.92	0.000	-1.242019	-.9841628

The reported standard errors in this model are probably meaningless. Women are observed repeatedly, and so the observations are not independent. Looking at the coefficients, we find a large southern effect against unionization and little time trend. The `robust` and `cluster()` options provide a way to fit this model and obtain correct standard errors:

```
. probit union age grade not_smsa south southXt, robust cluster(id)

Iteration 0:  log pseudolikelihood =  -13864.23
Iteration 1:  log pseudolikelihood = -13548.436
Iteration 2:  log pseudolikelihood = -13547.308
Iteration 3:  log pseudolikelihood = -13547.308

Probit regression                          Number of obs   =      26200
                                           Wald chi2(5)    =     165.75
                                           Prob > chi2     =     0.0000
Log pseudolikelihood = -13547.308          Pseudo R2       =     0.0229

              (Std. Err. adjusted for 4434 clusters in idcode)
```

union	Coef.	Robust Std. Err.	z	P>|z|	[95% Conf. Interval]	
age	.0059461	.0023567	2.52	0.012	.001327	.0105651
grade	.02639	.0078378	3.37	0.001	.0110282	.0417518
not_smsa	-.1303911	.0404109	-3.23	0.001	-.209595	-.0511873
south	-.4027254	.0514458	-7.83	0.000	-.5035573	-.3018935
southXt	.0033088	.0039793	0.83	0.406	-.0044904	.0111081
_cons	-1.113091	.1188478	-9.37	0.000	-1.346028	-.8801534

These standard errors are roughly 50% larger than those reported by the inappropriate conventional calculation. By comparison, another model we could fit is an equal-correlation population-averaged probit model:

```
. xtprobit union age grade not_smsa south southXt, i(id) pa

Iteration 1: tolerance = .04796083
Iteration 2: tolerance = .00352657
Iteration 3: tolerance = .00017886
Iteration 4: tolerance = 8.654e-06
Iteration 5: tolerance = 4.150e-07
```

```
GEE population-averaged model                    Number of obs      =      26200
Group variable:                      idcode      Number of groups   =       4434
Link:                                probit      Obs per group: min =          1
Family:                            binomial                     avg =        5.9
Correlation:                   exchangeable                     max =         12
                                                 Wald chi2(5)       =     241.66
Scale parameter:                          1      Prob > chi2        =     0.0000
```

union	Coef.	Std. Err.	z	P>|z|	[95% Conf. Interval]	
age	.0031597	.0014678	2.15	0.031	.0002829	.0060366
grade	.0329992	.0062334	5.29	0.000	.020782	.0452163
not_smsa	-.0721799	.0275189	-2.62	0.009	-.1261159	-.0182439
south	-.409029	.0372213	-10.99	0.000	-.4819815	-.3360765
southXt	.0081828	.002545	3.22	0.001	.0031946	.0131709
_cons	-1.184799	.0890117	-13.31	0.000	-1.359259	-1.01034

The coefficient estimates are similar, but these standard errors are smaller than those produced by `probit, robust cluster()`, as we would expect. If the equal-correlation assumption is valid, the population-averaged probit estimator above should be more efficient.

Is the assumption valid? That is a difficult question to answer. The population-averaged estimates correspond to an assumption of exchangeable correlation within person. It would not be unreasonable to assume an AR(1) correlation within person or to assume that the observations are correlated but that we do not wish to impose any structure. See [XT] xtgee for full details.

◁

What is important to understand is that `probit, robust cluster()` is robust to assumptions about within-cluster correlation. That is, it inefficiently sums within cluster for the standard error calculation rather than attempting to exploit what might be assumed about the within-cluster correlation.

dprobit

A probit model is defined as

$$\Pr(y_j \neq 0 \mid \mathbf{x}_j) = \Phi(\mathbf{x}_j \mathbf{b})$$

where Φ is the standard cumulative normal distribution, and $\mathbf{x}_j \mathbf{b}$ is called the probit score or index.

Since $\mathbf{x}_j \mathbf{b}$ has a normal distribution, interpreting probit coefficients requires thinking in the Z (normal quantile) metric. For instance, say that we estimated the probit equation

$$\Pr(y_j \neq 0) = \Phi(.08233\, x_1 + 1.529\, x_2 - 3.139)$$

Interpreting the x_1 coefficient we see that each one-unit increase in x_1 increases the probit index by .08233 standard deviations. Learning to think in the Z metric takes practice, and, even if you do, communicating results to others who have not learned to think this way is difficult.

A transformation of the results helps some people better understand them. The change in the probability somehow feels more natural, but how big that change is depends on where we start. Why not choose as a starting point the mean of the data? If $\overline{x}_1 = 21.29$ and $\overline{x}_2 = .42$, then we would report something like .0257, meaning the change in the probability calculated at the mean. We could make the calculation as follows.

The mean normal index is $.08233 \times 21.29 + 1.529 \times .42 - 3.139 = -.7440$, and the corresponding probability is $\Phi(-.7440) = .2284$. Adding our coefficient of $.08233$ to the index and recalculating the probability, we obtain $\Phi(-.7440 + .08233) = .2541$. The change in the probability is thus $.2541 - .2284 = .0257$.

In practice, people make this calculation somewhat differently and produce a slightly different number. Rather than making the calculation for a one-unit change in x, they calculate the slope of the probability function. Doing a little calculus, they derive that the change in the probability for a change in x_1 ($\partial\Phi/\partial x_1$) is the height of the normal density multiplied by the x_1 coefficient; that is,

$$\frac{\partial\Phi}{\partial x_1} = \phi(\overline{\mathbf{x}}\mathbf{b})b_1$$

Going through this calculation, they obtain $.0249$.

The difference between $.0257$ and $.0249$ is small; they differ because the $.0257$ is the exact answer for a one-unit increase in x_1, whereas $.0249$ is the answer for an infinitesimal change extrapolated out.

▷ Example 4

dprobit with the classic option transforms results as an infinitesimal change extrapolated out. Consider the automobile data again:

```
. use http://www.stata-press.com/data/r9/auto, clear
(1978 Automobile Data)

. generate goodplus = rep78>=4 if rep78 < .
(5 missing values generated)

. dprobit foreign mpg goodplus, classic
Iteration 0:   log likelihood = -42.400729
Iteration 1:   log likelihood = -27.643138
Iteration 2:   log likelihood = -26.953126
Iteration 3:   log likelihood = -26.942119
Iteration 4:   log likelihood = -26.942114
```

Probit regression, reporting marginal effects				Number of obs	=	69
				LR chi2(2)	=	30.92
				Prob > chi2	=	0.0000
Log likelihood = -26.942114				Pseudo R2	=	0.3646

foreign	dF/dx	Std. Err.	z	P>|z|	x-bar	[95% C.I.]
mpg	.0249187	.0110853	2.30	0.022	21.2899	.003192 .046646
goodplus	.46276	.1187437	3.81	0.000	.42029	.230027 .695493
_cons	-.9499603	.2281006	-3.82	0.000	1	-1.39703 -.502891

obs. P	.3043478	
pred. P	.2286624	(at x-bar)

z and P>\|z\| are the test of the underlying coefficient being 0

After estimation with dprobit, we can see the untransformed coefficient results by typing probit without options:

```
. probit
Probit regression                           Number of obs   =        69
                                            LR chi2(2)      =     30.92
                                            Prob > chi2     =    0.0000
Log likelihood = -26.942114                 Pseudo R2       =    0.3646
```

foreign	Coef.	Std. Err.	z	P>|z|	[95% Conf. Interval]
mpg	.082333	.0358292	2.30	0.022	.0121091	.152557
goodplus	1.528992	.4010866	3.81	0.000	.7428771	2.315108
_cons	-3.138737	.8209689	-3.82	0.000	-4.747807	-1.529668

In one case, one can argue that the classic, infinitesimal-change based adjustment could be improved on, and that is in the case of a dummy variable. A dummy variable takes on the values 0 and 1 only — 1 indicates that something is true, and 0 indicates that it is not. goodplus is such a variable. To understand the effect of goodplus, we want to know how much its being true or false affects the outcome probability.

That is, "at the means", the predicted probability of foreign for a car with goodplus = 0 is $\Phi(.08233\,\overline{x}_1 - 3.139) = .0829$. For the same car with goodplus = 1, the probability is $\Phi(.08233\,\overline{x}_1 + 1.529 - 3.139) = .5569$. The difference is thus $.5569 - .0829 = .4740$.

When we do not specify the classic option, dprobit makes the calculation for dummy variables in this way. Even though we fitted the model with the classic option, we can redisplay results with classic omitted:

```
. dprobit
Probit regression, reporting marginal effects     Number of obs =       69
                                                  LR chi2(2)    =    30.92
                                                  Prob > chi2   =   0.0000
Log likelihood = -26.942114                       Pseudo R2     =   0.3646
```

foreign	dF/dx	Std. Err.	z	P>|z|	x-bar	[95% C.I.]
mpg	.0249187	.0110853	2.30	0.022	21.2899	.003192	.046646
goodplus*	.4740077	.1114816	3.81	0.000	.42029	.255508	.692508
obs. P	.3043478						
pred. P	.2286624	(at x-bar)					

```
(*) dF/dx is for discrete change of dummy variable from 0 to 1
    z and P>|z| are the test of the underlying coefficient being 0
```

◁

❑ Technical Note

at (*matname*) allows you to evaluate effects at points other than the means. Let's obtain the effects for the above model at mpg = 20 and goodplus = 1:

(Continued on next page)

```
. matrix myx = (20,1)

. dprobit, at(myx)
```

Probit regression, reporting marginal effects

```
Number of obs =       69
LR chi2(2)    =    30.92
Prob > chi2   =   0.0000
```

Log likelihood = -26.942114

```
Pseudo R2     =   0.3646
```

foreign	dF/dx	Std. Err.	z	P>|z|	x	[95% C.I.]
mpg	.0328237	.0144157	2.30	0.022	20	.004569 .061078
goodplus*	.4468843	.1130835	3.81	0.000	1	.225245 .668524

obs. P	.3043478		
pred. P	.2286624	(at x-bar)	
pred. P	.5147238	(at x)	

(*) dF/dx is for discrete change of dummy variable from 0 to 1
    z and P>\|z\| are the test of the underlying coefficient being 0

❑

Model identification

The probit command has one more feature, which is probably the most useful. It will automatically check the model for identification and, if the model is underidentified, drop whatever variables and observations are necessary for estimation to proceed.

▷ Example 5

Have you ever fitted a probit model where one or more of your independent variables perfectly predicted one or the other outcome?

For instance, consider the following small amount of data:

Outcome y	Independent Variable x
0	1
0	1
0	0
1	0

Say that we wish to predict the outcome on the basis of the independent variable. Notice that the outcome is always zero when the independent variable is one. In our data, $\Pr(y = 0 \mid x = 1) = 1$, which means that the probit coefficient on x must be minus infinity with a corresponding infinite standard error. At this point, you may suspect that we have a problem.

Unfortunately, not all such problems are so easily detected, especially if you have a lot of independent variables in your model. If you have ever had such difficulties, then you have experienced one of the more unpleasant aspects of computer optimization. The computer has no idea that it is trying to solve for an infinite coefficient as it begins its iterative process. All it knows is that, at each step, making the coefficient a little bigger, or a little smaller, works wonders. It continues on its merry way until either (1) the whole thing comes crashing to the ground when a numerical overflow error occurs or (2) it reaches some predetermined cutoff that stops the process. Meanwhile, you have been waiting. In addition, the estimates that you finally receive, if any, may be nothing more than numerical roundoff.

Stata watches for these sorts of problems, alerts you, fixes them, and then properly fits the model.

Let's return to our automobile data. Among the variables we have in the data is one called `repair` that takes on three values. A value of 1 indicates that the car has a poor repair record, 2 indicates an average record, and 3 indicates a better-than-average record. Here is a tabulation of our data:

```
. use http://www.stata-press.com/data/r9/repair, clear
(1978 Automobile Data)

. tabulate foreign repair
```

		repair		
Car type	1	2	3	Total
Domestic	10	27	9	46
Foreign	0	3	9	12
Total	10	30	18	58

Notice that all the cars with poor repair records (`repair==1`) are domestic. If we were to attempt to predict `foreign` on the basis of the repair records, the predicted probability for the `repair==1` category would have to be zero. This in turn means that the probit coefficient must be minus infinity, and that would set most computer programs buzzing.

Let's try using Stata on this problem. First, we make up two new variables, `rep_is_1` and `rep_is_2`, that indicate the `repair` category.

```
. generate rep_is_1 = repair==1

. generate rep_is_2 = repair==2
```

The statement `generate rep_is_1=repair==1` creates a new variable, `rep_is_1`, that takes on the value 1 when `repair` is 1 and zero otherwise. Similarly, the next `generate` statement creates `rep_is_2` that takes on the value 1 when `repair` is 2 and zero otherwise. We are now ready to fit our model:

```
. probit foreign rep_is_1 rep_is_2
note: rep_is_1 != 0 predicts failure perfectly
      rep_is_1 dropped and 10 obs not used
Iteration 0:   log likelihood = -26.992087
Iteration 1:   log likelihood = -22.276479
Iteration 2:   log likelihood = -22.229184
Iteration 3:   log likelihood = -22.229138

Probit regression                               Number of obs   =         48
                                                LR chi2(1)      =       9.53
                                                Prob > chi2     =     0.0020
Log likelihood = -22.229138                     Pseudo R2       =     0.1765
```

foreign	Coef.	Std. Err.	z	P>|z|	[95% Conf. Interval]	
rep_is_2	-1.281552	.4297324	-2.98	0.003	-2.123812	-.4392916
_cons	1.21e-16	.295409	0.00	1.000	-.578991	.578991

Remember that all the cars with poor repair records (`rep_is_1`) are domestic, so the model cannot be fitted. At least it cannot be fitted if we restrict ourselves to finite coefficients. Stata noted that fact reporting, "Note: rep_is_1 != 0 predicts failure perfectly". This is Stata's mathematically precise way of saying what we said in English. When `rep_is_1` is not equal to 0, the car is domestic.

Stata then went on to say, "rep_is_1 dropped and 10 obs not used" and eliminated the problem. First, the variable `rep_is_1` had to be removed from the model because it would have an infinite coefficient. Then the ten observations that led to the problem had to be eliminated as well, so as not to bias the remaining coefficients in the model. The ten observations that are not used are the ten domestic cars that have poor repair records.

Finally, Stata fitted what was left of the model, using the remaining observations.

◁

❑ Technical Note

Stata is pretty smart about catching these problems. It will catch "one-way causation by a dummy variable", as we demonstrated above.

Stata also watches for "two-way causation", that is, a variable that perfectly determines the outcome, both successes and failures. In this case, Stata says that the variable "predicts outcome perfectly" and stops. Statistics dictate that no model can be fitted.

Stata also checks your data for collinear variables; it will say "so-and-so dropped due to collinearity". No observations need to be eliminated in this case, and model fitting will proceed without the offending variable.

It will also catch a subtle problem that can arise with continuous data. For instance, if we were estimating the chances of surviving the first year after an operation, and if we included in our model age, and if all the persons over 65 died within the year, Stata will say, "age > 65 predicts failure perfectly". It will then inform us about how it resolves the issue and fit what can be fitted of our model.

probit (and logit and logistic) will also occasionally display messages such as

```
note: 4 failures and 0 successes completely determined.
```

The cause of this message and what to do if you see it are described in [R] **logit**.

❑

Saved Results

probit saves in e():

Scalars

e(N)	number of observations		e(ll_0)	log likelihood, constant-only model
e(df_m)	model degrees of freedom		e(N_clust)	number of clusters
e(r2_p)	pseudo-R-squared		e(chi2)	χ^2
e(ll)	log likelihood			

Macros

e(cmd)	probit		e(vce)	*vcetype* specified in vce()
e(depvar)	name of dependent variable		e(vcetype)	title used to label Std. Err.
e(wtype)	weight type		e(crittype)	optimization criterion
e(wexp)	weight expression		e(properties)	b V
e(title)	title in estimation output		e(estat_cmd)	program used to implement estat
e(clustvar)	name of cluster variable		e(predict)	program used to implement predict
e(chi2type)	Wald or LR; type of model χ^2 test			

Matrices

e(b)	coefficient vector		e(V)	variance–covariance matrix of the estimators

Functions

e(sample)	marks estimation sample

`dprobit` saves in `e()`:

Scalars

e(N)	number of observations	e(N_clust)	number of clusters
e(df_m)	model degrees of freedom	e(chi2)	χ^2
e(r2_p)	pseudo-R-squared	e(pbar)	fraction of successes observed in data
e(ll)	log likelihood	e(xbar)	average probit score
e(ll_0)	log likelihood, constant-only model	e(offbar)	average offset

Macros

e(cmd)	dprobit	e(vcetype)	title used to label Std. Err.
e(depvar)	name of dependent variable	e(dummy)	string of blank-separated 0s and 1s;
e(wtype)	weight type		0 means that the corresponding
e(wexp)	weight expression		independent variable is not a dummy;
e(title)	title in estimation output		1 means that it is
e(clustvar)	name of cluster variable	e(crittype)	optimization criterion
e(at)	predicted probability (at x)	e(predict)	program used to implement predict
e(chi2type)	Wald or LR; type of model χ^2 test	e(properties)	b V

Matrices

e(b)	coefficient vector	e(dfdx)	marginal effects
e(V)	variance–covariance matrix of the estimators	e(se_dfdx)	standard errors of the marginal effects

Functions

e(sample)	marks estimation sample

Methods and Formulas

Probit analysis originated in connection with bioassay, and the word probit, a contraction of "probability unit", was suggested by Bliss (1934). For an introduction to probit and logit, see, for example, Aldrich and Nelson (1984), Johnston and DiNardo (1997), Long (1997), Pampel (2000), or Powers and Xie (2000). Long and Freese (2003, chapter 4) provide an introduction to probit and logit, along with Stata examples.

The log-likelihood function for probit is

$$\ln L = \sum_{j \in S} w_j \ln \Phi(\mathbf{x}_j \boldsymbol{\beta}) + \sum_{j \notin S} w_j \ln\left\{ 1 - \Phi(\mathbf{x}_j \boldsymbol{\beta}) \right\}$$

where Φ is the cumulative normal and w_j denotes the optional weights. $\ln L$ is maximized, as described in [R] **maximize**.

If robust standard errors are requested, the calculation described in *Methods and Formulas* of [R] **regress** is carried forward with $\mathbf{u}_j = \{\phi(\mathbf{x}_j \mathbf{b})/\Phi(\mathbf{x}_j \mathbf{b})\}\mathbf{x}_j$ for the positive outcomes and $-[\phi(\mathbf{x}_j \mathbf{b})/\{1 - \Phi(\mathbf{x}_j \mathbf{b})\}]\mathbf{x}_j$ for the negative outcomes, where ϕ is the normal density. q_c is given by its asymptotic-like formula.

Turning to `dprobit`, which is implemented as an ado-file, let $\mathbf{b}$ and $\mathbf{V}$ denote the coefficients and variance matrix calculated by `probit`. Let b_i refer to the ith element of $\mathbf{b}$. For continuous variables, or for all variables if `classic` is specified, `dprobit` reports

$$b_i^* = \left. \frac{\partial \Phi(\mathbf{x}\mathbf{b})}{\partial x_i} \right|_{\mathbf{x}=\overline{\mathbf{x}}} = \phi(\overline{\mathbf{x}}\mathbf{b})b_i$$

The corresponding variance matrix is $\mathbf{DVD}'$, where $\mathbf{D} = \phi(\overline{\mathbf{x}}\mathbf{b})\{\mathbf{I} - (\overline{\mathbf{x}}\mathbf{b})\mathbf{b}\overline{\mathbf{x}}\}$.

For dummy variables taking on values 0 and 1 when `classic` is not specified, `dprobit` makes the discrete calculation associated with the dummy changing from 0 to 1, $b_i^* = \Phi(\overline{\mathbf{x}}_1\mathbf{b}) - \Phi(\overline{\mathbf{x}}_0\mathbf{b})$, where $\overline{\mathbf{x}}_0 = \overline{\mathbf{x}}_1 = \overline{\mathbf{x}}$ except that the ith elements of $\overline{\mathbf{x}}_0$ and $\overline{\mathbf{x}}_1$ are set to 0 and 1, respectively. The variance of b_i is given by $\mathbf{dVd}'$, where $\mathbf{d} = \phi(\overline{\mathbf{x}}_1\mathbf{b})\overline{\mathbf{x}}_1 - \phi(\overline{\mathbf{x}}_0\mathbf{b})\overline{\mathbf{x}}_0$.

Note that, in all cases, `dprobit` reports test statistics z_i based on the underlying coefficients b_i.

> Chester Ittner Bliss (1899–1979) was born in Ohio. He was educated as an entomologist, earning degrees from Ohio State and Columbia, and was employed by the United States Department of Agriculture until 1933. When he lost his job as a result of the Depression, Bliss then worked with R. A. Fisher in London and at the Institute of Plant Protection in Leningrad before returning to a post at the Connecticut Agricultural Experiment Station in 1938. He was also Lecturer at Yale for 25 years. Among many contributions to biostatistics, his development and application of probit methods to biological problems are outstanding.

References

Aldrich, J. H. and F. D. Nelson. 1984. *Linear Probability, Logit, and Probit Models.* Newbury Park, CA: Sage.

Berkson, J. 1944. Application of the logistic function to bio-assay. *Journal of the American Statistical Association* 39: 357–365.

Bliss, C. I. 1934. The method of probits. *Science* 79: 38–39, 409–410.

Cochran, W. G. 1979. Chester Ittner Bliss 1899–1979. *Biometrics* 35: 715–717.

Finney, D. J. 1979. Chester Ittner Bliss 1899–1979. *Biometrics* 35: 717.

Hilbe, J. 1996. sg54: Extended probit regression. *Stata Technical Bulletin* 32: 20–21. Reprinted in *Stata Technical Bulletin Reprints*, vol. 6, pp. 131–132.

Johnston, J. and J. DiNardo. 1997. *Econometric Methods.* 4th ed. New York: McGraw–Hill.

Judge, G. G., W. E. Griffiths, R. C. Hill, H. Lütkepohl, and T.-C. Lee. 1985. *The Theory and Practice of Econometrics.* 2nd ed. New York: Wiley.

Long, J. S. 1997. *Regression Models for Categorical and Limited Dependent Variables.* Thousand Oaks, CA: Sage.

Long, J. S. and J. Freese. 2003. *Regression Models for Categorical Dependent Variables Using Stata.* rev. ed. College Station, TX: Stata Press.

Pampel, F. C. 2000. *Logistic Regression: A Primer.* Thousand Oaks, CA: Sage.

Powers, D. A. and Y. Xie. 2000. *Statistical Methods for Categorical Data Analysis.* San Diego, CA: Academic Press.

Also See

Complementary:	[R] **probit postestimation**, [R] **roc**
Related:	[R] **asmprobit**, [R] **brier**, [R] **biprobit**, [R] **clogit**, [R] **glm**, [R] **hetprob**, [R] **ivprobit**, [R] **logistic**, [R] **logit**, [R] **mprobit**, [R] **scobit**, [SVY] **svy: probit**, [XT] **xtprobit**
Background:	[U] **11.1.10 Prefix commands**, [U] **20 Estimation and postestimation commands**, [R] **estimation options**, [R] **maximize**, [R] *vce_option*

Title

> **probit postestimation** — Postestimation tools for probit and dprobit

Description

The following postestimation commands are of special interest after `probit` and `dprobit`:

command	description
estat clas	estat classification reports various summary statistics, including the classification table
estat gof	Pearson or Hosmer–Lemeshow goodness-of-fit test
lroc	graphs the ROC curve and calculates the area under the curve
lsens	graphs sensitivity and specificity versus probability cutoff

For information about these commands, see [R] **logistic postestimation**.

In addition, the following standard postestimation commands are available:

command	description
adjust[1]	adjusted predictions of $\mathbf{x}\beta$ or probabilities
estat	AIC, BIC, VCE, and estimation sample summary
estimates	cataloging estimation results
hausman	Hausman's specification test
lincom	point estimates, standard errors, testing, and inference for linear combinations of coefficients
linktest	link test for model specification
lrtest	likelihood-ratio test
mfx	marginal effects or elasticities
nlcom	point estimates, standard errors, testing, and inference for nonlinear combinations of coefficients
predict	predictions, residuals, influence statistics, and other diagnostic measures
predictnl	point estimates, standard errors, testing, and inference for generalized predictions
suest	seemingly unrelated estimation
test	Wald tests for simple and composite linear hypotheses
testnl	Wald tests of nonlinear hypotheses

[1] `adjust` does not work with time-series operators.

See the corresponding entries in the *Stata Base Reference Manual* for details.

Syntax for predict

predict [*type*] *newvar* [*if*] [*in*] [, *statistic* <u>nooff</u>set <u>rules</u> asif]

statistic	description
<u>pr</u>	probability of a positive outcome; the default
xb	linear prediction $\mathbf{x}_j \mathbf{b}$
stdp	standard error of the prediction
<u>score</u>	first derivative of the log likelihood with respect to $\mathbf{x}_j \boldsymbol{\beta}$

These statistics are available both in and out of sample; type predict ... if e(sample) ... if wanted only for the estimation sample.

Options for predict

pr, the default, calculates the probability of a positive outcome.

xb calculates the linear prediction.

stdp calculates the standard error of the linear prediction.

score calculates the equation-level score, $\partial \ln L / \partial (\mathbf{x}_j \boldsymbol{\beta})$.

nooffset is relevant only if you specified offset(*varname*) for probit. It modifies the calculations made by predict so that they ignore the offset variable; the linear prediction is treated as $\mathbf{x}_j \mathbf{b}$ rather than as $\mathbf{x}_j \mathbf{b} + \text{offset}_j$.

rules requests that Stata use any "rules" that were used to identify the model when making the prediction. By default, Stata calculates missing for excluded observations.

asif requests that Stata ignore the rules and exclusion criteria and calculate predictions for all observations possible using the estimated parameter from the model.

Remarks

Remarks are presented under the headings

> *Obtaining predicted values*
> *Performing hypothesis tests*

Obtaining predicted values

Once you have fitted a probit model, you can obtain the predicted probabilities using the predict command for both the estimation sample and other samples; see [U] **20 Estimation and postestimation commands** and [R] **predict**. Here we will make only a few additional comments.

predict without arguments calculates the predicted probability of a positive outcome. With the xb option, predict calculates the linear combination $\mathbf{x}_j \mathbf{b}$, where $\mathbf{x}_j$ are the independent variables in the jth observation and $\mathbf{b}$ is the estimated parameter vector. This is known as the index function since the cumulative density indexed at this value is the probability of a positive outcome.

In both cases, Stata remembers any "rules" used to identify the model and calculates missing for excluded observations unless rules or asif is specified. This is covered in the following example.

With the `stdp` option, `predict` calculates the standard error of the prediction, which is *not* adjusted for replicated covariate patterns in the data.

You can calculate the unadjusted-for-replicated-covariate-patterns diagonal elements of the hat matrix, or leverage, by typing

```
. predict pred
. predict stdp, stdp
. generate hat = stdp^2*pred*(1-pred)
```

▷ Example 1

In example 5 of [R] **probit**, we fitted the probit model `probit foreign rep_is_1 rep_is_2`. To obtain predicted probabilities, we type

```
. predict p
(option p assumed; Pr(foreign))
(10 missing values generated)
. summarize foreign p
```

Variable	Obs	Mean	Std. Dev.	Min	Max
foreign	58	.2068966	.4086186	0	1
p	48	.25	.1956984	.1	.5

Stata remembers any "rules" used to identify the model and sets predictions to missing for any excluded observations. In the previous example, `probit` dropped the variable `rep_is_1` from our model and excluded ten observations. When we typed `predict p`, those same ten observations were again excluded and their predictions set to missing.

`predict`'s `rules` option uses the rules in the prediction. During estimation, we were told, "rep_is_1 != 0 predicts failure perfectly", so the rule is that when `rep_is_1` is not zero, we should predict 0 probability of success or a positive outcome:

```
. predict p2, rules
. summarize foreign p p2
```

Variable	Obs	Mean	Std. Dev.	Min	Max
foreign	58	.2068966	.4086186	0	1
p	48	.25	.1956984	.1	.5
p2	58	.2068966	.2016268	0	.5

`predict`'s `asif` option ignores the rules and the exclusion criteria and calculates predictions for all observations possible using the estimated parameters from the model:

```
. predict p3, asif
. summarize for p p2 p3
```

Variable	Obs	Mean	Std. Dev.	Min	Max
foreign	58	.2068966	.4086186	0	1
p	48	.25	.1956984	.1	.5
p2	58	.2068966	.2016268	0	.5
p3	58	.2931034	.2016268	.1	.5

Which is right? By default, `predict` uses the most conservative approach. If a large number of observations had been excluded due to a simple rule, we could be reasonably certain that the `rules` prediction is correct. The `asif` prediction is only correct if the exclusion is a fluke and we would be willing to exclude the variable from the analysis, anyway. In that case, however, we should refit the model to include the excluded observations.

◁

Performing hypothesis tests

After estimation with `probit`, you can perform hypothesis tests using the `test` or `testnl` commands; see [U] **20 Estimation and postestimation commands**.

Methods and Formulas

All postestimation commands listed above are implemented as ado-files.

Also See

Complementary:	[R] **probit**; [R] **logistic postestimation**; [R] **adjust**, [R] **estimates**, [R] **hausman**, [R] **lincom**, [R] **linktest**, [R] **lrtest**, [R] **mfx**, [R] **nlcom**, [R] **predictnl**, [R] **suest**, [R] **test**, [R] **testnl**
Background:	[U] **13.5 Accessing coefficients and standard errors**, [U] **20 Estimation and postestimation commands**, [R] **estat**, [R] **predict**

Title

> **proportion** — Estimate proportions

Syntax

proportion *varlist* [*if*] [*in*] [*weight*] [, *options*]

options	description
Model	
<u>stdize</u>(*varname*)	variable identifying strata for standardization
<u>stdweight</u>(*varname*)	weight variable for standardization
<u>nostdrescale</u>	do not rescale the standard weight variable
<u>nolabel</u>	suppress value labels from *varlist*
<u>miss</u>ing	treat missing values like other values
if/in/over	
<u>over</u>(*varlist*[, <u>nolabel</u>])	group over subpopulations defined by *varlist*; optionally, suppress group labels
SE/Cluster	
<u>vce</u>(*vcetype*)	*vcetype* may be <u>bootstrap</u> or <u>jackknife</u>
<u>c</u>luster(*varname*)	adjust standard errors for intragroup correlation
Reporting	
<u>level</u>(#)	set confidence level; default is level(95)
<u>noh</u>eader	suppress table header
<u>nol</u>egend	suppress table legend

svy may be used with proportion; see [SVY] **svy: proportion**.

fweights, iweights, and pweights are allowed; see [U] **11.1.6 weight**.

proportion shares the features of all estimation commands; see [U] **20 Estimation and postestimation commands**.

Description

proportion produces estimates of proportions, along with standard errors.

Options

    Model

stdize(*varname*) specifies that the point estimates be adjusted by direct standardization across the strata identified by *varname*. This option requires the stdweight() option.

stdweight(*varname*) specifies the weight variable associated with the strata identified in the stdize() option. The standardization weights must be constant within the strata identified in the stdize() option.

nostdrescale prevents the standardization weights from being rescaled within the over() groups. This option requires stdize() but is ignored if the over() option is not specified.

nolabel specifies that value labels attached to the variables in *varlist* be ignored.

missing specifies that missing values in *varlist* be treated as valid categories, rather than omitted from the analysis (the default).

⌐─────── if/in/over ⌐──

over(*varlist* [, nolabel]) specifies that estimates be computed for multiple subpopulations, which are identified by the different values of the variables in *varlist*.

When this option is supplied with a single variable name, such as over(*varname*), the value labels of *varname* are used to identify the subpopulations. If *varname* does not have labeled values (or there are unlabeled values), the values themselves are used, provided that they are non-negative integers. Noninteger values, negative values, and labels that are not valid Stata names are substituted with a default identifier.

When over() is supplied with multiple variable names, each subpopulation is assigned a unique default identifier.

nolabel requests that value labels attached to the variables identifying the subpopulations be ignored.

⌐─────── SE/Cluster ⌐──

vce(*vcetype*); see [R] *vce_option*.

cluster(*varname*); see [R] **estimation options**.

⌐─────── Reporting ⌐──

level(*#*); see [R] **estimation options**.

noheader prevents the table header from being displayed. This option implies nolegend.

nolegend prevents the table legend identifying the subpopulations from being displayed.

Remarks

▷ Example 1

We can estimate the proportion of each repair rating in the auto data.

```
. use http://www.stata-press.com/data/r9/auto
(1978 Automobile Data)
. proportion rep78
Proportion estimation              Number of obs    =      69
```

		Proportion	Std. Err.	Binomial Wald [95% Conf. Interval]	
rep78					
	1	.0289855	.0203446	−.0116115	.0695825
	2	.115942	.0388245	.0384689	.1934152
	3	.4347826	.0601159	.3148232	.554742
	4	.2608696	.0532498	.1546113	.3671278
	5	.1594203	.0443922	.070837	.2480036

Here we use the `missing` option to include missing values as a category of rep78.

```
. proportion rep78, missing
Proportion estimation                    Number of obs    =       74
        _prop_6: rep78 = .
```

	Proportion	Std. Err.	Binomial Wald [95% Conf. Interval]	
rep78				
1	.027027	.0189796	-.0107994	.0648534
2	.1081081	.0363433	.0356761	.1805401
3	.4054054	.0574637	.2908804	.5199305
4	.2432432	.0502154	.1431641	.3433224
5	.1486486	.0416364	.0656674	.2316299
_prop_6	.0675676	.0293776	.0090181	.1261171

◁

▷ Example 2

We can also estimate proportions over groups.

```
. proportion rep78, over(foreign)
Proportion estimation                    Number of obs    =       69
        _prop_1: rep78 = 1
        _prop_2: rep78 = 2
        _prop_3: rep78 = 3
        _prop_4: rep78 = 4
        _prop_5: rep78 = 5
    Domestic: foreign = Domestic
     Foreign: foreign = Foreign
```

Over	Proportion	Std. Err.	Binomial Wald [95% Conf. Interval]	
_prop_1				
Domestic	.0416667	.0291477	-.0164966	.0998299
Foreign	(no observations)			
_prop_2				
Domestic	.1666667	.0543607	.0581916	.2751417
Foreign	(no observations)			
_prop_3				
Domestic	.5625	.0723605	.4181069	.7068931
Foreign	.1428571	.0782461	-.0132805	.2989948
_prop_4				
Domestic	.1875	.0569329	.0738921	.3011079
Foreign	.4285714	.1106567	.2077595	.6493834
_prop_5				
Domestic	.0416667	.0291477	-.0164966	.0998299
Foreign	.4285714	.1106567	.2077595	.6493834

◁

Saved Results

proportion saves in e():

Scalars

e(N)	number of observations	e(df_r)	sample degrees of freedom
e(N_over)	number of subpopulations	e(N_clust)	number of clusters
e(N_stdize)	number of standard strata		

Macros

e(cmd)	proportion	e(over_labels)	labels from over() variables
e(varlist)	*varlist*	e(over_namelist)	names from e(over_labels)
e(stdize)	*varname* from stdize()	e(namelist)	proportion identifiers
e(stdweight)	*varname* from stdweight()	e(label#)	labels from #th variable in *varlist*
e(wtype)	weight type	e(vce)	*vcetype* specified in vce()
e(wexp)	weight expression	e(vcetype)	title used to label Std. Err.
e(title)	title in estimation output	e(estat_cmd)	program used to implement estat
e(cluster)	name of cluster variable	e(properties)	b V
e(over)	*varlist* from over()		

Matrices

e(b)	vector of proportion estimates
e(V)	(co)variance estimates
e(_N)	vector of numbers of nonmissing observations
e(_N_stdsum)	number of nonmissing observations within the standard strata
e(_p_stdize)	standardizing proportions

Functions

e(sample)	marks estimation sample

Methods and Formulas

proportion is implemented as an ado-file.

Proportions are means of indicator variables; see [R] **mean**.

References

Cochran, W. G. 1977. *Sampling Techniques*. 3rd ed. New York: Wiley.

Stuart, A. and J. K. Ord. 1994. *Kendall's Advanced Theory of Statistics, Vol. I*. 6th ed. London: Arnold.

Also See

Complementary:	[R] **proportion postestimation**
Related:	[SVY] **svy: proportion**;
	[R] **mean**, [R] **ratio**, [R] **total**
Background:	[U] **20 Estimation and postestimation commands**,
	[R] **estimation options**, [R] *vce_option*

Title

> **proportion postestimation** — Postestimation tools for proportion

Description

The following postestimation commands are available for `proportion`:

command	description
estat	VCE
estimates	cataloging estimation results
lincom	point estimates, standard errors, testing, and inference for linear combinations of coefficients
nlcom	point estimates, standard errors, testing, and inference for nonlinear combinations of coefficients
test	Wald tests for simple and composite linear hypotheses
testnl	Wald tests of nonlinear hypotheses

See the corresponding entries in the *Stata Base Reference Manual* for details.

Methods and Formulas

All postestimation commands listed above are implemented as ado-files.

Also See

Complementary: [R] **proportion**; [R] **estimates**, [R] **lincom**, [R] **nlcom**, [R] **test**, [R] **testnl**

Related: [SVY] **svy: proportion postestimation**

Background: [U] **13.5 Accessing coefficients and standard errors**,

                [U] **13.6 Accessing results from Stata commands**,

                [R] **estat**

Title

> **prtest** — One- and two-sample tests of proportions

Syntax

Test that a variable has a specified proportion

> prtest *varname* == #$_p$ $\lceil if \rceil$ $\lceil in \rceil$ $\lceil$, $\underline{l}$evel(#) $\rceil$

Test that two variables have the same proportion

> prtest *varname*$_1$ == *varname*$_2$ $\lceil if \rceil$ $\lceil in \rceil$ $\lceil$, $\underline{l}$evel(#) $\rceil$

Test that a variable has the same proportion within two groups

> prtest *varname* $\lceil if \rceil$ $\lceil in \rceil$, by(*groupvar*) $\lceil\underline{l}$evel(#) $\rceil$

Immediate form of one-sample test of proportion

> prtesti #$_{obs1}$ #$_{p1}$ #$_{p2}$ $\lceil$, $\underline{l}$evel(#) $\underline{c}$ount $\rceil$

Immediate form of two-sample test of proportion

> prtesti #$_{obs1}$ #$_{p1}$ #$_{obs2}$ #$_{p2}$ $\lceil$, $\underline{l}$evel(#) $\underline{c}$ount $\rceil$

by may be used with prtest (but not prtesti); see [D] **by**.

Description

prtest performs tests on the equality of proportions using large-sample statistics.

In the first form, prtest tests that *varname* has a proportion of #$_p$. In the second form, prtest tests that *varname*$_1$ and *varname*$_2$ have the same proportion. In the third form, prtest tests that *varname* has the same proportion within the two groups defined by *groupvar*.

prtesti is the immediate form of prtest; see [U] **19 Immediate commands**.

The bitest command is a better version of the first form of prtest in that it gives exact *p*-values. Researchers should use bitest when possible, especially for small samples; see [R] **bitest**.

Options

> Main

by(*groupvar*) specifies a numeric variable that contains the group information for a given observation. This variable must have only two values. Do not confuse the by() option with the by prefix; both may be specified.

level(#) specifies the confidence level, as a percentage, for confidence intervals. The default is level(95) or as set by set level; see [U] **20.6 Specifying the width of confidence intervals**.

`count` specifies that integer counts instead of proportions be used in the immediate forms of `prtest`. In the first syntax, `prtesti` expects that $\#_{obs1}$ and $\#_{p1}$ are counts—$\#_{p1} \leq \#_{obs1}$—and $\#_{p2}$ is a proportion. In the second syntax, `prtesti` expects that all four numbers are integer counts, that $\#_{obs1} \geq \#_{p1}$, and that $\#_{obs2} \geq \#_{p2}$.

Remarks

The `prtest` output follows the output of `ttest` in providing a lot of information. Each proportion is presented along with a confidence interval. The appropriate one- or two-sample test is performed, and the two-sided and both one-sided results are included at the bottom of the output. In the case of a two-sample test, the calculated difference is also presented with its confidence interval. This command may be used for both large-sample testing and large-sample interval estimation.

▷ Example 1: One-sample test of proportion

In the first form, `prtest` tests whether the mean of the sample is equal to a known constant. Assume that we have a sample of 74 automobiles. We wish to test whether the proportion of automobiles that are foreign is different from 40 percent.

```
. use http://www.stata-press.com/data/r9/auto
(1978 Automobile Data)

. prtest foreign == .4
One-sample test of proportion                  foreign: Number of obs =       74
```

Variable	Mean	Std. Err.	[95% Conf. Interval]
foreign	.2972973	.0531331	.1931583 .4014363

```
    p = proportion(foreign)                               z =  -1.8034
Ho: p = 0.4
      Ha: p < 0.4              Ha: p != 0.4                 Ha: p > 0.4
  Pr(Z < z) = 0.0357      Pr(|Z| > |z|) = 0.0713        Pr(Z > z) = 0.9643
```

The test indicates that we cannot reject the hypothesis that the proportion of foreign automobiles is .40 at the 5% significance level.

◁

▷ Example 2: Two-sample test of proportion

We have two headache remedies that we give to patients. Each remedy's effect is recorded as 0 for failing to relieve the headache and 1 for relieving the headache. We wish to test the equality of the proportion of people relieved by the two treatments.

(Continued on next page)

```
. use http://www.stata-press.com/data/r9/cure
. prtest cure1 == cure2
```
```
Two-sample test of proportion                     cure1: Number of obs =      50
                                                  cure2: Number of obs =      59
```

Variable	Mean	Std. Err.	z	P>|z|	[95% Conf. Interval]
cure1	.52	.0706541			.3815205 .6584795
cure2	.7118644	.0589618			.5963013 .8274275
diff	-.1918644	.0920245			-.372229 -.0114998
	under Ho:	.0931155	-2.06	0.039	

```
         diff = prop(cure1) - prop(cure2)                      z =  -2.0605
     Ho: diff = 0

    Ha: diff < 0                 Ha: diff != 0                 Ha: diff > 0
 Pr(Z < z) = 0.0197         Pr(|Z| < |z|) = 0.0394         Pr(Z > z) = 0.9803
```

We find that the proportions are statistically different from each other at any level greater than 3.9%.

◁

▷ Example 3: Immediate form of one-sample test of proportion

prtesti is like prtest, except that you specify summary statistics rather than variables as arguments. For instance, we are reading an article that reports the proportion of registered voters among 50 randomly selected eligible voters as .52. We wish to test whether the proportion is .7:

```
. prtesti 50 .52 .70
```
```
One-sample test of proportion                         x: Number of obs =      50
```

Variable	Mean	Std. Err.	[95% Conf. Interval]
x	.52	.0706541	.3815205 .6584795

```
           p = proportion(x)                                   z =  -2.7775
      Ho: p = 0.7

    Ha: p < 0.7                  Ha: p != 0.7                 Ha: p > 0.7
 Pr(Z < z) = 0.0027         Pr(|Z| > |z|) = 0.0055         Pr(Z > z) = 0.9973
```

◁

▷ Example 4: Immediate form of two-sample test of proportion

To judge teacher effectiveness, we wish to test whether the same proportion of people from two classes will answer an advanced question correctly. In the first classroom of 30 students, 40% answered the question correctly, whereas in the second classroom of 45 students, 67% answered the question correctly.

```
. prtesti 30 .4 45 .67
Two-sample test of proportion                          x: Number of obs =        30
                                                       y: Number of obs =        45
```

Variable	Mean	Std. Err.	z	P>|z|	[95% Conf. Interval]
x	.4	.0894427			.2246955 .5753045
y	.67	.0700952			.532616 .807384
diff	-.27	.1136368			-.4927241 -.0472759
	under Ho:	.1169416	-2.31	0.021	

```
        diff = prop(x) - prop(y)                              z =  -2.3088
   Ho: diff = 0

   Ha: diff < 0                 Ha: diff != 0                 Ha: diff > 0
Pr(Z < z) = 0.0105        Pr(|Z| < |z|) = 0.0210        Pr(Z > z) = 0.9895
```

◁

Saved Results

prtest saves in r():

Scalars

r(z)	z statistic	r(N_#)	number of observations for variable #
r(P_#)	proportion for variable #		

Methods and Formulas

prtest and prtesti are implemented as ado-files.

A large-sample $(1 - \alpha)100\%$ confidence interval for a proportion p is

$$\widehat{p} \pm z_{1-\alpha/2}\sqrt{\frac{\widehat{p}\,\widehat{q}}{n}}$$

and a $(1 - \alpha)100\%$ confidence for the difference of two proportions is given by

$$(\widehat{p}_1 - \widehat{p}_2) \pm z_{1-\alpha/2}\sqrt{\frac{\widehat{p}_1\widehat{q}_1}{n_1} + \frac{\widehat{p}_2\widehat{q}_2}{n_2}}$$

where $\widehat{q} = 1 - \widehat{p}$ and z is calculated from the inverse normal distribution.

The one-tailed and two-tailed tests of a population proportion use a normally distributed test statistic calculated as

$$z = \frac{\widehat{p} - p_0}{\sqrt{p_0 q_0/n}}$$

where p_0 is the hypothesized proportion. A test of the difference of two proportions also uses a normally distributed test statistic calculated as

$$z = \frac{\widehat{p}_1 - \widehat{p}_2}{\sqrt{\widehat{p}_p \widehat{q}_p (1/n_1 + 1/n_2)}}$$

where

$$\widehat{p}_p = \frac{x_1 + x_2}{n_1 + n_2}$$

and x_1 and x_2 are the total number of successes in the two populations.

References

Sincich, T. 1987. *Statistics by Example*. 3rd ed. San Francisco: Dellen.

Wang, D. 2000. sg154: Confidence intervals for the ratio of two binomial proportions by Koopman's method. *Stata Technical Bulletin* 58: 16–19. Reprinted in *Stata Technical Bulletin Reprints*, vol. 10, pp. 244–247.

Also See

Related: [R] **bitest**, [R] **ci**, [R] **oneway**, [R] **proportion**, [R] **sdtest**,

                  [R] **signrank**, [R] **ttest**,

                  [MV] **hotelling**

Background: [U] **19 Immediate commands**

Title

qc — Quality control charts

Syntax

Draw a c chart

    cchart *defect_var unit_var* [, *cchart_options*]

Draw a p (fraction-defective) chart

    pchart *reject_var unit_var ssize_var* [, *pchart_options*]

Draw an R (range or dispersion) chart

    rchart *varlist* [*if*] [*in*] [, *rchart_options*]

Draw an $\overline{X}$ (control line) chart

    xchart *varlist* [*if*] [*in*] [, *xchart_options*]

Draw vertically aligned $\overline{X}$ and R charts

    shewhart *varlist* [*if*] [*in*] [, *shewhart_options*]

cchart_options	description
Plot	
marker_options	change look of markers (color, size, etc.)
marker_label_options	add marker labels; change look or position
connect_options	affect rendition of the plotted points
Control limits	
clopts(*cline_options*)	affect rendition of the control limits
Add plot	
addplot(*plot*)	add other plots to the generated graph
Y-Axis, X-Axis, Title, Caption, Legend, Overall	
twoway_options	any options other than by() documented in [G] ***twoway_options***

(Continued on next page)

pchart_options	description
Main	
<u>stabi</u>lized	stabilize the p chart when sample sizes are unequal
Plot	
marker_options	change look of markers (color, size, etc.)
marker_label_options	add marker labels; change look or position
connect_options	affect rendition of the plotted points
Control limits	
<u>clop</u>ts(*cline_options*)	affect rendition of the control limits
Add plot	
addplot(*plot*)	add other plots to the generated graph
Y-Axis, X-Axis, Title, Caption, Legend, Overall	
twoway_options	any options other than by() documented in [G] ***twoway_options***

rchart_options	description
Main	
<u>std</u>(#)	user-specified standard deviation
Plot	
marker_options	change look of markers (color, size, etc.)
marker_label_options	add marker labels; change look or position
connect_options	affect rendition of the plotted points
Control limits	
<u>clop</u>ts(*cline_options*)	affect rendition of the control limits
Add plot	
addplot(*plot*)	add other plots to the generated graph
Y-Axis, X-Axis, Title, Caption, Legend, Overall	
twoway_options	any options other than by() documented in [G] ***twoway_options***

(Continued on next page)

xchart_options	description
Main	
<u>std</u>(#)	user-specified standard deviation
<u>mean</u>(#)	user-specified mean
<u>lower</u>(#) <u>upper</u>(#)	lower and upper limits of the X-bar limits
Plot	
marker_options	change look of markers (color, size, etc.)
marker_label_options	add marker labels; change look or position
connect_options	affect rendition of the plotted points
Control limits	
<u>clopts</u>(cline_options)	affect rendition of the control limits
Add plot	
<u>addplot</u>(plot)	add other plots to the generated graph
Y-Axis, X-Axis, Title, Caption, Legend, Overall	
twoway_options	any options other than by() documented in [G] **twoway_options**

shewhart_options	description
Main	
<u>std</u>(#)	user-specified standard deviation
<u>mean</u>(#)	user-specified mean
Plot	
marker_options	change look of markers (color, size, etc.)
marker_label_options	add marker labels; change look or position
connect_options	affect rendition of the plotted points
Control limits	
<u>clopts</u>(cline_options)	affect rendition of the control limits
Y-Axis, X-Axis, Title, Caption, Legend, Overall	
combine_options	any options documented in [G] **graph combine**

Description

These commands provide standard quality-control charts. cchart draws a c chart; pchart, a p (fraction-defective) chart; rchart, an R (range or dispersion) chart; xchart, an $\overline{X}$ (control line) chart; and shewhart, vertically aligned $\overline{X}$ and R charts.

Options

> Main

stabilized stabilizes the p chart when sample sizes are unequal.

std(#) specifies the standard deviation of the process. The R chart is calculated (based on the range) if this option is not specified.

mean(#) specifies the grand mean, which is calculated if not specified.

lower(#) and upper(#) must be specified together or not at all. They specify the lower and upper limits of the $\overline{X}$ chart. Calculations based on the mean and standard deviation (whether specified by option or calculated) are used otherwise.

> Plot

marker_options affect the rendition of markers drawn at the plotted points, including their shape, size, color, and outline; see [G] ***marker_options***.

marker_label_options specify if and how the markers are to be labeled; see [G] ***marker_label_options***.

connect_options affect whether lines connect the plotted points and the rendition of those lines; see [G] ***connect_options***.

> Control limits

clopts(*cline_options*) affect the rendition of the control limits; see [G] ***cline_options***.

> Add plot

addplot(*plot*) provides a way to add other plots to the generated graph. See [G] ***addplot_option***.

> Y-Axis, X-Axis, Title, Caption, Legend, Overall

twoway_options are any of the options documented in [G] ***twoway_options***, excluding by(). These include options for titling the graph (see [G] ***title_options***) and options for saving the graph to disk (see [G] ***saving_option***).

combine_options (shewhart only) are any of the options documented in [G] **graph combine**. These include options for titling the graph (see [G] ***title_options***) and options for saving the graph to disk (see [G] ***saving_option***).

Remarks

Control charts may be used to define the goal of a repetitive process, to control that process, and to determine if the goal has been achieved. Walter A. Shewhart of Bell Telephone Laboratories devised the first control chart in 1924. In 1931, Shewhart published *Economic Control of Quality of Manufactured Product*. According to Burr, "Few fields of knowledge have ever been so completely explored and charted in the first exposition" (1976, 29). Shewhart states that "a phenomenon will be said to be controlled when, through the use of past experience, we can predict, at least within limits, how the phenomenon may be expected to vary in the future. Here it is understood that prediction within limits means that we can state, at least approximately, the probability that the observed phenomenon will fall within given limits" (1931, 6).

For more information on quality-control charts, see Burr (1976), Duncan (1986), Harris (1999), or Ryan (2000).

▷ Example 1: cchart

cchart graphs a c chart showing the number of nonconformities in a unit, where *defect_var* records the number of defects in each inspection unit and *unit_var* records the unit number. The unit numbers need not be in order. For instance, consider the following example dataset from Ryan (2000, 156):

```
. use http://www.stata-press.com/data/r9/ncu

. describe

Contains data from http://www.stata-press.com/data/r9/ncu.dta
  obs:            30
  vars:            2                           31 Mar 2005 03:56
  size:          360 (99.9% of memory free)

              storage   display    value
variable name   type    format     label      variable label

day            float    %9.0g                 Days in April
defects        float    %9.0g                 Numbers of Nonconforming Units

Sorted by:

. list in 1/5
```

	day	defects
1.	1	7
2.	2	5
3.	3	11
4.	4	13
5.	5	9

```
. cchart defects day, title(c Chart for Nonconforming Transistors)
```

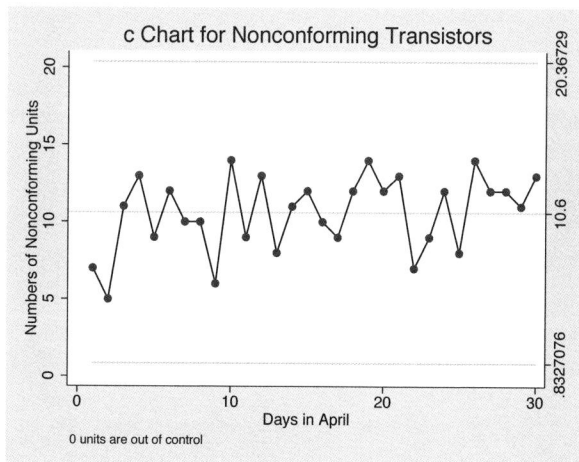

The expected number of defects is 10.6, with lower- and upper-control limits of .8327 and 20.36, respectively. No units are out of control.

◁

▷ Example 2: pchart

pchart graphs a p chart, which shows the fraction of nonconforming items in a subgroup, where *reject_var* records the number rejected in each inspection unit, *unit_var* records the inspection unit number, and *ssize_var* records the number inspected in each unit.

Consider the example dataset from Ryan (2000, 156) of the number of nonconforming transistors out of 1,000 inspected each day during the month of April:

```
. use http://www.stata-press.com/data/r9/ncu2

. describe

Contains data from http://www.stata-press.com/data/r9/ncu2.dta
  obs:           30
  vars:           3                              31 Mar 2005 14:13
  size:         480 (99.9% of memory free)

              storage  display    value
variable name   type   format     label    variable label

day            float   %9.0g               Days in April
rejects        float   %9.0g               Numbers of Nonconforming Units
ssize          float   %9.0g               Sample size

Sorted by:

. list in 1/5

      day    rejects    ssize

  1.    1         7     1000
  2.    2         5     1000
  3.    3        11     1000
  4.    4        13     1000
  5.    5         9     1000

. pchart rejects day ssize
```

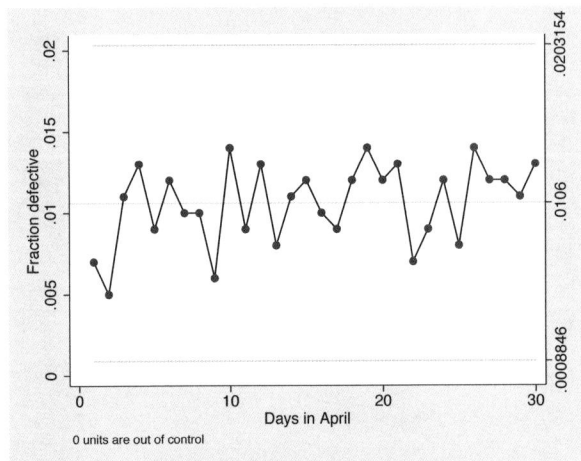

0 units are out of control

All the points are within the control limits, which are .0009 for the lower limit and .0203 for the upper limit.

In our example, the sample sizes are fixed at 1,000, so the ssize variable contains 1,000 for each observation. Sample sizes need not be fixed, however. Say that our data were slightly different:

```
. use http://www.stata-press.com/data/r9/ncu3
. list in 1/5
```

	day	rejects	ssize
1.	1	7	920
2.	2	5	920
3.	3	11	920
4.	4	13	950
5.	5	9	950

```
. pchart rejects day ssize
```

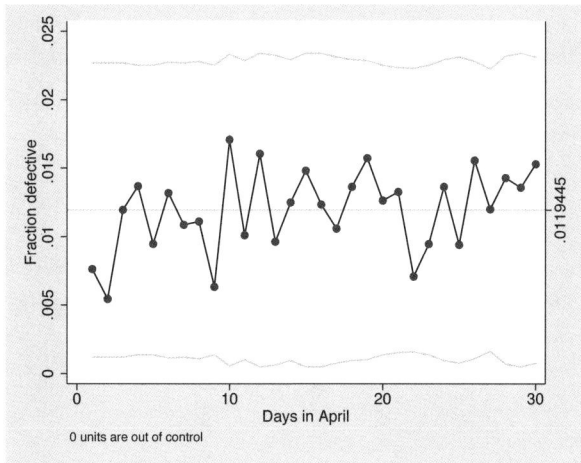

0 units are out of control

In this case, the control limits are, like the sample size, no longer constant. The stabilize option will stabilize the control chart:

```
. pchart rejects day ssize, stabilize
```

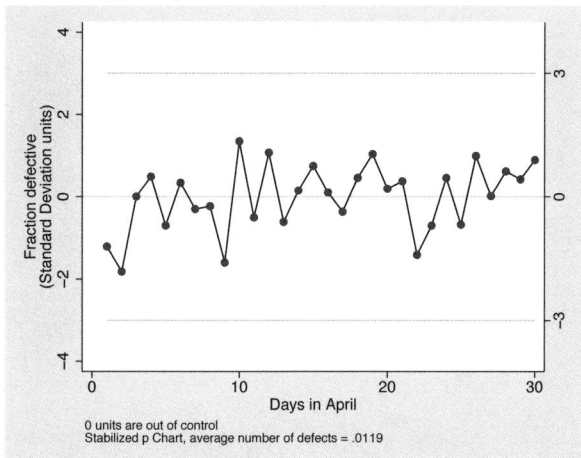

0 units are out of control
Stabilized p Chart, average number of defects = .0119

◁

▷ Example 3: rchart

rchart displays an R chart showing the range for repeated measurements at various times. Variables within observations record measurements. Observations represent different samples.

For instance, say that we take five samples of five observations each. In our first sample, our measurements are 10, 11, 10, 11, and 12. The data are

```
. list
```

	m1	m2	m3	m4	m5
1.	10	11	10	11	12
2.	12	10	9	10	9
3.	10	11	10	12	10
4.	9	9	9	10	11
5.	12	12	12	12	13

```
. rchart m1-m5, connect(l)
```

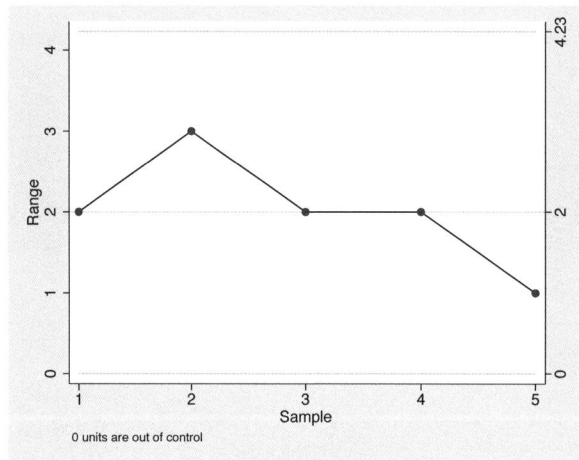

0 units are out of control

The expected range in each sample is 2 with lower- and upper-control limits of 0 and 4.23, respectively. If we knew that the process standard deviation is 0.3, we could specify

(Continued on next page)

```
. rchart m1-m5, connect(l) std(.3)
```

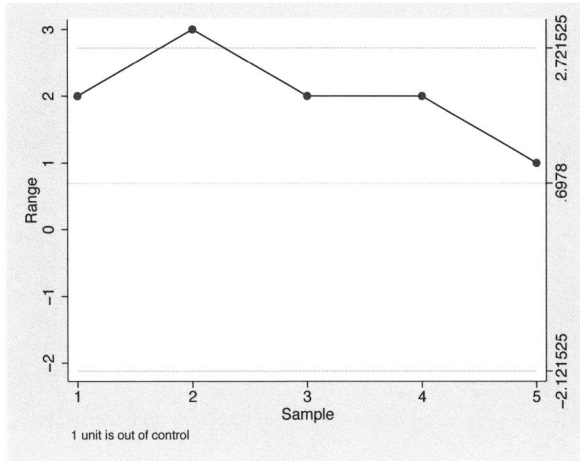

1 unit is out of control

◁

▷ Example 4: xchart

xchart graphs an $\overline{X}$ chart for repeated measurements at various times. Variables within observations record measurements, and observations represent different samples. Using the same data as in the previous example, we type

```
. xchart m1-m5, connect(l)
```

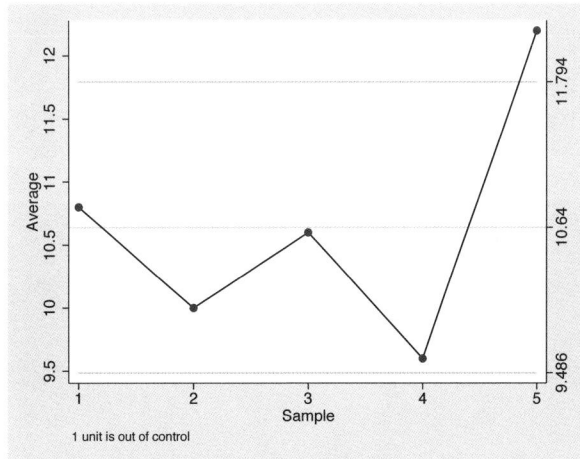

1 unit is out of control

The average measurement in the sample is 10.64, and the lower- and upper-control limits are 9.486 and 11.794, respectively. Suppose that we knew from prior information that the mean of the process is 11. Then we would type

```
. xchart m1-m5, connect(l) mean(11)
```

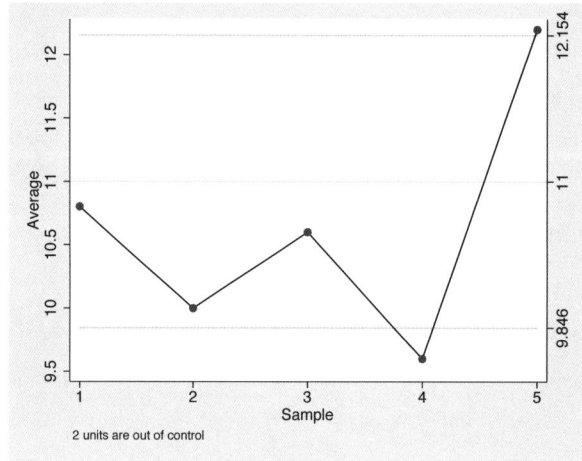

If we also knew that the standard deviation of the process is 0.3, we could type

```
. xchart m1-m5, connect(l) mean(11) std(.3)
```

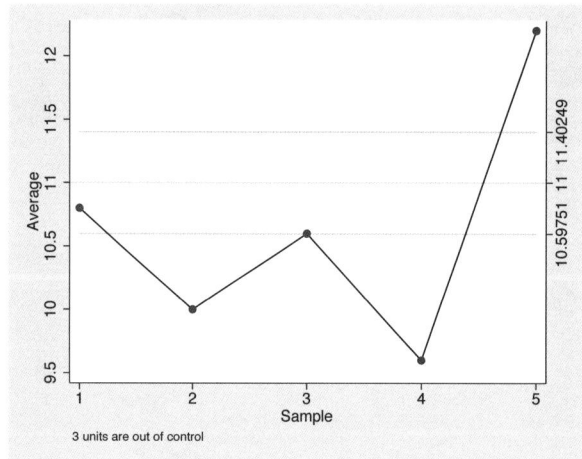

Finally, xchart allows us to specify our own control limits:

```
. xchart m1-m5, connect(l) mean(11) lower(10) upper(12)
```

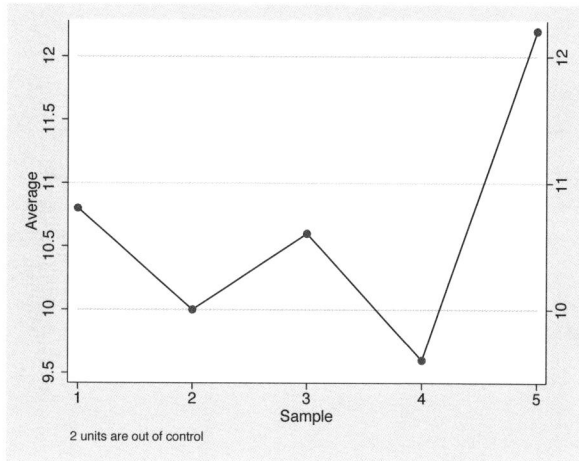

2 units are out of control

◁

▷ Example 5: shewhart

shewhart displays a vertically aligned $\overline{X}$ and R chart in the same image. To produce the best-looking combined image possible, you will want to use the xchart and rchart commands separately and then combine the graphs. shewhart, however, is more convenient.

Using the same data as previously, but realizing that the standard deviation should have been 0.4, we type

Walter Andrew Shewhart (1891–1967) was born in Illinois and educated as a physicist, with degrees from the Universities of Illinois and California. After a brief period teaching physics, he worked for the Western Electric Company and (from 1925) the Bell Telephone Laboratories. His name is most associated with control charts used in quality controls, but his many other interests ranged from quality assurance generally to the philosophy of science.

(Continued on next page)

```
. shewhart m1-m5, connect(1) mean(11) std(.4)
```

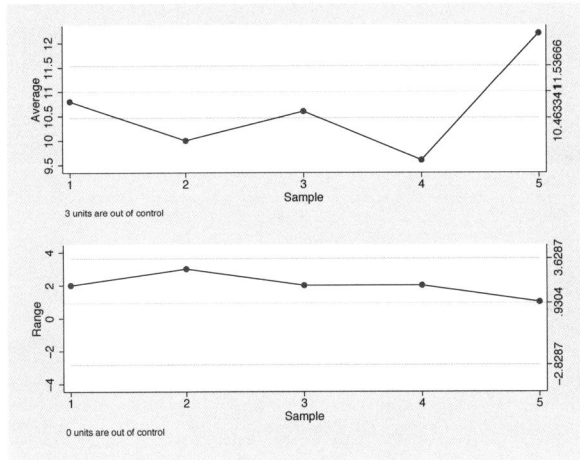

Methods and Formulas

cchart, pchart, rchart, xchart, and shewhart are implemented as ado-files.

For the c chart, the number of defects per unit, C, is taken to be a value of a random variable having a Poisson distribution. If k is the number of units available for estimating λ, the parameter of the Poisson distribution, and if C_i is the number of defects in the ith unit, then λ is estimated by $\overline{C} = \sum_i C_i/k$. Then

$$\text{central line} = \overline{C}$$

$$\text{UCL} = \overline{C} + 3\sqrt{\overline{C}}$$

$$\text{LCL} = \overline{C} - 3\sqrt{\overline{C}}$$

Control limits for the p chart are based on the sampling theory for proportions, using the normal approximation to the binomial. If k samples are taken, the estimator of p is given by $\overline{p} = \sum_i \widehat{p}_i/k$, where $\widehat{p}_i = x_i/n_i$, and x_i is the number of defects in the ith sample of size n_i. The central line and the control limits are given by

$$\text{central line} = \overline{p}$$

$$\text{UCL} = \overline{p} + 3\sqrt{\overline{p}(1 - \overline{p})/n_i}$$

$$\text{LCL} = \overline{p} - 3\sqrt{\overline{p}(1 - \overline{p})/n_i}$$

Control limits for the R chart are based on the distribution of the range of samples of size n from a normal population. If the standard deviation of the process σ is known,

$$\text{central line} = d_2\sigma$$

$$\text{UCL} = D_2\sigma$$

$$\text{LCL} = D_1\sigma$$

where d_2, D_1, and D_2 are functions of the number of observations in the sample and are obtained from the table published in Beyer (1976).

When σ is unknown,

$$\text{central line} = \overline{R}$$
$$\text{UCL} = (D_2/d_2)\overline{R}$$
$$\text{LCL} = (D_1/d_2)\overline{R}$$

where $\overline{R} = \sum_i R_i/k$ is the range of the k sample ranges R_i.

Control limits for the $\overline{X}$ chart are given by

$$\text{central line} = \overline{x}$$
$$\text{UCL} = \overline{x} + (3/\sqrt{n})\sigma$$
$$\text{LCL} = \overline{x} - (3/\sqrt{n})\sigma$$

if σ is known. If σ is unknown,

$$\text{central line} = \overline{x}$$
$$\text{UCL} = \overline{x} + A_2\overline{R}$$
$$\text{LCL} = \overline{x} - A_2\overline{R}$$

where $\overline{R}$ is the average range as defined above and A_2 is a function (*op. cit.*) of the number of observations in the sample.

References

Bayart, D. 2001. Walter Andrew Shewhart. In *Statisticians of the Centuries*, ed. C. C. Heyde and E. Seneta, 398–401. New York: Springer.

Beyer, W. H., ed. 1976. Factors for computing control limits. In *Handbook of Tables for Probability and Statistics*. 2nd ed. 451–465. Cleveland, OH: The Chemical Rubber Company.

Burr, I. W. 1976. *Statistical Quality Control Methods*. New York: Dekker.

Caulcutt, R. 2004. Control charts in practice. *Significance* 1: 81–84.

Duncan, A. J. 1986. *Quality Control and Industrial Statistics*. 5th ed. Homewood, IL: Irwin.

Harris, R. L. 1999. *Information Graphics: A Comprehensive Illustrated Reference*. New York: Oxford University Press.

Ryan, T. P. 2000. *Statistical Methods for Quality Improvement*. 2nd ed. New York: Wiley.

Saw, S. L. C. and T. W. Soon. 1994. sqc1: Estimating process capability indices with Stata. *Stata Technical Bulletin* 17: 18–19. Reprinted in *Stata Technical Bulletin Reprints*, vol. 3, pp. 174–175.

Shewhart, W. A. 1931. *Economic Control of Quality of Manufactured Product*. New York: Van Nostrand.

Also See

Related: [R] **serrbar**

Background: *Stata Graphics Reference Manual*

Title

> **qreg** — Quantile (including median) regression

Syntax

Quantile regression

> qreg *depvar* [*indepvars*] [*if*] [*in*] [*weight*] [, *qreg_options*]

Interquantile range regressions

> iqreg *depvar* [*indepvars*] [*if*] [*in*] [, *iqreg_options*]

Simultaneous-quantile regression

> sqreg *depvar* [*indepvars*] [*if*] [*in*] [, *sqreg_options*]

Quantile regression with bootstrap standard errors

> bsqreg *depvar* [*indepvars*] [*if*] [*in*] [, *bsqreg_options*]

Internal estimation command for quantile regression

> _qreg [*depvar* [*indepvars*] [*if*] [*in*] [*weight*]] [, *_qreg_options*]

qreg_options	description
Model	
quantile(*#*)	estimate *#* quantile; default is quantile(.5)
Reporting	
level(*#*)	set confidence level; default is level(95)
Opt options	
optimization_options	control the optimization process; seldom used
wlsiter(*#*)	attempt *#* weighted least-squares iterations before doing linear programming iterations

iqreg_options	description
Model	
quantiles(*# #*)	interquantile range; default is quantiles(.25 .75)
reps(*#*)	perform *#* bootstrap replications; default is reps(20)
Reporting	
level(*#*)	set confidence level; default is level(95)
nodots	suppress display of the replication dots

sqreg_options	description
Model	
<u>q</u>uantiles(# [# [# ...]])	estimate # quantiles; default is quantiles(.5)
<u>reps</u>(#)	perform # bootstrap replications; default is reps(20)
Reporting	
<u>l</u>evel(#)	set confidence level; default is level(95)
<u>nod</u>ots	suppress display of the replication dots

bsqreg_options	description
Model	
<u>q</u>uantile(#)	estimate # quantile; default is quantile(.5)
<u>reps</u>(#)	perform # bootstrap replications; default is reps(20)
Reporting	
<u>l</u>evel(#)	set confidence level; default is level(95)

_qreg_options	description
<u>q</u>uantile(#)	estimate # quantile; default is quantile(.5)
<u>l</u>evel(#)	set confidence level; default is level(95)
<u>ac</u>curacy(#)	relative accuracy required for linear programming algorithm; should not be specified
optimization_options	control the optimization process; seldom used
<u>wls</u>iter(#)	attempt # weighted least-squares iterations before doing linear programming iterations

by, rolling, statsby, and xi are allowed by qreg, iqreg, sqreg, and bsqreg; stepwise is allowed with qreg; see [U] **11.1.10 Prefix commands**.

qreg and _qreg allow aweights and fweights; see [U] **11.1.6 weight**.

See [U] **20 Estimation and postestimation commands** for additional capabilities of estimation commands.

Description

qreg fits quantile (including median) regression models, also known as least-absolute value models (LAV or MAD) and minimum L1-norm models.

iqreg estimates interquantile range regressions, regressions of the difference in quantiles. The estimated variance–covariance matrix of the estimators (VCE) is obtained via bootstrapping. iqreg has a limit of 336 *indepvars*.

sqreg estimates simultaneous-quantile regression. It produces the same coefficients as qreg for each quantile. Reported standard errors will be similar, but sqreg obtains an estimate of the VCE via bootstrapping, and the VCE includes between-quantiles blocks. Thus you can test and construct confidence intervals comparing coefficients describing different quantiles. sqreg has a limit of 336/q *indepvars*, where q is the number of quantiles specified.

bsqreg is equivalent to sqreg with one quantile. sqreg is faster than bsqreg, but bsqreg is not limited to 336 coefficients.

_qreg is the internal estimation command for quantile regression. _qreg is not intended to be used directly; see *Methods and Formulas* below.

Options for qreg

_____| Model |_____

quantile(*#*) specifies the quantile to be estimated and should be a number between 0 and 1, exclusive. Numbers larger than 1 are interpreted as percentages. The default value of 0.5 corresponds to the median.

_____| Reporting |_____

level(*#*); see [R] **estimation options**.

_____| Opt options |_____

optimization_options: <u>iter</u>ate(*#*), [<u>no</u>]log, <u>trace</u>. iterate() specifies the maximum number of iterations; log/nolog specifies whether or not to show the iteration log; and trace specifies that the iteration log should include the current parameter vector. These options are seldom used.

wlsiter(*#*) specifies the number of weighted least-squares iterations that will be attempted before the linear programming iterations are started. The default value is 1. If there are convergence problems—something we have never observed—increasing this number should help.

Options for iqreg

_____| Model |_____

quantiles(*# #*) specifies the quantiles to be compared. The first number must be less than the second, and both should be between 0 and 1, exclusive. Numbers larger than 1 are interpreted as percentages. Not specifying this option is equivalent to specifying quantiles(.25 .75), meaning the interquartile range.

reps(*#*) specifies the number of bootstrap replications to be used to obtain an estimate of the variance–covariance matrix of the estimators (standard errors). reps(20) is the default and is arguably too small. reps(100) would perform 100 bootstrap replications. reps(1000) would perform 1,000 replications.

_____| Reporting |_____

level(*#*); see [R] **estimation options**.

nodots suppresses display of the replication dots.

Options for sqreg

___Model___

quantiles(# [# [# ...]]) specifies the quantiles to be estimated and should contain numbers between 0 and 1, exclusive. Numbers larger than 1 are interpreted as percentages. The default value of 0.5 corresponds to the median.

reps(#) specifies the number of bootstrap replications to be used to obtain an estimate of the variance–covariance matrix of the estimators (standard errors). reps(20) is the default and is arguably too small. reps(100) would perform 100 bootstrap replications. reps(1000) would perform 1,000 replications.

___Reporting___

level(#); see [R] **estimation options**.

nodots suppresses display of the replication dots.

Options for bsqreg

___Model___

quantile(#) specifies the quantile to be estimated and should be a number between 0 and 1, exclusive. Numbers larger than 1 are interpreted as percentages. The default value of 0.5 corresponds to the median.

reps(#) specifies the number of bootstrap replications to be used to obtain an estimate of the variance–covariance matrix of the estimators (standard errors). reps(20) is the default and is arguably too small. reps(100) would perform 100 bootstrap replications. reps(1000) would perform 1,000 replications.

___Reporting___

level(#); see [R] **estimation options**.

Option for _qreg

quantile(#) specifies the quantile to be estimated and should be a number between 0 and 1, exclusive. The default value of 0.5 corresponds to the median.

level(#); see [R] **estimation options**.

accuracy(#) should not be specified; it specifies the relative accuracy required for the linear programming algorithm. If the potential for improving the sum of weighted deviations by deleting an observation from the basis is less than this on a percentage basis, the algorithm will be said to have converged. The default value is 10^{-10}.

optimization_options: iterate(#), [no]log, trace. iterate() specifies the maximum number of iterations; log/nolog specifies whether or not to show the iteration log; and trace specifies that the iteration log should include the current parameter vector. These options are seldom used.

wlsiter(#) specifies the number of weighted least-squares iterations that will be attempted before the linear programming iterations are started. The default value is 1. If there are convergence problems—something we have never observed—increasing this number should help.

Remarks

Remarks are presented under the headings

> *Median regression*
> *Generalized quantile regression*
> *Estimated standard errors*
> *Interquantile and simultaneous-quantile regression*

Median regression

qreg without options fits quantile regression models. The most common form is median regression, where the object is to estimate the median of the dependent variable, conditional on the values of the independent variables. This is very similar to ordinary regression, where the objective is to estimate the mean of the dependent variable. Simply put, median regression finds a line through the data that minimizes the sum of the *absolute* residuals rather than the sum of the *squares* of the residuals, as in ordinary regression.

▷ Example 1

Consider a two-group experimental design with five observations per group:

```
. use http://www.stata-press.com/data/r9/twogrp
. list
```

	x	y
1.	0	0
2.	0	1
3.	0	3
4.	0	4
5.	0	95
6.	1	14
7.	1	19
8.	1	20
9.	1	22
10.	1	23

```
. qreg y x
Iteration  1:  WLS sum of weighted deviations =        161.2

Iteration  1: sum of abs. weighted deviations =         111
Iteration  2: sum of abs. weighted deviations =         110
Median regression                               Number of obs =         10
    Raw sum of deviations       157 (about 14)
    Min sum of deviations       110             Pseudo R2     =     0.2994
```

y	Coef.	Std. Err.	t	P>|t|	[95% Conf. Interval]
x	17	3.924233	4.33	0.003	7.950702 26.0493
_cons	3	2.774852	1.08	0.311	-3.39882 9.39882

We have estimated the equation

$$y_{median} = 3 + 17\,x$$

Looking back at our data, x takes on the values 0 and 1, so the median for the x==0 group is 3, while for x==1 it is $3 + 17 = 20$. The output reports that the raw sum of absolute deviations about 14 is 157; that is, the sum of $|y - 14|$ is 157. 14 is the unconditional median of y, although in these data, any value between 14 and 19 could also be considered an unconditional median (we have an even number of observations, so the median is bracketed by those two values). In any case, the raw sum of deviations of y about the median would be the same no matter what number we choose between 14 and 19. (With a "median" of 14, the raw sum of deviations is 157. Now think of choosing a slightly larger number for the median and recalculating the sum. Half the observations will have larger negative residuals, but the other half will have smaller positive residuals, resulting in no net change.)

Turning now to the actual estimated equation, the sum of the absolute deviations about the solution $y_{\text{median}} = 3 + 17x$ is 110. The pseudo-R^2 is calculated as $1 - 110/157 \approx 0.2994$. This is based on the idea that the median regression is the maximum likelihood estimate for the double-exponential distribution.

◁

❏ Technical Note

qreg is an alternative to regular regression or robust regression—see [R] **regress** and [R] **rreg**. Let's compare the results:

```
. regress y x
```

Source	SS	df	MS		Number of obs =	10
					F(1, 8) =	0.00
Model	2.5	1	2.5		Prob > F =	0.9586
Residual	6978.4	8	872.3		R-squared =	0.0004
					Adj R-squared =	-0.1246
Total	6980.9	9	775.655556		Root MSE =	29.535

y	Coef.	Std. Err.	t	P>|t|	[95% Conf. Interval]	
x	-1	18.6794	-0.05	0.959	-44.07477	42.07477
_cons	20.6	13.20833	1.56	0.157	-9.858465	51.05847

Unlike qreg, regress fits ordinary linear regression and is concerned with predicting the mean rather than the median, so both results are, in a technical sense, correct. Putting aside those technicalities, however, we tend to use either regression to describe the central tendency of the data, of which the mean is one measure and the median another. Thus we can ask which method better describes the central tendency of these data?

Means—and therefore ordinary linear regression—are sensitive to outliers, and our data were purposely designed to contain two such outliers: 95 for x==0 and 14 for x==1. These two outliers dominated the ordinary regression and produced results that do not reflect the central tendency well—you are invited to enter the data and graph y against x.

Robust regression attempts to correct the outlier-sensitivity deficiency in ordinary regression:

```
. rreg y x, genwt(wt)
     Huber iteration 1:  maximum difference in weights = .7311828
     Huber iteration 2:  maximum difference in weights = .17695779
     Huber iteration 3:  maximum difference in weights = .03149585
  Biweight iteration 4:  maximum difference in weights = .1979335
  Biweight iteration 5:  maximum difference in weights = .23332905
  Biweight iteration 6:  maximum difference in weights = .09960067
  Biweight iteration 7:  maximum difference in weights = .02691458
  Biweight iteration 8:  maximum difference in weights = .0009113
```

```
Robust regression                                    Number of obs =      10
                                                     F(  1,    8) =   80.63
                                                     Prob > F      =  0.0000

           y │    Coef.   Std. Err.      t    P>|t|     [95% Conf. Interval]
─────────────┼────────────────────────────────────────────────────────────
           x │  18.16597   2.023114    8.98   0.000     13.50066    22.83128
       _cons │  2.000003   1.430558    1.40   0.200     -1.298869   5.298875
```

In this case, `rreg` discarded the first outlier completely. (We know this because we included the `genwt()` option on `rreg` and, after fitting the robust regression, examined the weights.) For the other "outlier", `rreg` produced a weight of 0.47.

In any case, the answers produced by `qreg` and `rreg` to describe the central tendency are similar, but the standard errors are quite different. In general, robust regression will have smaller standard errors since it is not as sensitive to the exact placement of observations near the median. Also, some authors (Rousseeuw and Leroy 1987, 11) have noted that quantile regression, unlike the median, may be sensitive to even a single outlier, if its leverage is sufficiently high. ❑

▷ Example 2

Let us now consider a less artificial example using the automobile data described in [U] **1.2.1 Sample datasets**. Using median regression, we will regress each car's price on its weight and length and whether it is of foreign manufacture:

```
. use http://www.stata-press.com/data/r9/auto, clear
(1978 Automobile Data)

. qreg price weight length foreign
Iteration  1:  WLS sum of weighted deviations =    114043.7

Iteration  1: sum of abs. weighted deviations =   114998.67
Iteration  2: sum of abs. weighted deviations =   111786.7
 (output omitted )
Iteration  9: sum of abs. weighted deviations =  108822.59

Median regression                                    Number of obs =      74
   Raw sum of deviations    142205 (about 4934)
   Min sum of deviations 108822.6                     Pseudo R2     =  0.2347

       price │    Coef.   Std. Err.      t    P>|t|     [95% Conf. Interval]
─────────────┼────────────────────────────────────────────────────────────
      weight │  3.933588   .8602185    4.57   0.000     2.217936    5.64924
      length │ -41.25191   28.86931   -1.43   0.157    -98.82993    16.3261
     foreign │  3377.771   577.3392    5.85   0.000     2226.304    4529.238
       _cons │  344.6494   3260.245    0.11   0.916    -6157.704    6847.003
```

The estimated equation is

$$\text{price}_{\text{median}} = 3.93 \, \text{weight} - 41.25 \, \text{length} + 3377.8 \, \text{foreign} + 344.65$$

The output may be interpreted in exactly the same way as linear regression output; see [R] **regress**. The variables `weight` and `foreign` are significant, but `length` is not significant. The median price of the cars in these data is \$4,934. Note that this is a median (one of the two center observations), not *the* median, which would typically be defined as the midpoint of the two center observations.

◁

Generalized quantile regression

Generalized quantile regression is similar to median regression in that it estimates an equation describing a quantile other than the .5 (median) quantile. For example, specifying `quant(.25)` estimates the 25th percentile or the first quartile.

▷ Example 3

Again we will begin with the ten-observation artificial dataset we used at the beginning of *Median regression* above. We will estimate the .6667 quantile:

```
. use http://www.stata-press.com/data/r9/twogrp

. qreg y x, quant(0.6667)
Iteration  1:  WLS sum of weighted deviations =   176.41631

Iteration  1: sum of abs. weighted deviations =   138.0054
Iteration  2: sum of abs. weighted deviations =   136.6714

.6667 Quantile regression                      Number of obs =        10
  Raw sum of deviations 159.3334 (about 20)
  Min sum of deviations 136.6714                Pseudo R2     =    0.1422
```

y	Coef.	Std. Err.	t	P>|t|	[95% Conf. Interval]	
x	18	54.21918	0.33	0.748	-107.0297	143.0297
_cons	4	38.33875	0.10	0.919	-84.40932	92.40932

The 0.6667 quantile in the data is 20. The estimated values are 4 for x==0 and 22 for x==1. These values are appropriate since the usual convention is to "count in" $(n + 1) \times$ quantile observations.

◁

▷ Example 4

Returning to real data, the equation for the 25th percentile of `price` based on `weight`, `length`, and `foreign` in our automobile data is

```
. use http://www.stata-press.com/data/r9/auto
(1978 Automobile Data)

. qreg price weight length foreign, quant(.25)
Iteration  1:  WLS sum of weighted deviations =   84014.501

Iteration  1: sum of abs. weighted deviations =   84083.651
Iteration  2: sum of abs. weighted deviations =   70700.381
  (output omitted )
Iteration  6: sum of abs. weighted deviations =   69603.554

.25 Quantile regression                        Number of obs =        74
  Raw sum of deviations  83825.5 (about 4187)
  Min sum of deviations 69603.55                Pseudo R2     =    0.1697
```

price	Coef.	Std. Err.	t	P>|t|	[95% Conf. Interval]	
weight	1.831789	.6680931	2.74	0.008	.4993193	3.164259
length	2.845558	24.78057	0.11	0.909	-46.57774	52.26885
foreign	2209.925	434.631	5.08	0.000	1343.081	3076.77
_cons	-1879.775	2808.067	-0.67	0.505	-7480.287	3720.738

Note that, in comparison with our previous median regression, the coefficient on length now has a positive sign, and the coefficients on foreign and weight are reduced. The actual lower quantile is $4,187, substantially less than the median $4,934. It appears that the factors are weaker in this part of the distribution.

We can also estimate the upper quartile as a function of the same three variables:

```
. qreg price weight length foreign, quant(.75)
Iteration  1:  WLS sum of weighted deviations =   102589.61

Iteration  1: sum of abs. weighted deviations =    104678.5
Iteration  2: sum of abs. weighted deviations =   101523.41
  (output omitted)
Iteration  7: sum of abs. weighted deviations =   98395.936

.75 Quantile regression                           Number of obs =        74
   Raw sum of deviations 159721.5 (about 6342)
   Min sum of deviations 98395.94                  Pseudo R2    =     0.3840
```

price	Coef.	Std. Err.	t	P>|t|	[95% Conf. Interval]	
weight	9.22291	2.653578	3.48	0.001	3.930515	14.51531
length	-220.7833	80.13906	-2.76	0.007	-380.6156	-60.95098
foreign	3595.133	1727.704	2.08	0.041	149.336	7040.93
_cons	20242.9	8534.528	2.37	0.020	3221.325	37264.48

This tells a very different story: weight is much more important, and length is now significant—with a negative coefficient! The prices of high-priced cars seem to be determined by different factors than the prices of low-priced cars.

◁

❏ Technical Note

One explanation for having substantially different regression functions for different quantiles is that the data are heteroskedastic, as we will demonstrate below. The following statements create a sharply heteroskedastic set of data:

```
. drop _all
. set obs 10000
obs was 0, now 10000
. set seed 50550
. gen x = .1 + .9 * uniform()
. gen y = x * uniform()^2
```

Let us now fit the regressions for the 5th and 95th quantiles:

```
. qreg y x, quant(.05)
Iteration  1:  WLS sum of weighted deviations =  453.41879

Iteration  1: sum of abs. weighted deviations =  453.44691
Iteration  2: sum of abs. weighted deviations =  344.26222
 (output omitted)
Iteration  8: sum of abs. weighted deviations =  182.25244

.05 Quantile regression                        Number of obs =      10000
  Raw sum of deviations  182.357 (about .0009234)
  Min sum of deviations 182.2524                Pseudo R2     =     0.0006
```

y	Coef.	Std. Err.	t	P>|t|	[95% Conf. Interval]	
x	.002601	.0002737	9.50	0.000	.0020646	.0031374
_cons	-.0001393	.0001666	-0.84	0.403	-.000466	.0001874

```
. qreg y x, quant(.95)
Iteration  1:  WLS sum of weighted deviations =  615.69745

Iteration  1: sum of abs. weighted deviations =  616.14813
Iteration  2: sum of abs. weighted deviations =  496.24163
 (output omitted)
Iteration  8: sum of abs. weighted deviations =  338.43889

.95 Quantile regression                        Number of obs =      10000
  Raw sum of deviations 554.6889 (about .61326343)
  Min sum of deviations 338.4389                Pseudo R2     =     0.3899
```

y	Coef.	Std. Err.	t	P>|t|	[95% Conf. Interval]	
x	.8898259	.006085	146.23	0.000	.8778981	.9017536
_cons	.0021514	.0036897	0.58	0.560	-.0050812	.0093839

The coefficient on x, in particular, differs markedly between the two estimates. For the mathematically inclined, it is not too difficult to show that the theoretical lines are $y = .0025\,x$ for the 5th percentile and $y = .9025\,x$ for the 95th, numbers in close agreement with our numerical results.

❏

Estimated standard errors

qreg estimates the variance–covariance matrix of the coefficients using a method of Koenker and Bassett (1982) and Rogers (1993). This is described in *Methods and Formulas* below. Rogers (1992) reports that, while this method seems adequate for homoskedastic errors, it appears to understate the standard errors for heteroskedastic errors. The irony is that exploring heteroskedastic errors is one of the major benefits of quantile regression. Gould (1992, 1997) introduced generalized versions of qreg that obtain estimates of the standard errors using bootstrap resampling (see Efron and Tibshirani 1993 or Wu 1986 for an introduction to bootstrapped standard errors). The iqreg, sqreg, and bsqreg commands provide a bootstrapped estimate of the entire variance–covariance matrix of the estimators.

▷ Example 5

The first example of qreg on real data above was a median regression of price on weight, length, and foreign using the automobile data. Here is the result of repeating the estimation using bootstrapped standard errors:

```
. set seed 1001
```

```
. use http://www.stata-press.com/data/r9/auto, clear
(1978 Automobile Data)

. bsqreg price weight length foreign
(fitting base model)
(bootstrapping ...................)
Median regression, bootstrap(20) SEs              Number of obs =       74
  Raw sum of deviations    142205 (about 4934)
  Min sum of deviations 108822.6                  Pseudo R2      =   0.2347
```

price	Coef.	Std. Err.	t	P>|t|	[95% Conf. Interval]	
weight	3.933588	3.12446	1.26	0.212	-2.297951	10.16513
length	-41.25191	83.71266	-0.49	0.624	-208.2116	125.7077
foreign	3377.771	1040.209	3.25	0.002	1303.14	5452.402
_cons	344.6494	7053.301	0.05	0.961	-13722.72	14412.01

The coefficient estimates are the same—indeed, they are obtained using the same technique. Only the standard errors differ. Therefore, the t statistics, significance levels, and confidence intervals also differ.

Since `bsqreg` (as well as `sqreg` and `iqreg`) obtains standard errors by randomly resampling the data, the standard errors it produces will not be the same from run to run, unless we first set the random-number seed to the same number; see [D] **generate**.

By default, `bsqreg`, `sqreg`, and `iqreg` use 20 replications. We can control the number of replications by specifying the `reps()` option:

```
. set seed 1001

. bsqreg price weight length foreign, reps(1000)
(fitting base model)
(bootstrapping .................(output omitted)...)
Median regression, bootstrap(1000) SEs            Number of obs =       74
  Raw sum of deviations    142205 (about 4934)
  Min sum of deviations 108822.6                  Pseudo R2      =   0.2347
```

price	Coef.	Std. Err.	t	P>|t|	[95% Conf. Interval]	
weight	3.933588	2.670425	1.47	0.145	-1.392407	9.259583
length	-41.25191	69.64623	-0.59	0.556	-180.1569	97.65311
foreign	3377.771	1095.176	3.08	0.003	1193.511	5562.031
_cons	344.6494	5944.856	0.06	0.954	-11511.99	12201.29

A comparison of the standard errors is informative:

variable	qreg	bsqreg reps(20)	bsqreg reps(1000)
weight	.8602	3.124	2.670
length	28.87	83.71	69.65
foreign	577.3	1040.2	1095.
_cons	3260.	7053.	5945.

The results shown above are typical for models with heteroskedastic errors. (Note that our dependent variable is `price`; if our model had been in terms of $\ln(\text{price})$, the standard errors estimated by `qreg` and `bsqreg` would have been nearly identical.) Also note that, even in the case of heteroskedastic errors, 20 replications is generally sufficient for hypothesis tests against 0.

◁

Interquantile and simultaneous-quantile regression

Consider a quantile-regression model where the qth quantile is given by

$$Q_q(y) = a_q + b_{q,1}x_1 + b_{q,2}x_2$$

For instance, the 75th and 25th quantiles are given by

$$Q_{.75}(y) = a_{.75} + b_{.75,1}x_1 + b_{.75,2}x_2$$
$$Q_{.25}(y) = a_{.25} + b_{.25,1}x_1 + b_{.25,2}x_2$$

The difference in the quantiles is then

$$Q_{.75}(y) - Q_{.25}(y) = (a_{.75} - a_{.25}) + (b_{.75,1} - b_{.25,1})x_1 + (b_{.75,2} - b_{.25,2})x_2$$

qreg fits models such as $Q_{.75}(y)$ and $Q_{.25}(y)$. iqreg fits interquantile models, such as $Q_{.75}(y) - Q_{.25}(y)$. The relationships of the coefficients estimated by qreg and iqreg are exactly as shown: iqreg reports coefficients that are the difference in coefficients of two qreg models, and, of course, iqreg reports the appropriate standard errors, which it obtains by bootstrapping.

sqreg is like qreg in that it estimates the equations for the quantiles

$$Q_{.75}(y) = a_{.75} + b_{.75,1}x_1 + b_{.75,2}x_2$$
$$Q_{.25}(y) = a_{.25} + b_{.25,1}x_1 + b_{.25,2}x_2$$

The coefficients it obtains are the same that would be obtained by estimating each equation separately using qreg. sqreg differs from qreg in that it estimates the equations simultaneously and obtains an estimate of the entire variance–covariance matrix of the estimators by bootstrapping. Thus you can perform hypothesis tests concerning coefficients both within and across equations.

For example, to fit the above model, you could type

```
. qreg y x1 x2, q(.25)
. qreg y x1 x2, q(.75)
```

Doing this, you would obtain estimates of the parameters, but you could not test whether $b_{.25,1} = b_{.75,1}$ or, equivalently, $b_{.75,1} - b_{.25,1} = 0$. If your interest really is in the difference of coefficients, you could type

```
. iqreg y x1 x2, q(.25 .75)
```

The "coefficients" reported would be the difference in quantile coefficients. Alternatively, you could estimate both quantiles simultaneously and then test the equality of the coefficients:

```
. sqreg y x1 x2, q(.25 .75)
. test [q25]x1 = [q75]x2
```

Whether you use iqreg or sqreg makes no difference in terms of this test. sqreg, however, because it estimates the quantiles simultaneously, allows you to test other hypotheses. iqreg, by focusing on quantile differences, presents results in a way that is easier to read.

Finally, sqreg can estimate quantiles singly,

```
. sqreg y x1 x2, q(.5)
```

and can thereby be used as a substitute for the slower bsqreg. (Gould [1997] presents timings demonstrating that sqreg is faster than bsqreg.) sqreg can also estimate more than two quantiles simultaneously:

```
. sqreg y x1 x2, q(.25 .5 .75)
```

▷ Example 6

In demonstrating qreg, we performed quantile regressions using the automobile data. We discovered that the regression of price on weight, length, and foreign produced vastly different coefficients for the .25, .5, and .75 quantile regressions. Here are the coefficients that we obtained:

Variable	25th percentile	50th percentile	75th percentile
weight	1.83	3.93	9.22
length	2.85	−41.25	−220.8
foreign	2209.9	3377.8	3595.1
_cons	−1879.8	344.6	20242.9

All we can say, having estimated these equations separately, is that price seems to depend differently on the weight, length, and foreign variables depending on the portion of the price distribution we examine. We cannot be more precise because the estimates have been made separately. With sqreg, however, we can estimate all the effects simultaneously:

```
. set seed 1001

. sqreg price weight length foreign, q(.25 .5 .75) reps(100)
(fitting base model)
(bootstrapping ................  (output omitted ) .....)
Simultaneous quantile regression            Number of obs =         74
  bootstrap(100) SEs                         .25 Pseudo R2 =     0.1697
                                             .50 Pseudo R2 =     0.2347
                                             .75 Pseudo R2 =     0.3840
```

price	Coef.	Bootstrap Std. Err.	t	P>|t|	[95% Conf. Interval]	
q25						
weight	1.831789	1.574777	1.16	0.249	-1.309005	4.972583
length	2.845558	38.63523	0.07	0.941	-74.20999	79.9011
foreign	2209.925	1008.521	2.19	0.032	198.494	4221.357
_cons	-1879.775	3665.184	-0.51	0.610	-9189.753	5430.204
q50						
weight	3.933588	2.529541	1.56	0.124	-1.111423	8.978599
length	-41.25191	68.62258	-0.60	0.550	-178.1153	95.6115
foreign	3377.771	1017.422	3.32	0.001	1348.586	5406.956
_cons	344.6494	6199.257	0.06	0.956	-12019.38	12708.68
q75						
weight	9.22291	2.483676	3.71	0.000	4.269374	14.17645
length	-220.7833	86.17422	-2.56	0.013	-392.6524	-48.91422
foreign	3595.133	1147.216	3.13	0.003	1307.083	5883.184
_cons	20242.9	9414.242	2.15	0.035	1466.79	39019.02

The coefficient estimates above are the same as those previously estimated, although the standard error estimates are a little different. sqreg obtains estimates of variance by bootstrapping. Rogers (1992) provides evidence that, in the case of quantile regression, the bootstrapped standard errors are better than those calculated analytically by Stata.

The important thing here, however, is that the full covariance matrix of the estimators has been estimated and stored, and, thus, it is now possible to perform hypothesis tests. Are the effects of weight the same at the 25th and 75th percentiles?

```
. test [q25]weight = [q75]weight

 ( 1)  [q25]weight - [q75]weight = 0

       F(  1,    70) =     8.97
            Prob > F =   0.0038
```

It appears that they are not. We can obtain a confidence interval for the difference using `lincom`:

```
. lincom [q75]weight-[q25]weight

 ( 1)  - [q25]weight + [q75]weight = 0
```

price	Coef.	Std. Err.	t	P>|t|	[95% Conf. Interval]
(1)	7.391121	2.467548	3.00	0.004	2.469752	12.31249

Indeed, we could test whether the `weight` and `length` sets of coefficients are equal at the three quantiles estimated:

```
. quietly test [q25]weight = [q50]weight

. quietly test [q25]weight = [q75]weight, accum

. quietly test [q25]length = [q50]length, accum

. test [q25]length = [q75]length, accum

 ( 1)  [q25]weight - [q50]weight = 0
 ( 2)  [q25]weight - [q75]weight = 0
 ( 3)  [q25]length - [q50]length = 0
 ( 4)  [q25]length - [q75]length = 0

       F(  4,    70) =     2.43
            Prob > F =   0.0553
```

`iqreg` focuses on one quantile comparison but presents results that are more easily interpreted:

```
. set seed 1001

. iqreg price weight length foreign, q(.25 .75) reps(100) nodots
```

.75-.25 Interquantile regression				Number of obs =	74
bootstrap(100) SEs				.75 Pseudo R2 =	0.3840
				.25 Pseudo R2 =	0.1697

price	Coef.	Bootstrap Std. Err.	t	P>|t|	[95% Conf. Interval]
weight	7.391121	2.467548	3.00	0.004	2.469752	12.31249
length	-223.6288	83.09868	-2.69	0.009	-389.3639	-57.89376
foreign	1385.208	1193.557	1.16	0.250	-995.2672	3765.683
_cons	22122.68	9009.159	2.46	0.017	4154.478	40090.88

Looking only at the .25 and .75 quantiles (the interquartile range), the `iqreg` command output is easily interpreted. Increases in `weight` correspond significantly to increases in `price` dispersion. Increases in `length` correspond to decreases in `price` dispersion. The `foreign` variable does not significantly change `price` dispersion.

Do not make too much of these results; the purpose of this example is simply to illustrate the `sqreg` and `iqreg` commands and to do so in a context that suggests why analyzing dispersion might be of interest.

Note that lincom after sqreg produced the same t statistic for the interquartile range of weight, as did the iqreg command above. In general, they will not agree exactly due to the randomness of bootstrapping, unless the random-number seed is set to the same value before estimation (as was done in this case).

◁

Gould (1997) presents simulation results showing that the coverage—the actual percentage of confidence intervals containing the true value—for iqreg is appropriate.

Saved Results

qreg saves in e():

Scalars

e(N)	number of observations	e(sum_adev)	sum of absolute deviations
e(df_m)	model degrees of freedom	e(sum_rdev)	sum of raw deviations
e(df_r)	residual degrees of freedom	e(f_r)	residual density estimate
e(q)	quantile requested	e(convcode)	0 if converged; otherwise,
e(q_v)	value of the quantile		return code for why nonconvergence

Macros

e(cmd)	qreg	e(predict)	program used to implement predict
e(depvar)	name of dependent variable	e(properties)	b V

Matrices

e(b)	coefficient vector	e(V)	variance–covariance matrix of the estimators

Functions

e(sample)	marks estimation sample

iqreg saves in e():

Scalars

e(N)	number of observations	e(sumrdev0)	lower quantile sum of raw deviations
e(df_r)	residual degrees of freedom	e(sumrdev1)	upper quantile sum of raw deviations
e(reps)	number of replications	e(sumadev0)	lower quantile sum of absolute deviations
e(q0)	lower quantile requested	e(sumadev1)	upper quantile sum of absolute deviations
e(q1)	upper quantile requested	e(convcode)	0 if converged; otherwise, return code for why nonconvergence

Macros

e(cmd)	iqreg	e(properties)	b V
e(depvar)	name of dependent variable	e(predict)	program used to implement predict
e(vcetype)	title used to label Std. Err.		

Matrices

e(b)	coefficient vector	e(V)	variance–covariance matrix of the estimators

Functions

e(sample)	marks estimation sample

sqreg saves in e():

Scalars

e(N)	number of observations	e(reps)	number of replications
e(n_q)	number of quantiles requested	e(sumrdv#)	sum of raw deviations for q#
e(q#)	the quantiles requested	e(sumadv#)	sum of absolute deviations for q#
e(df_r)	residual degrees of freedom	e(convcode)	0 if converged; otherwise, return code for why nonconvergence

Macros

e(cmd)	sqreg	e(properties)	b V
e(depvar)	name of dependent variable	e(predict)	program used to implement predict
e(eqnames)	names of equations		
e(vcetype)	title used to label Std. Err.		

Matrices

e(b)	coefficient vector	e(V)	variance–covariance matrix of the estimators

Functions

e(sample)	marks estimation sample

bsqreg saves in e():

Scalars

e(N)	number of observations	e(q_v)	value of the quantile
e(df_r)	residual degrees of freedom	e(sum_adev)	sum of absolute deviations
e(reps)	number of replications	e(sum_rdev)	sum of raw deviations
e(q)	quantile requested	e(convcode)	0 if converged; otherwise, return code for why nonconvergence

Macros

e(cmd)	bsqreg	e(properties)	b V
e(depvar)	name of dependent variable	e(predict)	program used to implement predict

Matrices

e(b)	coefficient vector	e(V)	variance–covariance matrix of the estimators

Functions

e(sample)	marks estimation sample

_qreg saves in r():

Scalars

r(N)	number of observations	r(ic)	number of iterations
r(df_m)	model degrees of freedom	r(f_r)	residual density estimate
r(sum_w)	sum of the weights	r(q)	quantile requested
r(sum_adev)	sum of absolute deviations	r(q_v)	value of the quantile
r(sum_rdev)	sum of raw deviations	r(convcode)	1 if converged; 0 otherwise

Methods and Formulas

`qreg`, `iqreg`, `sqreg`, and `bsqreg` are implemented as ado-files.

According to Stuart and Ord (1991, 1084), the method of minimum absolute deviations was first proposed by Boscovich in 1757 and was later developed by Laplace; Stigler (1986, 39–55) and Hald (1998, 97–103, 112–116) provide historical details. According to Bloomfield and Steiger (1980), Harris (1950) subsequently observed that the problem of minimum absolute deviations could be turned into the linear programming problem that was first implemented by Wagner (1959). Interest has grown in this method due to interest in robust methods. Statistical and computational properties of minimum absolute deviation estimators are surveyed by Narula and Wellington (1982).

Define q as the quantile to be estimated; the median is $q = .5$. For each observation i, let r_i be the residual

$$r_i = y_i - \sum_j \beta_j x_{ij}$$

Define the multiplier h_i

$$h_i = \begin{cases} 2q & \text{if } r_i > 0 \\ 2(1-q) & \text{otherwise} \end{cases}$$

The quantity being minimized with respect to β_j is $\sum_i |r_i| h_i$, so quantiles other than the median are estimated by weighting the residuals. For example, if we want to estimate the 75th percentile, we weight the negative residuals by 0.50 and the positive residuals by 1.50. It can be shown that the criterion is minimized when 75 percent of the residuals are negative.

This is set up as a linear programming problem and is solved via linear programming techniques, as suggested by Armstrong, Frome, and Kung (1979) and used by courtesy of Marcel Dekker, Inc. The definition of convergence is exact in the sense that no amount of added iterations could improve the solution. Each step is described by a set of observations through which the regression plane passes, called the *basis*. A step is taken by replacing a point in the basis if the sum of weighted absolute deviations can be improved. If this occurs, a line is printed in the iteration log. The linear programming method is started by doing a weighted least-squares (WLS) regression to identify a good set of observations to use as a starting basis.

The variances are estimated using a method suggested by Koenker and Bassett (1982). This method can be put into a form recommended by Huber (1967) for M-estimates, where

$$\text{cov}(\boldsymbol{\beta}) = \mathbf{R}_2^{-1} \mathbf{R}_1 \mathbf{R}_2^{-1}$$

$\mathbf{R}_1 = \mathbf{X}'\mathbf{W}\mathbf{W}'\mathbf{X}$ (in the Huber formulation), $\mathbf{W}$ is a diagonal matrix with elements

$$W_{ii} = \begin{cases} q/f_{\text{residuals}}(0) & \text{if } r > 0 \\ (1-q)/f_{\text{residuals}}(0) & \text{if } r < 0 \\ 0 & \text{otherwise} \end{cases}$$

and $\mathbf{R}_2$ is the design matrix $\mathbf{X}'\mathbf{X}$. This is derived from formula 3.11 in Koenker and Bassett, although their notation is much different. $f_{\text{residuals}}()$ refers to the density of the true residuals. Koenker and Bassett leave much unspecified, including how to obtain a density estimate for the errors in real data. At this point, we offer our contribution (Rogers 1993).

We first sort the residuals and locate the observation in the residuals corresponding to the quantile in question, taking into account weights if they are applied. We then calculate w_n, the square root of the sum of the weights. Unweighted data are equivalent to weighted data in which each observation has weight 1, resulting in $w_n = \sqrt{n}$. For analytically weighted data, the weights are rescaled so that the sum of the weights is the number of observations, resulting in $\sqrt{n}$ again. For frequency-weighted data, w_n literally is the square root of the sum of the weights.

We locate the closest observation in each direction, such that the sum of weights for all closer observations is w_n. If we run off the end of the dataset, we stop. We calculate w_s, the sum of weights for all observations in this middle space. Typically, w_s is slightly greater than w_n.

If there are k parameters, then exactly k of the residuals must be zero. Thus we calculate an adjusted weight $w_a = w_s - k$. The density estimate is the distance spanned by these observations divided by w_a. Because the distance spanned by this mechanism converges toward zero, this estimate of density converges in probability to the true density.

The pseudo-R^2 is calculated as

$$1 - \frac{\text{sum of weighted deviations about estimated quantile}}{\text{sum of weighted deviations about raw quantile}}$$

This is based on the likelihood for a double-exponential distribution $e^{h_i |r_i|}$.

References

Armstrong, R. D., E. L. Frome, and D. S. Kung. 1979. Algorithm 79-01: A revised simplex algorithm for the absolute deviation curve fitting problem. In *Communications in Statistics, Simulation and Computation* B8(2), 175–190. New York: Dekker.

Bloomfield, P. and W. Steiger. 1980. Least absolute deviations curve-fitting. *SIAM Journal on Scientific and Statistical Computing* 1: 290–301.

Efron, B. and R. Tibshirani. 1993. *An Introduction to the Bootstrap.* New York: Chapman & Hall.

Gould, W. W. 1992. sg11.1: Quantile regression with bootstrapped standard errors. *Stata Technical Bulletin* 9: 19–21. Reprinted in *Stata Technical Bulletin Reprints*, vol. 2, pp. 137–139.

———. 1997. sg70: Interquantile and simultaneous-quantile regression. *Stata Technical Bulletin* 38: 14–22. Reprinted in *Stata Technical Bulletin Reprints*, vol. 7, pp. 167–176.

Gould, W. W. and W. H. Rogers. 1994. Quantile regression as an alternative to robust regression. *1994 Proceedings of the Statistical Computing Section.* Alexandria, VA: American Statistical Association.

Hald, A. 1998. *A History of Mathematical Statistics from 1750 to 1930.* New York: Wiley.

Harris, T. 1950. Regression using minimum absolute deviations. *The American Statistician* 4: 14–15.

Huber, P. J. 1967. The behavior of maximum likelihood estimates under nonstandard conditions. *Proceedings of the Fifth Berkeley Symposium on Mathematical Statistics and Probability* 1: 221–233.

———. 1981. *Robust Statistics.* New York: Wiley.

Jolliffe, D., B. Krushelnytskyy, and A. Semykina. 2000. sg153: Censored least absolute deviations estimator: CLAD. *Stata Technical Bulletin* 58: 13–16. Reprinted in *Stata Technical Bulletin Reprints*, vol. 10, pp. 240–244.

Koenker, R. and G. Bassett, Jr. 1982. Robust tests for heteroscedasticity based on regression quantiles. *Econometrica* 50: 43–61.

Koenker, R. and K. Hallock. 2001. Quantile regression. *Journal of Economic Perspectives* 15: 143–156.

Narula, S. C. and J. F. Wellington. 1982. The minimum sum of absolute errors regression: A state of the art survey. *International Statistical Review* 50: 317–326.

Rogers, W. H. 1992. sg11: Quantile regression standard errors. *Stata Technical Bulletin* 9: 16–19. Reprinted in *Stata Technical Bulletin Reprints*, vol. 2, pp. 133–137.

———. 1993. sg11.2: Calculation of quantile regression standard errors. *Stata Technical Bulletin* 13: 18–19. Reprinted in *Stata Technical Bulletin Reprints*, vol. 3, pp. 77–78.

Rousseeuw, P. J. and A. M. Leroy. 1987. *Robust Regression and Outlier Detection.* New York: Wiley.

Stigler, S. M. 1986. *The History of Statistics.* Cambridge, MA: Belknap Press of Harvard University Press.

Stuart, A. and J. K. Ord. 1991. *Kendall's Advanced Theory of Statistics, Vol. 2.* 5th ed. New York: Oxford University Press.

Wagner, H. M. 1959. Linear programming techniques for regression analysis. *Journal of the American Statistical Association* 54: 206–212.

Wu, C. F. J. 1986. Jackknife, bootstrap and other resampling methods in regression analysis. *Annals of Statistics* 14: 1261–1350 (including comments and reply).

Also See

Complementary:	[R] **qreg postestimation**
Related:	[R] **bootstrap**, [R] **regress**, [R] **rreg**
Background:	[U] **11.1.10 Prefix commands**,
	[U] **20 Estimation and postestimation commands**,
	[R] **estimation options**, [R] **maximize**

Title

qreg postestimation — Postestimation tools for qreg, iqreg, sqreg, and bsqreg

Description

The following postestimation commands are available for `qreg`, `iqreg`, `bsqreg`, and `sqreg`:

command	description
adjust	adjusted predictions of $\mathbf{x}\beta$
estat	VCE and estimation sample summary
estimates	cataloging estimation results
lincom	point estimates, standard errors, testing, and inference for linear combinations of coefficients
linktest	link test for model specification
mfx	marginal effects or elasticities
nlcom	point estimates, standard errors, testing, and inference for nonlinear combinations of coefficients
predict	predictions, residuals, influence statistics, and other diagnostic measures
predictnl	point estimates, standard errors, testing, and inference for generalized predictions
test	Wald tests for simple and composite linear hypotheses
testnl	Wald tests of nonlinear hypotheses

See the corresponding entries in the *Stata Base Reference Manual* for details.

Syntax for predict

For qreg, iqreg, *and* bsqreg

> predict [*type*] *newvar* [*if*] [*in*] [, [xb | stdp | <u>r</u>esiduals]]

For sqreg

> predict [*type*] *newvar* [*if*] [*in*] [, <u>e</u>quation(*eqno* [,*eqno*]) *statistic*]

statistic	description
xb	linear prediction; the default
stdp	standard error of the linear prediction
stddp	standard error of the difference in linear predictions
<u>r</u>esiduals	residuals

These statistics are available both in and out of sample; type `predict ... if e(sample) ...` if wanted only for the estimation sample.

Options for predict

xb, the default, calculates the linear prediction.

stdp calculates the standard error of the linear prediction.

stddp is allowed only after you have fitted a model using sqreg. The standard error of the difference in linear predictions $(\mathbf{x}_{1j}\mathbf{b} - \mathbf{x}_{2j}\mathbf{b})$ between equations 1 and 2 is calculated.

residuals calculates the residuals, that is, $y_j - \mathbf{x}_j\mathbf{b}$.

equation(*eqno* [*,eqno*]) specifies the equation to which you are making the calculation.

> equation() is filled in with one *eqno* for options xb, stdp, and residuals. equation(#1) would mean that the calculation is to be made for the first equation, equation(#2) would mean the second, and so on. Alternatively, you could refer to the equations by their names. equation(income) would refer to the equation named "income" and equation(hours) to the equation named "hours".

> If you do not specify equation(), results are the same as if you had specified equation(#1).

> To use stddp, you must specify two equations. You might specify equation(#1, #2) or equation(q80, q20) to indicate the 80th and 20th quantiles.

Methods and Formulas

All postestimation commands listed above are implemented as ado-files.

Also See

Complementary:	[R] **qreg**; [R] **adjust**, [R] **estimates**, [R] **lincom**, [R] **linktest**, [R] **mfx**, [R] **nlcom**, [R] **predictnl**, [R] **test**, [R] **testnl**
Background:	[U] **13.5 Accessing coefficients and standard errors**, [U] **20 Estimation and postestimation commands**, [R] **estat**, [R] **predict**

Title

query — Display system parameters

Syntax

query [memory | output | interface | graphics | efficiency | network |

    update | trace | mata | other]

Description

query displays the settings of various Stata parameters.

Remarks

query provides more system information than you will ever want to know. You do not need to understand every line of output query produces if all you need is one piece of information. Here is what happens when you type query:

```
. query

    Memory settings (also see query memory for a more complete report)
        set memory        10M
        set maxvar        5000
        set matsize       400

    Output settings
        set more          on
        set rmsg          off
        set dp            period    may be period or comma
        set linesize      79        characters
        set pagesize      23        lines

        set logtype       smcl      may be smcl or text

    Interface settings
        set dockable      on
        set dockingguides on
        set locksplitters off
        set persistfv     off
        set persistvtopic off
        set xptheme       on

        set linegap       1         pixels
        set scrollbufsize 32000     characters
        set varlabelpos   32        column number
        set reventries    100       lines

        set maxdb         50        dialog boxes
        set smalldlg                (not relevant)
```

531

```
Graphics settings
    set graphics         on
    set scheme           s2color
    set printcolor       automatic   may be automatic, asis, gs1, gs2, gs3
    set copycolor        automatic   may be automatic, asis, gs1, gs2, gs3

Efficiency settings
    set adosize          500         kilobytes
    set virtual          off

Network settings
    set checksum         off
    set timeout1         120         seconds
    set timeout2         300         seconds

    set httpproxy        off
    set httpproxyhost
    set httpproxyport    80

    set httpproxyauth    off
    set httpproxyuser
    set httpproxypw

Update settings
    set update_query     on
    set update_interval  7
    set update_prompt    on

Trace (programming debugging) settings
    set trace            off
    set tracedepth       32000
    set traceexpand      on
    set tracesep         on
    set traceindent      on
    set tracenumber      off
    set tracehilite

Mata settings
    set matastrict       off
    set matalnum         off
    set mataoptimize     on
    set matafavor        space       may be space or speed
    set matacache        400         kilobytes
    set matalibs         lmataado;lmatabase
    set matamofirst      off

Other settings
    set type             float       may be float or double
    set level            95          percent confidence intervals
    set maxiter          16000       max iterations for estimation commands
    set searchdefault    local       may be local, net, or all
    set seed             X075bcd151f123bb5159a55e50022865746ad
    set varabbrev        on
```

The output is broken into several divisions: memory, output, interface, graphics, efficiency, network, update, trace, mata, and other settings, and we will discuss each one in turn.

We generated the output above using Stata/SE for Windows. Here is what happens when we type query and we are running Intercooled Stata for Linux(GUI):

```
. query
```

Memory settings
    set memory   1M
    set maxvar   2048   (not settable in this version of Stata)
    set matsize   200

Output settings
    set more   on
    set rmsg   off
    set dp   period   may be period or comma
    set linesize   85   characters
    set pagesize   22   lines

    set logtype   smcl   may be smcl or text

Interface settings

    set linegap   1   pixels
    set scrollbufsize 32000   characters
    set fastscroll   on
    set varlabelpos   32   column number

    set maxdb   50   dialog boxes

Graphics settings
    set graphics   on
    set scheme   s2color
    set printcolor   automatic may be automatic, asis, gs1, gs2, gs3

Efficiency settings
    set adosize   500   kilobytes
    set virtual   off

Network settings
    set checksum   off
    set timeout1   120   seconds
    set timeout2   300   seconds

    set httpproxy   off
    set httpproxyhost
    set httpproxyport 8080

    set httpproxyauth off
    set httpproxyuser
    set httpproxypw

Update settings
    set update_query   (not relevant)
    set update_interval   (not relevant)
    set update_prompt   (not relevant)

Trace (programming debugging) settings
    set trace   off
    set tracedepth   32000
    set traceexpand   on
    set tracesep   on
    set traceindent   on
    set tracenumber   off
    set tracehilite

```
Mata settings
     set matastrict       off
     set matalnum         off
     set mataoptimize     on
     set matafavor        space       may be space or speed
     set matacache        400         kilobytes
     set matalibs         lmataado;lmatabase
     set matamofirst      off

Other settings
     set type             float       may be float or double
     set level            95          percent confidence intervals
     set maxiter          16000       max iterations for estimation commands
     set searchdefault    local       may be local, net, or all
     set seed             X075bcd151f123bb5159a55e50022865746ad
     set varabbrev        on
```

Memory settings

Memory settings indicate how much memory is allocated to Stata, the maximum number of variables, and the maximum size of a matrix.

For more information, see

```
memory    [D] memory
maxvar    [D] memory
matsize   [R] matsize
```

Output settings

Output settings show how Stata displays output on the screen and in log files.

For more information, see

```
    more    [R] more
    rmsg    [P] rmsg
      dp    [D] format
linesize    [R] log
pagesize    [R] more
 logtype    [R] log
 eolchar    [R] set
   icmap    [R] set
```

(Continued on next page)

Interface settings

Interface settings control how Stata's interface works.

For more information, see

dockable	[R]	set
dockingguides	[R]	set
locksplitters	[R]	set
persistfv	[R]	set
persistvtopic	[R]	set
xptheme	[R]	set
smoothfonts	[R]	set
smoothsize	[R]	set
linegap	[R]	set
revwindow	[R]	set
varwindow	[R]	set
scrollbufsize	[R]	set
fastscroll	[R]	set
varlabelpos	[R]	set
reventries	[R]	set
maxdb	[R]	**db**
smalldlg	[R]	set

Graphics settings

Graphics settings indicate how Stata's graphics are displayed.

For more information, see

graphics	[G]	**set graphics**
scheme	[G]	**set scheme**
printcolor	[G]	**set printcolor**
copycolor	[G]	**set printcolor**
macgphengine	[R]	set
piccomments	[R]	set

Efficiency settings

The efficiency settings set the maximum amount of memory allocated to automatically loaded do-files, the maximum number of remembered-contents dialog boxes, and the use of virtual memory.

For more information, see

adosize	[P]	**sysdir**
virtual	[D]	**memory**

Network settings

Network settings determine how Stata interacts with the Internet.

For more information, see [R] **netio**.

Update settings

Update settings determine how Stata performs updates.

For more information, see [R] **update**.

Trace settings

Trace settings adjust Stata's behavior and are particularly useful in debugging code.

For more information, see [P] **trace**.

Mata settings

Mata settings affect Mata's system parameters.

For more information, see [M-3] **mata set**.

Other settings

The other settings are a miscellaneous collection.

For more information, see

type	[D] **generate**
level	[R] **level**
maxiter	[R] **maximize**
searchdefault	[R] **search**
seed	[D] **generate**
varabbrev	[R] **set**

In general, the parameters displayed by query can be changed by set; see [R] **set**.

Also See

Complementary: [R] **set**

Related: [R] **db**, [R] **level**, [R] **log**, [R] **matsize**, [R] **maximize**, [R] **more**, [R] **netio**, [R] **search**, [R] **view**, [D] **format**, [D] **generate**, [D] **memory**, [G] **set graphics**, [G] **set printcolor**, [G] **set scheme**, [P] **rmsg**, [P] **sysdir**, [P] **trace**, [M-3] **mata set**